ORIGIN OF MATTER AND EVOLUTION OF GALAXIES

ORIGIN OF MATTER AND EVOLUTION OF GALAXIES

The 10th International Symposium on Origin of Matter and Evolution of Galaxies: From the Dawn of Universe to the Formation of Solar System

Sapporo, Japan 4 – 7 December 2007

EDITORS
Takuma Suda
Takaya Nozawa
Akira Ohnishi
Kiyoshi Kato
Masayuki Y. Fujimoto
Hokkaido University, Japan
Toshitaka Kajino
National Astronomical Observatory of Japan, Tokyo
Shigeru Kubono
University of Tokyo, Japan

All papers have been peer reviewed

SPONSORING ORGANIZATIONS

Department of Cosmosciences, Hokkaido University
National Astronomical Observatory of Japan
Center for Nuclear Study, University of Tokyo
RIKEN
Research Center for Nuclear Physics, Osaka University
KEK
Department of Astronomy, University of Tokyo
The Science Council of Japan
The Physical Society of Japan
The Astronomical Society of Japan

Melville, New York, 2008
AIP CONFERENCE PROCEEDINGS ■ VOLUME 1016

Editors:

Takuma Suda
Takaya Nozawa
Akira Ohnishi
Kiyoshi Kato
Masayuki Y. Fujimoto

Department of Cosmosciences
Hokkaido University
Kita 10 Nishi 8, Kita-ku, Sapporo
Hokkaido 060-0810
Japan

E-mail: suda@astro1.sci.hokudai.ac.jp
tnozawa@mail.sci.hokudai.ac.jp
ohnishi@nucl.sci.hokudai.ac.jp
kato@nucl.sci.hokudai.ac.jp
fujimoto@astro1.sci.hokudai.ac.jp

Toshitaka Kajino
National Astronomical Observatory
2-21-1 Osawa, Mitaka
Tokyo 181-8588
Japan

E-mail: kajino@nao.ac.jp

Shigeru Kubono
Center for Nuclear Study
University of Tokyo
Wako Branch at RIKEN
Hirosawa 2-1, Wako,
Saitama 351-0198
Japan

E-mail: kubono@cns.s.u-tokyo.ac.jp

L.C. Catalog Card No. 2008926914
ISBN 978-0-7354-0537-0
ISSN 0094-243X
Printed in the United States of America

CONTENTS

INVITED and ORAL TALKS

1. OPENING

2. BIG BANG COSMOLOGY and PARTICLE ASTROPHYSICS

3. SPECIAL SESSION: FIRST STARS (1)
NUCLEAR ABUNDANCES IN FIRST STARS

4. SPECIAL SESSION: FIRST STARS (2)
CHEMICAL EVOLUTION and STRUCTURE FORMATION IN THE UNIVERSE

5. NUCLEOSYNTHESIS IN STARS, NOVAE and SUPERNOVAE (1)

6. NUCLEOSYNTHESIS IN STARS, NOVAE and SUPERNOVAE (2)

7. NUCLEAR DATA FOR ASTROPHYSICS

8. NUCLEAR STRUCTURE and REACTIONS FOR ASTROPHYSICS (1)

9. NUCLEAR STRCUTURE and REACTIONS FOR ASTROPHYSICS (2)

10. EXPLOSION MECHANISM OF SUPERNOVAE

11. NEUTRON STAR and HIGH DENSITY MATTER

12. SPECIAL SESSION: RI BEAMS FOR NUCLEAR ASTROPHYSICS

13. WEAK INTERACTION and NEUTRINO PHYSICS

14. NUCLEAR ABUNDANCES IN SUPERNOVAE

15. STELLAR and METEORITIC ABUNDANCES

POSTER SESSION

PREFACE

Chemical elements are fundamental matter to comprise the universe and hold a great deal of interest for astronomers and nuclear physicists, for these play an important role in understanding the dawn of the universe to the formation of solar system. Astronomers look for a clue to understand the formation and evolution of Universe, Galaxy, and solar system through the composition of element species in stars and meteorites. On the other hand, nuclear physicists have interests in the general properties and behaviors of atomic nuclei including unstable ones created by accelerator facilities.

In order to integrate and/or share our knowledge on various fields of astronomy and nuclear physics, a series of symposia entitled "Origin of Matter and Evolution of Galaxies" have been held almost every two years since 1988. These symposia have been successful in encouraging the discussion and collaboration between astronomers and nuclear physicists, which has contributed to the progresses in the field of nuclear astrophysics.

Now, the symposium, held at Hokkaido University, Sapporo, from December 4 to 7, is the tenth in a series of OMEG symposia. This is the first attempt to move to the area other than Tokyo district and to arrange a session focusing on the specific topics on astronomy and nuclear physics: "First Stars" for the former and "RI beams for Nuclear Astrophysics" for the latter. In addition, this symposium covers a wide range of scientific topics such as the cosmic ray, particle physics and the formation of solar system as well as those focused in the previous symposia. It assembled more than 120 active researchers coming from different countries and different area of expertise. The present volume contains the reviews and contributions of this symposium made by both oral and poster presentations. We wish that it gives a fair cross section of our present knowledge on the nuclear astrophysics and will fruitfully contribute to future research efforts.

The symposium is hosted by Department of Cosmosciences (DoC), Hokkaido University, National Astronomical Observatory of Japan (NAOJ), Center for Nuclear Study, University of Tokyo (CNS), RIKEN, Research Center for Nuclear Physics, Osaka University (RCNP), KEK, Department of Astronomy, University of Tokyo, Science Council of Japan, and The Physical Society of Japan. This symposium is sponsored by The Astronomical Society of Japan.

We are very grateful to the administration officers of the host institutes, students, and young researchers for their services provided to this symposium, especially to Ms. Hitomi Yoshida and Ms. Asako Sato for their secretarial assistance. We are also grateful to the referees selected from participants who carefully reviewed the manuscripts and gave us useful comments. We believe that the manuscripts will be very much improved by these peer review processes. Finally we would like to thank all the participants for their lively contributions to the discussions and for their help for prompt publication of this book.

March, 2008

T. Suda (Editor in Chief)
T. Nozawa, A. Ohnishi, K. Kato, M. Y. Fujimoto, T. Kajino, S. Kubono

List of Committees

International Advisory Committee:

H. Ando (NAOJ/Tokyo),
A. Balantekin (Wisconsin),
R. Boyd (LLNL),
L. Buchmann (TRIUMF),
R. Gallino (Torino),
W. Haxton (Washington, Seattle),
K. Koyama (Kyoto),
K. Langanke (GSI),
C. S. Lee (Korea),
W. P. Liu (CIAE, Beijing),
G. J. Mathews (Notre Dame),
A. Mengoni (IAEA, Vienna/CERN),
Y. Nagai (JAEA),
M. Nakahata (Tokyo),
P. Nissen (Aarhus),
T. Otsuka (UT/UT-CNS, Tokyo),
N. Prantzos (Paris),
C. Rolfs (Bochum),
K. Sagara (Kyusyu),
K. Sato (Tokyo),
H. Schatz (Michigan State),
J. Silk (Cambridge),
M. S. Smith (Oak Ridge),
C. Spitaleri (Catania),
F.-K. Thielemann (Basel),
H. Toki (RCNP),
J. Truran (Chicago),
M. Wiescher (Notre Dame),
H. Yurimoto (Hokkaido)

Local Organizing Committee:

* M.-Y. Fujimoto (Hokkaido Univ.),
** K. Kato (Hokkaido Univ.),
** S. Kubono (UT-CNS),
** T. Kajino (NAOJ/GUAS/UT),
W. Aoki (NAOJ/GUAS),
T. Motobayashi (RIKEN/UT),
T. Suda (RIKEN),
T. Kishimoto (RCNP),
H. Miyatake (KEK),
K. Nomoto (Univ. of Tokyo),
T. Kozasa (Hokkaido Univ.),
S. Watanabe (Hokkaido Univ.),
K. Ishikawa (Hokkaido Univ.),
% A. Ohnishi (Hokkaido Univ.),
% T. Suda (Hokkaido Univ.),
% T. Nozawa (Hokkaido Univ.)

* Chairperson, ** Vice-Chair, % Scientific Secretary

Program Committee:

* H. Utsunomiya (Konan Univ.),
T. Hayakawa (JAEA),
T. Teranishi (Kyusyu Univ.),
H. Umeda (Univ. of Tokyo),
Y. Fujita (Osaka),
K. Kuramoto (Hokkaido Univ.),
and LOC members

T. Shima (RCNP),
A. Ozawa (Tsukuba Univ.),
K. Sumiyoshi (Numazu),
K. Terada (Hiroshima),
T. Nakamura (TITech),
A. Habe (Hokkaido Univ.),

* Chairperson

Host Institutes / Organizations:

- Department of Cosmosciences, Graduate School of Science, Hokkaido University
- National Astronomical Observatory of Japan (NAOJ)
- Center for Nuclear Study, University of Tokyo (UT-CNS)
- RIKEN Nishina Center
- Research Center of Nuclear Physics, Osaka Univsersity (RCNP)
- High Energy Accelerator Research Organization (KEK)
- Department of Astronomy, Graduate School of Science, University of Tokyo (UT)

- The Science Council of Japan
- The Physical Society of Japan (JPS)

Sponsor:

- The Astronomical Society of Japan

OMEG07
Origin of Matter and Evolution of Galaxies
Dec. 4-7, 2007, Sapporo, Japan
http://nucl.sci.hokudai.ac.jp/~omeg07

INVITED and ORAL TALKS

1. OPENING

Opening Address

Masayuki Y. Fujimoto
Department of Cosmosciences, Graduate School of Science, Hokkaido University

Dear Symposium Participants,

I would like to give a short opening address on behalf of the organizing committee. It is our pleasure to welcome all of you and we are happy to have this International Symposium on the Origin of Matter and Evolution of Galaxies, from the dawn of Universe to the formation of solar system here in Sapporo.

This is the 10th meeting of a series of international conferences, which started from 1988, and hence, continues almost 20 years. The former meetings were all held in Tokyo and adjacent prefectures and this is the first time for the symposium to be held outside near Tokyo area. I would like to thank all of the participants for traveling a long distance, especially those who come from foreign countries. It is our pleasure to have so many participants.

The aim of this symposium is, in my understanding, to get together physicists and astronomers, and to encourage the discussion and collaboration to understand the basic properties and states of matter and elements in the Universe. As you know, there has been a long history of interactions between nuclear physics and astronomy, which traces back to the first half of the last century: CNO cycles and pp chain reactions as the stellar energy sources by Bethe and Weizecker, the big bang nucleosynthesis by Gamov, and the prediction of resonance state of carbon 12 by Hoyle. Recently, there have been great progresses in the observations and experiments, which is along with the theoretical endeavors. They reveal new aspects of our Universe and material properties and are expected to renew the prospects to these collaborations. Put aside the elusive dark matter and enigmatic dark energy, it is amazing to know how such a world, consisting of baryons, and full of diversities and wonders as we observe, has been produced from the featureless, uniformly distributed, matter produced by the Big Bang. On the astronomy side, the 8m class telescopes has brought us the information on not only the far and faint galaxies, but also the oldest stars, formed at the very beginning of the formation of our galaxy. On the physics side, the colliding experiments with heavy radioactive nuclei beams have contributed to the disclosure unknown aspects of nuclei. These two topics are featured as the special sessions, first stars and RI beams, in this symposium. In addition, many interesting papers will be presented for the various topics of astronomy, astrophysics, and nuclear physics, and also of the cosmic ray, particle physics, and the meteorite and planetary sciences. We hope these contributions provide stimulus for the research and cooperation to understand the origin of matter and evolution of Universe.

CP1016, *Origin of Matter and Evolution of Galaxies,*
edited by T. Suda, T. Nozawa, A. Ohnishi, K. Kato, M. Y. Fujimoto, T. Kajino, and S. Kubono

I would like to take an advantage of this opportunity to introduce a new enterprise of our Institute. Our graduate school of science, Hokkaido university, was reorganized last year to form a new interdisciplinary department of Cosmosciences. This aims to bring the people together from the various related fields of sciences including physics, astronomy, and planetary sciences, to promote the joint-research and education to investigate the evolution of matter "from the big bang to the origin of life" seamlessly. The motto of our department is in line with the purpose of this symposium, and this is one of the reasons that our department hosts this symposium. Our department also has the nuclear reaction data center founded this year in collaboration with Japan Atomic Energy Agency. This is another reason to have this symposium.

We hope for all of you to enjoy this 3-day meeting and 1-day excursion here in Hokkaido in the atmosphere and climate quite different from the other parts of Japan. We wish this symposium is stimulating, productive and fruitful for you.

Thank you for your attention.

2007.12.4

2. BIG BANG COSMOLOGY and PARTICLE ASTROPHYSICS

Big Bang Cosmology

G. J. Mathews

Center for Astrophysics, Department of Physics,
University of Notre Dame, Notre Dame, IN 46556, USA

Abstract. The origin of matter and many interesting phenomena occurred during the big bang between the first instants of cosmic expansion time of the surface of photon last scattering. Even so, there are only two means by which to probe the physics of the big bang epoch. One is the observed power spectrum of temperature fluctuations in the cosmic microwave background. This spectrum derives from a combination of physics from the first instants of cosmic inflation, and the physics of matter and radiation at the photon last scattering surface. The other probe is the observed ashes of primordial nucleosynthesis which occurred during the radiation dominated epoch from about 1 sec to 10^3 sec into the big bang. This paper summarizes these two cosmic probes and their roles in motivating and constraining new cosmological paradigms. Among the topics discussed are the limits which these probes place upon the nature of dark matter and dark energy, along with possible insights into physics beyond the standard model of particle physics.

Keywords: Big Bang; Early Universe; Extra Dimensions; Brane World Cosmology
PACS: 98.80.Es, 95.35.+d, 95.30.Cq

INTRODUCTION

Almost every form of matter which we are aware of (photons, leptons, baryons, light elements, dark matter, dark energy, etc.) formed and began its evolution during the big bang. Indeed, this is a very vigorous time in the study of big bang cosmology. I count over 800 papers with the words "big bang" in the title or abstract within the past year alone. It would be an enormous task to overview all of the exciting developments in big bang cosmology in the last few years. For the purposes of this overview, however, I will try simply outline some of what I think are the most intriguing problems.

For our purposes, the big bang spans the time frame from the first instant of creation until the surface of last photon scattering some 2.5×10^5 years later. Within this time frame, a lot of interesting things occur including the Planck epoch, the birth of space-time, inflation, reheating, a variety of cosmic phase transitions (e.g. supersymmetry breaking, baryogenesis, the electroweak transition, and the QCD transition), the epoch of big bang nucleosynthesis (BBN) , and of course the photon last scattering to produce the cosmic microwave background (CMB).

The dynamics of the universe during the big bang are quite simple. Beginning from the Einstein equation, $G^{\mu\nu} + \Lambda g^{\mu\nu} = 8\pi G T^{\mu\nu}$ and adopting a homogeneous, isotropic Lemaitre-Friedmann-Robertson-Walker (LFRW) metric, the evolution of the scale factor a or Hubble parameter H with time is simply given from the time-time component of the Einstein equation:

$$\left(\frac{\dot{a}}{a}\right)^2 = H^2 = \frac{8}{3}\pi G\rho - \frac{k}{a^2} - \frac{\Lambda}{3} = H_0^2\left[\frac{\Omega_\gamma}{a^4} + \Omega_m a^3 + \Omega_k a^2 + \Omega_\Lambda\right] , \quad (1)$$

CP1016, *Origin of Matter and Evolution of Galaxies,*
edited by T. Suda, T. Nozawa, A. Ohnishi, K. Kato, M. Y. Fujimoto, T. Kajino, and S. Kubono

where $\Omega_\gamma = 8\pi G\rho_\gamma/(3H_0^2)$, $\Omega_m = (8\pi G\rho_m)/(3H_0^2)$, $\Omega_k = -k/(a^2H_0^2)$, and $\Omega_\Lambda = \Lambda/(3H_0^2)$ are the present effective contributions toward closure. For most of the big bang only the radiation term in Eq. (1) is important. As we shall see, however, there are interesting variants of big bang cosmology where this is not the case.

However, there are really only two observational cosmological probes of the big bang. The only probe of the radiation dominated epoch is the yield of light elements from BBN in the temperature regime from 10^8 to 10^{10} K and times of about 1 to 10^3 sec into the big bang. The other probe is the spectrum of temperature fluctuations in the CMB which contains information of the first quantum fluctuations in the universe, and the details of the distribution and evolution of dark matter, baryonic matter and photons near the surface of photon last scattering. It is worthwhile, therefore to begin with a brief review of the present status of these observational probes.

Light Element Abundances

One of the powers of standard-homogeneous BBN [1, 2, 3] is that all of the light element abundances are determined in terms of a single parameter η_{10} which is the baryon-to-photon ratio in units of 10^{-10}. The crucial test of the standard BBN is, therefore, whether a single value of η_{10} can be found which reproduces all of the observed primordial abundances.

Primordial deuterium is best determined from its absorption line in high redshift Lyman α clouds. The average of measurements of six absorption-line systems towards five QSOs gives [4] $D/H = 2.78^{+0.44}_{-0.38} \times 10^{-5}$. This would imply an value of $\eta_{10} = 5.9 \pm 0.5$. This is an important result because it is also very close to the value $\Omega_b h^2 = 0.0219 \pm 0.0007$ ($\eta_{10} = 6.00 \pm 0.25$) deduced [17] from the *WMAP* analysis described below. Because of this concordance the *WMAP* determination of η_{10} is generally accepted as the most accurate determination of η_{10}.

The primordial lithium abundance, on the other hand, is inferred from old low-metallicity halo stars which exhibit an approximately constant ("Spite plateau") lithium abundance as a function of surface temperature. There is, however, some controversy [5] concerning the depletion of ^{7}Li on the surface of such halo stars and/or during the big bang itself [6]. Here, we adopt the value from [7] $^7\text{Li} = 1.23^{+0.68}_{-0.32} \times 10^{-10}$, where the errors are 95% confidence limits.

The primordial helium abundance is obtained by measuring extragalactic HII regions in low-metallicity irregular galaxies. The primordial helium abundance Y_p so deduced has tended to have one of two possible values (a low value, e.g. $Y_p = 0.238 \pm 0.002 \pm 0.005$, [10] and a high value $Y_p 0.2452 \pm 0.0015$ [8]). There is also, however, a dilemma [9] regarding the uncertainty in the observationally determined primordial helium abundance. Many recent evaluations (e.g. [8]) give a rather narrow range of abundance uncertainty. The extent of systematic errors is still being debated. Ref. [9] has concluded that correlations in various uncertainties could stretch the error in the inferred primordial abundance to a range of $0.232 \leq Y_p \leq 0.258$.

While this is being sorted out, however, it has been deduced by several authors (cf. [6, 11]) that the combined deuterium and *WMAP* constraints on the baryon-to-photon ratio implies that the primordial helium abundance should be $Y_p = 0.2484^{0.0004}_{0.0005}$ [11] or $Y_p = 0.2479 \pm 0.0004$ [6]. If we adopt the narrow helium abundance of [8] and the *WMAP* constraint of [11] there is, a possible 2-3σ discrepancy between the ^{4}He + ^{7}Li and the D + *WMAP* results.

If this dilemma persists it may provide insight into new physics beyond the minimal BBN model, for example, brane-world effects [12], cosmic quintessence [13], time varying constants [14], etc. [3]. It has been noted, however [15] that much of the possible discrepancy could be accounted for simply by adopting a new shorter neutron decay lifetime [16].

CMB Observations

The best current constraints on the CMB come from the third year data of the *WMAP* (Wilkinson Microwave Anisotropy Probe) data [17]. This mission obtained a full sky map of the temperature anisotropy of the cosmic microwave background radiation at 13 arcminute resolution.

When the *WMAP* data are combined with data from other CMB observations along with that from type Ia supernovae, and large scale structure, the following parameters have been deduced [17] for a variety of cosmological parameters in a standard LFRW ΛCDM cosmology: $(\Omega_\Lambda, \Omega_m, \Omega_m h^2, \Omega_b h^2, h, n_s, \tau, \sigma_8) = (0.732 \pm 0.018, 0.268 \pm 0.018, 0.1324^{+0.0042}_{-0.0041}, 0.02186 \pm 0.068, 0.704^{+0.015}_{-0.016}, 0.947 \pm 0.015, 0.073^{+0.27}_{-0.28}, 0.776^{+0.031}_{-0.032})$, where h is the Hubble parameter in units of 100 km s^{-1} Mpc^{-1}, τ is the ionization optical depth, and σ_8 is the amplitude of galaxy cluster fluctuations, and n_s is the slope for the scalar perturbation spectrum from inflation. The precision with which these parameters are now known is much better than a decade ago. However, these parameters are based upon the simplest possible ΛCDM cosmology, and the analysis must be redone for more complicated cosmologies such as those described below. Also, there are still some discrepancies between the best fit cosmology and the data. For example there is evidence of a suppression of the lowest multipole moments of the CMB. This has been suggested [18, 19] as possible evidence for a compact topology for the universe. Another curiosity is a possible excess of power on the largest multipoles (smallest scales) as measured by *ACBAR* [20] and CBI [21]. This excess might suggest new physics, some examples of which are described below.

WHAT ARE THE QUESTIONS?

The important questions regarding the big bang are something like the following: 1)How did the universe begin? 2)Why are there 3 large dimensions? 3) What drives inflation? 4) Are there observable effects from: supersymmetric particles, string excitations, etc. 5) Is there evidence for large extra dimensions? 6) How does the universe reheat? 7) How and when was the net baryon number generated? 8) When and how was the dark matter generated? 9) When and how was the dark energy created? 10) Are there observable effects from cosmic phase transitions such as Supersymmetry breaking, the Electroweak transition, and the QCD transition? 11) Is there a primordial magnetic field?

Although this is a far from exhaustive list, I now summarize some intriguing developments toward answering some of these questions.

How Did the Universe Begin?

The universe most likely was born out of the chaos at the Planck epoch. Hence, tone must have a complete theory of quantum gravity. The best candidate for such a theory is string theory [22] which requires a radical departure from our perceptions of space-time. At the simplest level, particles are not point-like but represent excitations of a one dimensional string. Quantizing this string, however, requires more than four space-time dimensions. A theory with gauge particles of no more than spin 2 requires 10 dimensions. A generalization in which all of the string theories are equivalent (M-theory) requires the introduction of one more dimension [23, 24].

Another radical change in our view of space-time in string theory comes from what is called *holography*. Roughly speaking this is the notion that physics in a region can be described by the fundamental degrees of freedom living on the boundary. This leads to what is called the AdS/CFT correspondence [25]. Simply put, there is a coupling constant describing the strength of the interaction among strings. In the limit of a large coupling constant, the strings can be described as an anti-deSitter (AdS) space-time and a compact space. In the other limit, the strings are described by a conformal gauge theory living on the boundary. Thus, there is a duality ($t - duality$) between a gauge theory and a space-time. Thus, space-time may not be fundamental. Instead the universe begins as a simple string describable by a quantum field theory. Only after the coupling constant begins to grow, do strings form a separate large space time. This bizarre conclusion has led to the introduction of many papers (e.g. [26]). Here, we consider a few popular examples of how the universe might have emerged.

In the low-energy limit, the heterotic M-theory is an 11-dimensional supergravity coupled to two 10-dimensional $E_8 \times E_8$ gauge theories. The universe can be thought of as two smooth 10-dimensional manifolds plus a bulk dimension between them. This is the basis for a large number of papers in "brane-world" cosmology. As a practical phenomenological model [27] our universe can be represented as a thin three-brane embedded in an infinite five-dimensional bulk anti-de Sitter space. Physical particles are trapped on a three-dimensional brane via curvature in the bulk dimension. Gravitons can reside as fluctuations in the background gravitational field living in both the brane and bulk dimension.

Why are there only three large dimensions?

There is no *a priori* explanation as to why only three dimensions are large while the others have compactified to be small. In the brane-world phenomenology there is an interesting conjecture [28] based upon a "relaxation principle" as to how only 3 spatial dimensions became large. Initially, all possible dimensions and universes exist, however, branes collide and annihilate. Since the volume of the total n-dimensional space occupied by the branes of high spatial dimension is large, they annihilate most easily. It is argued [28] that for a space with 9 spatial dimensions, only branes of dimension $d \leq 3$ become sufficiently diluted that they can survive annihilation and grow to become large.

Another approach is that of the *string gas cosmology* [29]. In this picture strings wind around branes and resist expansion. However, winding and anti-winding modes of the strings can annihilate allowing expansion. For strings of dimension d (d = 1 for strings), the winding modes only cancel for volumes with a dimension of $2d+1$. Hence, a universe with three large spatial dimensions is singled out.

A third intriguing possibility is suggested [30] by gravity in higher dimensional space. In a four-dimensional space-time there are only two unique geometric tensors which satisfy vanishing of the covariant derivative (Bianchi identities). In higher dimensions, however, there is a new tensor which also has the desired properties, the Gauss-Bonnet tensor $H_{\mu\nu}$. The Einstein equation then becomes: $G^{\mu\nu} + \Lambda g^{\mu\nu} + \alpha H_{\mu\nu} = \kappa^2 T^{\mu\nu}$ where κ^2 and Λ are the gravitational coupling and cosmological constant in the higher dimensional space-time. This leads to a modified LFRW equation which is quadratic equation in H^2 and allows two possible solutions. Small αH^2 corresponds to the normal LFRW expansion (1). Large αH^2 corresponds to an exponentially inflating cosmology. If the universe begins in the large solution there is no graceful exit from the inflating cosmology unless [30], the higher-dimensions stabilize (i.e. their Hubble parameter vanishes). Then it can be shown that with p large dimensions, the expansion becomes [30]

$$\frac{1}{2}p(p-1)H^2 = \frac{1}{2}p(p-1)(p-2)(p-3)|\alpha|H^4 + \kappa^2\rho \ . \qquad (2)$$

Unless $p \leq 3$ an initially inflating universe is unable to exit. However, the H^4 contribution vanishes for $p \leq 3$ and a natural transition to an LFRW universe can be made.

What drives Inflation?

The simplest explanation for the fact that the universe is so nearly flat today ($\Omega_{tot} = 1.00 \pm 0.018$ [17]) and the near isotropy of CMB almost demands [31] that the universe has gone through a rapid inflation. In the previous section we introduced one mechanism by which such inflation could be driven by changes in the LFRW expansion rate due to the influence of higher order dimensions on the Einstein equation. The traditional view, however, is that some vacuum energy $V(\phi)$ drives inflation due to the existence of a self-interacting scalar field ϕ. That is, the energy density of the cosmic fluid in the early universe becomes dominated by that of the scalar field for which $\rho_\phi = \dot{\phi}^2/2 + (\nabla\phi)^2 + V(\phi)$, and the inflaton field ϕ evolves according to a damped harmonic-oscillator-like equation of motion: $\ddot{\phi} + 3H\dot{\phi} - \nabla^2\phi + dV/d\phi = 0$. As the universe expands, H is large and $\dot{\phi}$ becomes small. It is trapped in the slowly varying $V(\phi)$ dominated regime and the scale factor grows exponentially.

The biggest unknown in this paradigm, however, the form of $V(\phi)$. Dozens of forms have been proposed and more are sure to come. There is at least the suggestion [32] that the simplest phenomenological form $V(\phi) = (m/2)\phi^2$ is motivated by a slight modification of the Kahler potential in string theory. In a more phenomenological approach, all possible forms of the potential which can be approximated by a polynomial have been compared with available data [33] with the conclusion that almost any form for the potential works except a fourth order potential [$V(\phi) = (\lambda/4)\phi^4$]. Clearly, this remains a mystery to be solved in coming generations.

Observable effects from supersymmetric particles, string excitations, etc?

Before leaving this topic of the beginning of the universe we should address the question as to whether there could be an imprint on the big bang from the existence of Planck-mass particles. Planck-scale mass particles generically exist in the Planck-scale compactification schemes of string theory from the Kaluza-Klein states, winding modes, and the massive (excited) string modes. The possibility of Planck-scale mass particles which couple to inflation is a generic situation. During inflation such a coupling could have left an imprint [34]on the CMB and the matter power spectrum. Indeed, there continue to remain puzzling features of the power spectrum [17] at high multipolarity which might be evidence of such a coupling.

Is there Evidence for Large Extra Dimensions?

If the universe could consist of two manifolds separated by a large extra dimension. Is it possible that such an extra dimension could manifest itself on the dynamics of the universe? There is at least one possibility [35]. For example, in a Randall-Sundrum II [27] brane-world cosmology, the cosmic expansion for a 3-space embedded in a higher dimensional space can be written as

$$\left(\frac{\dot{a}}{a}\right)^2 = \frac{8\pi G_{\rm N}}{3}\rho - \frac{k}{a^2} + \frac{\Lambda_4}{3} + \frac{\kappa_5^4}{36}\rho^2 + \frac{\mu}{a^4} \;, \tag{3}$$

where, the first term on the right hand side is obtained by relating the four-dimensional gravitational constant $G_{\rm N}$ to the five-dimensional gravitational constant, κ_5 $G_{\rm N} = \kappa_5^4\lambda/48\pi$, with λ the intrinsic tension of the brane. The four-dimensional cosmological constant Λ_4 is related to its five-dimensional counterpart $\Lambda_4 = \kappa_5^4\lambda^2/12 + 3\Lambda_5/4$. Λ_5 should be negative in order for Λ_4 to obtain its presently observed small value.

Standard big-bang cosmology does not contain the fourth and fifth terms of Eq. (3). The fourth term arises from the imposition of a junction condition for the scale factor on the surface of the brane, and is not likely to be significant for most of during the big bang epoch of interest here. The fifth term, however, scales just like radiation with a constant μ and is called the *dark radiation.* It derives from the electric (Coulomb) part of the five-dimensional Weyl tensor [36]. The coefficient μ is a constant of integration whose magnitude and sign depend on initial conditions. Because the dark radiation scales as a^{-4} it can affect both BBN and the CMB. For example, it can significantly improve [35] the fit to BBN abundances and the CMB if μ is negative.

When and how was the dark energy created?

There are a variety of models for the dark energy besides that of a simple cosmological constant. Dark energy is often attributed to a vacuum energy in the form of a "quintessence" scalar field or *k*-essence [37] which must be slowly evolving along an effective potential. A quintessence or *k*-essence field is of interest as it can be constrained by both BBN and the CMB [13].

However, the simple coincidence that both of dark matter and dark energy currently contribute comparable mass energy toward the closure of the universe begs the question as to whether they could be different manifestations of the same physical phenomenon [38, 39, 40, 41].

The expansion rate with brane-bulk energy exchange leads to another modified LFRW cosmology to an observer on the brane [38]:

$$H^2 = \frac{\dot{a}^2}{a^2} = \frac{8}{3}\pi G_N(\rho+\bar{\rho}) - \frac{k}{a^2} + \Lambda_4 + \frac{\kappa^4}{36}(\rho+\bar{\rho})^2 + \chi \;, \tag{4}$$

while the (5,5) component leads to a modified dark radiation term χ, $\dot{\chi}+4(\dot{a}/a)(\chi+(\kappa^2/6)T_5^5)=(\kappa^4/18)(\rho+\bar{\rho}+\tau)T$. This cosmology has been shown [38] to reproduce remarkably well constraints from the CMB and large scale structure without a manifest cosmological constant.

Is there evidence of supersymmetric matter in the early universe?

There has been considerable interest in observations [42] indicating that the abundance of ^{6}Li in metal poor stars appears to be primordial, and at an abundance much higher than that expected from BBN. The ^{6}Li abundance as a function of metallicity exhibits a plateau similar to that for ^{7}Li in very metal-poor stars, This abundance, however, is a factor of $\sim 10^3$ larger than that predicted by BBN. At the same time ^{7}Li abundance is a factor of 3 below the BBN expectation. The two lithium abundance anomalies might be a manifestation of the existence of new unstable particles during the big bang ([43] and Refs. therein).

Several recent papers [44, 45] have considered heavy negatively charged decaying X^- particles that modify BBN and may be identified as the supersymmetric partner of the tau. In this paradigm, the heavy X^- particles bind to the nuclei produced in BBN to form X-nuclei. The massive X^- particles reduce the reaction Coulomb barriers thereby enhancing the thermonuclear reaction rates and extending the duration of BBN to lower temperatures. This can lead to a a large enhancement of the ^{6}Li abundance [45] (for example by the $^4\mathrm{He}_X(d,X^-)^6\mathrm{Li}$ reaction, while depleting ^{7}Li, thus solving both the ^{6}Li and ^{7}Li problems simultaneously.

Is there a primordial magnetic field?

The existence of a primordial magnetic field (PMF) $\sim$1 nG whose field lines collapse as structure forms is one possible explanation for the magnetic fields observed in galactic clusters. Such a PMF, however, could have influenced a variety of phenomena in the early universe [46] such as the cosmic microwave background (CMB), e.g. [47]. In [48] we reported on correlations between the calculated CMB power spectrum (influenced by a PMF) and the primary curvature perturbations. We found that the PMF affects the CMB on both small and large angular scales in the TT and TE modes. The introduction of a PMF leads to a better fit to the CMB power spectrum for the higher multipoles, and the fit to the lowest multipoles can be used to constrain the correlation of the PMF with the density fluctuations. Our prediction for the BB mode for a PMF of average field strength $|B_\lambda| = 4.0$ nG is consistent with the upper limit deduced from the latest CMB observations. We found that the BB mode is dominated by the vector mode of the PMF for higher multipoles. We also showed that by fitting the complete power spectrum one can break the degeneracy between the PMF amplitude and its power spectral index.

ACKNOWLEDGMENTS

Work at the University of Notre Dame supported by the U.S. Department of Energy under Nuclear Theory Grant DE-FG02-95-ER40934.

REFERENCES

1. R.V. Wagoner, *Ap.J.* **179** (1973) 343.
2. J. Yang, M.S. Turner, G. Steigman, D.N. Schramm, and K.A. Olive, *Ap.J.* **281** (1984) 493.
3. R. A. Malaney and G. J. Mathews, G. J. , Phys. Rep. **229** , 145 (1993).
4. D. Kirkman, et al., Astrophys. J. Suppl., **149**, 1 (2003).
5. M.H. Pinsonneault, et al., Astrophys. J., **527**, 180 (1999).
6. A. Coc, et al., Astrophys. J., **600**, 544 (2004).
7. S. G. Ryan, et al., Astrophys. J., **532**, 654 (2000).
8. Y. Izatov, et al. Astrophys. J., **527**, 757 (1999).
9. K. A. Olive and E. D. Skillman, Astrophys. J., 617, 29 (2004).
10. B. D. Fields and K. A. Olive, Astrophys. J., **506**, 177 (1998).
11. R. H. Cyburt, B. D. Fields and K. A. Olive, Phys. Lett., **567**, 227 (2003).
12. K. Ichiki, P. M. Garnavich, T. Kajino, G. J. Mathews and M. Yahiro, Phys. Rev. **D68**, 083518 (2003).
13. M. Yahiro, G. J. Mathews, K. Ichiki, T. Kajino and M. Orito, Phys. Rev. **D65** 063502 (2002).
14. K. Ichikawa and M. Kawasaki, Phys. Rev. **D69**, 123506, (2004).
15. G. J. Mathews T. Kajino, and T. Shima, Phys. Rev. **D71**, 021302 (2005).
16. A. Serebrov, et al., Phys. Lett., **B605**, 72 (2005).
17. A. Kogut, et al., ApJ, 665, 355 (2007); D. Spergel, et al., ApJ, 665, 377 (2007); L. Page, et al., ApJ, 665, 335 (2007); G. Hinshaw, et al., ApJ, 665, 288 (2007); N. Jarosik, et al., ApJ, 665, 263 (2007).
18. M. Tegmark, A. de Oliveira-Costa and A. Hamilton, Phys. Rev., D68, 123523 (2003).
19. D. Menzies, and G. J. Mathews, J. Cosmo. and Astropart. Phys. **10**, p. 008 (2005).
20. C.-L. Kuo et al., ApJ, **664**, 687 (2007).
21. A. C. S. Readhead et al., Astrophys. J., 609, 498 (2004); J. L. Sievers et al., ApJ, **660**, 976 (2007).
22. G. T. Horowitz, New Journal of Physics, **7**, 201 (2005).
23. Horava & E. Witten, NPB, 506 (1996).
24. E. Witten, Adv. Theor. Math. Phys., **2** 253, (1998).
25. J. Maldacena, Adv. Theor. Math. Phys. **2**, 231 (1998).
26. D. Marlof, Am. J. Phys., **73**, 730 (2004).
27. L. Randall and R. Sundrum, Phys. Rev. Lett. **83**, 3370 (1999); **83**, 4690 (1999).
28. A. Karch and L. Randall, Phys. Rev. Lett. **95**,161601 (2005).
29. R. H. Brandenberger and C. Vafa, Nucl. Phys., **B316**, 391, (1989); S. Alexander, R. H. Brandenberger and D. Esson, Phys. Rev. **D62**, 103509 (2000).
30. F. Ferrer and S. Räsänen, JHEP (2007) submitted, hep-th/0707.0499.
31. E. W. Kolb and M. S. Turner, *The Early Universe*, (Addison-Wesley, Menlo Park, Ca., 1990).
32. S. Kachru, R. Kallosch, A. Linde, and S. P. Trivedi, Phys. ReV. **D66**, 046005 (2003).
33. W. H. Kinney, E. W. Kolb, A. Melchiorri and A. Riotto, Phys. Rev. **D74**, 023502 (2006).
34. G. J. Mathews, D. J. Chung, K. Ichiki, K., T. Kajino, M. Orito, Physical Review **D70**, 083505 (2004).
35. K. Ichiki, M. Yahiro, T. Kajino, M. Orito, G. J. Mathews, Phys. Rev. **D66**, 043521 (2002).
36. T. Shiromizu, K.I. Maeda, and M. Sasaki, Phys. Rev. **D62**, 024012 (2000).
37. V. Barger, E. Guarnaccia, and D. Marfatia Phy. Lett. **B635**, 61 (2006).
38. K. Umezu, K. Ichiki, T. Kajino, G. J. Mathews, R. Nakamura, M. Yahiro, Phys. Rev. D **73**, 063527 (2006).
39. J. R. Wilson, G. J. Mathews, and G. M. Fuller, PRD, 75, 043521 (2007)
40. G. J. Mathews, N. Q. Lan, and C. Kolda, PRD Submitted (2008).
41. R. Bean, S. Carroll, and M. Trodden, astro-ph/0510059.
42. M. Asplund *et al.*, Astrophys. J. 644, 229 (2006).
43. M. Kusakabe, T. Kajino and G. J. Mathews, Phys. Rev. D 74, 023526 (2006).
44. M. Pospelov, arXiv:hep-ph/0605215; R. H. Cyburt *et al.*, JCAP, 0611, 014 (2006); M. Kaplinghat and A. Rajaraman, Phys. Rev. D 74, 103004 (2006); K. Kohri and F. Takayama, arXiv:hep-ph/0605243; C. Bird, K. Koopmans and M. Pospelov, arXiv:hep-ph/0703096.
45. M. Kusakabe, T. Kajino, R. N. Boyd, T. Yoshida and G. J. Mathews, Phys. Rev. D, 76, 121302 (2007); ApJ (2008) in press.
46. D. Grasso and H. R. Rubinstein, Phys. Rep, **348**,163 (2001).
47. D. G. Yamazaki, K. Ichiki, T. Kajino and G. J. Mathews, Astrophys. J., **646**, 719 (2006).
48. D. G. Yamazaki, K. Ichiki, T. Kajino and G. J. Mathews, PRD, Submitted (2007).

Big-Bang Nucleosynthesis with Negatively-Charged Massive Particles as a Cosmological Solution to the ^{6}Li and ^{7}Li Problems

Motohiko Kusakabe[a,b,1], Toshitaka Kajino[a,b,c], Richard N. Boyd[a,d], Takashi Yoshida[a] and Grant J. Mathews[e]

[a] *Division of Theoretical Astronomy, National Astronomical Observatory of Japan, Mitaka, Tokyo 181-8588, Japan*

[b] *Department of Astronomy, Graduate School of Science, University of Tokyo, Hongo, Bunkyo-ku, Tokyo 113-0033, Japan*

[c] *Department of Astronomical Science, The Graduate University for Advanced Studies, Mitaka, Tokyo 181-8588, Japan*

[d] *Present Address: Lawrence Livermore National Laboratory, 7000 East Avenue, Livermore, CA 94550, USA*

[e] *Department of Physics and Center for Astrophysics, University of Notre Dame, Notre Dame, IN 46556, USA*

Abstract. Observations of metal poor halo stars exhibit a possible plateau of ^{6}Li abundance as a function of metallicity similar to that for ^{7}Li, suggesting a big bang origin. However, the inferred primordial abundance of ^{6}Li is $\sim$1000 times larger than that predicted by standard big bang nucleosynthesis (BBN) for the baryon-to-photon ratio inferred from the WMAP data. On the other hand, the inferred ^{7}Li primordial abundance is about 3 times smaller than the prediction. We study a possible simultaneous solution to both the problems of underproduction of ^{6}Li and overproduction of ^{7}Li in BBN. This solution involves a hypothetical massive, negatively-charged leptonic particle that would bind to the light nuclei produced in BBN, but would decay long before it could be detected. Because the particle gets bound to the existing nuclei after the cessation of the usual big bang nuclear reactions, a second longer epoch of nucleosynthesis can occur among X-nuclei which have reduced Coulomb barriers. We numerically carry out a fully dynamical BBN calculation, simultaneously solving the recombination and ionization processes of negatively-charged particles by normal and X-nuclei as well as many possible nuclear reactions among them. We confirm that a reaction in which the hypothetical particle is transferred can occur that greatly enhance the production of ^{6}Li while a reaction through an atomic excited state of X-nucleus depletes ^{7}Li. It is confirmed that BBN in the presence of these hypothetical particles, together with or without an event of stellar burning process, can simultaneously solve the two Li abundance problems.

Keywords: Big Bang nucleosynthesis, dark matter, early universe, cosmology
PACS: 26.35.+c, 95.35.+d, 98.80.Cq, 98.80.-k

INTRODUCTION

The precise observational cosmology has started with a recent mission of the Wilkinson Microwave Anisotropy Probe (WMAP) of the cosmic microwave background (CMB).

---

[1] Research Fellow of the Japan Society for the Promotion of Science.

CP1016, *Origin of Matter and Evolution of Galaxies*,
edited by T. Suda, T. Nozawa, A. Ohnishi, K. Kato, M. Y. Fujimoto, T. Kajino, and S. Kubono

The CMB-WMAP data strongly suggest that our universe is most likely flat and cosmic expansion is accelerating, which indicates the finite values of energy densities of dark energy (cosmological constant) $\Omega_\Lambda = 0.74$ and cold dark matter (CDM) $\Omega_{\rm CDM} = 0.22$ as well as baryon $\Omega_{\rm B} = 0.04$. It is the biggest challenge to resolve what is the nature of the dark energy and the dark matter in modern cosmology and particle and nuclear astrophysics.

Recent spectroscopic observations of metal poor halo stars (MPHSs) have indicated a primordial plateau of the ^{6}Li abundance as a function of the metallicity [1] which is similar to that for ^{7}Li [1, 2, 3]. The standard Big Bang Nucleosynthesis (BBN) model predicts primordial ^{6}Li abundance about three orders of magnitude smaller than the observed abundance level and ^{7}Li abundance a factor of $\sim$ three overproduced when one adopts a value of the universal baryonic density parameter, $\Omega_{\rm B} = 0.04$, inferred from the recent WMAP-3 data [4].

A number of solutions have been suggested for Li abundance anomalies. Some relate the Li anomalies to the possible existence of unstable heavy particles in the early Universe [5, 6, 7, 8, 9, 10, 11, 12] which lead to non-thermal nucleosynthesis induced by the particle decay. Some have suggested a possible early epoch of enhanced cosmic ray nucleosynthesis via the $\alpha + \alpha$ reactions [13].

Several recent works [14, 15, 16, 17, 18, 19, 20, 21] study a suggestion that heavy negatively charged unstable particles could modify BBN and lead to ^{6}Li production. In this paradigm, the heavy particles (here denoted as X^-) would bind to the nuclei synthesized during BBN to produce exotic nuclei (hereafter denoted as X-nuclei). These massive X^- particles would be bound in orbits that would have radii comparable to those of the nuclei to which they were attached. Their presence, however, would reduce the Coulomb barrier seen by incident charged particles, thereby enhancing the thermonuclear reaction rates [21].

Pospelov [16] suggested that a large enhancement of the ^{6}Li abundance could result from an X^- bound to ^{4}He (denoted ^{4}He$_X$). This allows for the X^- transfer reaction of ^{4}He$_X$$(d,X^-)$^{6}Li which can enhance the production of ^{6}Li by as much as seven orders of magnitude. Subsequently, Hamaguchi et al. [18] carried out a theoretical calculation of the cross section for this X^- transfer reaction in a quantum three-body model. Their value was about an order of magnitude smaller than that of Pospelov [16]. This difference can be traced to their exact treatment of the quantum tunneling in the fusion process and the use of a better nuclear potential.

A resonant reaction ^{7}Be$_X$$(p,\gamma)^{8}B_X$ has recently been proposed [19] that further destroys ^{7}Be$_X$ through an atomic excited state of ^{8}B$_X$.

In this paper we study the potential consequences of X^- particles on BBN, including recombination of the X^- particles by the existing normal nuclei. Several new reaction processes have been included that can produce or destroy ^{6}Li and ^{7}Li in the early universe. We utilize a fully dynamical description which is solved numerically, and which includes relevant kinetic and chemical equilibrium. The purpose of the present work is to better quantify whether there are observable consequences of the resulting reaction network. The ultimate goal is to better explain the observed overproduction (1000 times) and underproduction (3 times) of the abundances of ^{6}Li and ^{7}Li, respectively.

MODEL

First, we estimate binding energies of nuclides with X^-. We assume that the X^- particle has spin 0, charge $-e$, and mass m_X much larger than nuclear mass of the order of 1 GeV. The charge distribution of nuclides is assumed to be Gaussian. We then solved the two-body Shrödinger equation by a variational calculation (Gaussian expansion method [22]), and obtained binding energies. When experimental data exist, we used charge radii determined from experiment. Note that since the X^- particles can bind only electromagnetically to nuclei, their binding energies are typically small ($\sim 0.1-1$ MeV), and are largest for heavy nuclei. Hence, they are not appreciably bound to nuclei until the temperature becomes low enough.

Second, we estimate the thermonuclear reaction rates of X-bound nuclei [21]. We take only non-resonant reactions. As for radiative neutron capture reactions, we take rates of standard BBN. As for reaction rates for charged particle reactions, we take astrophysical S factors for standard BBN reactions, and corrected nuclear charges and reduced mass. Reaction Q-values are corrected taking account of calculated binding energies.

Third, Pospelov [16] suggested that the rate of X^- transfer reaction $^4\mathrm{He}_X(d,X^-)^6\mathrm{Li}$ is enhanced by 7 orders of magnitude more than that of $^4\mathrm{He}(d,\gamma)^6\mathrm{Li}$. Hamaguchi et al. [18] carried out a theoretical calculation of the cross section for $^4\mathrm{He}_X(d,X^-)^6\mathrm{Li}$ in a quantum three-body model. We adopt the result of Hamaguchi et al. [18].

Bird et al. [19] has recently suggested that the $^7\mathrm{Be}_X(p,\gamma)^8\mathrm{B}_X$ resonant reaction could occur through an atomic excited state of $^8\mathrm{B}_X$ with a threshold energy of 167 keV. This channel does destroy a significant amount of $^7\mathrm{Be}_X$ and is included in our present study. To check the nuclear flow to the higher mass region we added the $^8\mathrm{Be}_X+p \rightarrow {}^9\mathrm{B}_X^{*a} \rightarrow {}^9\mathrm{B}_X+\gamma$ reaction, where $^9\mathrm{B}_X^{*a}$ indicates an atomic excited state of $^9\mathrm{B}_X$. However this reaction was now found to be unimportant because its threshold energy is relatively large.

We have put both recombination and ionization processes of X^- particles into our BBN network code and have dynamically solved the associated set of rate equations [17] to find when the X-nuclei decoupled from the cosmic expansion precisely.

RESULT

The panels in Fig. 1 show light element abundances as a function of $T_9 \equiv T/(10^9\ \mathrm{K})$ with the cosmic photon temperature T, in a calculation with certain input parameters, where $m_X = 50$ GeV, X abundance is $Y_X = n_X/n_b = 0.1$ with number densities of X^- and baryon, n_X and n_b, respectively, and decay lifetime of X is $\tau_X = \infty$. The left part of the panels (high temperature region) is the same as in standard BBN. However, when the temperature of the universe decreases to $T_9 \sim 0.5$, the recombination of $^7\mathrm{Be}$ proceeds, and other nuclides also start recombining with X^- particles. One can find the destruction of $^7\mathrm{Be}_X$ at $T_9 \sim 0.3-0.2$. The destruction mechanism is $^7\mathrm{Be}_X(p,\gamma)^8\mathrm{B}_X$ through an atomic resonance [19]. In our previous calculation with rough inputs on the charge radii of nuclides, the same $^7\mathrm{Be}_X(p,\gamma)^8\mathrm{B}_X$ reaction but through the 1^+ nuclear excited state of $^8\mathrm{B}$ is found to have a possibility of destruction of $^7\mathrm{Be}_X$ [20]. But we confirmed that

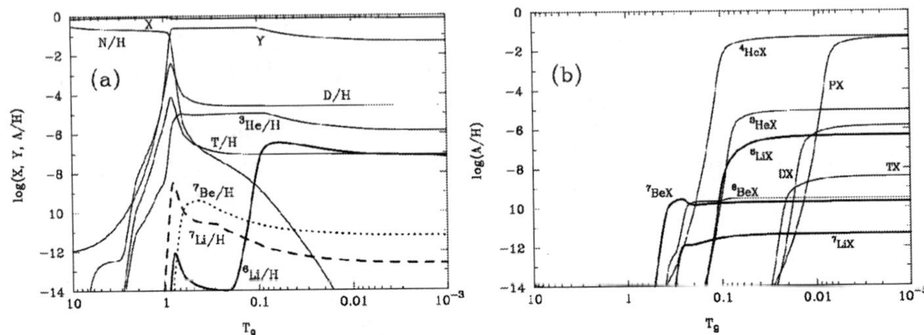

FIGURE 1. Calculated abundances of normal nuclei (left) and X-nuclei (right panel) as a function of T_9. For this figure we have taken the abundance of negatively charged particles X^- to be $Y_X = n_X/n_b = 0.1$, and its lifetime is taken to be long $\tau_X = \infty$. The figure is reprinted from Ref. [21].

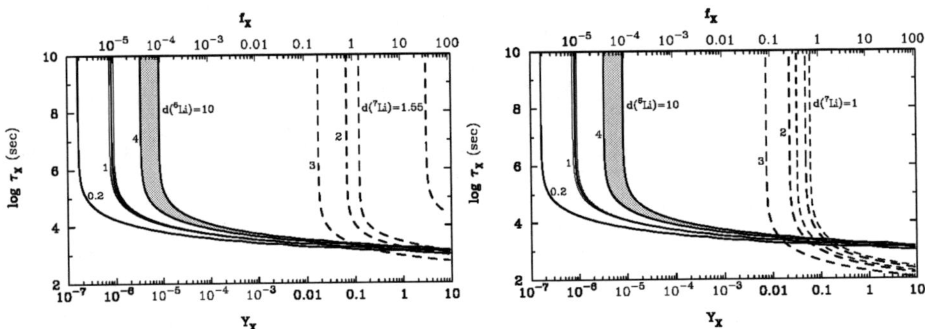

FIGURE 2. Contours of constant lithium abundance relative to the value observed in MPHSs, i.e., $d(^6\text{Li})$ = $^6\text{Li}^{\text{Calc}}/^6\text{Li}^{\text{Obs}}$ (solid curves) and $d(^7\text{Li})$ = $^7\text{Li}^{\text{Calc}}/^7\text{Li}^{\text{Obs}}$ (dashed curves) when the charged-current decay of $^7\text{Be}_X \rightarrow {}^7\text{Li}+X^0$ is not included (left) and is included (right panel). The adopted values of observed abundances are $^6\text{Li/H}= (7.1 \pm 0.7) \times 10^{-12}$ [1] and $^7\text{Li/H}= (1.23^{+0.68}_{-0.32}) \times 10^{-10}$ [2], respectively. Thin solid (dashed) lines around the line of $d(^6\text{Li}) = 1$ [$d(^7\text{Li}) = 1$] enclose the 1 σ uncertainty in the adopted observational constraint based upon the dispersion of the observed plateau. In the gray region, the condition $d(^6\text{Li}) > d(^7\text{Li})$ is satisfied. The figure is reprinted from Ref. [21].

this resonant channel is not important in this calculation using more realistic charge radii, or binding energies [21]. We confirmed that ^{6}Li is produced abundantly by the $^4\text{He}_X(d,X^-)^6\text{Li}$ reaction [16].

We calculate nucleosynthesis with parameters of the X^- particle; abundance Y_X and decay lifetime τ_X. We fixed baryon-to-photon ratio to be $\eta = 6.1 \times 10^{-10}$. For a set of parameters, light element abundances are obtained. This X-catalyzed BBN has differences in abundances of ^{6}Li and ^{7}Li, and other nuclear abundances were not influenced significantly. We obtain the calculated Li abundances relative to the observed abundances, which we designate as d-factors. These factors should be explained by a stellar depletion in atmospheres of MPHSs from the levels at the cosmological nucleosynthesis to those observed in MPHSs in order to obtain the observed abundances. d=1 satisfies the requirement that resulting lithium isotopic abundances are consistent with observations.

In the left panel of Fig. 2, the contours of $d(^6\text{Li})$ [solid curves] and $d(^7\text{Li})$ [dashed curves] are shown. The upper and lower solid curves correspond to the abundance level which satisfies our adopted constraint ($1.7 \times 10^{-12} \leq {}^6\text{Li/H} \leq 7.1 \times 10^{-11}$) as discussed in Ref. [21] from the abundance of MPHSs, $(^6\text{Li/H})_{\text{MPHS}} = (7.1 \pm 0.7) \times 10^{-12}$ [1]. Thus the upper right region of the figure is excluded for ^{6}Li overproduction. The right-upper side from the lower solid curve indicates the region for which the ^{6}Li abundance is higher than that observed in MPHSs. In the right-upper side from the second right dashed curves, the resulting ^{7}Li abundance is lower than $^7\text{Li/H} \approx 2.5 \times 10^{-10}$. Therefore, we conclude that BBN with negatively charged particles provides a simultaneous solution to the ^{7}Li overproduction problem and the ^{6}Li underproduction problem, (as also deduced by Ref. [19]) in the parameter region $Y_X \geq 0.9$, $\tau_X \approx (1.0 - 1.8) \times 10^3$ s.

Next, we consider another case where $^7\text{Be}_X$ converts to ^{7}Li by a weak charged current transition from X^- to X^0, i.e. $^7\text{Be}_X \rightarrow {}^7\text{Li}+X^0$. This quickly transforms $^7\text{Be}_X$ to ^{7}Li (type II model in Ref. [19]). In this model, the rate of recombination effectively determines the $^7\text{Be} \rightarrow {}^7\text{Li}$ conversion rate induced by X^-. The results of this case are shown in the right panel of Fig. 2. The general features are very similar to the previous case. However, due to a slightly stronger destruction rate due to the $^7\text{Be}_X \rightarrow {}^7\text{Li}+X^0$ decay followed by the $^7\text{Li}(p,\alpha)^4\text{He}$ and $^7\text{Li}(X^-,\gamma)^7\text{Li}_X(p,\alpha)^4\text{He}_X$ reactions, the contours of the ^{7}Li abundance are systematically shifted toward smaller values of Y_X. The parameter region which solves both the ^{6}Li and ^{7}Li problems also slightly shifts to $Y_X \approx 0.04 - 0.2$, $\tau_X \approx (1.4 - 2.6) \times 10^3$ s.

SUMMARY

We calculate light element nucleosynthesis during BBN with negatively-charged X^- particles in a fully dynamical manner. The ^{6}Li/H and ^{7}Li/H observed in MPHSs can constrain the lifetime and abundance of an X^- particle. These observational constraints require the lifetime and abundance to be in the ranges of $\tau_X \approx (1.0 - 1.8) \times 10^3$ s and $Y_X \geq 0.9$. When the reaction $^7\text{Be}_X \rightarrow {}^7\text{Li}+X^0$ is also taken into account, these ranges change to $\tau_X \approx (1.4 - 2.6) \times 10^3$ s and $Y_X \approx 0.04 - 0.2$. Therefore, introducing X^- particles with an adequate lifetime and abundance can be a solution for both of the factor of ~ 1000 underproduction of ^{6}Li and the factor of 3-4 overproduction of ^{7}Li in standard BBN.

A constraint on the X^- particle mass can be made from the dark-matter content deduced from the WMAP analysis of the CMB if X^- finally decays to a dark matter Y^0 and any residues. If the abundance of X^- particles is $Y_X \geq 0.1 - 1$ as summarized above, the mass of dark matter particles Y^0, which are produced from the decay of X^- particles, turns out to be $m_Y \leq 10 - 100$ GeV, thus leading to the constraint $m_X \leq O(100 \text{ GeV})$ when $m_X \sim m_Y$.

ACKNOWLEDGMENTS

We are very grateful to Professor Masayasu Kamimura for enlightening suggestions on the nuclear reaction rates for transfer and radiative capture reactions. This work has been supported in part by the Mitsubishi Foundation, the Grant-in-Aid for Scientific Research (17540275) of the Ministry of Education, Science, Sports and Culture of Japan, and the JSPS Core-to-Core Program, International Research Network for Exotic Femto Systems (EFES). MK acknowledges the support by the Japan Society for the Promotion of Science. Work at the University of Notre Dame was supported by the U.S. Department of Energy under Nuclear Theory Grant DE-FG02-95-ER40934. RNB gratefully acknowledges the support of the National Astronomical Observatory of Japan during his stay there.

REFERENCES

1. M. Asplund, D. L. Lambert, P. E. Nissen, F. Primas and V. V. Smith, *Astrophys. J.*, **644**, 229–259 (2006).
2. S. G. Ryan, T. C. Beers, K. A. Olive, B. D. Fields and J. E. Norris, *Astrophys. J.*, **530**, L57–L60 (2000).
3. J. Melendez and I. Ramirez, *Astrophys. J.*, **615**, L33–L36 (2004).
4. D. N. Spergel *et al.* [WMAP Collaboration], *Astrophys. J. Suppl.*, **170**, 377–408 (2007).
5. S. Dimopoulos, R. Esmailzadeh, G. D. Starkman, and L. J. Hall, *Phys. Rev. Lett.*, **60**, 7–10 (1988).
6. S. Dimopoulos, R. Esmailzadeh, G. D. Starkman, and L. J. Hall, *Astrophys. J.*, **330**, 545–568 (1988).
7. S. Dimopoulos, R. Esmailzadeh, L. J. Hall, and G. D. Starkman, *Nucl. Phys.*, **B311**, 699–718 (1989).
8. K. Jedamzik, *Phys. Rev. Lett.*, **84**, 3248–3251 (2000).
9. R. H. Cyburt, J. R. Ellis, B. D. Fields and K. A. Olive, *Phys. Rev.*, **D67**, 103521 (2003).
10. K. Jedamzik, *Phys. Rev.*, **D70**, 063524 (2004).
11. M. Kawasaki, K. Kohri and T. Moroi, *Phys. Rev.*, **D71**, 083502 (2005).
12. M. Kusakabe, T. Kajino and G. J. Mathews, *Phys. Rev.*, **D74**, 023526 (2006).
13. E. Rollinde, E. Vangioni and K. A. Olive, *Astrophys. J.*, **651**, 658–666 (2006).
14. R. H. Cyburt, J. R. Ellis, B. D. Fields, K. A. Olive and V. C. Spanos, *JCAP*, **0611**, 014 (2006).
15. M. Kaplinghat and A. Rajaraman, *Phys. Rev.*, **D74**, 103004 (2006).
16. M. Pospelov, *Phys. Rev. Lett.*, **98**, 231301 (2007).
17. K. Kohri and F. Takayama, *Phys. Rev.*, **D76**, 063507 (2007).
18. K. Hamaguchi, T. Hatsuda, M. Kamimura, Y. Kino and T. T. Yanagida, *Phys. Lett.*, **B650**, 268–274 (2007).
19. C. Bird, K. Koopmans and M. Pospelov, arXiv:hep-ph/0703096
20. M. Kusakabe, T. Kajino, R. N. Boyd, T. Yoshida and G. J. Mathews, *Phys. Rev.*, **D76**, 121302 (2007).
21. M. Kusakabe, T. Kajino, R. N. Boyd, T. Yoshida and G. J. Mathews, arXiv:0711.3858 [astro-ph].
22. E. Hiyama, Y. Kino, and M. Kamimura, *Prog. Part. Nucl. Phys.*, **51**, 223–307 (2003).

^{7}Li/^{6}Li Ratio in Distant Interstellar Media and Primordial ^{7}Li

– Implementation in Big-Bang Cosmology and Galactic Chemical Evolution –

S. Kawanomoto*, T. Kajino†, T. K. Suzuki**, H. Ando*, W. Aoki* and Subaru HDS collaborators*

**Optical and Infrared Astronomy Division, National Astronomical Observatory, Mitaka, Tokyo 181-8588, Japan*
†Division of Theoretical Astronomy, National Astronomical Observatory, Mitaka, Tokyo 181-8588, Japan
***Graduate School of Arts & Sciences, University of Tokyo, Komaba, Meguro, Tokyo, 153-8902, Japan*

Abstract. We have detected the isotopic abundance ratio of ^{7}Li/^{6}Li in the interstellar media (ISMs) along the line of sight to HD169454 and HD250290 using High Dispersion Spectrograph equipped with the Subaru Telescope. We also observed ζ Oph for the comparison purpose with the previous data. The first background star HD169454 is at the distance 1.7kpc inward from the solar sysytem in the Galactic plane, and the absorbing clouds are estimated to locate at 0.125kpc $\leq d \leq$ 0.7kpc and/or 1kpc $\leq d \leq$ 1.7kpc. If the observed ^{7}Li/^{6}Li ratio is for the gas cloud at the larger distance 1kpc $\leq d \leq$ 1.7kpc, this could be the first observation of the ISM ^{7}Li/^{6}Li ratio beyond the solar neighborhood. As for the second background star HD250290, the distance from the solar system is uncertain, 2.2 ± 0.2 kpc or $0.439^{+0.446}_{-0.147}$ kpc. The observed abundance ratios turn out to be ^{7}Li/^{6}Li $= 8.1^{+3.6}_{-1.8}$, and $6.3^{+3.0}_{-1.7}$ for HD169454 and HD250290, respectively. These values are in reasonable agreement with those observed previously in the ISMs of solar neighborhood within $\pm 1\sigma$ error bars and also consistent with our measurement of ^{7}Li/^{6}Li $= 7.1^{+2.9}_{-1.6}$ for nearby cloud along the line of sight to ζ Oph. This suggests a piece of evidence for homogeneous mixing and instantaneous recycling of gas component in the Galactic disk. We cannot claim a slope of the ^{7}Li/^{6}Li ratio within $\pm 1\sigma$ error bars as a function of Galactocentric distance R_G. Decomposing the observed ISM ^{7}Li/^{6}Li ratio into several possible source of ^{7}Li, we discuss three different contributions to the ^{7}Li from the Galactic cosmic-ray interactions, the stellar nucleosynthesis, and the Big-Bang nucleosynthesis.

Keywords: Big-Bang Nucleosynthesis, Light element abundances, Interstellar Media

1. INTRODUCTION

In the scheme of the Big-Bang cosmology, elemental abundances of light elements preserve rich information from the early Universe to the present Milky Way Galaxy. Significant amount of ^{2}H(D), ^{3}He, ^{4}He, and ^{7} Li were synthesized during the first three minits in the Universe. Sinse primordial abundances of these elements are quite sensitive to the cosmological parameter $\Omega_b \equiv \rho_b/\rho_C$, where ρ_b is universal baryon density and ρ_C is the critical density to close the Universe, a number of observations have been devoted to determine these primordial abundances. Recently independent Ω_b have been estimated from the WMAP data of CMB anisotropies [17, 18], however, there still serious debate concerning the extent of systematic errors in the observed primordial

CP1016, *Origin of Matter and Evolution of Galaxies,*
edited by T. Suda, T. Nozawa, A. Ohnishi, K. Kato, M. Y. Fujimoto, T. Kajino, and S. Kubono

abundances (see figure 22 in refs. [7]). Thus precise determination of the primordial abundance of the ^{7}Li is urgently sought.

^{7}Li, however, has a disadvantage because of its considerably smaller primordial abundance than D and ^{4}He. Another disadvantage has been the fact that determination of the primordial ^{7}Li abundance has been limited to observations of the photospheres of nearby metal-deficient dwarf stars whose Li abundances are subject to uncertainties due to poorly understood processes of convective mixing [15, 10] that can result in overionization and also nuclear destruction of Li in stellar envelopes.

An independent method was proposed [4, 6] to make use of the isotopic abundance ratio ^{7}Li/^{6}Li in the ISM. This method has several advantages: First and foremost, both ^{7}Li and ^{6}Li in the ISM are totally free from stellar depletion effects and therefore the ^{7}Li/^{6}Li ratio is not affected by the destruction sink of stellar processing. Second, since both isotopes ^{7}Li and ^{6}Li have the same chemical properties, any effects due to overionization and dust condensation are canceled in their ratio.

The ^{7}Li/^{6}Li ratio presumably depends on Galactic chemical evolution (GCE) which affects each isotope in a different manner. This complication should be untangled when one decomposes the observed ISM–^{7}Li/^{6}Li ratio into the primordial abundance component and the other production components. The only way to produce the ^{6}Li is the Galactic Cosmic Ray interactions with the ISMs. However, ^{7}Li has at least three different stellar production sites in addition to the GCR-ISM interactions. These are nucleosynthesis in type II supernova explosion, nova nucleosynthesis, and nucleosynthesis in asymptotic giant branch stars. These production sources have different intensities and efficiencies which depend on the star formation rate and on the stellar-mass vs. gas-mass fraction as a function of Galactocentric distance, R_G.

We propose our results of the observed lithium isotopic ratio ^{7}Li/^{6}Li of the ISMs probable far from the Sun and introduce the application the isotopic ratio to the Big-Bang Cosmology.

2. OBSERVATION AND DATA REDUCTION

In order to obtain data covering wide range of R_G, we chose HD169454 and HD250290 as background stars located near the Galactic center and anticenter direction, respectively. The distances of these stars are estimated at 1.7kpc and 2.2kpc from the Sun, respectively. Although distance to the absorbing cloud can not be presumed exactly, in the most optimistic case, R_G range of absorber is widen to $\sim$3 kpc around the Sun. In addition to these two stars, we chose one more nearby star ζ Oph as a reference because it was most frequently studies by several groups. It is located at a distance 0.15kpc from the Sun. The basic parameters for the background stars are summerized in TABLE 1.

TABLE 1. Observed data for background stars.

Star	Sp. T.	V (mag)	$B-V$ (mag)	l (deg)	b (deg)	$E(B-V)$ (mag)	d (kpc)	R_G (kpc)
ζ Oph	O9V	2.58	+0.02	6.3	+23.6	+0.30	0.15	8.36
HD169454	O+	6.65	+0.74	17.5	−0.7	+1.02	1.7	6.8
HD250290	B3Ib	7.41	+0.49	186.6	+0.1	+0.62	2.2	10.7

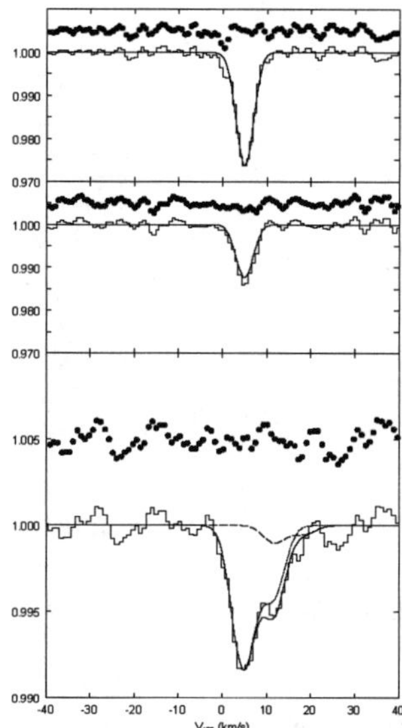

FIGURE 1. Absorption spectra toward HD169454. Upper and Middle: The K I absorption line at 4044Å and 4047Åwith the best fit model (solid line). Residuals are shown as filled circles with offset value of 1.005. Bottom: The Li I 6708Å absorption lines with the best fit (solid line). ^{7}Li and ^{6}Li components are shown as dotted and dashed lines, respectively. Residuals are shown as filled circles with offset value of 1.005.

All observations were taken with the High Dispersion Spectrograph (HDS) [14] on the optical-Nasmyth platform of the 8.2m Subaru Telescope of National Astronomical Observatory of Japan at the summit of Mauna Kea. In order to resolve isotopic shift between ^{7}Li and ^{6}Li, we set the spectral resolving power at $\lambda/\Delta\lambda \sim 100,000$. Typical signal-to-noise ratio was higher than 1,000 per pixel element. The observational data were collected in two spectral regions: one a blue region including K I 4044 Å and 4047 Å for resolving the velocity structure of the absorbing cloud, the other a red region including the Li I 6708 Å for investigating the isotopic ratio of Li. For ζ Oph, only a red spectrum was obtained. We adopted the velocity component estimated by previous work [2, 11, 13, 3]. Our objects were sufficiently bright on the Subaru Telescope, that all exposures were limited to 10 minutes at most to avoid saturation. We took tens of exposures for each object to attain the required S/N ratio.

Observational data were reduced in a standard way using Image Reduction and Analysis Facility (IRAF) [1].

[1] IRAF is distributed by National Optical Astronomy Observatory, which is operated by the Association of Universities for Research in Astronomy, Inc., under cooperative agreement with the National Science Foundation.

3. ANALYSIS

The ^{7}Li- and ^{6}Li-absorption lines overlap each other. In addition, the line profile of each Li isotope itself is complicated due to doublets. Moreover, the observed Li line profile is presumed to be the superposition of a certain number of interstellar clouds with a variety of doppler shifts. The K I line is invaluable for decomposing the contributions of individual clouds because K I 4044 Å and 4047 Å lines are singlets at our wavelength resolving power.

Assuming that an absorption line can be represented by a simple Gaussian profile, there are three free parameters used to reproduce absorption line profile, equivalent width, line center wavelength and Gaussian FWHM (full width at the half maximum). We used an alternative equivalent parameter set to determine each line; equivalent width, radial velocity and velocity spread parameter. In addition to these three parameters, we introduced a fourth parameter to define the Li isotopic ratio, ^{7}Li/^{6}Li. In order to find the absorption line parameters that minimize the χ^2 between observations and synthetic profiles, we use an iterative process.

An example of the observational data and line profile fitting are shown in FIGURE 1. And all results of the profile fitting are summarized in TABLE 2.

TABLE 2. Results of profile fitting.

Star	Species	V_{LSR} (kms^{-1})	b (km/s)	E.W. (mÅ)	N (10^9cm^{-2})	^{7}Li/^{6}Li
ζ Oph	^{7}Li I	-0.47 ± 0.08	1.26 ± 0.19	1.06 ± 0.04	3.56 ± 0.14	
	^{6}Li I	-0.47 ± 0.08	1.26 ± 0.19	0.15 ± 0.04	0.50 ± 0.14	$7.1^{+2.9}_{-1.6}$
HD169454	^{39}K I	$+4.92\pm0.06$	1.97 ± 0.12	2.53 ± 0.07	$4.63\pm0.13\times10^4$	
	^{7}Li I	$+4.76\pm0.14$	3.15 ± 0.16	1.81 ± 0.06	6.08 ± 0.20	
	^{6}Li I	$+4.76\pm0.14$	3.15 ± 0.16	0.22 ± 0.06	0.74 ± 0.20	$8.1^{+3.6}_{-1.8}$
HD250290	^{39}K I	$+0.90\pm0.11$	3.52 ± 0.14	4.80 ± 0.14	$8.78\pm0.26\times10^4$	
	^{7}Li I	$+0.64\pm0.09$	3.35 ± 0.11	4.21 ± 0.10	14.15 ± 0.33	
	^{6}Li I	$+0.64\pm0.09$	3.35 ± 0.11	0.69 ± 0.23	2.32 ± 0.77	$6.3^{+3.0}_{-1.7}$

4. RESULTS AND DISCUSSIONS

In FIGURE 2, we show our data together with the previously observed data. Our new data are in reasonable agreement with the mean value of those observed previously in the solar neighborhood within $\pm1\sigma$ error bars except for ρ Oph.

As mentioned in section 1, ^{6}Li is entirely synthesized by GCR-ISM interactions and ^{7}Li has three different production source; (1) GCR-ISM interaaction; (2) stellar nucleosynthesis (SNe, novae, AGB stars); and (3) BBN. The observed ^{7}Li/^{6}Li ratio can thus be written

$$\left(\frac{^{7}\mathrm{Li}}{^{6}\mathrm{Li}}\right)_{\mathrm{obs}} = \frac{^{7}\mathrm{Li}_{\mathrm{GCR}}}{^{6}\mathrm{Li}_{\mathrm{GCR}}} + \frac{^{7}\mathrm{Li}_{\mathrm{STAR}}}{^{6}\mathrm{Li}_{\mathrm{GCR}}} + \frac{^{7}\mathrm{Li}_{\mathrm{BBN}}}{^{6}\mathrm{Li}_{\mathrm{GCR}}} . \tag{1}$$

Since the average trend of the observed $(^{7}\mathrm{Li}/^{6}\mathrm{Li})_{\mathrm{obs}}$ ratios as shown in FIGURE 2 does not describe a clear gradient as a function of Galactocentric distance, we assume that

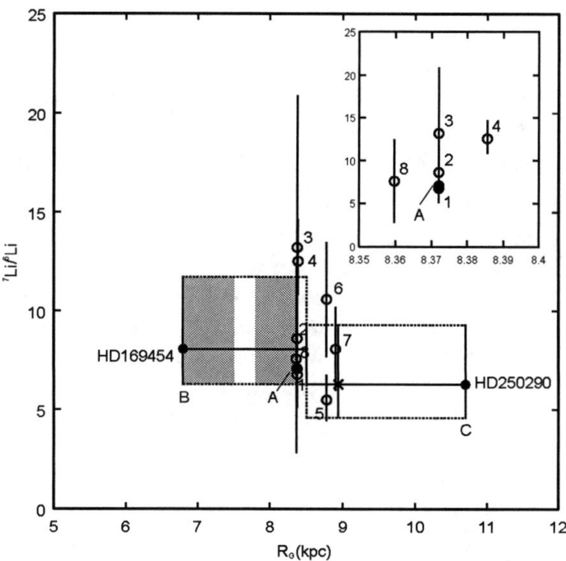

FIGURE 2. The lithium isotopic ratio, ^{7}Li/^{6}Li, vs. Galactocentric distance, R_G. Filled circles are our data for the ISM ^{7}Li/^{6}Li toward ζ Oph (denoted by A), HD169454(B) and HD250290(C). Open circles are the previous data from Meyer et al. [11], Lemoine et al. [13] and Howarth et al. [3] for ζ Oph (1 to 3), Meyer et al. [11] and Knauth et al. [8] for ζ Per (5 and 6), Knauth et al. [9] for o Per (7), Lemoine et al. [12] for ρ Oph (4) and Knauth et al. [9] for χ Oph (8), respectively

the $(^7\mathrm{Li}/^6\mathrm{Li})_{\mathrm{obs}}$ is independent of R_G. The mean weighted average of all observed data including ours is $(^7\mathrm{Li}/^6\mathrm{Li})_{\mathrm{obs}} = 7.6 \pm 2.3(1\sigma)$. Let us try to decompose this into several production sources of ^{7}Li.

The first term on the RHS of Equation (1) takes a constant value of $^7\mathrm{Li_{GCR}}/^6\mathrm{Li_{GCR}} = 1.5 \pm 0.3$ almost independent of the GCE model [16, 1]. The second term is calculated theoretically in our GCE model by using stellar nucleosynthesis yields of ^{7}Li from several different sources divided by the ISM ^{6}Li abundance of $^6\mathrm{Li/H} = 3.4 \times 10^{-10}$ [19, 5]. The stellar term is estimated to be $^7\mathrm{Li_{STAR}}/^6\mathrm{Li_{GCR}} = 5.4 \pm 1.6$. The third primordial term can be determined by subtracting the first and second terms from the observed value $(^7\mathrm{Li}/^6\mathrm{Li})_{\mathrm{obs}} = 7.6 \pm 2.3(1\sigma)$: $^7\mathrm{Li_{BBN}}/^6\mathrm{Li_{GCR}} = 0.7^{+2.3}_{-0.7}(1\sigma)$. Multiplying by the ISM ^{6}Li abundance, we can estimate the corresponding primordial abundance of ^{7}Li to be $^7\mathrm{Li/H} = 2.4^{+7.8}_{-2.4}(1\sigma) \times 10^{-10}$.

To determine more precise primordial abundance than this value, we should observe the outer region of the Galaxy. Because there are less effective stellar activities in the outer region of the Galaxy, the ^{7}Li/^{6}Li ratio should be dominated by the BBN term.

5. CONCLUSIONS

We reported observations of the interstellar 6708Å Li I absorption lines obtained using the Subaru Telescope with spectral resolution of $\sim 100,000$ and signal-to-noise ratio $> 1,000$. We obtained values of the ^{7}Li/^{6}Li isotopic ratio at 1σ error, $7.1^{+2.9}_{-1.6}$, $8.1^{+3.6}_{-1.8}$ and $6.3^{+3.0}_{-1.7}$ for ζ Oph, HD169454 and HD250290, respectively. If the gas clouds are located near the background stars, this would be the first observational data of ^{7}Li/^{6}Li in the ISMs beyond the nearby gas clouds of the solar neighborhood.

These values were compared with previous data for the local gas clouds near the Sun, and we found our observed ratios to be in reasonable agreement within $\pm 1\sigma$ error bars.

Assuming an R_G–independence of each component of ^{7}Li/^{6}Li from the GCR-ISM interactions ($^7\mathrm{Li}_{\mathrm{GCR}}/^6\mathrm{Li}_{\mathrm{GCR}}$), stellar nucleosynthesis ($^7\mathrm{Li}_{\mathrm{STAR}}/^6\mathrm{Li}_{\mathrm{GCR}}$), and Big-Bang nucleosynthesis ($^7\mathrm{Li}_{\mathrm{BBN}}/^6\mathrm{Li}_{\mathrm{GCR}}$), we decompose the observed ratio into these three source components. We find that the primordial ^{7}Li abundance is likely to be $^7\mathrm{Li/H} = 2.4^{+7.8}_{-2.4}(1\sigma) \times 10^{-10}$ or $^7\mathrm{Li/H} \leq 10 \times 10^{-10}$ which are estimated from our inferred ratio $^7\mathrm{Li}_{\mathrm{BBN}}/^6\mathrm{Li}_{\mathrm{GCR}} = 0.7^{+2.3}_{-0.7}(1\sigma)$.

In the next giant telescope era, we can find the isotopic ratio in outer region of the Galaxy and more strict constraint to the primordial ^{7}Li abundance will be defined.

REFERENCES

1. Austin, S. M. 1981, Progr. Part. Nucl. Phys. 7, 1
2. Ferlet, R., Dennefeld, M. 1984, A&A. 138, 303
3. Howarth, I. D., Price, R. J., Crawford, I. A., Hawkins, I. 2002, MNRAS. 335, 267
4. Kajino, T., Suzuki, T.-K., Kawanomoto, S. Ando, H. 2000, IAU Circ. 198, 344
5. Kawanomoto, S. 2002, PhD thesis University of Tokyo
6. Kawanomoto, S., Suzuki, K. T. Ando, H., Kajino, T. 2003, Nucl. Phys. A 718. 659
7. Kirkman, D., Tytler, D., Suzuki, N., O'Meara, J. M., Lubin, D. 2003, ApJS. 149, 1
8. Knauth, D. C., Federman, S. R., Lambert, D. L., Crane, P. 2000, Nature 405, 656
9. Knauth, D. C., Federman, S. R., Lambert, D. L. 2003, ApJ. 586, 268
10. Kurucz, R. L. 1995, ApJ. 452, 102
11. Meyer, D. M., Hawkins, I., Wright, E. L. 1993, ApJL. 409, L61
12. Lemoine, M., Ferlet, R., Vidal-Madjar, A., Emerich, C., Bertin, P. 1993, A&A. 269, 469
13. Lemoine, M., Ferlet, R., Vidal-Madjar, A. 1995, A&A. 298, 879
14. Noguchi, K., Aoki, W., Kawanomoto, S. Ando, H., Honda, S., Izumiura, H., Kambe, E., Okita, K., Sadakane, K., Sato, B., Tajitsu, A., Takada-Hidai, M., Tanaka, W., Watanabe, E., Yoshida, M. 2002, PASJ. 54, 855
15. Pinsonneault, M. H., Walker, T. P., Steigman, G., Narayanan, Vijay K. 1999, ApJ. 527, 180
16. Ramaty, R., Scully, S. T., Lingenfelter, R. E., & Kozlovsky, B. 2000, ApJ, 534, 747
17. Spergel, D. N., Verde, L., Peiris, H. V., Komatsu, E., Nolta, M. R., Bennett, C. L., Halpern, M., Hinshaw, G., Jarosik, N., Kogut, A., Limon, M., Meyer, S. S., Page, L., Tucker, G. S., Weiland, J. L., Wollack, E., Wright, E. L. 2003, ApJS. 148, 175
18. Spergel, D. N., Bean, R., Doré, O., Nolta, M. R., Bennett, C. L., Dunkley, J., Hinshaw, G., Jarosik, N., Komatsu, E., Page, L., Peiris, H. V., Verde, L., Halpern, M., Hill, R. S., Kogut, A., Limon, M., Meyer, S. S., Odegard, N., Tucker, G. S., Weiland, J. L., Wollack, E., Wright, E. L. 2007, ApJS. 170, 377
19. Suzuki, T. K., Yoshii, Y., & Kajino, T. 1999, ApJL, 522, L125

3. SPECIAL SESSION: FIRST STARS (1) NUCLEAR ABUNDANCES IN FIRST STARS

Observations of Very Metal-Poor Stars in the Galaxy

Timothy C. Beers

Department of Physics and Astronomy, Center for the Study of Cosmic Evolution, and Joint Institute for Nuclear Astrophysics, Michigan State University, E. Lansing, MI 48824 USA

Abstract.
I report on recent results from observations of stars with metallicities [Fe/H] ≤ -2.0. These include a substantial new sample of objects with high-resolution observations obtained as part of a follow-up of the HK Survey, The Hamburg/ESO Survey, and the ongoing survey SEGUE: Sloan Extension for Galactic Understanding and Exploration. Perspectives on the next directions are also provided.

Keywords: Galaxy: abundances – nuclear reactions, nucleosynthesis, abundances – stars: abundances – stars: Population II
PACS: 26.30.Hj, 97.10.Tk, 97.20.Tr, 97.30.Hk

THE IMPORTANCE OF VERY METAL-POOR STARS

It is widely recognized that stars with metallicities less than 1% of the solar value ([Fe/H] < -2.0; the Very Metal-Poor (VMP) stars in the nomenclature of Beers & Christlieb [1] provide invaluable information on the origin of the elements in the Universe, and on the nature of the neutron-capture processes that were the dominant source of the elements beyond the iron peak. Perhaps less appreciated is the fact that these same stars place strong constraints on the formation and evolution of individual galaxies, and thereby provide a direct connection with cosmological studies. Within the Milky Way, the shape of the low-metallicity tail of the Metallicity Distribution Function (MDF) is beginning to reveal the characteristic abundances of elements associated with the major epochs of star formation in the early Galaxy. Changes in the MDF as a function of distance are revealing the assembly history of the Galaxy (see, e.g., the article by Carollo & Beers in this volume). The frequency of various abundance signatures, e.g., the [C/Fe] and [α-element/Fe] ratios, are being used to investigate the dominant production sites, and the Initial Mass Function (IMF) of the stellar populations in the proto-Galactic fragments that theory suggests were involved in galaxy assembly. In this short review, I summarize a few of the recent observational results, and indicate the directions which the field will be going in the next few years.

PAST AND ONGOING SURVEYS

Past searches for VMP stars include those based on the selection of high proper-motion stars in the local volume [e.g., 2, 3], and in particular, stars identified from non-kinematically biased *in-situ* objective-prism surveys, such as the HK survey of Beers

CP1016, *Origin of Matter and Evolution of Galaxies,*
edited by T. Suda, T. Nozawa, A. Ohnishi, K. Kato, M. Y. Fujimoto, T. Kajino, and S. Kubono

and colleagues [4, 5], and the more recent Hamburg/ESO Survey (HES) of Christlieb and collaborators [6]. These two surveys form the basis for the majority of the high-resolution spectroscopic studies that have been carried out over the past decade, such as the First Stars program of Cayrel et al. [7], the HERES project of Barklem et al. [8], and the 0Z Project of Cohen et al. [9].

Although work continues on the medium-resolution follow-up of HK and HES candidate VMP stars, to date the efforts have yielded roughly 2500 firm identifications. However, it should be emphasized that as these stars are selected in the volume sorrounding the Sun, and explore no more than 10 kpc (in the case of the HK survey) to 15 kpc (in the case of the HES) away, these samples are dominated by the now-recognized inner-halo population of the Galaxy [10]. The outer-halo population, which Carollo et al. suggest is comprised of a MDF with peak metallicity a factor of three times lower ([Fe/H]$= -2.2$) than that associated with the inner halo ([Fe/H]$= -1.6$), dominates outside of 15-20 kpc from the Galactic center. This may well account for the difficulty in the identification of significant numbers of stars at the lowest metallicities, the Ultra Metal-Poor (UMP) and Hyper Metal-Poor (HMP) stars, with [Fe/H] < -4.0 and [Fe/H] < -5.0, respectively. If progress is to be made in finding large numbers of such stars, different tactics must be explored.

Fortunately, this process has already begun. Already, the largest numbers of VMP stars revealed to date has emerged from medium-resolution spectroscopic spectra taken during the course of the SDSS (primarily calibration stars) and SEGUE. SEGUE is a survey directed at studies of Galactic structure, but it includes a number of target categories (such as F turnoff stars, K giants, and a low-metallicity category extending across all spectral types) that is successful in the identification of large numbers of VMP stars. Even though SEGUE is not yet complete, the total list of VMP stars already exceeds 10,000 stars, quadruple the number from the HK/HES efforts combined. Of greatest importance, SDSS/SEGUE can explore to much larger distances, beyond the 10-15 kpc region where the inner-halo population dominates. Furthermore, the availability of reasonably accurate proper motions for the subset of nearby stars (out of 4-5 kpc from the Sun) makes it feasible to identify candidate VMP, EMP, UMP, and HMP stars based on the characteristic retrograde signature of the outer-halo population. This mode of selection has not been implemented in SEGUE, but it is one of the target categories that will be explored in the SEGUE-2 project, which will be executed from July 2008 to July 2009 as part of the proposed next extension of the Sloan Survey, known as SDSS-III (see `http://www.sdss3.org/outermilkyway.php`). We are hopeful that this dedicated effort will finally be able to break through the UMP/HMP barrier.

Space precludes a comprehensive discussion of even the most recent results coming from high-resolution spectroscopic studies of VMP stars, so below I simply summarize a few of the results of greatest relevance to this conference. Other examples can be gleaned from the online version of my talk.

RECENT RESULTS

Carbon-Enhanced Metal-Poor (CEMP) stars

The CEMP stars (and their sub-categories) are defined by Beers & Christlieb [1] based on the criteria listed in Table 1. Their clear signficance to understanding the chemical evolution of the Galaxy results from the apparently large fraction of such stars that are found among the VMP stars (which varies between 10% and 20%, according to investigations carried out to date). At the lowest iron abundances, CEMP stars represent 40% of stars with [Fe/H] < -3.5 [1], and 100% (3 of 3) of stars known with [Fe/H] < -4.0 [11, 12, 13]. A number of articles in this volume discuss attempts to understand the origin of the UMP and HMP CEMP stars.

TABLE 1. Definition of sub-classes of metal-poor stars (based on Beers & Christlieb [1]

Carbon-enhanced metal-poor stars	
CEMP	[C/Fe] $> +1.0$
CEMP-r	[C/Fe] $> +1.0$ and [Eu/Fe] $> +1.0$
CEMP-s	[C/Fe] $> +1.0$, [Ba/Fe] $> +1.0$, and [Ba/Eu] $> +0.5$
CEMP-r/s	[C/Fe] $> +1.0$ and $0.0 <$ [Ba/Eu] $< +0.5$
CEMP-no	[C/Fe] $> +1.0$ and [Ba/Fe] < 0
Neutron-capture-rich stars	
r-I	$+0.5 \leq$ [Eu/Fe] $\leq +1.0$ and [Ba/Eu] < 0
r-II	[Eu/Fe] $> +1.0$ and [Ba/Eu] < 0
s	[Ba/Fe] $> +1.0$ and [Ba/Eu] $> +0.5$
r/s	$0.0 <$ [Ba/Eu] $< +0.5$

Aoki et al. [14] reports new observations (obtained with the Subaru telescope) and summarizes results from the high-resolution studies of CEMP stars reported in the literature. Among the most intriguing results from this study concerns the apparent difference in the MDFs (see Figure 1) between the CEMP-s (those exhibiting abundance signatures characteristic of the s-process, and which likely originated from mass-tranfer events from a now-deceased intermediate mass AGB companion) stars, and the CEMP-no (those exhibiting no neutron-capture elements, and whose origin remains the subject of current discussion) stars.

Aoki et al. [15] presents additional Subaru results for a small sample (seven) of CEMP stars selected from SDSS/SEGUE to be located at the main-sequence halo turnoff. These stars are important because they lie in the region of the H-R diagram where there is no possibility of extensive mixing of their outer atmospheres having taken place (at least via usual mechanisms); they thus represent an essentially pure signature of the elemental abundance patterns produced by their progenitors. It is of interest that the majority of the CEMP stars in this study exhibit kinematics that may associate them with the outer-halo population reported by Carollo et al. [10].

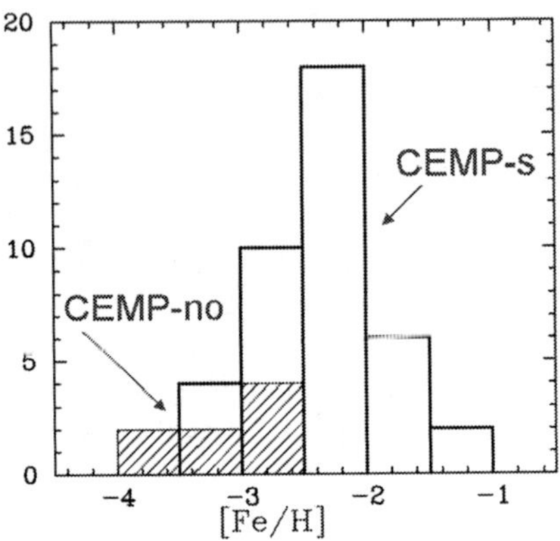

FIGURE 1. Distribution of [C/Fe] for CEMP stars discussed by Aoki et al. [14]. Note the predominance of the CEMP-no stars (cross-hatched bars) at the lowest [Fe/H].

Another recent study has measure the abundance of flourine (F) in a CEMP-s star [16, see Figure 2], using the Phoenix near-IR high-resolution spectrograph on the Gemini-S telescope. This observation represents the first such detection in a CEMP star (and revealed this star to have an over-abundance [F/Fe] $= +2.9$, nearly 1000 times that of the solar ratio). This element provides a senstive probe of the operation of the s-process in AGB stars [e.g., 17], and may provide information on the IMF from which the AGB progenitor was drawn [18]. Additional high-resolution spectroscopy of CEMP stars for the study of this element are clearly required.

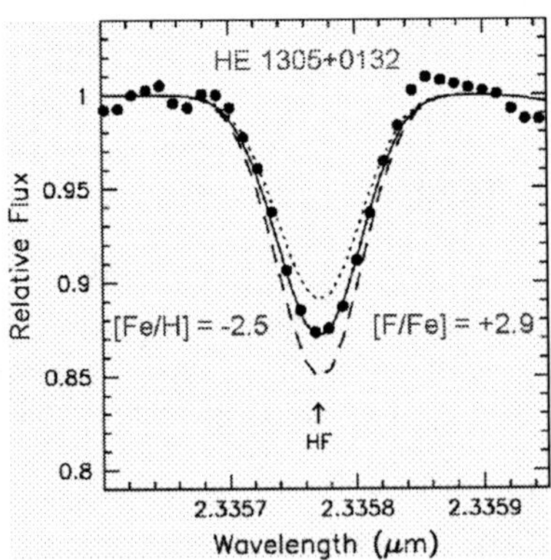

FIGURE 2. The detection of F in a VMP CEMP star by Schuler et al. [16]; the HF molecular feature is very prominent.

Among the recent theoretical studies of the significance of the CEMP stars, I draw attention to the work of Tumlinson [19], which has argued that the critical mass of early star formation in the Universe was shifted to favor the formation of intermediate-mass stars (essentially due to the "floor" set by the cosmic microwave background temperature

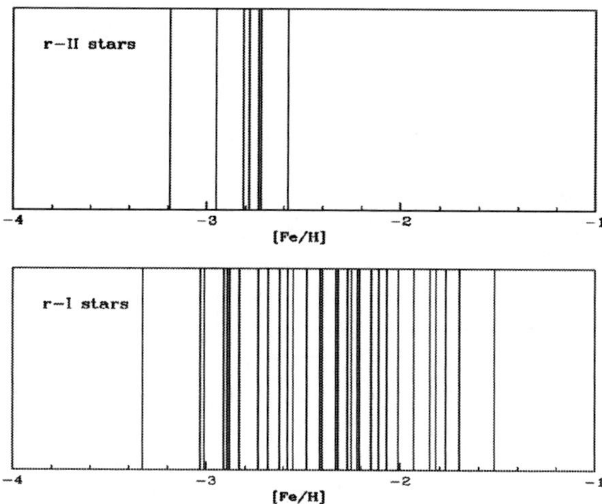

FIGURE 3. Distribution of [Fe/H] for r-process-enhanced stars identified by Barklem et al. [8]. Note the obvious contrast between the behavior of the r-II stars with that of the r-I stars.

at the time) that are copius carbon producers. Others [e.g., 20, 21] have argued that significant carbon (and nitrogen) may be produced from the winds of massive Mega Metal-Poor (MMP; [Fe/H] < -6), rapidly rotating stars that might have formed at the very earliest epochs of the Universe. Such stars would not be expected to produce s-process elements, but their association with the origin of the CEMP-no stars has yet to be established.

Highly r-process-enhanced stars

One of the goals of the HERES survey of Barklem et al. [8] was to dramatically increase the numbers of VMP stars with enhanced r-process elements. This goal was met; eight new examples of the phenomenon were identified, and a critical difference was noted between the MDF of the so-called r-II stars (those with [Eu/Fe] $> +1.0$, and low Ba, according to Table 1), and the moderately r-process-enhanced stars ($+0.5 \leq$ [Eu/Fe] $< +1.0$). As is clear from Figure 3, while the r-I stars exhibit an MDF that covers a wide range, the MDF of the r-II stars is restricted to roughly $-3.2 <$ [Fe/H] < -2.6).

An additional handful of r-II stars have since been identified by other studies, including at least one with detected uranium ([22]; HE 1523-0901, with [Fe/H] $= -2.95$); others are still unpublished. It is presumably no coincidence that *all* of the newly discovered r-II stars fall in the same low metallicity interval as those from the Barklem et al. study. One possible explanation is that the main astrophysical r-process arises from from stars of a relatively small mass range (many have suggested 10-15 $M_{\odot}$), which may have been the dominant contributor of heavy elements at the time when the Galaxy reached a mean metallicity around [Fe/H] ~ -3.0. Other alternatives surely exist.

Observations of light n-capture elements

Although the stars with large enhancements of n-capture elements are of great interest, it is equally important to consider the abundances of n-capture elements for the great majority of metal-poor stars with "normal" levels, as these help constrain the question of the universality of the r-process, and address whether multiple r-process sites must be considered. Recently, Francois et al. [23] reported abundance determinations for some 16 n-capture elements for a sample of 32 VMP and EMP stars observed during the course of the Cayrel et al. First Stars program. These data greatly enlarge the numbers of stars with [Fe/H] < -2.8 with such measurements available. The star-to-star scatter among these elements reaches a peak at around [Fe/H] $= -3.0$, but below this metallicity, too few stars exist with measurements to be certain of the behavior.

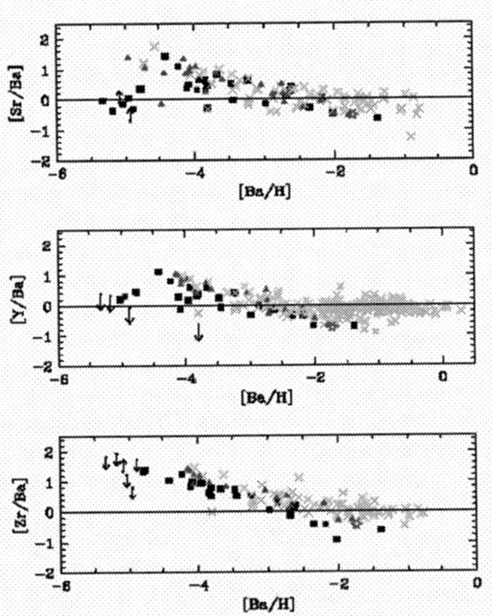

FIGURE 4. Reported abundances for Sr, Y, and Zr, relative to Ba, as a function of [Ba/H], from Francois et al. [23]. The new data are shown as the black squares (or upper limits). Other data are from the literature.

By adopting the element Ba as a reference element, Francois et al. demonstrate the existence of strong anti-correlations of the lighter n-capture ratios [Sr/Ba], [Y/Ba], and [Zr/Ba] with the [Ba/H] abundance from [Ba/H] ~ -1.5 down to [Ba/H] ~ -4.5 (Figure 4). It remains unclear if this behavior changes below [Ba/H] $= -4.5$, as only upper limits on the lighter element ratios are available for the majority of stars in this region; the few measurements that *do* exist suggest a possible reversal of the trends observed above this limit. In any case, the results confirm that there must exist a second n-capture process to account for the synthesis of the first r-process peak elements, variously referred to in the literature as the "weak" r-process [24] or the Light Element Production Process [LEPP 25]. Subtraction of the predicted contributions to these elements by the main r-process suggests that this mechanism is responsible for the production of 90-95% of the observed amount of Sr, Y, and Zr in stars with [Ba/H] > -4.5. See Montes et al. [26] for additional discussion of the LEPP.

NEW AND ANTICIPATED HIGH-RESOLUTION SPECTROSCOPIC STUDIES

There exist a number of large high-resolution spectroscopic surveys that are just now getting underway, which should greatly enlarge the numbers of VMP stars with available elemental abundance information. The Chemical Abundances of Halo Stars (CASH) survey is making use of the Hobby-Eberly Telescope to obtain moderately high-resolution ($R = 15,000$) spectroscopy for up to 1000 VMP stars identified during the course of the HK-I, HK-II, and HES efforts, with a large number of additional stars from SDSS/SEGUE. Aoki et al. have recently been awarded a Key Project status on the Subaru Telescope, with the aim of obtaining $R = 50,000$ spectroscopic observations for up to 200 VMP stars, the majority of which will be drawn from SDSS/SEGUE targets. Both of these surveys have as one of their primary aims to test for the existence (or not) of chemical signatures that might be associated with the inner/outer halo dichotomy reported by Carollo et al. [10].

In the near future, the proposed SDSS-III project will undertake a massive survey of some 100,000 red giants with resolving power $R = 20,000$, concentrating on the H-band region in the near-IR. This survey, known as APOGEE (APO Galactic Evolution Experiment; see `http://www.sdss3.org/innermilkyway.php`) is expected to take place between 2011 and 2014, and will target objects in the Galactic bulge, bar, disk, and halo components. It is expected that on the order of 15 individual elements will be obtained per star.

Of course, we all look forward to the possible execution of the WFMOS (Wide Field Multi-Object Spectrograph) survey of on the order of one million stars at resolving power $R = 50,000$. Currently, this Gemini instrument is expected to be mounted on the prime focus of the Subaru telescope, in order to take advantage of its wide field of view. The hope, and expectation, is that this survey will finally reveal the elemental abundances of stars that probe the entire history of chemical evolution throughout the Galaxy.

ACKNOWLEDGMENTS

The author is grateful to the organizers for providing assistance with travel and accomodation expenses. This work also received support from grants AST 07-07776 and PHY 02-15783; Physics Frontier Center / Joint Institute for Nuclear Astrophysics (JINA), awarded by the US National Science Foundation.

REFERENCES

1. T. C. Beers & N. Christlieb, "The Discovery and Analysis of Very Metal-Poor Stars in the Galaxy," in *Annual Reviews of Astronomy & Astrophysics*, **43**, 2005, pp. 531–580.
2. S. G. Ryan & J. E. Norris, *Astron. J.* **101**, 1835–1864 (1991).
3. B. W. Carney et al., *Astron. J.* **112**, 668–692 (1996).
4. T. C. Beers, G. W. Preston, & S. A. Shectman, *Astron. J.* **90**, 2089–2102 (1985).
5. T. C. Beers, G. W. Preston, & S. A. Shectman *Astron. J.* **103**, 1987–2034 (1992).
6. N. Christlieb, "Finding the Most Metal-Poor Stars of the Galactic Halo with the Hamburg/ESO Objective-Prism Survey," in *Reviews of Modern Astronomy*, **16**, 2003, pp. 191–206.
7. R. Cayrel et al., *Astron. & Astrophys.* **416**, 1117–1138 (2004).
8. P. S. Barklem et al., *Astron. & Astrophys.* **439**, 129–151 (2005).
9. J. G. Cohen et al., arXiv: 0709.1279 (2007).
10. D. Carollo et al., *Nature* **450**, 1020–1025 (2007).
11. N. Christlieb et al., *Nature* **419**, 904–906 (2002).
12. A. Frebel et al., *Nature* **434**, 871–873 (2005).
13. J. E. Norris et al., *Astrophys. J.* **670**, 774–788 (2007).
14. W. Aoki et al., *Astrophys. J.* **655**, 492–521 (2007).
15. W. Aoki et al., *Astrophys. J.*, in press (2008)
16. S. C. Schuler et al., *Astrophys. J.* **667**, L81–L84 (2007).
17. S. W. Campbell & J. C. Lattanzio, arXiv: 0709:4567 (2007).
18. M. Lugaro et al., *Astrophys. J.*, submitted (2008).
19. J. Tumlinson, *Astrophys. J.* **664**, L63–L66 (2007).
20. G. Meynet, S. Ekstrom, & A. Maeder, *Astron. & Astrophys.* **447**, 623–639 (2006)
21. C. Chiappini et al., arXiv: 0712.3434 (2007).
22. A. Frebel et al., *Astrophys. J.* **660**, L117–L120 (2007).
23. P. Francois et al. *Astron. & Astrophys.* **476**, 935–950 (2007).
24. Y.-Z. Qian & G. J. Wasserburg, *Phys. Rep.* **333**, 77–108 (2000).
25. C. Travaglio et al., *Astrophys. J.* **601**, 864–884 (2004).
26. F. Montes et al. 2007, *Astrophs. J.* **671**, 1685–1695 (2007).

Lithium Abundances in Extremely Metal-Poor Turn-Off Stars [1]

W. Aoki[*,†], P. Barklem[**], T. C. Beers[‡], N. Christlieb[**] and S. Inoue[*]

*National Astronomical Observatory, Mitaka, Tokyo, 181-8588 Japan
†Department of Astronomical Science, The Graduate University of Advansed Stidies, Mitaka, Tokyo, 181-8588 Japan
**Department of Astronomy and Space Physics, Uppsala University, Box 515, 751-20 Uppsala, Sweden
‡Department of Physics and Astronomy, CSCE: Center for the Study of Cosmic Evolution, and JINA: Joint Institute for Nuclear Astrophysics, Michigan State University, East Lansing, MI 48824-1116, USA

Abstract.
The Lithium (Li) abundances measured for very metal-poor turn-off (unevolved) stars have been interpreted as the result of Big Bang nucleosynthesis. However, the value is lower by a factor of two or three than the prediction of standard Big Bang nucleosynthesis models, adopting the cosmological parameters determined by the measurements of cosmic microwave background radiation with the WMAP satellite. Moreover, the recent measurements for extremely metal-poor stars (objects having iron abundances less than 1/1000th solar) suggest a scatter of the Li abundance, or a possible decreasing trend with decreasing metallicity. In order to further investigate the Li production and destruction processes in the very early universe, we have determined Li abundances for extremely metal-poor stars based on high-resolution spectra for the resonance line of neutral Li. The result of our analysis, combined with previous measurements, indicates that the Li abundances of extremely metal-poor stars are, on average, lower than those of stars with higher metallicity, while the scatter or trend of the Li abundance remains unclear. We discuss possible reasons for the lower Li abundances in extremely metal-poor stars, such as depletion of Li in low-mass unevolved stars, or destruction of Li by the first generations of massive progenitors.

Keywords: Galaxy: abundances – nuclear reactions, nucleosynthesis, abundances – stars: abundances – stars: Population II
PACS: 97.10.Tk, 97.20.Jg, 97.20.Jg, 97.20.Wt, 98.80.Ft

INTRODUCTION

Lithium abundances in metal-deficient stars in the Galactic halo have been investigated as a constraint on Big Bang nucleosynthesis models that predict the primordial ^{7}Li abundance as a function of baryon density. The primordial Li abundance is expected to be recorded in stars that formed in the very early universe. Although ^{7}Li is destroyed at a temperature higher than 2×10^6 K, it is preserved on the surfaces of unevolved stars with low metallicity and relatively high effective temperature ($T_{\rm eff} > 6000$ K) in which the surface convection zone is quite thin. These stars are called turn-off stars

[1] Based on data collected at the Subaru Telescope, which is operated by the National Astronomical Observatory of Japan.

CP1016, *Origin of Matter and Evolution of Galaxies*,
edited by T. Suda, T. Nozawa, A. Ohnishi, K. Kato, M. Y. Fujimoto, T. Kajino, and S. Kubono

(or main-sequence turn-off stars), because main-sequence stars evolve to giants via the turning point in the so-called HR diagram that displays stellar luminosity as a function of effective temperature. Lithium production in the Big Bang is discussed by Mathews [1], Kusakabe et al. [2], and Kawanomoto et al. [3] in this conference.

The Li abundances of metal-deficient turn-off stars have been measured since the 1980s. Spite & Spite [4] found a uniform Li abundance (the so-called Spite plateau) for metal-deficient main-sequence stars, concluding that is provided evidence of primordial Li production at low baryon density. Since that time, additional measurements have been intensively made for a number of halo turn-off stars (e.g. Ryan et al.[5], Asplund et al. [6]). The Spite plateau value is estimated to be A(Li)=$\log[n(\mathrm{Li})/n(\mathrm{H})]+12=2.0$–2.2, which corresponds to a baryon density of $\eta=2\text{–}4\times10^{-10}$. This value is, however, significantly lower than that inferred by measurements of the cosmic microwave background radiation (CMB) with the WMAP satellite ($\eta=6\times10^{-10}$ [7], which yields A(Li)$=2.6$), and the implication of the observed Li abundances in metal-deficient stars is now in controversy.

Measurements of Li abundances in turn-off stars were recently extended to the lowest metallicity range ([Fe/H]<-3)[2] by Bonifacio et al. [8]. While a decreasing trend of Li abundances with decreasing metallicity is suggested by previous studies (e.g., Ryan et al. [5]), the most recent data clearly exhibited lower Li abundances for stars with [Fe/H]<-3 than for stars with higher metallicity. This result might provide a hint to solve the discrepancy between the Li abundances measured for halo stars and the value expected from the standard Big Bang nucleosynthesis model with the baryon density determined from the CMB observation.

Another problem with the Li abundances observed in metal-deficient stars is the "depletion" of Li in HE 1327–2326, the most metal(iron)-deficient star known to date ([Fe/H]~-5.5 [9, 10]). The evolutionary status of this object is most likely a subgiant (another possibility is a main-sequence star) with an effective temperature higher than 6000 K. While such stars usually have Li abundances close to the Spite plateau value, there is no Li absorption feature detected in HE 1327–2326; the upper limit of the inferred Li abundance is an order of magnitude lower than the Spite plateau value (Aoki et al. [10]; Frebel et al. [11]). No physical reason for the Li depletion in this object is identified yet.

One important and possible observational approach to such Li problems is to determine the Li abundances for a larger sample of extremely metal-poor ([Fe/H]<-3) stars to confirm the lower Li abundances in such stars found by Bonifacio et al.[8], and to search for the dependence of Li abundance on metallicity or on other stellar parameters.

OBSERVATIONS AND ANALYSES

In order to investigate the Li abundances for extremely metal-poor stars, we collected high resolution spectra of turn-off stars obtained with the Subaru Telescope High Dispersion Spectrograph (HDS; Noguchi et al.[12]) in three different programs: (1) A program

---

[2] $[\mathrm{A/B}]=\log(N_\mathrm{A}/N_\mathrm{B})-\log(N_\mathrm{A}/N_\mathrm{B})_\odot$

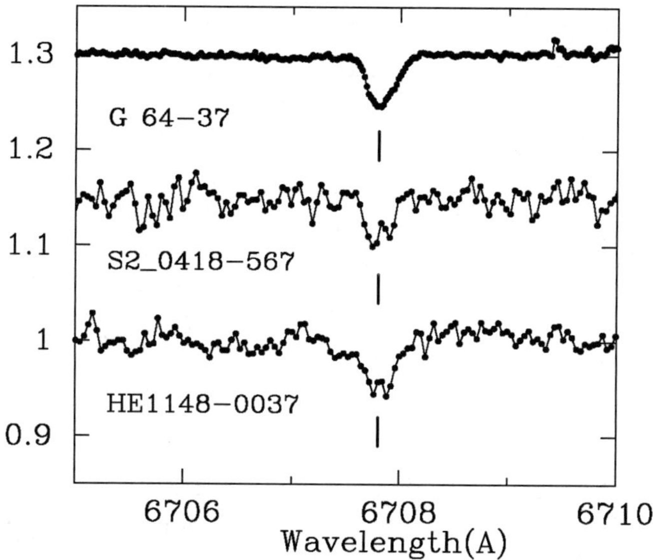

FIGURE 1. Examples of spectra for the Li 6708 Å range. The top shows the high quality spectrum of the bright metal-poor star G 64–37, while the other two are the spectra of two faint objects newly studied by the present work.

to determine Li isotope ratios (P.I.: Inoue); (2) A program to study extremely metal-poor stars (P.I.: Aoki); (3) A program to investigate turn-off stars with carbon-excesses selected from the SDSS sample (P.I.: Aoki). Although the data quality (resolving power and signal-to-noise ratios) of these data covers quite wide ranges, depending on the observational program, measurements of Li abundances are still possible with reasonable precision. Figure 1 shows examples of the spectra for the range of the Li resonance line at 6708 Å for three stars obtained in the three programs. We note that, although the SDSS sample was selected to be candidate carbon-enhanced objects, the stars studied here exhibit no excess of carbon and their Li abundance is not expected to be affected by special processes such as mass transfer from an evolved companion across a binary system. Two bright, well studied metal-poor stars with [Fe/H]= −2.3 are also analyzed for comparison purposes.

The Li abundances are determined by fitting synthetic spectra calculated based on model atmospheres [13, 14] to the observed one by searching for the χ^2 minimum. An important parameter of the model atmospheres that affects the Li abundance determination is the effective temperature. Among several methods to estimate effective temperatures for turn-off stars, an analysis of hydrogen Balmer line profiles might be the most reliable technique. We adopted the effective temperature determined from the Hα profile [15], as has been done by recent other work [6, 8]. Metallicities of our objects are determined by an LTE analyses of neutral and singly ionized species of iron.

The derived Li abundances as a function of the iron abundance are shown in Figure 2. The 2σ range of the χ^2 analysis is adopted as the fitting error of the Li absorption

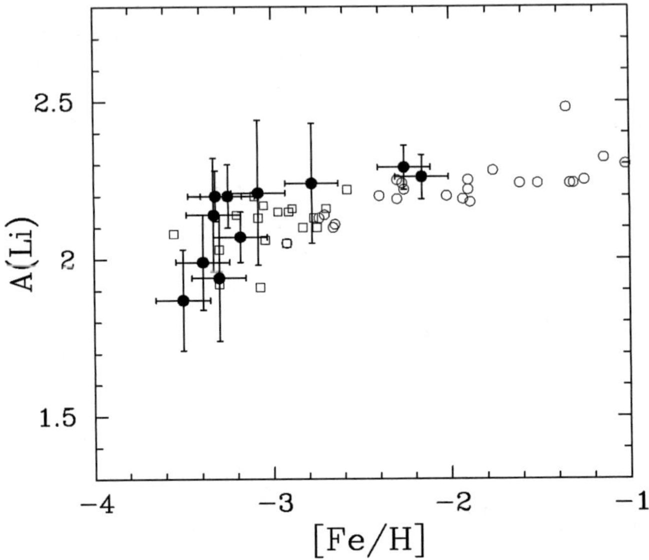

FIGURE 2. Li abundances as a function of [Fe/H]. Filled circles with error bars indicate our results, while open circles and squares mean those of Asplund et al. [6] and Bonifacio et al. [8], respectively. The so-called Spite plateau value found for $-2.5 <$[Fe/H]< -1 is A(Li)~ 2.2, while the Li abundance predicted by the standard Big Bang nucleosynthesis model is 2.6. The Li abundances for stars with [Fe/H] < -3 is lower, on average, by 0.2 dex than those with higher metallicity.

feature. In addition, the errors due to uncertainties in the stellar parameters, in particular that of the effective temperature (100-200 K, depending on the object), are included in the errors indicated in the figure. The fitting error ranges between 0.02 and 0.15 dex, while the typical error due to the uncertainty on the stellar parameters is 0.1–0.2 dex.

DISCUSSION AND CONCLUDING REMARKS

Figure 2 shows the Li abundances as a function of [Fe/H] for our sample, along with those of Asplund et al. [6] and Bonifacio et al. [8]. The Li abundances of the two comparison stars with [Fe/H] ~ -2.3 determined by our analysis agree well with the measurements by Asplund et al. [6], confirming no or little systematic differences between these studies.

The average of the A(Li) of the eight stars with [Fe/H] < -3 ($<A$(Li)$>$) is 2.08 and the standard deviation (σ) is 0.12 dex. The standard deviation is comparable with the measurement errors of our analysis (0.08–0.23 dex). Hence, we can claim nothing about the scatter of the A(Li) in these stars. The $<A$(Li)$>$ is lower by 0.2 dex than the values at higher metallicity (2.27, the average of our two reference stars; 2.23, the average of the Asplund et al.'s sample for $-2.5 <$ [Fe/H] < -2.0). The discrepancy of 0.2 dex is significant, compared to the $\sigma N^{-1/2}$=$0.12/\sqrt{8}$=0.04, where N is the number of objects of our sample for [Fe/H] < -3. Our conclusion here is that the Li abundances in stars with [Fe/H] < -3 are 0.2 dex lower than those of stars with higher metallicity, while no significant scatter or trend to metallicity is found in our sample. This result confirms the study by Bonifacio et al. [8].

The effective temperatures of our targets are between 5900 and 6550 K. The sample includes both main-sequence and subgiant stars, which are estimated from the surface gravity determined from the analyses of iron lines. No clear correlation is found between the Li abundance and the effective temperature nor the evolutionary status.

Several hypotheses for depletion of Li in metal-deficient stars have been proposed in the past few years. One is the depletion of Li at the surface of low-mass stars we are currently observing. Korn et al. [16] carefully investigated the abundances of Li and other elements for turn-off, subgiant, and giant stars in the globular cluster NGC 6397. They found a dependence of the abundances of iron and other metals on evolutionary status, and took this as evidence for the effect of diffusion. They estimated the Li abundance at the surface of old, turn-off (unevolved) stars decreased by 0.25 dex from its initial value. If this process is the reason for the low Li abundances in extremely metal-poor stars, one question is why the Li abundances for stars with [Fe/H] > -2.5 are so uniform, while they are lower in the lower metallicity range.

A scenario of Li depletion by the first generation of stars have been proposed by Piau et al. [17]. Assuming that a part of Li depletion in metal-poor stars from the primordial value is due to the destruction by the first generation of massive stars, they suggested that a significant amount of primordial (metal-free) gas in the Galactic halo was processed by such massive stars. Since stars with the lowest metallicity should be most severely affected by first generations of stars, their low Li abundances are naturally explained by this scenario, and the non-detection of Li in HE 1327–2326 might be the extreme case. However, it is still unclear whether such a large fraction of primordial material was actually processed through the first generation objects.

The reason for the low Li abundances in extremely metal-poor stars, as well as for the discrepancy between the observed Li abundance and the value predicted by standard Big Bang nucleosynthesis models, is not identified yet, although the abundance behavior in [Fe/H]< -3 provides a strong constraint on the models or scenarios for the Li processes in the early universe. Further observational studies, in particular for stars with lower metallicity ([Fe/H]< -3.5) are strongly desired. A new observing program with high resolution spectroscopy for candidates extremely metal-poor stars discovered by SDSS/SEGUE (Beers et al. [18]) will start from March, 2008 using the Subaru Telescope. Such observations will reveal the scatter of Li abundances, or their trend to metallicity, for stars in the lowest metallicity range.

ACKNOWLEDGMENTS

W. A. is supported by a Grant-in-Aid for Science Research from JSPS (grant 18104003). T. C. B. is grateful to the organizers for providing assistance with travel and accommodation expenses. This work also received support from grants AST 07-07776 and PHY 02-15783; Physics Frontier Center / Joint Institute for Nuclear Astrophysics (JINA), awarded by the US National Science Foundation.

REFERENCES

1. Mathews, G. J., this volume
2. Kusakabe, M. et al., this volume
3. Kawanomoto, S. et al., this volume
4. Spite, M., & Spite, F. 1982, Nature, 297, 483
5. Ryan, S. G., Norris, J. E., & Beers, T. C. 1999, Astrophysical Journal, 523, 654
6. Asplund, M., Lambert, D. L., Nissen, P. E., Primas, F., & Smith, V. V. 2006, Astrophysical Journal, 644, 229
7. Spergel, D. N., et al. 2003, Astrophysical Journals, 148, 175
8. Bonifacio, P., et al. 2007, Astronomy & Astrophysics, 470, 153
9. Frebel, A., Aoki, W., Christlieb, N. et al. 2005, Nature, 434, 871
10. Aoki, W., et al. 2006, Astrophysical Journal, 639, 897
11. Frebel, A., Eriksson, K., Collet, R., Christlieb, N., Aoki, W., submitted to Astrophysical Journal
12. Noguchi, K., Aoki, W., Kawanomoto, S. et al. 2002, Publication of Astronomical Observatory of Japan, 54, 855
13. Kurucz, R. L. 1993, CD-ROM 13, ATLAS9 Stellar Atmospheres Programs and 2 km/s Grid (Cambridge: Smithsonian Astrophys. Obs.)
14. Castelli, F., & Kurucz, R. L. 2003, Modelling of Stellar Atmospheres, 210, 20P
15. Barklem, P. S., Stempels, H. C., Allende Prieto, C., Kochukhov, O. P., Piskunov, N., & O'Mara, B. J. 2002, Astronomy & Astrophysics, 385, 951
16. Korn, A. J., Grundahl, F., Richard, O., Barklem, P. S., Mashonkina, L., Collet, R., Piskunov, N., & Gustafsson, B. 2006, Nature, 442, 657
17. Piau, L., Beers, T. C., Balsara, D. S., Sivarani, T., Truran, J. W., & Ferguson, J. W. 2006, Astrophysical Journal, 653, 300
18. Beers, T. C. 2008, this volume

The Nucleosynthetic Signatures of the First Star Survivors Among Hyper Metal-Poor Stars with [Fe/H] < -4.5

T. Suda*, Y. Komiya†, T. Nishimura*, N. Iwamoto**, M. Aikawa‡ and M. Y. Fujimoto*

*Department of Cosmosciences, Graduate School of Science, Hokkaido University, Sapporo 060-0810, Japan
†Department of Astronomy, Graduate School of Science, Tohoku University, Sendai 980-8578, Japan
**Japan Atomic Energy Agency, 2-4 shirakata-shirane, Tokai, Ibaraki 319-1195, Japan
‡Open Course Ware, Hokkaido University, Sapporo 060-0811, Japan

Abstract. The first stars in our Universe (Population III) are the useful probes for the star formation history in the very early Universe. These stars were born in the primordial gas cloud produced soon after the Big Bang and considered to be metal-free, i.e., comletely lack of elements heavier than lithium. In order to identify these survivors, we should consider the effect of changing surface abundances during their long lives. The surface abundances are modified by the accretion of gas from the interstellar matter and/or the binary mass transfer. The former possibly affects the iron group elements through the pollution of the parent cloud with the ejecta of supernovae. On the other hand, the latter can affect the abundance pattern of light and s-process elements through the evolution of primary star that experienced the internal mixing and dredge-up during thermally pulsating AGB before the final mass loss. The top three of the iron deficient stars HE1327-2326, HE0107-5240, and HE0557-4840 are reported as the candidates of the first stars by the current observations of extremely metal-poor stars in the Galactic halo. These stars have [Fe/H] < -4.5 and share the common feature of large enhancement of C. We argue that these abundance patterns are testified to the evolutionary characteristics of the first stars with low- and intermediate-mass by trying to constrain the mass of primary under the assumuption that they were the suvivors of secondary stars in the binary system when they were born. We also argue that the binary scenario for these stars are validated with the merging history of the Galaxy. The considerations of stellar evolution, nucleosynthesis, and Galactic chemical evolution reveal that these three stars are promising candidate for the first stars.

Keywords: Population III;stellar evolution;metal-poor stars;giants;Galactic halo
PACS: 97.10.Cv,97.20.Wt,97.10.Tk,97.80.-d,98.35.Gi

INTRODUCTION

The role of the first stars of our universe (Population III stars) is important for our understanding of the evolution of our universe. They are theoretically devoid of elements heavier than lithium at their birth in the primordial gas cloud. Because of the lifetimes as long as ~ 10 Gyr for the survivors, we should take into account the possible modifications of their surface chemical abundances. The low-mass first stars, if they were born, can have suffered from the accretion of metal-rich gas, polluted by the supernova ejecta of massive stars of the first and subsequent generations in the mother clouds where they were formed. The small amount of iron observed among stars in the Galactic halo

CP1016, *Origin of Matter and Evolution of Galaxies*,
edited by T. Suda, T. Nozawa, A. Ohnishi, K. Kato, M. Y. Fujimoto, T. Kajino, and S. Kubono

is readily explained in terms of this external pollution if their mother clouds remained undisrupted in one Gyr or so [1]. In addition, the first stars can be polluted by the gas, transferred from the binary companion if the internal mixing at their later phase of evolution modifies the surface abundances of primary stars.

The consideration of the effect of these internal and external pollution in stars is crucial for the identification of the Pop. III objects. This is because the currently observed extremely iron-poor stars [2, 3, 4] with [Fe/H] < -4.5 show peculiar abundance patterns that cannot be explained by the standard theory of single star evolution with low-mass. These three stars, HE0107-5240, HE1327-2326, and HE0557-4840 share the common feature of large enhancement of carbon.

The origin of C-enhancement can be explained by the stellar evolution models at very low metallicity because the first stars share the peculiarity in the nucleosynthesis with the extremely metal-poor stars of [Fe/H] < -2.5, and hence, [CNO/H] < -2.5. Such stars experience the mixing of hydrogen into the helium convective zone, driven by the runaway of helium burning, differently from the stars of younger populations I and II with larger metal abundances [5]. For stars of mass less than or equal to $1.1 M_{\odot}$, this occurs during the first off-center core flash, which develops in a full amplitude of helium burning rate $L_{\mathrm{He}} = 10^9 L_{\odot}$. For stars of masses between 1.2 and $3 M_{\odot}$, it occurs in the beginning of helium shell flashes on the asymptotic giant branch [1]. Hydrogen, engulfed and carried inward by the helium convection, is captured by ^{12}C, nuclear product of helium burning, to produce ^{13}C via beta decay in a half way, and further transported into the region of high temperatures to produce neutrons with the reaction $^{13}\mathrm{C}(\alpha,n)^{16}\mathrm{O}$. With this neutron source, neutron-capture nucleosynthesis proceeds in the helium convective zone as discussed in the paper in this volume by Nishimura et al. [6].

In this work, we explore the possibility of the binary orgin for the most iron-deficient stars by considering the stellar evolution, nucleosynthesis, and the Galactic chemical evolution.

ORIGIN OF HYPER- (ULTRA-) METAL-POOR STARS

Figure 1 shows the comparisons of abundance distribution obtained by the stellar models with the observed abundance patterns of HE0107-5240, HE1327-2326 and HE0557-4840. We explore the ensuing neutron-capture nucleosynthesis in the helium convective shell with the one-zone approximation [7]. Our nuclear network is an extension of that used by Aikawa et al. [8] to include 67 isotopes of 17 light species from the neutron and ^{1}H through ^{35}S, combined with s-process nucleosynthesis code for heavier elements [see, e.g. 9]. Since the amount of ^{13}C mixed into the inner helium convective zone varies with the strength of the helium shell flash, and hence, with the stellar mass, and also depends on the treatment of convection, we regard it as parameter and assume constant rate of mixing for simplicity. As soon as ^{13}C is mixed, neutrons are produced and captured primarily by ^{12}C to form ^{13}C and then reappear in consequence of additional $^{13}\mathrm{C}(\alpha,n)^{16}\mathrm{O}$ reactions. This neutron cycle reactions continue until $^{16}\mathrm{O}(n,\gamma)^{17}\mathrm{O}$ reaction has converted most of the injected ^{13}C into ^{17}O, and then, follow the reactions $^{17}\mathrm{O}(\alpha,n)^{20}\mathrm{Ne}$. Subsequently, neutron capture reactions to produce s-process elements begin with the seed nuclei such as neon and magnesium, which synthesize enough

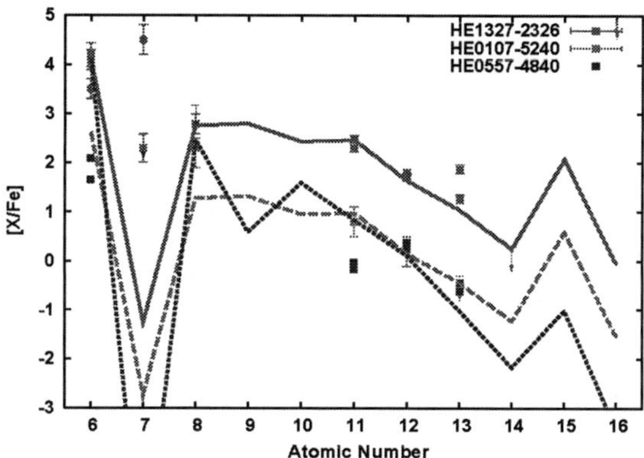

FIGURE 1. Abundance distribution as a result of the helium flash driven deep mixing at the helium shell flash. Derived abundances for HE0107-5240, HE1327-2326 and HE0557-4840 are also plotted.

amount of Sr and still heavier elements successively without iron initially.

For HE0107-5240, the reproduction of the abundances is thoroughly discussed in Suda et al. [1]. For HE1327-2326, the total amount of mixing into the helium convective zone is taken as $4 \times 10^{-5} M_\odot$, of ^{13}C (corresponding to the abundance ratio $X_{13}/X_{12} = 0.01$) at a constant rate for 10^9 s. This choice of mixing parameter is ten times of that for the model of HE0107-5240. The result for HE0557-4840 shows the consequence of the third dredge-up during the thermal pulses. For comparisons with observation, we normalized these abundance pattern so as to give the observed carbon abundance [C/H] = -1.7 and -3.15 with taking account of dilution in the surface convection after the dredge-up. This gives a good agreement of elements with HE1327-2326 through Na, Mg, Al and Sr. The abundance of Ba is barely consistent within the upper limit in this case [see, 10]. The model for HE0557-4840 also agrees well with observation for light elements. When the surface carbon abundance satisfies [C/H] > -2.5 by the mixing and dredge-up, neutron-rich isotopes made in the convective zone are dredged up to the surface along with carbon during third dredge-up episodes [5, 9].

For stars of mass $M < 1.5 M_\odot$, the third dredge-up will not occur and the surface enrichment of [CN/H] $= -1.7$ for HE1327-2326 is solely due to the He-FDDM, which is consistent with observed large nitrogen to carbon ratio N/C $\sim$ 1, differently from HE0107-5240 (N/C $\sim$ 1/150). In the case of EMP dwarfs, the size of surface convection is an order of a few times $10^{-3} M_\odot$, the accreted mass of $\sim 0.01 M_\odot$ is sufficient to disguise the surface abundances with accreted materials. Accordingly, the binary separation can be as large as a few ten times the distance between the sun and the earth, and the period can be as large as 100 years, similar to those predicted for HE0107-5240 [1]. The abundance patterns of these elements near the iron-peak are similar between three stars, and more or less common to other extremely metal-poor stars of [Fe/H] ~ -3.

We compute the model stars of $3 M_\odot$ and [Fe/H] $= -4.8$ to search for the possibility of reproducing abundance pattern of HE0557-4840. The result is shown in Figure 2. We find the third dredge-up during the thermal pulses in shell helium burning without the

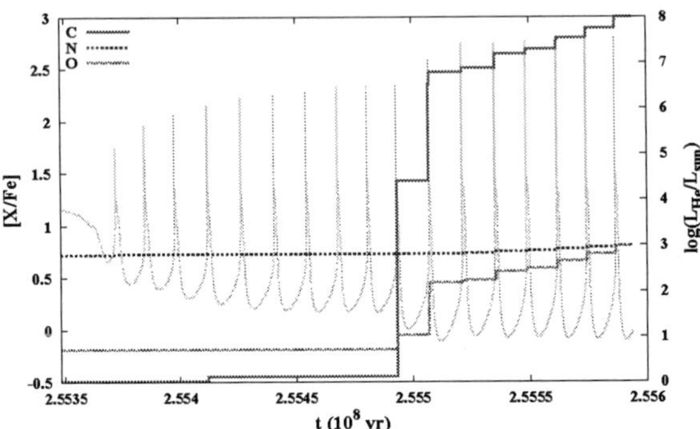

FIGURE 2. Time evolution of the shell helium burning in asymptotic giant branch phase. The evolution is shown for helium burning rate (scales in the rightside) and abundances (scales in the leftside) of carbon, nitrogen, and oxygen, where $t = 0$ is set at the zero-age main sequence.

He-FDDM. For mass greater than 3 $M_{\odot}$, the stars with [Fe/H] < -4.5 do not experience strong He-FDDM at the AGB [11]. The only small amount of hydrogen mixing events are found, but it is not enough to promote neutron capture nucleosynthesis and the subsequent dredge-up. On the other hand, these models have opportunities of carbon enhancement in the surface through the third dredge-up.

The occurence of the third dredge-up in Population III AGB models may be possible if we consider the effect of surface pollution. The surface pollution causes the enhancement of CNO elements in the hydrogen burning shell that brings about the occurence of the third dredge-up. The models with mass greater than $2.5M_{\odot}$ can be polluted its hydrogen burning shell with the accreted materials during the second dredge-up [1].

For the origin of these elements, there are two possibilities from our binary scenario, one is acquired pollution that the stars were born of metal-free gas and accreted metal-rich gas and the other is the innate origin that the stars were born from the gas already polluted by the first and subsequent supernovae. These two cases make little differences in the nucleosynthesis leading to the enhancement of light elements through Si and Ba; for the scaled-solar metal abundances of [Fe/H] $= -5.6$ with the 0.4 dex enrichment of alpha elements, O, Ne, Mg and Si, the stars of mass $M \leq 1.2M_{\odot}$ ignite helium burning at the center and meet with the hydrogen mixing during the helium shell burning on asymptotic giant branch. Only difference is the Pb abundance since all the pristine metals were converted into the heaviest s-process elements, which amounts the enhancement of [Pb/Fe] $\simeq 3$. The former possibility is explored extensively by Suda et al. [1], who argue the pollution in the primordial clouds of smaller masses than our Milky Way where they were born.

The binary scenario is also supported by the previous work on the origin of carbon-enhanced metal-poor stars that the initial mass function (IMF) of EMP stars are different for [Fe/H] < -2.5 and peaked at $\sim 10M_{\odot}$ [12]. Figure 3 shows the comparison of the theoretical metallicity distribution function (MDF) with the observations. Theoretical

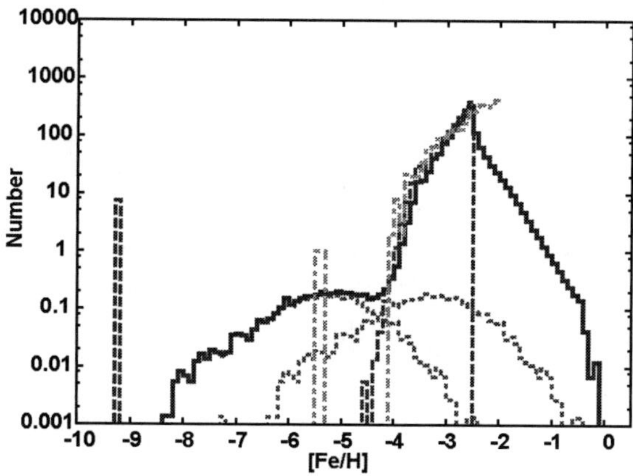

FIGURE 3. Comparison of observed (dash-dotted lines) and simulated metallicity distribution function. Models without the accretion are shown by dashed lines, while those with accretion are shown by solid line. The distribution of dwarf (right) and giants (left) are shown by the short-dashed lines, which is according to the mass of surface convection.

MDFs are the results of the simulation of the merging tree for our Galaxy starting with the primordial mini-halo of $10^6 M_\odot$ based on the extended Press-Schechter theory. We assume the virial temperature of ~ 300 K by the cooling of interstellar gas. The lifetime of mini halo is set at 0.1 Gyr during which the moving stars of 0.8 $M_\odot$ with virial velocity accrete gas according to the Bondi's formalism. The pollution of mini-haloes is approximated by the one-zone model and amounts to [Fe/H] $= -3.5$ by single event of supernova. The clear cut-off at [Fe/H] ~ -4 is reproduced with the first stars shoulder below the cut-off that represents the survivors of the first generation stars. With this figure, the most of currently known EMP stars are in binary system that are affected by a few events of supernovae and by accretion of matter in their host cloud of star formation. The stars in the lowest metallicity range come from the massive binary whose primary stars have $\sim 10 M_\odot$ and ended their lives as supernovae. It should be investigated in more detail whether such high value of binary mass ratio is possible in star formation. In conclusion, it is predicted that the binary star formation in high-mass IMF was dominant in the early universe and the survivor of Population III stars will be found among [Fe/H] < -4. The high-mass IMF based on binary scenario can also be a explanation for the scarcity of stars in [Fe/H] < -4, so called G-dwarf problem in the the Galactic halo [13].

CONCLUSIONS AND DISCUSSION

We followed the evolution of Population III model stars with use of stellar evolution and nucleosynthesis program to reproduce the most iron-deficient stars in the Galactic halo. We also derived the metallicity distribution function based on the merger tree of our

Galaxy with an assumption of high mass peaked initial mass function that comes from the comparison of stellar models with the observed abundances of extremely metal-poor stars. It is suggested that the three stars discussed here should be the secondary stars of Population III binary. The initial mass M of primary stars for these stars will be $1.5 < M < 3$ for HE0107-5240, $1 < M < 1.5$ for HE1327-2326, and $3 < M < 4$ for HE0557-4840 where M in units of $M_{\odot}$.

The results obtained here is not dependent on the formation mechanism of low-mass stars in the primordial cloud. It is to be noted that the existence of the survivor of Population III stars does not necessarily requires the formation of low-mass stars in the metal-free circumstances. The binary scenario does not conflict with the current understandings of star formation theory that the initial mass of the Population III stars should be more than $\sim 10-100 M_{\odot}$, if most of them are formed in binary system. Within the binary scenario, we do not need any contribution of pair-instability supernovae in the early universe as suggested by the observations of EMP stars (see the discussion in this volume by Komiya et al. [14]). It will be more and more important in considering the effect of pollution by the internal mixing in AGB stars and the mass transfer in the binary system. It will also be important to investigate the final fate of supernova binaries on whether the accretion of metal-rich gas increases after the supernovae.

Why all three stars with [Fe/H] < -4.5 show carbon enhancement? The answer to this question will be related to the formation limit of binary system. If the binary system is too wide in the star forming primordial cloud, the fragmentation does not occur due to the lack of cooling element. This corresponds to the lower limit of pollution with carbon-rich matter by wind accretion in binary mass transfer.

REFERENCES

1. T. Suda, M. Aikawa, M. N. Machida, M. Y. Fujimoto, and I. I. Jr., *The Astrophysical Journal* **611**, 476–493 (2004).
2. N. Christlieb, M. Bessell, T. Beers, B. Gustafsson, A. Korn, P. Barklem, T. Karlsson, M. Mizuno-Wiedner, and S. Rossi, *Nature* pp. 904–906 (2002).
3. A. Frebel, W. Aoki, N. Christlieb, H. Ando, M. Asplund, P. Barklem, T. Beers, K. Eriksson, C. Fechner, M. Fujimoto, et al., *Nature* **434**, 871–873 (2005).
4. J. E. Norris, et al., *The Astrophysical Journal* **670**, 774–788 (2007).
5. M. Y. Fujimoto, Y. Ikeda, and I. I. Jr., *The Astrophysical Journal Letters* **529**, L25–L28 (2000).
6. T. Nishimura, M. Aikawa, N. Iwamoto, T. Suda, M. Y. Fujimoto, and I. J. Iben, "Interpretation of extremely metal-poor stars as candidates of first generation stars," in [15], in press.
7. M. Y. Fujimoto, M. Aikawa, and K. Katō, *The Astrophysical Journal* **519**, 733–744 (1999).
8. M. Aikawa, M. Y. Fujimoto, and K. Katō, *The Astrophysical Journal* **560**, 937–956 (2001).
9. N. Iwamoto, T. Kajino, G. J. Mathews, M. Y. Fujimoto, and W. Aoki, *The Astrophysical Journal* **602**, 377–387 (2004).
10. T. Nishimura, M. Aikawa, T. Suda, and M. Y. Fujimoto, *The Astrophysical Journal* (2008), submitted.
11. T. Suda, and M. Y. Fujimoto, *The Astrophysical Journal* (2008), submitted.
12. Y. Komiya, T. Suda, H. Minaguchi, T. Shigeyama, W. Aoki, and M. Y. Fujimoto, *The Astrophysical Journal* **658**, 367–390 (2007).
13. H. E. Bond, *The Astrophysical Journal* **248**, 606–611 (1981).
14. Y. Komiya, T. Suda, A. Habe, and M. Y. Fujimoto, "Near Field Cosmology with Binary," in [15], in press.
15. T. Suda, et al., editors, *Origin of Matter and Evolution of Galaxies 2007*, AIP Conference Proceedings, Melville, New York, 2008, in press.

Supernova Nucleosynthesis and Extremely Metal-Poor Stars

Nozomu Tominaga*, Hideyuki Umeda*, Keiichi Maeda†,
Nobuyuki Iwamoto** and Ken'ichi Nomoto*,†

*Department of Astronomy, University of Tokyo, Bunkyo-ku, Tokyo 113-0033, Japan
†Institute for the Physics and Mathematics of the Universe, University of Tokyo, Kashiwa-shi, Chiba, 277-8582, Japan
**Nuclear Data Center, Nuclear Science and Engineering Directorate, Japan Atomic Energy Agency, Tokai, Ibaraki 319-1195, Japan

Abstract. We investigate hydrodynamical and nucleosynthetic properties of the jet-induced explosion of a population III $40M_\odot$ star with a two-dimensional special relativistic hydrodynamical code and compare the abundance patterns of the yields with those of the metal-poor stars. We conclude that (1) the ejection of Fe-peak products and the fallback of unprocessed materials can account for the abundance patterns of the extremely metal-poor (EMP) stars and that (2) the jet-induced explosion with different energy deposition rates can explain the difference of the abundance patterns of the metal-poor stars. Furthermore, the abundance distribution after the explosion and the angular dependence of the yield are obtained for the models with high and low energy deposition rates $\dot{E}_{\rm dep} = 120 \times 10^{51}$erg s^{-1} and 1.5×10^{51}erg s^{-1}. We also find that the peculiar abundance pattern of a Si-deficient metal-poor star HE 1424–0241 can be reproduced by the angle-delimited yield for $\theta = 30^\circ - 35^\circ$ of the model with the energy deposition rate of $\dot{E}_{\rm dep} = 120 \times 10^{51}$ erg s^{-1}.

Keywords: Galaxy: halo — gamma rays: bursts — nuclear reactions, nucleosynthesis, abundances — stars: abundances — stars: Population II — supernovae: general
PACS: 26.30.-k, 26.30.Ef, 26.50.+x, 97.10.Tk, 97.20.Tr, 97.20.Wt, 97.60.Bw

INTRODUCTION

In the early universe, the enrichment by a single supernova (SN) can dominate the pre-existing metal contents (e.g., [1]). The Pop III SN shock compresses the SN ejecta consisting of heavy elements, e.g., O, Mg, Si, and Fe, and the circumstellar materials consisting of H and He. The abundance pattern of the enriched gas reflects nucleosynthesis in the SN. The SN compression will initiate a SN-induced star formation (e.g., [2]); the second-generation stars will be formed from the enriched gases. Thus the abundances of the second generation stars preserve the information of nucleosynthesis in Pop III SNe. The low-mass ($\sim 1M_\odot$) stars among them have long lifetimes, survive until present days, and might be observed as the metal-poor stars. Therefore, some of the metal-poor stars can be used to make a constraint on the yields of the Pop III SN.

The abundance patterns of extremely metal-poor (EMP) stars with [Fe/H] < -3[1] suggest the aspherical SN explosions in the early universe. The C-enhanced type of the

[1] Here [A/B] $= \log_{10}(N_{\rm A}/N_{\rm B}) - \log_{10}(N_{\rm A}/N_{\rm B})_\odot$, where the subscript $\odot$ refers to the solar value and $N_{\rm A}$ and $N_{\rm B}$ are the abundances of elements A and B, respectively.

CP1016, *Origin of Matter and Evolution of Galaxies*,
edited by T. Suda, T. Nozawa, A. Ohnishi, K. Kato, M. Y. Fujimoto, T. Kajino, and S. Kubono

TABLE 1. Jet-induced explosion models.

Name	M_0	R_0	$\dot{E}_{dep}$	E_{dep}	θ_{jet}	Γ_{jet}	f_{th}	$M(^{56}Ni)$	M_{rem}
	$[M_\odot]$	[km]	$[10^{51}$ergs s$^{-1}]$	$[10^{51}$ergs]	[degrees]			$[M_\odot]$	$[M_\odot]$
A	1.4	900	120	15	15	100	10^{-3}	2.1×10^{-1}	9.1
B	1.4	900	1.5	15	15	100	10^{-3}	3.9×10^{-6}	16.9

EMP stars have been well explained by the faint SNe [3, 4, 5, 6], except for their large Co/Fe and Zn/Fe ratios (e.g., [7, 8]). The enhancement of Co and Zn in low metallicity stars requires explosive nucleosynthesis under high entropy. In a *spherical* model, a high entropy explosion corresponds to a high energy explosion that inevitably synthesizes a large amount of ^{56}Ni. Thus, it was suggested that some faint SNe are associated with a narrow jet within which a high entropy region is confined [3].

In this paper, we present hydrodynamical and nucleosynthetic models of the jet-induced explosions of a 40 $M_\odot$ star and show that the jet-induced SN explosions are responsible for the formation of the EMP stars. Further, we investigate the angular dependence of the yield to compare the yields with the abundance patterns of the metal-poor stars.

MODELS

We investigate a jet-induced SN explosion of a Pop III $40M_\odot$ star [3, 6] by means of a two-dimensional relativistic Eulerian hydrodynamic and nucleosynthesis calculation with the gravity [9, 10, 11]. The nucleosynthesis calculation is performed as a post-processing with the reaction network including 280 isotopes up to ^{79}Br ([12, 13], see Table 1 in [3]). The thermodynamic histories are traced by maker particles representing Lagrangian mass elements (e.g., [14, 15]).

The explosion mechanism of GRB-associated SNe is still under debate (e.g., a neutrino annihilation, [16]; and a magneto-rotation, [17]). Thus, we do not consider how the jet is launched, but we deal the jet parametrically with the following five parameters [11]: energy deposition rate ($\dot{E}_{dep}$), total deposited energy (E_{dep}), initial half angle of the jets (θ_{jet}), initial Lorentz factor (Γ_{jet}), and the ratio of thermal to total deposited energies (f_{th}).

We inject the jets at a radius $R_0 \sim 900$ km, corresponding to an enclosed mass of $M_0 \sim 1.4M_\odot$, and follow the jet propagation. We investigate the dependence of nucleosynthesis outcome on $\dot{E}_{dep}$ for a range of $\dot{E}_{dep,51} \equiv \dot{E}_{dep}/10^{51}\mathrm{ergs\,s^{-1}} = 0.3 - 1500$ and in particular show the abundance patterns of the yields of two models; (A) a model with $\dot{E}_{dep,51} = \dot{E}_{dep}/(10^{51}\mathrm{ergs\ s^{-1}}) = 120$, (B) a model with $\dot{E}_{dep,51} = 1.5$. Here, we fix the other parameters as $E_{dep} = 1.5\times10^{52}$ergs, $\theta_{jet} = 15°$, $\Gamma_{jet} = 100$, and $f_{th} = 10^{-3}$ in all models. The mass of jets is $M_{jet} \sim 8\times10^{-5}M_\odot$. The model parameters, the ejected Fe mass [M(Fe)], and the central remnant mass (M_{rem}) are summarized in Table 1.

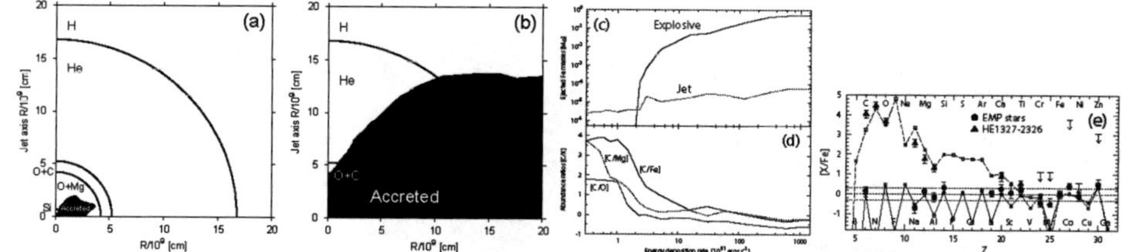

FIGURE 1. Initial locations of the mass elements which are finally accreted (*black*), for (a) model A and (b) model B. The background circles represent the boundaries between the layers in the progenitor star; the H, He, O+C, O+Mg, and Si layers from the outside. (c) Ejected ^{56}Ni mass (*solid line*: explosive nucleosynthesis products, *dashed line*: the jet contribution) as a function of the energy deposition rate. (d) Dependence of abundance ratios, [C/Fe] (*solid line*), [C/Mg] (*dashed line*), and [C/O] (*dotted line*), on the energy deposition rate. (e) Comparisons of the abundance patterns of metal-poor stars and of models. The abundance patterns of the normal EMP stars ([20], *ccircles*) and HE 1327–2326 ([21, 22], *triangles*) are reproduced by models with $\dot{E}_{\rm dep,51} = 120$ (*solid line*) and $\dot{E}_{\rm dep,51} = 1.5$ (*dashed line*).

RESULTS

The hydrodynamical calculations are followed until the homologously expanding structure is reached ($v \propto r$). Then, the ejected mass elements are identified from whether their radial velocities exceed the escape velocities at their position.

Fallback

Figures 1a and 1b show "accreted" regions for models A and B, where the accreted mass elements initially located in the progenitor [10, 11]. The O layer is separated into the two layers: (1) the O+Mg layer with $X(^{24}\mathrm{Mg}) > 0.01$ and (2) the O+C layer with $X(^{12}\mathrm{C}) > 0.1$. The inner matter is ejected along the jet-axis but not along the equatorial plane. On the other hand, the outer matter is ejected even along the equatorial plane, since the lateral expansion of the shock terminates the infall as the shock reaches the equatorial plane.

The accreted mass is larger for lower $\dot{E}_{\rm dep}$. This stems from the balance between the ram pressures of the injecting jet ($P_{\rm jet}$) and the infalling matter ($P_{\rm fall}$) (e.g., [18, 19]). The critical energy deposition rate ($\dot{E}_{\rm dep,cri}$) giving $P_{\rm jet} = P_{\rm fall}$ is lower for the outer layer. Thus, the jet injection with lower $\dot{E}_{\rm dep}$ is realized at a later time when the central remnant becomes more massive. Additionally, the lateral expansion of the jet is more efficiently suppressed for lower $\dot{E}_{\rm dep}$. As a result, the accreted region is larger and $M_{\rm rem}$ is larger for lower $\dot{E}_{\rm dep}$.

Figures 1c and 1d show the dependence of M(Fe) and the abundance ratios [C/O], [C/Mg], and [C/Fe] on the energy deposition rate $\dot{E}_{\rm dep}$, respectively [9, 10]. A model with lower $\dot{E}_{\rm dep}$ has larger $M_{\rm rem}$, higher [C/O], [C/Mg], and [C/Fe], and smaller M(Fe) because of the larger amount of fallback (Figs. 1ab). The larger amount of fallback decreases the mass of the inner core (e.g., Fe, Mg, and O) relative to the mass of the outer layer (e.g., C, Figs. 1ab). Since O and Mg are synthesized in the inner layers than

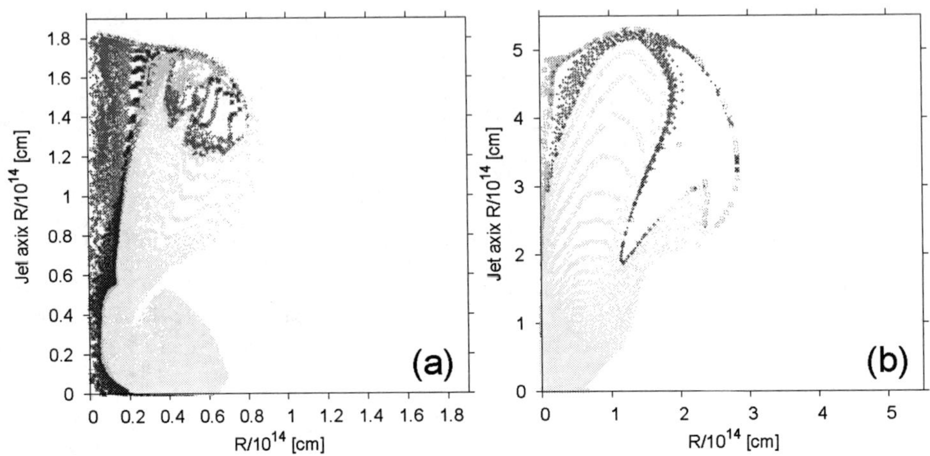

FIGURE 2. Positions of the mass elements at $t = 10^5$ s for (a) model A and (b) model B. Symbols of the marks represents the abundance of the mass element (H: *circle*, He: *triangles*, O+C: *squares*, O+Mg: *stars*, Si: *crosses*, and Fe: *pluses*). Size of the marks represents the origin of the mass element (the jet: *dots*, and the shocked stellar mantle: *filled circles*).

C, [C/O] and [C/Mg] are larger for the larger infall of the O layer. Also, the fallback of the O layer decreases $M(\mathrm{Fe})$ because Fe is mainly synthesized explosively in the Si and O+Mg layers. Therefore, the variation of $\dot{E}_{\mathrm{dep}}$ in the jet-induced SN explosions predicts that the variations of [C/O], [C/Mg], and [C/Fe] corresponding to the variation of $M(\mathrm{Fe})$.

Figure 1e shows that the abundance patterns of the normal EMP stars [20], and the HMP star, HE 1327–2326 [21, 22], are reproduced by models with $\dot{E}_{\mathrm{dep},51} = 120$, and 1.5, respectively (see Table 1 for model parameters). The model for the normal EMP stars ejects $M(\mathrm{Fe}) \sim 0.2 M_{\odot}$. On the other hand, the model for HE 1327–2326 ejects $M(\mathrm{Fe}) \sim 4 \times 10^{-6} M_{\odot}$.

Abundance distribution

Figures 2a and 2b show the abundance distributions and density structures at $t = 10^5$ s for models A and B [11]. We classify the mass elements by their abundances as follows: (1) Fe with $X(^{56}\mathrm{Ni}) > 0.04$, (2) Si with $X(^{28}\mathrm{Si}) > 0.08$, (3) O+Mg with $X(^{16}\mathrm{O}) > 0.6$ and $X(^{24}\mathrm{Mg}) > 0.01$, (4) O+C with $X(^{16}\mathrm{O}) > 0.6$ and $X(^{12}\mathrm{C}) > 0.1$, (5) He with $X(^{4}\mathrm{He}) > 0.7$, and (6) H with $X(^{1}\mathrm{H}) > 0.3$. If a mass element satisfies two or more conditions, the mass element is classified into the class with the smallest number.

The abundance distribution and thus the composition of the ejecta depend on the direction. In model A, the O+Mg, O+C, He, and H mass elements locate in the all direction. On the other hand, most of the Fe and Si mass elements locate at $\theta < 10^\circ$ and stratify in this order from the jet axis and a part of them locate at $15^\circ < \theta < 35^\circ$. Interestingly, the Fe mass elements surround the Si mass elements at $15^\circ < \theta < 35^\circ$. In model B, most of the O+C and He mass elements locate at $\theta < 3^\circ$, while the Fe mass

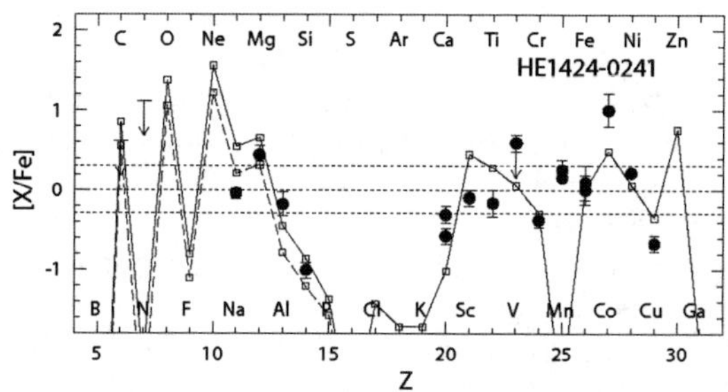

FIGURE 3. Comparison between the abundance pattern of HE 1424–0241 (*filled circles*) and the angle-delimited yields of model A for $30^\circ \leq \theta < 40^\circ$ (*solid line*) and $30^\circ \leq \theta < 35^\circ$ (*dashed line*).

elements injected as a jet expand laterally up to $\theta \sim 50^\circ$ and the H mass elements are distributed in the all directions. The lateral expansion of the Fe mass elements in models A and B are led by the collision with the stellar mantle and the internal pressure of the jet.

A very peculiar, Si-deficient, metal-poor star HE 1424–0241 was observed [23]. Its abundance pattern with high [Mg/Si] (~ 1.4) and normal [Mg/Fe] (~ 0.4) is difficult to be reproduced by previous SN models. This is because $\log\{[X(^{24}\mathrm{Mg})/X(^{28}\mathrm{Si})]/[X(^{24}\mathrm{Mg})/X(^{28}\mathrm{Si})]_\odot\} \lesssim 1.6$ is realized in the O+Mg layer at the presupernova stage (e.g., [24, 3]). Thus, in order to reproduce the abundance pattern of HE 1424–0241, the SN yield is required to consist of explosively-synthesized Fe but not explosively-synthesized Si.

The angle-delimited yield may possibly explain the high [Mg/Si] and normal [Mg/Fe]. Figure 3 shows that the yields integrated over $30^\circ \leq \theta < 40^\circ$ and $30^\circ \leq \theta < 35^\circ$ of model A reproduce the abundance pattern of HE 1424–0241. The yield consist of Mg in the inner region and Fe in the outer region (Fig. 2a). Although there are some elements to be improved, the elusive feature of HE 1424–0241 could be explained by taking into account the angular dependence of the yield. The high [Mg/Si] and normal [Mg/Fe] can be realized with an appropriate integration range if the Fe mass elements penetrate the stellar mantle (i.e., the duration of the jet injection is long) and if the O+Mg mass elements are ejected in all directions (i.e., $\dot{E}_{\mathrm{dep}}$ is high).

CONCLUSIONS AND DISCUSSION

We perform two-dimensional hydrodynamical and nucleosynthesis calculations of the jet-induced explosions of a Pop III $40M_\odot$ star and show two jet-induced explosion models A and B as summarized in Table 1. We have shown that (1) the yields of the explosions with large $\dot{E}_{\mathrm{dep}}$ explain the abundances of the normal EMP stars, and (2) the explosions with small $\dot{E}_{\mathrm{dep}}$ are responsible for the formation of the HMP star, HE 1327–2326.

(1) **Fallback**: the dynamics and the abundance distributions depend sensitively on the energy deposition rate $\dot{E}_{\mathrm{dep}}$. The explosion with lower $\dot{E}_{\mathrm{dep}}$ leads to a larger amount of

fallback, and consequently smaller M(Fe) and higher [C/O], [C/Mg], and [C/Fe]. Such dependences of [C/Fe] and M(Fe) on $\dot{E}_{\rm dep}$ predict that the higher [C/Fe] tends to be realized for lower [Fe/H]. Note, however, the formation of star with low [C/Fe] and [Fe/H] is possible because [Fe/H] depends on the swept-up H mass, i.e., the interaction between the SN ejecta and ISM (e.g., [2]).

(2) **Angular dependence**: we present the aspherical abundance distributions and investigate the angular dependence of the yield. The angle-delimited yield could reproduce the extremely peculiar abundance pattern of HE 1424–0241. However, we note that the angle-delimited yield depends strongly on which mass elements are included into the integration. This would be determined by the abundance mixing in the SN ejecta and by the region where the next-generation star takes in the metal-enriched gas. Although it is necessary to calculate three-dimensional evolution of the supernova remnant (e.g., [25]), these would be interesting issues.

REFERENCES

1. J. Audouze, & J. Silk, *ApJ* **451**, L49–L52 (1995).
2. D. F. Cioffi, C. F. McKee, E. Bertschinger, *ApJ* **334**, 252–265 (1988).
3. H. Umeda, & K. Nomoto, *ApJ* **619**, 427–445 (2005).
4. N. Iwamoto, H. Umeda, N. Tominaga, K. Nomoto, & K. Maeda, *Science* **309** 451–453 (2005).
5. K. Nomoto, N. Tominaga, H. Umeda, C. Kobayashi, & K. Maeda, *Nucl. Phys. A* **777**, 424–458 (2006).
6. N. Tominaga, H. Umeda, & K. Nomoto, *ApJ* **660**, 516–540 (2007).
7. E. Depagne, et al., *A&A* **390**, 187–198 (2002).
8. M. S. Bessell, & N. Christlieb, 2005, "From Lithium to Uranium: Elemental Tracers of Early Cosmic Evolution", edited by V. Hill, et al., IAU Symposium Proceedings of the international Astronomical Union 228, Cambridge University Press, Cambridge, 2005, pp.237-238
9. N. Tominaga, et al., *ApJ* **657**, L77–L80 (2007).
10. N. Tominaga, *PhD thesis*, University of Tokyo, (2007).
11. N. Tominaga, *ApJ* submitted (arXiv:0711.4815) (2007).
12. W. R. Hix, & F.-K. Thielemann, *ApJ* **460**, 869–894 (1996).
13. W. R. Hix, & F.-K. Thielemann, *ApJ* **511**, 862–875 (1999).
14. I. Hachisu, T. Matsuda, K. Nomoto, & T. Shigeyama, *ApJ* **358**, L57–L61 (1990).
15. K. Maeda, & K. Nomoto, *ApJ* **598**, 1163–1200 (2003).
16. S. E. Woosley, *ApJ* **405**, 273–277 (1993).
17. G. E. Brown, et al. *New Astronomy* **5**, 191–210 (2000).
18. C. Fryer, & P. Mészáros, *ApJ* **588**, L25–L28 (2003).
19. K. Maeda, & N. Tominaga, in preparation (2008).
20. R. Cayrel, et al., *A&A* **416**, 1117–1138 (2004).
21. A. Frebel, et al., *Nature* **434**, 871–873 (2005).
22. W. Aoki, et al., *ApJ* **639**, 897–917 (2006).
23. J. G. Cohen, et al., *ApJ* **659**, L161–L164 (2007).
24. S. E. Woosley, & T. A. Weaver, *ApJS* **101**, 181–235 (1995).
25. N. Nakamura, & T. Shigeyama, *ApJ* **541**, L59–L62 (2000).

Evolution of Dust in Primordial Supernova Remnants and Its Influence on the Elemental Composition of Hyper-Metal-Poor Stars

Takaya Nozawa*, Takashi Kozasa*, Asao Habe*, Eli Dwek†, Hideyuki Umeda**, Nozomu Tominaga**, Keiichi Maeda‡,§ and Ken'ichi Nomoto**,‡

*Department of Cosmosciences, Graduate School of Science, Hokkaido University, Sapporo 060-0810, Japan
†Laboratory for Astronomy and Solar Physics, NASA Goddard Space Flight Center, Greenbelt, MD 20771
**Department of Astronomy, School of Science, University of Tokyo, Bunkyo-ku, Tokyo 113-0033, Japan
‡Institute for the Physics and Mathematics of the Universe, University of Tokyo, Kashiwa, Chiba 277-8568, Japan
§Max-Planck-Institut für Astrophysik, Karl-Schwarzschild Strasse 1, 85741 Garching, Germany

Abstract.
The calculations for the evolution of dust within Population III supernova remnants (SNRs) are presented, based on the models of dust formed in the unmixed ejecta of Type II SNe. We show that once dust grains collide with the reverse shock penetrating into the ejecta, their fates strongly depend on the initial radius $a_{\rm ini}$. For SNRs expanding into the interstellar medium (ISM) with $n_{\rm H,0} = 1$ cm^{-3}, grains of $a_{\rm ini} < 0.05$ μm are trapped in the hot gas to be completely destroyed; grains of $a_{\rm ini} = 0.05$–0.2 μm are piled up in the dense shell formed behind the forward shock; grains of $a_{\rm ini} > 0.2$ μm are injected into the ISM without being eroded significantly. The total mass of surviving dust is 0.01 to 0.8 $M_\odot$ for $n_{\rm H,0} = 10$ to 0.1 cm^{-3}. We also investigate the influence of the piled-up dust on the elemental abundances of the second-generation stars formed in the dense shell of Population III SNRs. The comparison of the calculated elemental abundances with those observed in hyper-metal-poor (HMP) and ultra-metal-poor (UMP) stars indicates that the transport of dust separated from metal-rich gas can be an important process in determining the abundance patterns of Mg and Si in HMP and UMP stars.

Keywords: dust; extinction, early universe, supernovae: general, stars; abundances, Population III
PACS: 95.30.Lz, 95.30.Wi, 97.10.Tk, 97.20.Wt, 97.60.Bw, 98.38.Mz

INTRODUCTION

Dust grains in the early universe have great impacts on the evolutional history of the universe: Dust grains control the energy balance in interstellar space through absorption of stellar light and re-emission of it, and play critical roles in the formation processes of stars and galaxies. In particular, their thermal emission greatly increases the efficiency of cooling of gas in metal-poor circumstances, which causes the star-forming gas cloud to fragment into low-mass gas clumps of 0.1–1 $M_\odot$ for metallicity of 10^{-6} to 10^{-5} $Z_\odot$ [1, 2, 3]. The absorption and thermal emission by dust grains heavily depend on their composition, size, and amount. Thus, it is one of the most important subjects in astrophysics to elucidate the origin and evolution of dust grains.

CP1016, *Origin of Matter and Evolution of Galaxies*,
edited by T. Suda, T. Nozawa, A. Ohnishi, K. Kato, M. Y. Fujimoto, T. Kajino, and S. Kubono

The dominant sources of dust in the early universe are considered to be supernovae (SNe) evolving from the early generations of massive stars. Theoretical studies have investigated the formation of dust in primordial Type II SNe (SNe II) [4, 5] and pair-instability SNe [5, 6], and have revealed the composition, size, and mass of dust formed in the ejecta; for SNe II, dust grains of 0.1–2 $M_{\odot}$ are formed, which is in good agreement with the amount required to account for dust content in high-redshift galaxies [7, 8]. However, the reverse shock induced by the interaction of the SN ejecta with the surrounding medium reprocesses the newly condensed dust grains and makes their size and mass significantly different from those at the time of dust formation [9, 10].

In this proceedings, we present the calculations for the processing of dust through the collisions with the reverse shocks and its transport within Population III SN remnants (SNRs), in order to reveal the size and amount of dust injected from SNe into the interstellar medium (ISM). The results of calculations show that a part of the surviving dust grains are accumulated in the dense SN shell. These piled-up grains may have significant influences on the elemental abundances of the second-generation stars formed there. Therefore, assuming that the composition of the piled-up grains reflects the elemental composition of those stars, we explore the abundance patterns of the stars formed in the dense shell of Population III SNRs and compare to those observed in hyper-metal-poor (HMP) and ultra-metal-poor (UMP) stars.

EVOLUTION OF DUST IN PRIMORDIAL SNRS

The calculations of the dust evolution in SNRs are described in detail by Nozawa et al. (2007) [10]. By taking the size distribution and spatial distribution of each dust species from the dust formation model by [5] as the initial conditions, the transport of dust and its destruction by sputtering are calculated. The time evolution of the gas temperature and density in SNRs expanding into the uniform ISM with the hydrogen number density of $n_{\rm H,0} = 0.1$, 1, and 10 cm^{-3} are calculated by adopting the hydrodynamic models of Population III SNe by [11]. In what follows, we present the results for the evolution of the dust formed in the unmixed ejecta of SNe II with the progenitor masses of $M_{\rm pr} = 13$, 20, 25, and 30 $M_{\odot}$ and explosion energy of 10^{51} ergs.

In Figure 1*a*, we show the time evolutions of positions of C, Mg_2SiO_4, and Fe grains within the SNR for $M_{\rm pr} = 20$ $M_{\odot}$ and $n_{\rm H,0} = 1$ cm^{-3}, and the time evolutions of their radii are given in Fig 1*b*. The thick solid lines in Fig. 1*a* depict the positions of the forward shock, the reverse shock, and the surface of the He core. C, Mg_2SiO_4, and Fe grains comove with the gas until they encounter the reverse shock at 3650 yr, 6300 yr, and 13000 yr, respectively. Once dust grains intrude into the reverse shock, their fates strongly depend on their initial radii $a_{\rm ini}$ and compositions. Because the deceleration rate of a grain is inversely proportional to its radius, the relatively small grains with $a_{\rm ini} = 0.01$ μm (*dotted lines*) are efficiently decelerated by the gas drag and are trapped in the hot gas of $\geq 10^6$ K created by the passage of the reverse and forward shocks. These small grains captured in the hot gas are completely destroyed by the thermal sputtering. C, Mg_2SiO_4, and Fe grains with $a_{\rm ini} = 0.1$ μm (*dashed lines*) penetrating into the hot gas are eroded due to the kinetic and/or thermal sputterings, and their radii are reduced by 43%, 69%, and 52%, respectively. C and Mg_2SiO_4 grains of $a_{\rm ini} = 0.1$ μm are finally

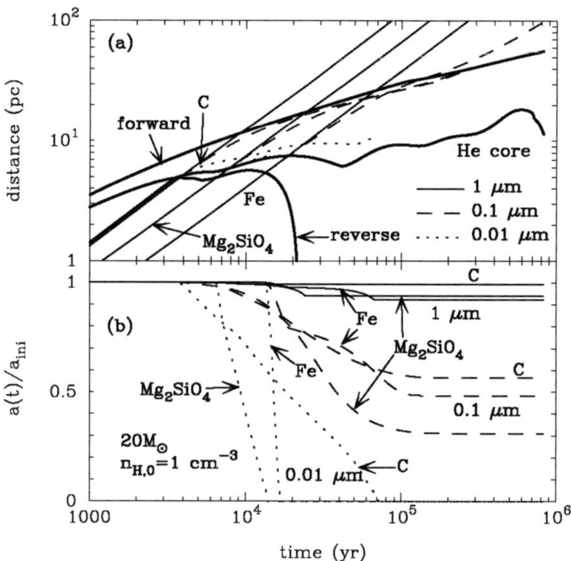

FIGURE 1. Time evolutions of (*a*) positions of C, Mg_2SiO_4, and Fe grains within the SNR for $M_{pr} = 20$ $M_\odot$ and $n_{H,0} = 1$ cm^{-3} and (*b*) ratios of their radii to the initial ones. The evolution of dust with $a_{ini} =$ 0.01, 0.1, and 1 μm is denoted by the dotted, dashed, and solid lines, respectively. In (*a*), the positions of the forward shock, the reverse shocks, and the surface of the He core are indicated by the thick solid lines.

trapped in the dense SN shell formed at $\sim 2 \times 10^5$ yr, where the dust grains are no longer eroded by the thermal sputtering because the gas temperature drops down quickly below 10^5 K [12]. The 0.1 μm-sized Fe grains with higher bulk density are ejected into the ISM. Since the deceleration by the gas drag is inefficient, grains with $a_{ini} = 1$ μm (*solid lines*) also go across the outwardly expanding shock front and are injected into the ISM without being processed significantly,

The behaviors of the transport and destruction of dust in a SNR heavily depend on the density of gas in the ISM; as the ISM gas density is higher, the density of the shocked gas is higher, which leads to the efficient deceleration and erosion of dust through more frequent collisions with the hot gas. As a result, the lower limit of the initial radius of dust injected into the ISM increases with increasing the ISM density and is 0.03, 0.2, and 0.5 μm for $n_{H,0} = 0.1$, 1, and 10 cm^{-3}, respectively. Furthermore, the initial radius below which dust is completely destroyed is enhanced by the increase of the density in the ISM, and is 0.01, 0.05, and 0.2 μm for $n_{H,0} = 0.1$, 1, and 10 cm^{-3}. Accordingly, as shown in Figure 2, the total mass of the surviving dust decreases with increasing the ambient density and spans the range of 0.01–0.8 $M_\odot$ for $n_{H,0} = 10$ to 0.1 cm^{-3}, depending on the progenitor mass. In any cases, the resulting size distribution of the dust that survives the destruction shows a serious lack of small-sized grains, in comparison with that at the time of dust formation. Note that the evolution of dust within a SNR does not depend on the progenitor mass considered here because the same explosion energy results in the similar time evolution of the gas temperature and density.

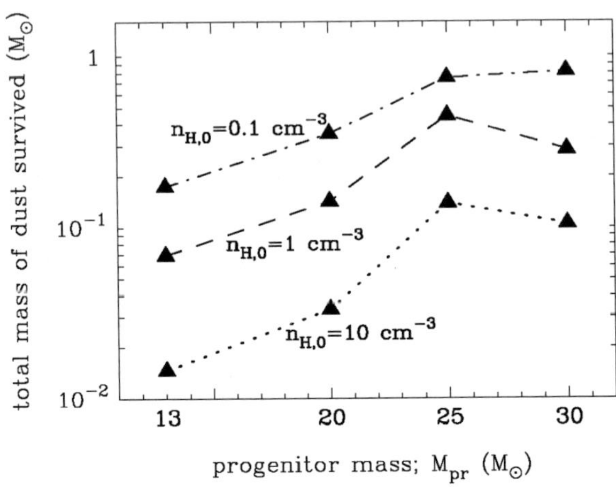

FIGURE 2. Total mass of surviving dust vs. the progenitor mass. The results for $n_{\rm H,0} = 0.1$, 1, and 10 cm^{-3} are connected by the dot-dashed, dashed, and dotted lines, respectively.

ELEMENTAL ABUNDANCES IN THE DENSE SN SHELL

Currently known most iron-deficient stars, HE 0107-5240 with [Fe/H] $= -5.3$ ([13, 14]) and HE 1327-2326 with [Fe/H] $= -5.45$ ([15, 16]) show peculiar abundance patterns: more than 100 times excesses of C, N, and O and 1–100 times enhancements of Mg and Si relative to Fe. Also, UMP star, HE 0557-4880 with [Fe/H] $= -4.75$ ([17]) exhibits a large ($\sim$50 times) overabundance of C. Although the elemental compositions of these low-mass stars are expected to reflect the nucleosynthesis in the very early generations of stars, the origin of the abundance patterns in HMP and UMP stars is still an open question.

The results for the evolution of dust within a SNR presented in this paper show that the dust grains which are not injected into the ISM but survive the destruction through the collision with the reverse shock are accumulated in the dense SN shell in 10^5–10^6 yr. This transport of dust to the SN shell may not only enable the formation of stars with solar mass scales there [1, 2, 3] but also influence the elemental abundance of those stars. Therefore, based on the composition of the piled-up grains, we investigate the metal abundance patterns of the stars formed in the dense shell of SNRs and compare to those observed in HMP and UMP stars, assuming that they are the second-generation stars formed in the shell of Population III SNRs.

The results for the elemental composition and metallicity in the dense shell are summarized in Table 1 for various progenitor masses and ambient densities. Metallicity in the shell is in the range from 10^{-6} up to 10^{-4} $Z_\odot$, which is high enough to form low-mass stars with 0.1–1 $M_\odot$ [1, 2, 3]. We can also see that most of calculated [Fe/H] spans the range of -6 to -4.5, which are in good agreement with those for HMP and UMP stars.

In Figure 3a, we show the abundances of Mg and Si (*upper panel*) and C and O

TABLE 1. Elemental composition and metallicity in the dense shell of primordial SNe II for various progenitor masses and ambient densities.

M_{pr} ($M_{\odot}$)	[Fe/H]	[C/Fe]	[O/Fe]	[Mg/Fe]	[Si/Fe]	[Al/Fe]	[S/Fe]	$\log(Z/Z_{\odot})$
				$n_{H,0} = 0.1$ cm^{-3}				
13	−6.43	−0.274	−0.699	−0.230	1.92	−2.60	0.239	−5.89
20	−5.20	0.117	−0.595	0.034	0.410	−1.97	0.242	−5.44
25	−5.90	1.11	−1.42	−0.500	−0.552	−0.563	0.242	−5.55
30	−5.56	0.566	−0.043	0.739	0.866	0.905	0.242	−5.33
				$n_{H,0} = 1$ cm^{-3}				
13	−5.15	1.11	−0.555	−0.459	1.01	–	−2.18	−4.72
20	−5.53	0.992	0.585	1.16	1.87	–	0.200	−4.68
25	−5.23	1.09	−0.412	0.407	0.989	–	0.241	−4.79
30	−5.11	0.797	0.242	1.09	1.26	−5.72	0.242	−4.60
				$n_{H,0} = 10$ cm^{-3}				
13	−4.13	0.284	−2.54	−3.89	0.599	–	–	−4.40
20	−4.92	0.946	−2.15	−1.80	2.14	–	–	−4.09
25	−5.10	1.60	0.122	0.232	2.34	–	−1.45	−3.91
30	−5.11	−0.207	0.375	−1.23	2.66	–	−0.696	−3.84

(*lower panel*) in the dense shell as a function of [Fe/H] for the ISM density of 1 cm^{-3}. The results of calculations are drawn by the filled symbols, and the observational data of HMP and UMP stars are plotted by the open symbols. We can see that in addition to Fe, the calculated abundances of Mg and Si can also reproduce the observations of HMP stars with 1–100 times overabundances seen in HMP stars. Some models (such as $M_{pr} =$ 20 and 30 $M_{\odot}$ for $n_{H,0} = 0.1$ cm^{-3}) also show the modest overabundances for both Mg and Si (see Table 1). Because the elemental composition of dust piled up in the shell can reproduce the abundance patterns of refractory elements such as Mg, Si, and Fe in HMP and UMP stars, we can conclude that the transport of dust separated from metal-rich gas within SNRs can be responsible for the abundance patterns in HMP and UMP stars, if they are the second-generation stars formed in the dense shell of primordial SNRs. However, more than one hundred times excesses of C and O observed in HMP and UMP stars cannot be reproduced by any models considered here.

The reasons is that in the calculation we assumed that the metal-rich gas in the ejecta of SN does not mix with the gas in the shell. Then, as an extreme case, we examine the abundance patterns in the shell by considering that in addition to the piled-up grains, the gas outside the innermost Fe layer of the SN is incorporated into the shell. The results are shown by the filled symbols in Figure 3*b*. This case can produce the extreme overabundances ($\sim$1000 times) of C and O, but results in unreasonable excesses ($\geq$100 times) of Mg and Si, which is not in agreement with the observations. Nevertheless, it might be possible to reproduce the elemental abundances of HMP stars if the Si-Mg-rich layer is not mixed with the gas in the shell. Anyway, it is necessary to examine what extent of the gas in the SN ejecta can mix with the gas in the shell when the second-generation stars form in the SN shell. In the future works, we will consider this subject.

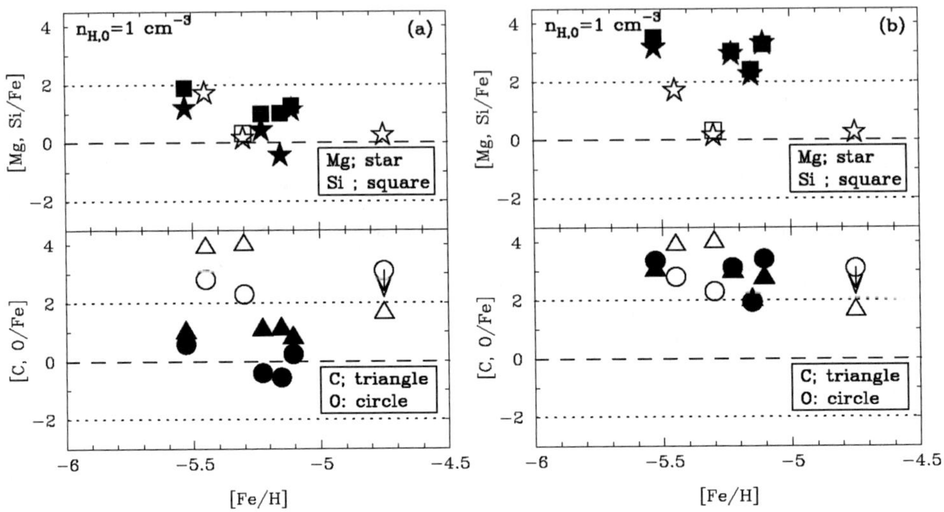

FIGURE 3. Abundances of Mg and Si (*upper panel*) and C and O (*lower panel*) in the shell of the SNR for $n_{\mathrm{H},0} = 1\ \mathrm{cm}^{-3}$; (*a*) the elemental abundances obtained from the grains piled up in the shell, and (*b*) the elemental abundances including the gas outside the innermost Fe layer as well as the piled-up grains. The results of calculations are shown by the filled symbols. The observational data for HMP and UMP stars are denoted by the open symbols and are taken from [14, 15, 16, 17, 18].

ACKNOWLEDGMENTS

This work has been supported in part by a Grant-in-Aid for Scientific Research from the Japan Society for the Promotion of Sciences (19740094, 18104003).

REFERENCES

1. Omukai, K., et al., *ApJ*, **626**, 627–643 (2005)
2. Tsuribe, T., & Omukai, K., *ApJ*, **642**, L61–L64 (2006)
3. Schneider, R., et al., *MNRAS*, **369**, 1437–1444 (2006)
4. Todini, P., & Ferrara, A., *MNRAS*, **325**, 726–736 (2001)
5. Nozawa, T., et al., *ApJ*, **598**, 785–803 (2003)
6. Schneider, R., Ferrara, A., & Salvaterra, R., *MNRAS*, **351**, 1379–1386 (2004)
7. Morgan, H. L., & Edmunds, M. G., *MNRAS*, **343**, 427–442 (2003)
8. Dwek, E., Galliano, F., & Jones, A. P., *ApJ*, **662**, 927–939 (2007)
9. Bianchi, S., & Schneider, R., *MNRAS*, **378**, 973–982 (2007)
10. Nozawa, T., et al., *ApJ*, **666**, 955–966 (2007)
11. Umeda, H., & Nomoto, K., *ApJ*, **565**, 385–404 (2002)
12. Nozawa, T., Kozasa, T., & Habe, A., *ApJ*, **648**, 435–451 (2006)
13. Christlieb, N., et al., *Nature*, **419**, 904–906 (2002)
14. Christlieb, N., et al., *ApJ*, **603**, 708–728 (2004)
15. Frebel, A., et al., *Nature*, **434**, 871–873 (2005)
16. Aoki, W., et al., *ApJ*, **639**, 897–917 (2006)
17. Norris, J. E., et al., *ApJ*, **670**, 774–788 (2007)
18. Frebel, A., et al., *ApJ*, **638**, L17–L20 (2006)

4. SPECIAL SESSION: FIRST STARS (2) CHEMICAL EVOLUTION and STRUCTURE FORMATION IN THE UNIVERSE

Formation of the first stars in the universe

Naoki Yoshida

Department of Physics, Nagoya University, Furocho, Chikusa, Nagoya 464-8602, Japan

Abstract. The standard cosmological model predicts that first cosmological objects are formed when the age of the universe is a few hundred million years. Recent theoretical works and numerical simulations consistently suggests that the first objects are very massive primordial stars. We introduce the key physics and explain why the first stars are thought to be massive, rather than to be low-mass stars. The state-of-the-art simulation includes all the relevant atomic and molecular physics to follow the thermal evolution of a prestellar gas cloud to very high densities. We also show the result of a simulation of the formation of primordial stars in a reionized gas.

Keywords: Population III Stars
PACS: 97.20.Wt

PHYSICS OF PRIMORDIAL STAR FORMATION

The study of primordial star formation has a long history. The formation of the first cosmological objects via gas condensation by molecular hydrogen cooling has been studied for many years since late 1960's. One-dimensional hydrodynamic simulations of spherical gas collapse were performed by a number of researchers [1]. Omukai & Nishi (1998) included a detailed treatment of all the relevant chemistry and radiative processes and thus were able to provide accurate results on the thermal evolution of a collapsing primordial gas cloud up to stellar densities. These authors found that, while the evolution of a spherical primordial gas cloud proceeds in a roughly self-similar manner, there are a number of differences in the thermal evolution from that of present-day, metal- and dust-enriched gas clouds.

Recently, three-dimensional hydrodynamic calculations were performed by several groups. Statistical properties of primordial star-forming clouds and the overall effect of cosmological bias have been studied in detail [2, 3]. Simulations of the formation of primordial proto-stars have been hampered by complexity of involved physics such as radiative transfer in very high-density primordial gases. A critical technique we describe in this contribution is computation of molecular line opacities and continuum opacities. With the implementation of optically thick cooling, the gas evolution can be accurately followed to much higher densities than was possible in the previous studies [4, 6]. We show that the method works well in problems of collapsing gas clouds, in terms of computation of radiative cooling rates and resulting density and temperature structure. Our method can be applied to a broad range of three-dimensional problems in non-trivial configurations. We apply this technique to a cosmological simulation of the formation of primordial star.

CP1016, *Origin of Matter and Evolution of Galaxies*,
edited by T. Suda, T. Nozawa, A. Ohnishi, K. Kato, M. Y. Fujimoto, T. Kajino, and S. Kubono

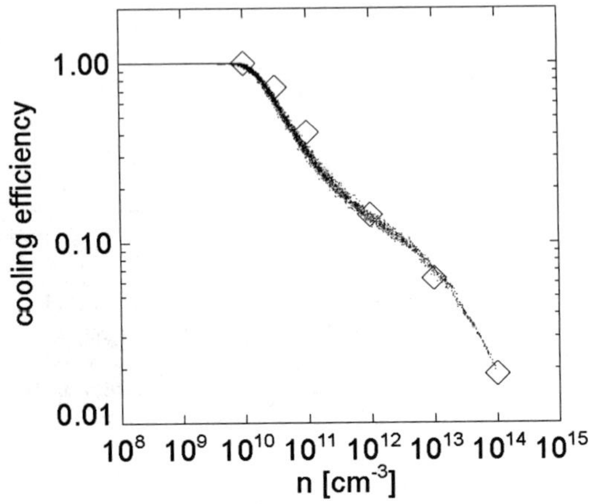

FIGURE 1. The effect of molecular line opacities to the net cooling rate.

OPTICALLY THICK COOLING RATE

Molecular hydrogen line cooling

When the gas density and the molecular fraction are high, the cloud becomes opaque to molecular lines and then H_2 line cooling becomes inefficient. The net cooling rate can be expressed as

$$\Lambda_{\mathrm{H_2,thick}} = \sum_{u,l} h\nu_{ul}\, \beta_{\mathrm{esc},ul}\, A_{ul}\, n_u\,, \tag{1}$$

where n_u is the population density of hydrogen molecules in the upper energy level u, A_{ul} is the Einstein coefficient for spontaneous transition, $\beta_{\mathrm{esc},ul}$ is the probability for an emitted line photon to escape without absorption, and $h\nu_{ul} = \Delta E_{ul}$ is the energy difference between the two levels.

In order to calculate the escape probability, we first evaluate the opacity for each molecular line as

$$\tau_{lu} = \alpha_{lu} L, \tag{2}$$

where L is the characteristic length scale. Since the absorption coefficients α_{lu} are computed in a straightforward manner, although somewhat costly, the remaining key task is the evaluation of the length scale L. By noting that the important quantity we need is the effective gas cooling rate, we can formulate a reasonable and well-motivated approximation. To this end, we decided to use the Sobolev method that is widely used in the study of stellar winds and planetary nebulae.

We calculate the Sobolev length along a line-of-sight as

$$L_r = \frac{v_{\mathrm{thermal}}}{|dV_r/dr|}, \tag{3}$$

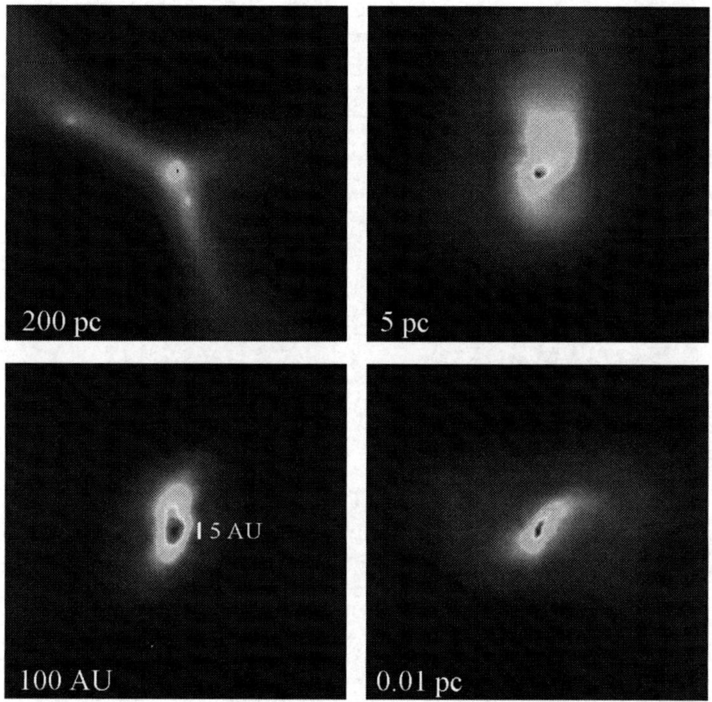

FIGURE 2. The projected density distributions up to the time when the central core becomes optically thick to continuum. From Yoshida et al. (2006, ApJ)

where $v_{\mathrm{thermal}} = \sqrt{kT/m_{\mathrm{H}}}$ is the thermal velocity of H_2 molecules, and V_r is the fluid velocity in the direction. A suitable angle-average must be computed in order to obtain the net escape probability. Details are found in [4].

We test our method by performing a three-dimensional calculation of spherical cloud collapse. FIGURE 1 shows the normalized H_2 line cooling rate against local density. We use an output at the time when the central density is $n_{\mathrm{c}} = 10^{14}\mathrm{cm}^{-3}$. In the figure, we compare our simulation results with those from the full radiative transfer calculations of [1] (open diamonds). Clearly our method works very well. The steepening of the slope at $n > 10^{12}\mathrm{cm}^{-3}$, owing to the velocity change where infalling gas settles gradually onto the center, is well-reproduced. We emphasize that the level of agreement shown in FIGURE 1 can be achieved only if *all* of the local densities (of chemical species), temperatures, and velocities are reproduced correctly.

FIGURE 2 shows the density field around a proto-star found in our cosmological simulation. Structures down-to $\sim$ AU scale is resolved in the simulation.

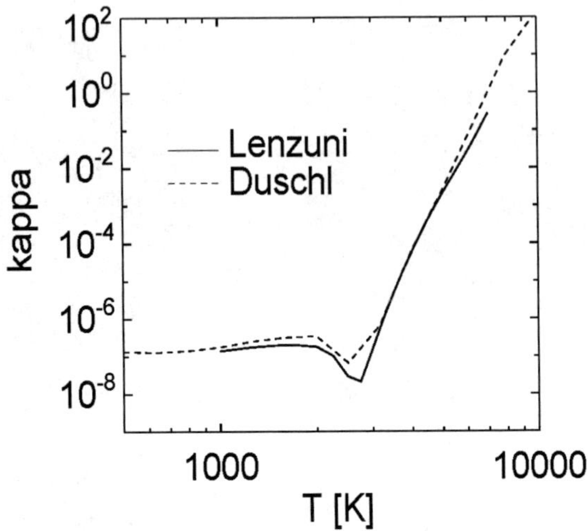

FIGURE 3. The Planck opacity as a function of temperature for a density of $10^{16}\mathrm{cm}^{-3}$.

Cooling by collision-induced emission

Next, we implement computation of local optical depth to continuum radiation using the Planck opacity table of Lenzuni et al. (1991). The continuum opacity is used to evaluate the net cooling rate by collision-induced emission (CIE). The optically thin cooling rate can be computed as in [4].

FIGURE 3 compares the opacities for a gas of primordial composition calculated by Lenzuni et al. (1991) and Mayer & Duschl (2005). We find no significant difference between the two opacity tables, and thus we use the one by Lenzuni et al. We calculate local optical depths in six orthogonal directions from a target point in a smoothed particle hydrodynamics manner (see, e.g., [5]), by projecting the SPH kernels of surrounding gas particles which have their own density and temperature. The net energy transfer rate in each direction scales as $\propto 1/(1+\tau)$ when the optical depth is small. Again, a suitable angle average must be calculated. We simply take the mean of six directions,

$$f_{\rm reduce} = \frac{1}{6}\sum\frac{1}{1+\tau_i}. \tag{4}$$

We then use this local reduction factor to evaluate the cooling rate [1] FIGURE 4 shows the net reduction factor for the CIE cooling rate against local density. We note that, although our method is still approximate, it takes into account local density, temperature, and their

[1] This estimate becomes incorrect when τ is large. However, the radiative cooling rate is extremely small for large τ and so the exact behavior at very optically-thick regime does not matter to the net cooling rate.

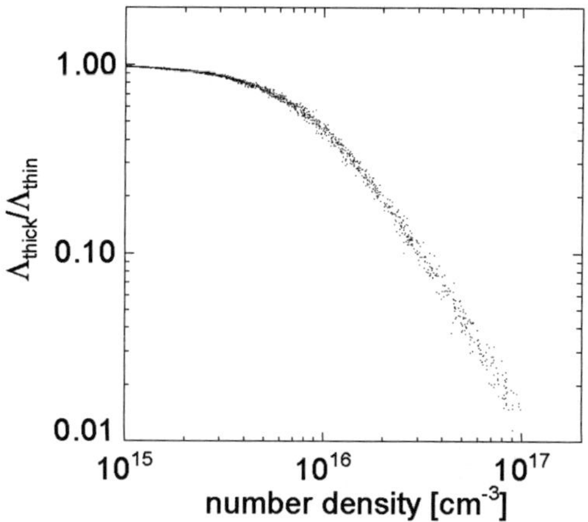

FIGURE 4. The ratio of $\Lambda_{CIE,thick}/\Lambda_{CIE,thin}$ as a function of density. We use an output when the central density reaches 10^{20}cm^{-3}.

structures (geometry) in a self-consistent manner. Our method can be applied to general problems, whereas simple functional fits that is based only on local density will fail in estimating the true cooling rate when the cloud core has a complex structure such as (weak) turbulence.

DISSOCIATION COOLING

When a gas cloud core becomes opaque to continuum radiation, the gas can still absorb thermal energy input owing to gravitational contraction by dissociating molecules. Ionization cooling is unimportant in the temperature regime we consider.

For densities much larger than 10^{15}cm^{-3}, the reaction time scale becomes significantly shorter than the dynamical time. Then, we can use equilibrium chemistry which would actually *simplify* our handling of chemistry evolution. The species abundances can be determined from the coupled Saha-Boltzmann equations:

$$\frac{n(\mathrm{H})^2}{n(\mathrm{H_2})} = \frac{z_{\mathrm{H}}^2}{z_{\mathrm{H_2}}} \left(\frac{\pi k m_{\mathrm{H}}}{h^2}\right)^{3/2} T^{3/2} \exp\left(-\frac{\chi_{\mathrm{H_2}}}{kT}\right), \tag{5}$$

$$\frac{n(\mathrm{H^+})^2}{n(\mathrm{H})} = \frac{2}{z_{\mathrm{H}}} \left(\frac{2\pi k m_{\mathrm{e}}}{h^2}\right)^{3/2} T^{3/2} \exp\left(-\frac{\chi_{\mathrm{H}}}{kT}\right). \tag{6}$$

We solve these equations at a given gas density iteratively. The associated cooling/heating due to dissociation/formation is implemented self-consistently as

$$\varepsilon_{\mathrm{H2}} = \chi_{\mathrm{H_2}} \Delta n_{\mathrm{H_2}} \tag{7}$$

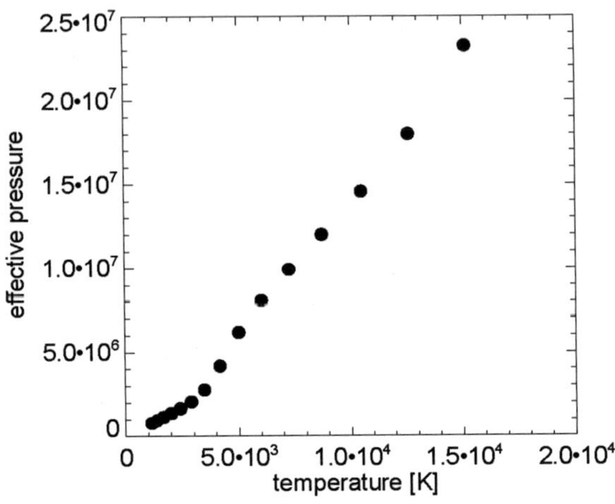

FIGURE 5. Pressure for a non-ideal gas, as calculated by [8], for a particle number density of $10^{19}\mathrm{cm}^{-3}$.

where $\chi_{\mathrm{H}_2} = 4.48\mathrm{eV}$ is the molecular binding energy.

When dissociation of hydrogen molecules is completed at $T \sim 5000$ K, there will be no further mechanisms that enable the gas to lose its thermal energy, and then the gas temperature increases following the so-called adiabatic track.

The equation of state in the high pressure regime must be modified to account for various non-ideal gas effects [8]. FIGURE 5 shows the effective gas pressure for which various non-ideal gas effects are included.

PRIMORDIAL STAR-FORMATION IN A REIONIZED GAS

We employ the technique described in the previous sections in a cosmological simulation. Earlier [5], we used a large cosmological simulation to study the evolution of early relic HII regions until second-generation gas clouds are formed. We further explore the evolution of these prestellar gas clouds. The highest density achieved by the simulation is $\sim 10^{18}\mathrm{cm}^{-3}$, at which point the central core is optically thick even to continuum. Full-scale dissociation of hydrogen molecules is taking place in the core, which works as an effective *cooling* mechanism. We investigate in detail the structure of such gas clouds. We then compute the gas mass accretion rate and use it as an input to a proto-stellar calculation.

FIGURE 6 shows the radial temperature profile around the proto-star and the gas mass accretion rate. The temperature structure can be understood by various atomic and molecular processes (see [4, 6]). HD line cooling brings the gas temperature below 100 Kelvin. Note that the minimum temperature is set by the that of cosmic microwave background.

The central proto-stellar 'seed' is accreting the surrounding gas at a rate $> 10^{-3} M_{\odot}/\mathrm{yr}$

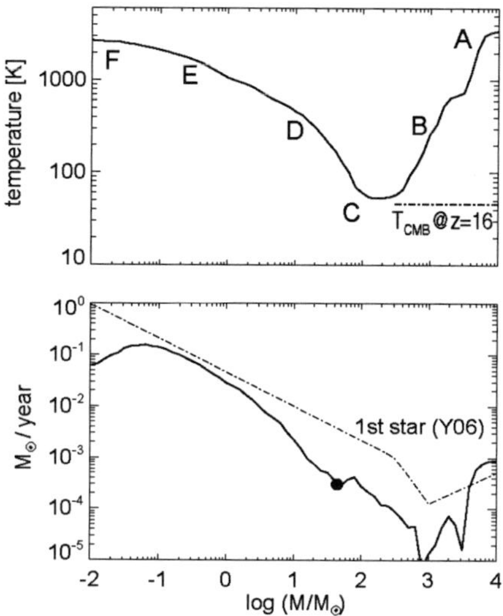

FIGURE 6. (Top) The radial temperature profile around the proto-star as a funciton of enclosed gas mass. (Bottom) The instantaneous gas mass accretion rate. We also show the accretion rate of [4] as the dot-dashed line.

and thus a star with mass $\sim 10M_\odot$ will form within 10^4 years. However, the final stellar mass is determined by processes such as radiative feedback from the protostar. We treat the evolution of a protostar as a sequence of a growing hydrostatic core with an accreting envelope. The ordinary stellar structure equations are applied to the hydrostatic core. The structure of the accreting envelope is calculated under the assumption that the flow is steady for a given mass accretion rate.

FIGURE 7 shows the resulting evolution of the protostar. After a transient phase and an adiabatic growth phase at $M_* < 10M_\odot$, the protostar enters the Kelvin-Helmholtz phase and contracts by radiating its thermal energy. When the central temperature reaches 10^8K, hydrogen burning by the CNO cycle begins with a slight amount of carbon synthesized by helium burning. This phase is marked by a solid circle in the figure. The energy generation by hydrogen burning halts contraction when the mass is $35M_\odot$ and its radius is ~ 2.8 solar radii. Soon after, the star reaches the zero-age main sequence (ZAMS). The protostar relaxes to a ZAMS star within about 10^5 years from the birth of the protostellar seed. Accretion is not halted by radiation from the protostar to the end of our calculation.

It is important to point out that the mass of the parent cloud from which the star formed is $M_{\rm cloud} \sim 40M_\odot$. The final stellar mass is likely limited by the mass of the gravitationally unstable parent cloud. We thus argue that primordial stars formed from an ionized gas are massive, with a characteristic mass of several tens of solar masses, allowing overall uncertainties in the accretion physics and also the dependence of the minimum gas temperature on redshift. They are smaller than the first stars formed from

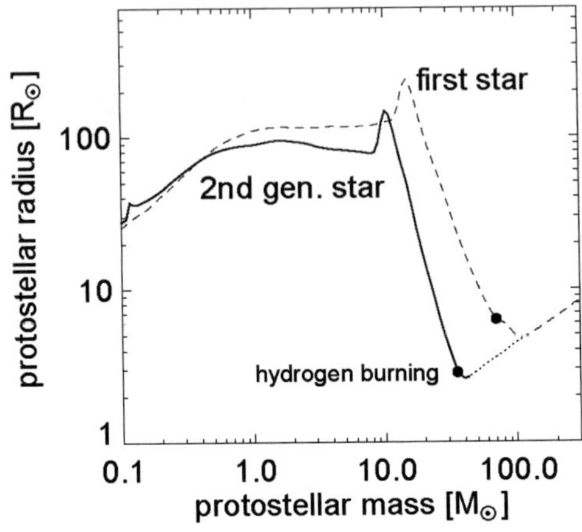

FIGURE 7. Evolution of the proto-stellar radius and the mass (solid line). The solid circle marks the time when efficient hydrogen burning begins. The dotted line shows the mass (and radius) growth which is calculated under the assumption that a larger amount of gas than the parent cloud can be accreted. For reference we also show the result from [4] for a first generation star that forms in an initially neutral gas cloud.

a neutral gas, but are not low-mass objects as suggested by earlier studies.

The elemental abundance patterns of recently discovered hyper metal-poor stars suggest that they might have been born from the interstellar medium that was metal-enriched by supernovae of these massive primordial stars[9].

ACKNOWLEDGMENTS

The work is supported in part by the Grants-in-Aid for Young Scientists 17684008 by the Ministry of Education, Culture, Science and Technology of Japan, and by The Mitsubishi Foundation.

REFERENCES

1. Omukai, K. & Nishi, R., 1998, ApJ, 508, 141
2. Yoshida, N., Abel, T, Hernquist, L., Sugiyama, N., 2003, ApJ, 592, 645
3. Gao, L., Yoshida, N., Abel, T., Frenk, C. S., Jenkins, A., Springel, V., 2007, MNRAS, 378, 449
4. Yoshida, N., Omukai, K., Hernquist, L., Abel, T., 2006, ApJ, 652, 6
5. Yoshida, N., Oh, S.-P., Kitayama, T., Hernquist, L., 2007, ApJ, 663, 687
6. Yoshida, N., Omukai, K., Hernquist, L., 2007, ApJL, 667, 117
7. Lenzuni, P., D., Chernoff, D. F., & Salpeter, E. E., 1991, ApJS, 76, 759
8. Saumon, D., Chabrier, G., & van Horn, H. M., 1995, ApJS, 99, 713
9. Iwamoto, N., Umeda, H., Tominaga, N., Nomoto, K., & Maeda, K., 2005, Science, 309, 451

Chemical Evolution of C–Zn Produced by the First Generation Stars

Yuhri Ishimaru*, Shinya Wanajo† and Nikos Prantzos**

* *Academic Support Center, Kogakuin University, 2665–1, Nakanomachi, Hachioji, Tokyo 192-0015, Japan*
† *Department of Astronomy, Graduate School of Science, University of Tokyo, 7–3–1 Hongo, Bunkyo-ku, Tokyo 113-8654, Japan*
** *Institut d'Astrophysique de Paris, 98 bis, Boulevard Arago, 75014, Paris, France.*

Abstract. Recent observational studies show considerable small dispersions for abundance ratios of C–Zn relative to iron in metal-poor halo stars. On the other hand, abundance ratios of neutron-capture elements show large scatters, suggesting incomplete mixing of the interstellar medium at the beginning of the Galaxy. We attempt to explain these contradictory observations, using an inhomogeneous chemical evolution model, which is shown to well reproduce large scatters in neutron-capture elements in our previous studies. In particular, taking into account of difference of observational trends among iron-peak elements, we discuss nucleosynthesis of the very first stars.

Keywords: nucleosynthesis, stars: abundances, Galaxy: evolution, Galaxy: halo
PACS: 26.30.+k; 97.10.Tk; 97.20.Tr; 97.20.Wt; 98.35.Bd; 98.35.Gi

INTRODUCTION

Metal-poor stars record enrichment history of the Galaxy at the early epoch. Abundance analysis of these stars reveals large star-to-star scatters in neutron-capture elements [1, 2]. This may be interpreted as a result of incomplete mixing of the interstellar medium (ISM) at the beginning of the Galaxy. On the other hand, recent studies also show dispersions of abundance ratios of C–Zn relative to Fe in extremely metal-poor stars are considerably small[3]. It seems to suggest that the ISM was already mixed completely, when those stars were formed, contrary to interpretation of neutron-capture elements.

Therefore in this study, we discuss whether a consistent scenario can explain distributions of abundance ratios of lighter to heavier elements. In the previous study, we have already shown that inhomogeneous enrichment scenario can well account for the observed dispersions of abundance ratios of neutron-capture elements and have discussed the site(s) of *r*-process from comparison of model predictions with observations [4]. Applying the same model to other elements, we attempt to explain the difference of scatters between lighter and heavier elements. We also discuss nucleosynthesis of the very first stars, and possible roles of hypernovae.

CP1016, *Origin of Matter and Evolution of Galaxies,*
edited by T. Suda, T. Nozawa, A. Ohnishi, K. Kato, M. Y. Fujimoto, T. Kajino, and S. Kubono

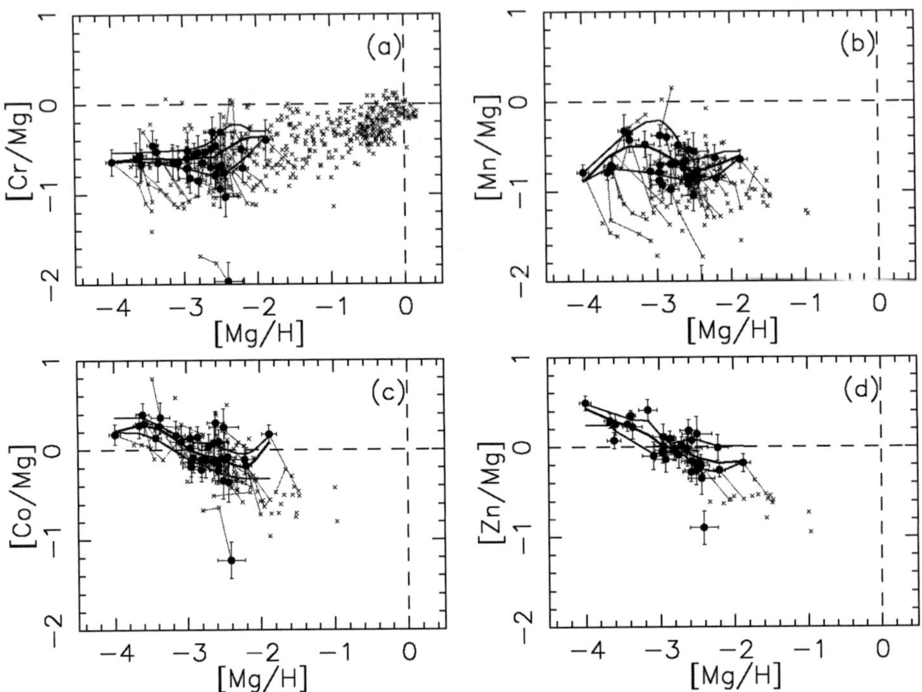

FIGURE 1. Relations between magnesium abundance and relative abundance ratios of iron-peak elements; a) Cr, b) Mn, c) Co, and d) Zn, over magnesium. Filled circles are taken from data of Cayrel et al. (2004) and Honda el al. (2004)[3, 2], and crosses are from other references. Thick and narrow lines indicate average and 50% confidence regions, respectively.

OBSERVATIONAL TRENDS OF IRON-PEAK ELEMENTS

Recent observations of metal-poor stars have shown remarkably small dispersions in abundance ratios of O–Zn relative to iron [3, 2], especially for iron-peak elements. In addition, abundance ratios of iron-peak elements seems to correlate with metallicity under [Fe/H]~ -3, i.e., those of Cr and Mn relative to Fe decrease, while those of Co and Zn increase with decreasing of metallicity. These observational trends possibly indicate variety of nucleosynthesis of supernovae (SNe) by different progenitors. However, because theoretical iron yields have some uncertainties by models, iron is not suitable to determine the progenitors which enriched observed metal-poor stars. Therefore as shown in figure 1, we use magnesium instead of iron, since yield of Mg is known to be correlated with progenitor mass regardless of SN models. As shown by averaged lines (thick lines), abundance ratios of Cr and Mn relative to Mg are almost constant irrespective of Mg abundance, while those for Co and Zn decrease with Mg abundance. As shown by the 50 % confidence regions (narrow lines), the width of scatters typically take the maximum value at [Mg/H]~ -3 and decrease towards lower metallicity. In the following sections, we discuss what are indicated from these observations, from points of view of

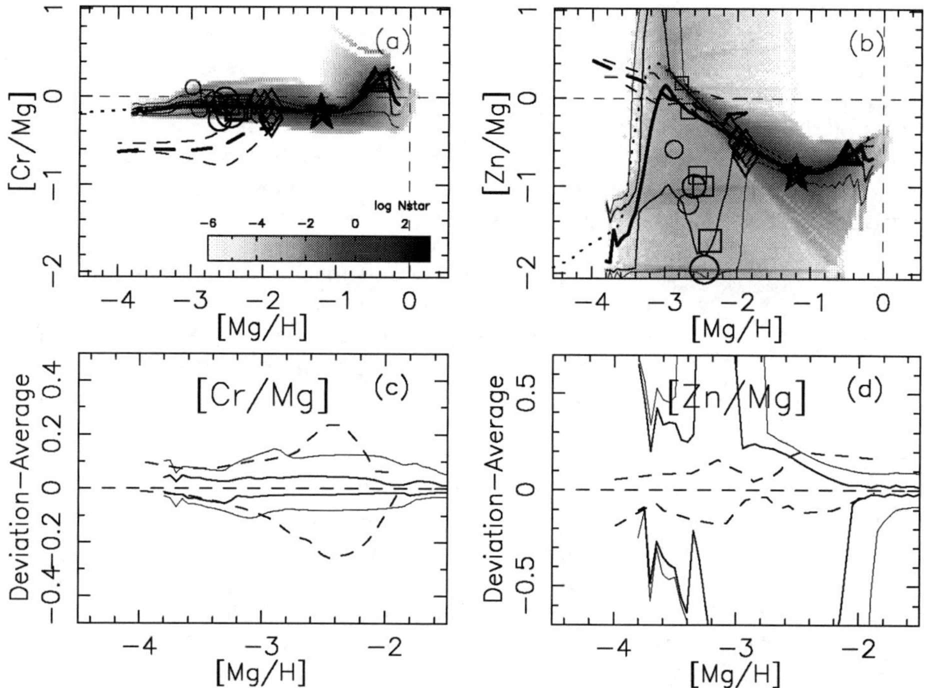

FIGURE 2. (a) [Cr/Mg] and (b) [Zn/Mg] as functions of [Mg/H]. Gray-scale indicates distribution of stellar fraction predicted by a model using the data of SN yield by Chieffi & Limongi (2004). Open symbols indicate stars formed through SN from progenitors of zero metal (circles), $5 \times 10^{-5} Z_{\odot}$ (squares), $5 \times 10^{-3} Z_{\odot}$ (diamonds), $5 \times 10^{-2} Z_{\odot}$ (stars), and $0.3 Z_{\odot}$ (triangles). The size of symbols increase with progenitor masses from $13 M_{\odot}$ to $35 M_{\odot}$. The average stellar distributions (thick-solid lines), 50% (solid lines) and 90% confidence intervals (thin-solid lines) are compared with observational average (thick-dashed lines) and 50% (solid lines) confidence intervals (thin-dashed lines). The deviations of 50% and 90% confidence lines from average as functions of [Mg/H] are also shown in (c) for [Cr/Mg] and in (d) for [Zn/Mg]. The lines are the same with (a) and (b) but indicate deviation from average value.

galactic chemical evolution.

EFFECTS OF SUPERNOVAE BY ZERO-METAL STARS

In our previous study, observed wide spread of Eu in metal-poor stars are shown to be well-reproduced by inhomogeneous enrichment scenario. In this model, star formation is assumed to be induced by individual SNe, and thus, chemical compositions of new stars are given by the mass average of the supernova remnant (SNR) and the surrounding inter-stellar medium swept up by expansion of the SNR. For comparison, we take three sets of SN yields; Woosley & Weaver (1995)[5], Chieffi & Limongi (2004)[6] and Nomoto et al. (2006)[7]. Using various SN yields, we calculate distributions of relative abundance ratios of [C–Zn/Mg] vs. [Mg/H], and discuss implications of observational scatters in abundance ratios of metal-poor stars.

Predicted distribution (gray-scale) of [Cr/Mg] and [Zn/Mg] as functions of [Mg/H] based on SN yields of Chieffi & Limongi (2004)[6] are compared with observational dispersions in figure 2ab. As shown in these figures, most of stars formed from the zero-metal stars (circles) distribute at around [Mg/H]~ -3, and that causes large scatters at this metallicity. The difference of the width of spread at around [Mg/H]~ -3 between Cr and Zn comes from dependence of SN yields on progenitor mass. This is also shown in Figs. 2cd, in which predicted deviations of upper and lower 50% confidence lines from average (thick lines) are compared with those of observations (dashed lines).

In this model, extremely metal-poor stars of [Mg/H] < -3 are formed from gas containing products of zero-metal stars but highly diluted with the ISM by chance. Thus, in the region of [Mg/H] < -3, stellar chemical compositions are determined mainly by the ISM, and lower metallicity stars are formed from gas containing more massive stars. In general, Mg yield is correlated with stellar mass, and thus, products of massive stars has lower [Fe-peak/Mg] as shown by circles in Figs. 2ab. Therefore, concerning stars of [Mg/H] < -3 formed from diluted ISM, [Fe-peak/Mg] more or less increases with metallicity, but never decreases by this model such as observed trend of [Zn/Mg].

In addition, stronger dependence of yields on stellar mass causes larger dispersion as the case of [Zn/Mg]. This model requires weak dependence on stellar mass such as the case of [Cr/Mg] to account for observed small dispersions, although weak mass dependence is hard to reproduce clear correlations of abundance ratios with metallicity. Models with other SN models also show similar results, since Mg yield generally depends on stellar mass. Thus, observed correlations with small dispersions possibly suggest other source besides normal SNe.

EFFECTS OF HYPERNOVAE

Discussions of the previous section point out difficulty explaining the observed trends of abundance ratios of extremely metal-poor stars with small dispersions only by normal SNe. Here we discuss possibility of hypernovae (HNe) as the other source for enrichment of the early galaxy.

Figure 3 show the cases of [Cr/Mg] and [Zn/Mg] predicted with a model using Nomoto et al. (2006)[7], in which all stars heavier than $20M_\odot$ are assumed to be HNe. As shown in this figure, stars formed from ejecta of HNe (double circles) mainly distribute at metallicity about 1 dex lower than those from normal SNe (circles). The explosion energy of HNe is more than 10 times larger than that of SNe. The remnant of HNe expands wider, sweeping up amount of ISM, and thus, it is diluted more efficiently than the case of SNe. As a result, although HNe produce much iron, stars formed from them distribute at lower metallicity; [Mg/H]~ -4, comparing to those from SNe.

Abundance ratios of stars formed from HNe are determined by nucleosynthesis of HNe. As shown by Figs. 3ab, the difference of yields between SNe and HNe causes variations of abundance ratios with metallicity for stars of [Mg/H]< -3. Anti-correlation between stellar abundance ratio [X/Mg] and metallicity, such as the case of Zn, can be reproduced, if abundance ratio of [X/Mg] of HNe products is relatively higher than those of SNe.

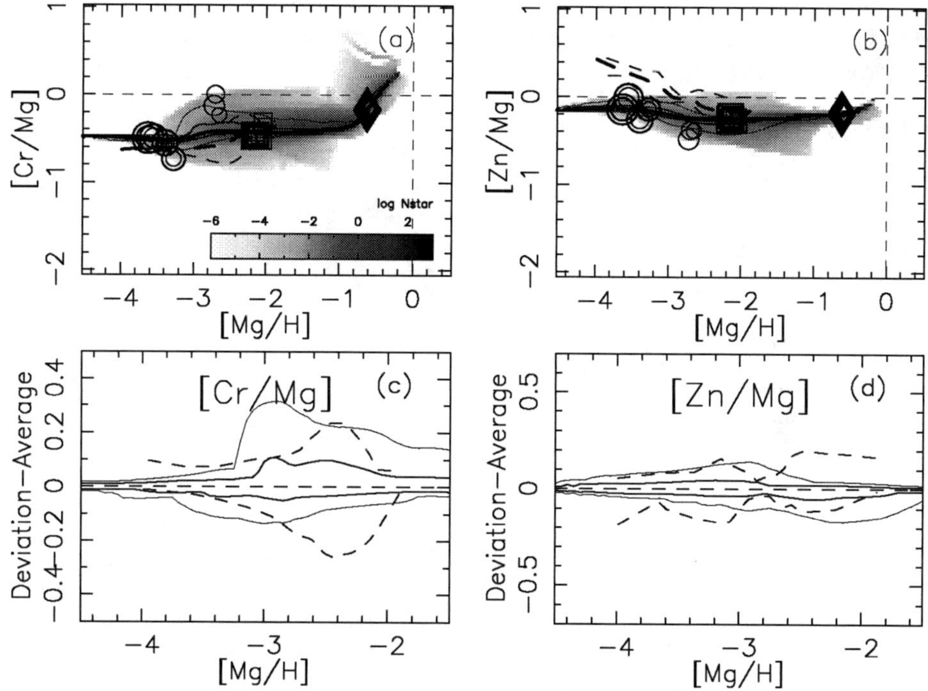

FIGURE 3. Same with Fig. 2 but with a model using the data of SN yield by Nomoto et al. (2006), in which all stars heavier than $20M_{\odot}$ are assumed to be hypernovae. Stars formed through SN from progenitors of zero metal, $5 \times 10^{-3}Z_{\odot}$, and $0.2Z_{\odot}$ are indicated by circles, squares, and diamonds, respectively. Double symbols denote stars formed from hypernovae. The size of symbols increase with progenitor masses from $13M_{\odot}$ to $40M_{\odot}$.

Figs. 3cd show that resulted dispersions are comparable to observations for both cases of Cr and Zn. HNe are regarded to come from stars massive than $20M_{\odot}$, and possibly enrich the Galaxy with amount of metal in advance of normal SNe. Therefore, when SNe start inducing star formation, the ISM has already been enriched by HNe, and few extremely metal-poor stars that causes large scatters are formed from SNe. As a result, at least HNe can be one of the candidates which account for the observed trend of iron-peak elements with small scatters, though even the latest model of HNe does not completely explain observed behaviors concerning all elements.

CONCLUSIONS

It is basically possible to account for o the observed relations between abundance ratios of [Fe-peak/Mg] and metallicity with small dispersion, through the consistent galactic evolution scenario which can explain scatters of abundances of neutron-capture elements. However, it is difficult to reproduce only by nucleosynthesis of normal supernova and process of gas mixing. Observed trends requires other source of metal which enrich the Galaxy at the very first stage. One of the possible candidates must be hypernovae. But our discussion does not exclude other possibilities such as population III stars. Future cumulative works of observational data of metal-poor stars must give strong constraints on nucleosynthesis models of SN at the early epoch.

REFERENCES

1. Ryan, S. G., Norris, J. E., & Beers, T. C. 1996, ApJ, 471, 254
2. Honda, S., Aoki, W., Kajino, T., et al., 2004, ApJ, 607, 474
3. Cayrel, R., et al., 2004, A&A, 416, 1117
4. Ishimaru, Y., Wanajo, S., Aoki, W., & Ryan, S. G., 2004, ApJ, 600, L47
5. Woosley, S. E., Weaver, T. A., 1995, ApJS, 101, 181
6. Chieffi, A., Limongi, M., 2004, ApJ, 608, 405
7. Nomoto, K., Umeda, H., Tominaga, N., et a., 2006, in preparation

Near-Field Cosmology with Binary

Yutaka. Komiya*, Takuma. Suda†, Asao. Habe† and Masayuki. Y. Fujimoto†

*Department of Astronomy, Graduate School of Science, Tohoku University, Sendai 980-8578, Japan
†Department of Cosmosciences, Graduate School of Science, Hokkaido University, Sapporo 060-0810, Japan

Abstract. Extremely metal-poor (EMP) stars in the Milky Way halo are the low-mass survivors of the first stars and their descendants that were born and first lit the universe during the "cosmological dark age". We use their properties to study the star formation and galaxy evolution process in the early universe. From the analysis of their surface abundances, it has been shown that EMP stars with metallicity $[Fe/H] \lesssim -2.5$ were born under the initial mass function (IMF) peaked around $10 M_{\odot}$ and that most of their relic stars currently observed are the secondary members of binary systems. We investigate the star formation and early chemical evolution with the derived IMF in the context of the hierarchical formation process of the Galaxy to demonstrate that the metallicity distribution function (MDF) of these halo stars is well reproduced in terms of the high-mass IMF and the binary origin of low-mass survivors. Our results suggest that the three most iron-poor stars of $[Fe/H] < -4.5$ known to date are Population III stars that were formed out of the primordial gas and have suffered the surface pollution due to the accretion of metal-rich gas. We also discuss the nature of the first stars and in particular the possible traces of pair-instability supernovae (PISNe).

Keywords: metal-poor star, Milky Way halo, Galaxy formation, chemical evolution, first star

INTRODUCTION

Recent large-scale surveys such as HK survey [1] and Hamburg/ESO (HES) survey [2] have observed more than 1000 stars with metallicity $[Fe/H] \lesssim -2.5$ in the halo of the Milky Way. They are low-mass stars of $M < 0.8 M_{\odot}$ that were formed in the proto-galaxies in the early universe. In this paper, we call all stellar population with metallicity $[Fe/H] \lesssim -2.5$ formed in the early universe as EMP population and their low-mass members still in nuclear burning stages as EMP survivors. The follow-up observations with high dispersion spectroscopy reveal the elemental abundances of many EMP survivors (e.g.[3, 4]). The properties of EMP stars may furnish a vital clue for understanding the formation and evolution of stars in the early universe and the nature of first stars and first supernovae. In this paper we investigate the formation process of the Galaxy with the halo EMP stars as a probe and work on "Near-Field Cosmology".

The abundances of EMP stars are useful probes into the chemical evolution of early universe but their surface abundances may have been modified from their original ones during their long lives from early universe to the present. In order to draw information from their observations, therefore, it is critical to take into account these changes, but there are few studies about them so far. There are three processes responsible for the changes; the internal mixing of EMP survivors themselves, and the external pollution by the binary mass transfer and by the accretion of interstellar matter (ISM). In particular,

CP1016, *Origin of Matter and Evolution of Galaxies*,
edited by T. Suda, T. Nozawa, A. Ohnishi, K. Kato, M. Y. Fujimoto, T. Kajino, and S. Kubono

binarity plays an important role in the evolution of EMP survivors [5, 6].

Observationally, it is known that 20-25% of EMP survivors show large carbon enhancement [7, 8]. The carbon enhancement is thought to be outcome of the binary evolution; in which the primary stars evolve to develop the surface carbon-enrichment during the AGB phase and transfer it onto the secondary through wind accretion. Komiya et al. [6] estimate the mass of primary stars of carbon-rich EMP stars from their observed abundance patterns and show that typical mass of EMP population is $\sim 10M_{\odot}$. Their result suggests that most of EMP survivors were formed as the low-mass members of binaries, and have been possibly affected by mass loss from AGB stars or by supernova explosions of primary stars.

It is remarkable that all of three ultra metal-poor (UMP) stars of $[Fe/H] < -4.5$ discovered to date show carbon enhancement [9, 10, 11]. Suda et al. [5] propose a binary scenario to explain the peculiar abundances of these stars and argue the importance of surface pollution. A Pop. III binary with intermediate-mass primary and low-mass secondary star is formed in the pristine gas without metals. After the birth, massive stars born in the same host cloud explode and pollute ISM by the supernova ejecta. ISM enriched with metal accretes onto the Pop. III binary, and their surface iron abundance increases to be $[Fe/H] \sim -5$. On the other hand, carbon synthesized in the primary star of Pop. III binary during AGB phase is transferred to the low-mass secondary star through stellar wind. Eventually, the low-mass star is observed as a carbon-enriched UMP star. Other scenarios are proposed that assert the UMP stars to be the second-generation of stars, born of gas polluted with supernova ejecta from the first generation star [12, 13]. Because of very low original metallicity and high accretion rate in the early universe, however, surface pollution can no longer be ignored in these stars.

In this paper, we take these peculiar effects to the stars in the early universe into account and try to investigate the formation of halo EMP stars as population and their evolution history. Our study is based on the current knowledge of structure formation, that the galaxies are formed through the merge of the smaller scaled clouds. We calculate the MDF of EMP survivors taking into account the effect of structure formation, high-mass IMF, binarity, and the surface pollution by accretion of metal-rich ISM and compare the results with the observations.

METHOD

It is known that galaxies are formed hierarchically, that is, many mini-halos merge and grow to constitute the Galactic halo. First, we construct a merger tree of the dark matter mini-halos according to the method by Somerville & Kolatt [14] based on the extended Press-Schechter theory[15, 16]. Then, we calculate the chemical evolution of mini-halos, and the formation and ISM accretion of EMP stars to study the MDF of EMP survivors.

In calculating the chemical evolution, we make the following assumptions: (1) Stars are formed stochastically in constant average rate of 10^{-10}/yr, in all mini-halos formed with virial temperature $T_{vir} > 10^3$K [17]: (2) The IMF of EMP population is a high mass one with the medium mass at $M_{\rm md} = 10M_{\odot}$ and the standard deviations of $\Delta_M = 0.4$ in a log-normal form: (3) The IMF of stars with $[Fe/H] > -2.5$ is taken to be the same as IMF of the Galactic spheroid ($M_{\rm md} = 0.2M_{\odot}$ and $\Delta_M = 0.33$): (4) Lifetime of massive stars is set at 10^7yr: (5) Iron yield of mass $0.07M_{\odot}$ is ejected by Type II supernovae and mixed in the host mini-halo instantaneously and uniformly: (6) When the mini-halos merge, gas and metal in the mini-halos are mixed immediately.

The first stars in the mini-halos are suggested to be very massive ($> 100M_{\odot}$) [18, 19], differently from the stars of EMP population. Very massive stars with mass $M = 130 - 260M_{\odot}$ explode as PISNe to yield a large amount of metals and huge energy deposited may expel all the gas in the host mini-halos [20]. In order to constrain the nature of first stars, we calculate three following cases: *Case I*: Pop. III stars are formed with the same formation rate and the same IMF as EMP stars. *Case II*: All the primordial mini-halos of the viral temperature $T_{\rm vir} < 10^4$K have a single PISN event to be destroyed. For the mini-halos of higher virial temperatures, Pop. III stars are formed in the same way as EMP stars. *Case III*: The PISN event take place only in mini-halos with $[\mathrm{Fe/H}] < -5.5$ and of $T_{vir} < 10^4$K. In other halos, stars are formed in the same way as Case I.

We assume that PISNe explode as soon as formed and eject the iron yield of $10M_{\odot}$, which is blown-off and uniformly mixed with all the galactic matter. In Case I, all the mini-halos are formed with primordial gas without metals, but in Cases II and III, mini-halos are not always metal-free but those formed later are polluted by the PISN ejecta.

We follow the metal-enrichment process through the supernovae and merging of all the mini-halos by keeping record of all the individual low-mass stars and the changes in their surface abundances. The accretion rate of ISM is much larger in the low-mass mini-halos than in the present-day Galaxy because of their low virial velocities. Accordingly, a considerable amount of gas with metal elements possibly accretes onto EMP stars in the early universe. We assume the Bondi-Hoyle accretion of ISM, $\dot{M} = 4\pi(GM)^2\rho(V^2+c_s^2)^{-3/2}$, where ρ and c_s are the density and sound velocity of ISM, V is relative velocity of stars to the ISM. To evaluate the surface metallicity after pollution, we assume that the mass of surface convective zone is $0.003M_{\odot}$ for a main sequence star and $0.3M_{\odot}$ for a giant.

The accretion rate strongly depends on c_s and V. In the host clouds where the first stars are formed, the relative velocity between the stars and ISM can be much lower than the virial velocities. Here, we may assume two models, Case A and Case B, with the lower and upper limits of ISM accretion rates. Case A assumes the virialized clouds; the density of ISM is set at an averaged density, $\rho_{\rm av}$, and the stellar relative velocity to ISM is at the virial velocity, $V_{\rm vir}$. Case B assumes that stars and gas are concentrated to the central region of mini-halos and their relative velocities reduce smaller than the virial velocities until their mother clouds suffer merging. Primordial gas is cooled down to ~ 200K due to H_2 molecular cooling. Accordingly we assume an isobaric contraction from the virial temperatures to have $\rho = \rho_{\rm av}(200\mathrm{K}/T_{\rm vir})$, and neglect the relative velocity of stars to the gas clouds they were born.

Furthermore, we take account of the time delay between the merging of dark halos and the mixing of stars and gas in Case B. We know the time of merging of dark halos from the merger tree but gas and the stellar systems located will not mix immediately. Lacey & Cole [16] calculate the angular momentum loss of satellite galaxy by dynamical friction and estimate this delay time. We use their formula. After the merging, stars are assumed to move with virial velocity of host halos. For case A, we simply assume that stellar system also merge when dark halos merge.

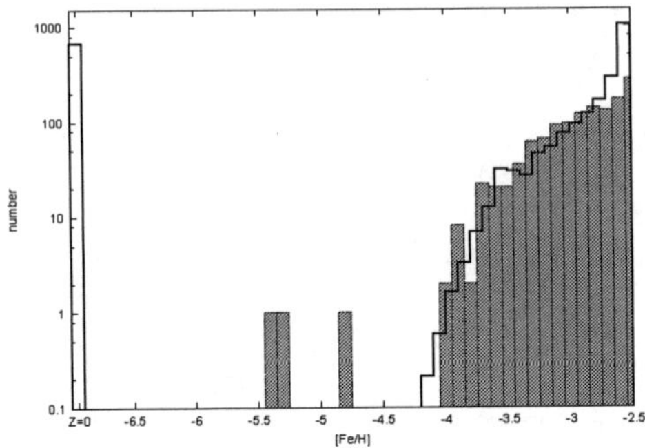

FIGURE 1. Metallicity distribution functions of EMP survivors in the Galactic halo. Shaded histogram shows the number of stars observed by HES survey and HK survey. Thick solid line shows the predicted number of stars in the detected area of the two surveys for Case I calculation without surface pollution.

As for the ISM accretion process, we assume that the members of binary systems accrete mass independently, i.e., we use the mass of low-mass members to evaluate an accretion rate for Case A. For Case B, on the other hand, we assume that the accretion rate is determined by the total mass of binary and all the gas accreted settles onto the secondary members.

NUMERICAL RESULTS

Figure 1 shows the resultant MDF of EMP survivors for Case I. The effect of surface pollution is not taken into account in Figs. 1 and 2. We also plot the number of stars obtained by HK survey and HES survey, which cover 2800+4100 and 8225 square degrees, respectively [1, 2]. Our numerical results in this figure are calculated for the same area as their fields; all giant stars are assumed to be observable to the end of the halo and the main sequence stars are assumed to be observable in comparable number of giants from the evolutionary consideration [see 6].

Theoretical result is quite similar to observed distribution, except for UMP stars. The cut-off of the MDF around $[\mathrm{Fe/H}] \sim -4$ results from the hierarchical galaxy formation. Because primordial mini-halos of $10^6 M_\odot$ are polluted by a single supernova to have $[\mathrm{Fe/H}] = -3.5 + \log(M_{yield}/0.07M_\odot)(10^6 M_\odot/M_{\mathrm{mh}}))$, the metallicity of second or later generation stars should be more metal-rich. It means that UMP stars should have been formed from the primordial gas without metals before the explosions of the first supernovae in the host mini-halos and that the metals currently observed from their surfaces have to be provided after their birth. This point will be discussed later in detail.

Calculated number of Pop. III stars (see Z=0 in Fig. 1) is inconsistent with observations. Case I model predicts several hundreds of Pop. III stars whereas only 3 UMP stars are known to date. It means that low-mass star formation is much more inefficient in the primordial clouds than in the clouds of EMP population. In order to be consistent with the observations, the formation rate of low-mass secondary stars in Pop. III binary is smaller by a factor of $\sim 1/100$ than in EMP binaries.

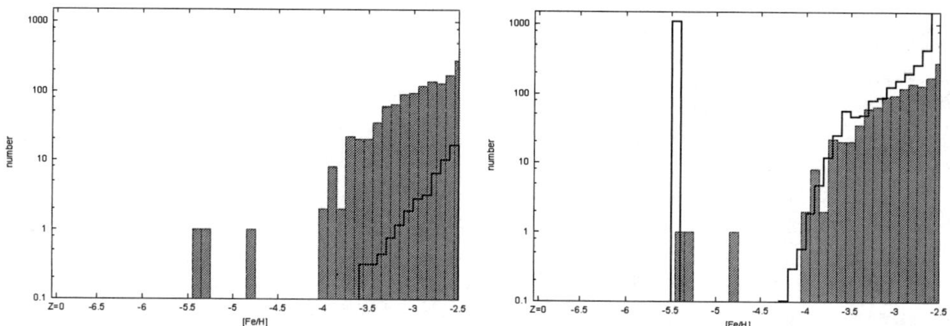

FIGURE 2. Same as figure 1, but for Case II (left) and Case III (right), respectively.

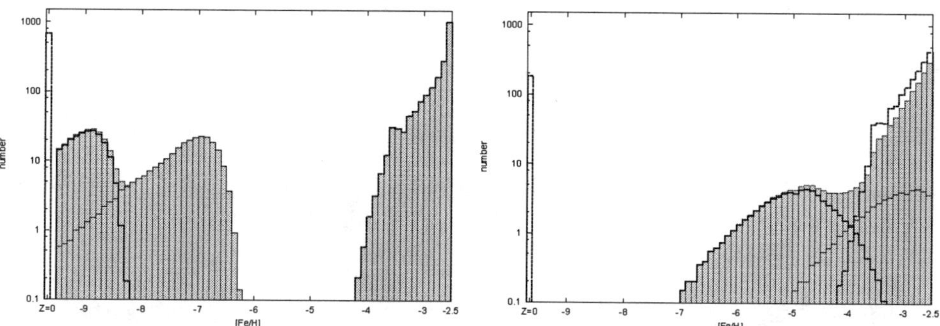

FIGURE 3. Effect of surface pollution on the MDFs for Case I. Dashed lines show MDFs of Pop. III and EMP stars. Shaded histograms show MDFs with surface pollution for Case A (left panel) and Case B (right panel), respectively. Thin and thick solid lines show the distributions of Pop. III stars with pollution for main sequence stars and giants, respectively. Accreted metals are mixed in the surface convection of mass $0.003M_{\odot}$ and $0.3M_{\odot}$ for main sequence stars and giants, respectively.

Left and right panels of Figure 2 show results for Case II and Case III, respectively. In Case II, PISNe eject too much metal and increase the metallicity of the Galaxy up to $[\mathrm{Fe/H}] \sim -3.5$ before the formation of the EMP stars. Predicted number of EMP stars is smaller and depends more strongly on the metallicity than observed MDF.

In Case III, MDF of stars with $[\mathrm{Fe/H}] \gtrsim -4$ is similar to Case I and consistent with observations of EMP stars. It also predicts similarly much larger number of stars with the critical metallicity of $[\mathrm{Fe/H}] \sim -5.5$ than observed. This means again that low mass star formation is inefficient at this small metallicity than at the metallicity of EMP population. These stars are pre-enriched by PISNe, and have to have the abundance patterns similar to the yields of PISNe. This is at variance with the observed abundance patterns of UMP stars. It indicates that PISNe don't take place, or the surface of UMP stars is covered by the accreted ISM polluted by later Type II supernovae. In conclusion, if PISNe take places in the early universe, the upper limit to the metallicity for PISNe should be lower than $[\mathrm{Fe/H}] \simeq -5.5$.

Figure 3 shows the MDFs with the effect of the accretion of ISM taken into account. We adopt Case I model about first stars. For Case A (left panel) MDF of $[Fe/H] \gtrsim -4$ is not affected by surface pollution. Surface metallicity of Pop. III stars increases up to $[Fe/H] \sim -7$ for main sequence and also increases only up to ~ -9 for giant. This difference is due to the difference of the mass of surface convective zone. In this case, the surface pollution is too small to explain the observation of UMP stars. For Case B (right panel), the accretion rate is higher and the metallicity of Pop. III giants is distributed around $[Fe/H] \simeq -5$. Main sequence Pop. III stars become $[Fe/H] \sim -3$ and MDF of EMP stars is also changed. This result suggests that UMP stars are polluted Pop. III stars and that some main sequence stars with $[Fe/H] \sim -3$ are polluted Pop. III stars.

SUMMARY

We studied a basic picture of the formation history of EMP and UMP stars. Our main conclusions are as follows. Hierarchical galaxy formation process has a critical effect on the MDF and causes a cut-off around metallicity $[Fe/H] \sim -4$. Fraction of low mass stars should be very low for Pop. III stars to meet observational constraint. PISNe give the overproduction of iron, which is hardly reconciled with the observed MDF. If PISNe take place, they are restricted only in mini-halos with metallicity less than the observed lowest metallicity $[Fe/H] = -5.4$ of HE1327-2326. All UMP stars observed to date are possibly polluted Pop. III stars.

The next generation of the projects of the search for EMP stars in the Galaxy (SDSS/SEGUE, LAMOST) will provide more insight into the structure formation of the Galaxy, promoting "Near-Field Cosmology".

REFERENCES

1. Beers, T. C., Preston, G. W., & Shectman, S. A., *AJ* **103**, 6, 1987-2034 (1992).
2. Christlieb, N., Green, P. J., Wisotzki, L., & Reimers, D., *A&A* **375**, 366-374 (2001).
3. Aoki, W., Norris, J. E., Ryan, S. G., Beers, T. C. & Ando, H., *ApJ* **567**,1166-1182 (2003).
4. Spite, M. et al., *A&A* **430**, 655-668 (2005).
5. Suda, T., Aikawa, M., Machida, M. N., Fujimoto, M. Y., & Iben, I. Jr., *ApJ* **611**, 476-493 (2004).
6. Komiya, Y., Suda, T., Minaguchi, H., Shigeyama, T., Aoki, W., & Fujimoto, M., *ApJ* **658**, 367-390, (2007).
7. Rossi, S. C. F., Beers, T. C., & Sneden, C., *ASP Conf. Ser.* **165**, 264-268 (1999).
8. Christlieb, N., *Rev Mod. Astron* **16**, 191C (2003).
9. Christlieb, N., Bessell, M. S., Beers, T. C., et al., *Nature* **419**, 904-906 (2002).
10. Frebel, A., Aoki, W., Christlieb, N., et al., *Nature* **434**, 871-873 (2005).
11. Norris, John E., Christlieb, N., Korn, A. J., Eriksson, K., et al., *ApJ* **670**, 774-788 (2007).
12. Umeda, H., & Nomoto, K., *Nature* **422**, 871-873 (2003).
13. Meynet, G., Ekström, S., & Maeder, A., *A&A* **447**, 623-639 (2006).
14. Somerville, R. S., & Kolatt, T. S., *MNRAS* **305**, 1 -14 (1999).
15. Bond, J. R., Cole, S., Efstathiou, G., & Kaiser, N., *ApJ* **379**, 440-460 (1991).
16. Lacey, C., & Cole, S., *MNRAS* **262**, 627-649 (1993).
17. Tegmark, M., Silk, J., Rees, M. J., Blanchard, A., Abel, T., & Palla, F., *ApJ* **474**, 1-12 (1997).
18. Abel, T, Bryan, G. L., & Norman, M. L., *Science* **295**, 93-98 (2002).
19. Yoshida, N., Omukai, K., Hernquist, L., & Abel, T., *ApJ* **652**, 6-25 (2006).
20. Heger, A., & Woosley, S. E., *ApJ* **567**, 532-543 (2002).

The dichotomy of the halo of the Milky Way

Daniela Carollo* and Timothy C. Beers†

*Research School of Astronomy and Astrophysics, Mount Stromlo Observatory, The Australian National University Mount Stromlo Observatory, Cotter Road, Weston, ACT 2611, Australia & INAF, Osservatorio Astronomico di Torino, 10025, Pino Torinese, Italy
†Department of Physics and Astronomy, Center for the Study of Cosmic Evolution, Joint Institute for Nuclear Astrophysics, Michigan State University, E. Lansing, Michigan 48824, USA

Abstract. We summarize evidence that the halo of the Milky Way comprises two different, and broadly overlapping, stellar components. The two structures exhibit different chemical compositions, spatial distributions, and kinematics. These results were obtained through an analysis of more than 20,000 calibration stars from the Sloan Digital Sky Survey (SDSS). The duality of the stellar halo directly impacts galaxy formation models, for the Milky Way and other large spirals.

Keywords: Astronomy, Astrophysics, Milky Way, Galactic Halo
PACS: 90,98.35.Ac, 98.35.Df, 98.35.Gi, 98.35.Ln

INTRODUCTION

The structure of the halo of the Milky Way has recently been revised by the work of Carollo et al. (2007) [1]. Confirming previous speculations based on much smaller data sets, the halo is indeed clearly divisible in two overlapping stellar components – the inner halo and the outer halo. The first structure dominates at R < 10-15 kpc, exhibits highly eccentric stellar orbits, is in slightly prograde rotation, and comprises stars with a peak metallicity around [Fe/H] ~ -1.6. The outer halo is dominant at R > 15-20 kpc, exhibits a much more uniform distribution in eccentricity, includes stars on highly retrograde orbits, and possesses a peak metallicity three times lower, i.e. [Fe/H] ~ -2.2. Previous work provided hints that the halo of the Milky Way may not comprise a single population, primarily based on analysis of the spatial profiles (or inferred spatial profiles) for halo objects, and possible indications of a net retrograde motion. In any case, the past samples of tracer objects were not sufficiently large to establish the dichotomy of the halo with confidence, and usually were suitable only for consideration of a limited number of the expected signatures of its presence. In Carollo et al. (2007) [1] all of the expected signals for the presence of two different stellar halo populations (different spatial distributions, kinematics, and chemical compositions) are now seen. This was made possible through the analysis of a homogeneously selected sample of more than 20,000 stars, originally obtained as calibration data during the course of the Sloan Digital Sky Survey (SDSS, [2]).

CP1016, *Origin of Matter and Evolution of Galaxies*,
edited by T. Suda, T. Nozawa, A. Ohnishi, K. Kato, M. Y. Fujimoto, T. Kajino, and S. Kubono

DERIVATION OF THE KINEMATIC AND ORBITAL PARAMETERS

The total number of unique stars in the sample is 20,366, and comprises mainly F and G main-sequence turnoff stars. The apparent magnitude range is $15.5 < g_0 < 17.0$ (spectrophotometric calibration stars), and $17.0 < g_0 < 18.5$ (telluric calibration stars). The color ranges are $0.6 < (u\text{-}g)_0 < 1.2$; $0 < (g\text{-}r)_0 < 0.6$ (see Figure 1).

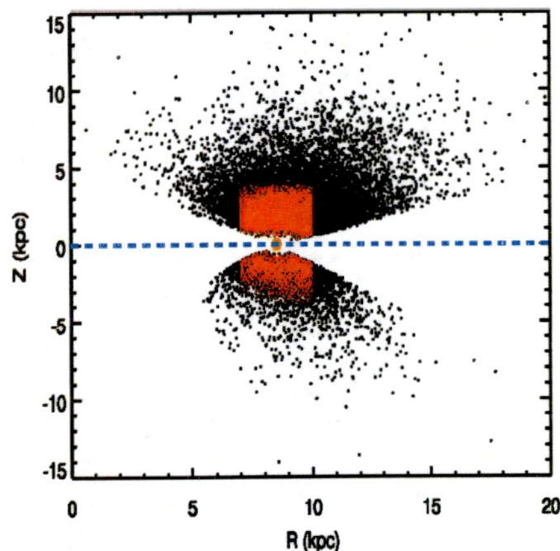

FIGURE 1. The spatial distribution of the SDSS-DR5 calibration stars in the Z-R plane, where Z is the distance from the Galactic plane, and R is the distance from the center of the Galaxy projected onto the Galactic plane. The dashed blue line represents the Galactic plane, while the filled orange dot is the position of the Sun, at Z = 0 kpc and R = 8.5 kpc. The "wedge shape" of the selection area is the result of limits of the SDSS footprint in Galactic latitude. The red points indicate the 11,458 stars that satisfy our criteria for a "local sample" of stars, having 7 kpc < R < 10 kpc, with distance estimates from the Sun < 4 kpc, and with viable measurements of stellar parameters and proper motions.

In order to derive the full space motions of the stars, we require astrometry (positions and proper motions), radial velocities, and distances. The SDSS provides positions with an accuracy of 0.1", while proper motions are provided by the re-calibrated USNO-B2 catalogue ([3]), with an accuracy of 3-4 mas/yr. The radial velocities are derived from matches to an external library of high-resolution spectral templates with accurately known velocities; its accuracy is around 5-20 km/s, depending on the S/N of the spectrum. The distances are evaluated using the cluster fiducials of Beers et al. (2000) [4]; the accuracy is around 10-20%. The SEGUE Stellar Parameter Pipeline (SSPP, [5, 6, 7]) provides the stellar physical parameters, i.e. effective temperature, surface gravity, and metallicity, with accuracies of 100 K, 0.25 dex, and 0.20 dex, respectively.

In order to obtain the best available estimates of the kinematic and orbital parameters for the stars in our sample, we consider only those stars satisfying several cuts: (1) A selection in the effective temperature range 5000 K < T_{eff} < 6800 K, over which the SSPP is expected to provide the highest accuracy, reducing the number of stars to 19,687, (2) A selection for stars in the sample with distances d < 4 kpc from the Sun, in order to restrict the kinematical and orbital analyses to a local volume (where the assumptions going into their calculation are best satisfied), reducing the number of stars to 15,435. The choice of stars in a local volume reduces errors in the derived transverse velocities, which scale with distance from the Sun. The number of stars in the remaining sample is 11,458. The proper motions, used in combination with radial velocities and the estimated distances, provide the information required to calculate the full space motions (U,V,W) of the stars relative to the Local Standard of Rest (LSR). We have also obtained the velocity components of the stars in a cylindrical reference frame with origin at the Galactic center (V_R, V_Φ, V_Z).

The orbital parameters are derived adopting a Stäckel-type gravitational potential that comprises a flattened oblate disk and a spherical massive dark halo ([8, 9]). The quantities obtained are r_{peri}, which correspond to the closest approach of an orbit to the Galactic center, and r_{apo}, which is the farthest extent of an orbit from the Galactic center. The orbital eccentricities and the maximum distance of stellar orbits above or below the Galactic plane, Z_{max}, are also evaluated.

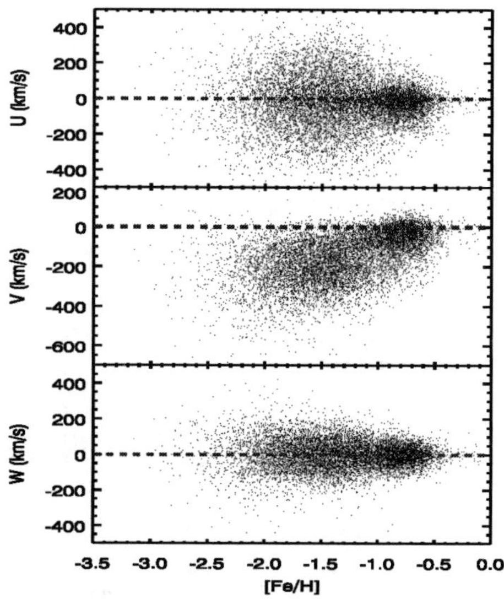

FIGURE 2. The U, V, and W components of the space motions of stars in our sample as a function of [Fe/H]. A blue dashed line at 0 km/s has been added in each panel for reference. In the metallicity range $-1.2 <$ [Fe/H] < -0.3, all three panels exhibit a low-velocity dispersion population of stars with mean values of the velocity components near 0 km/s, and dispersion in velocities of around 40-50 km/s. These are the stars of the thick-disk and metal-weak thick-disk populations. At metallicities below [Fe/H] ~ -1.2, the large velocity dispersions (on the order of 100 to 150 km/s) observed in each component are associated with the broadly overlapping inner- and outer-halo populations of stars.

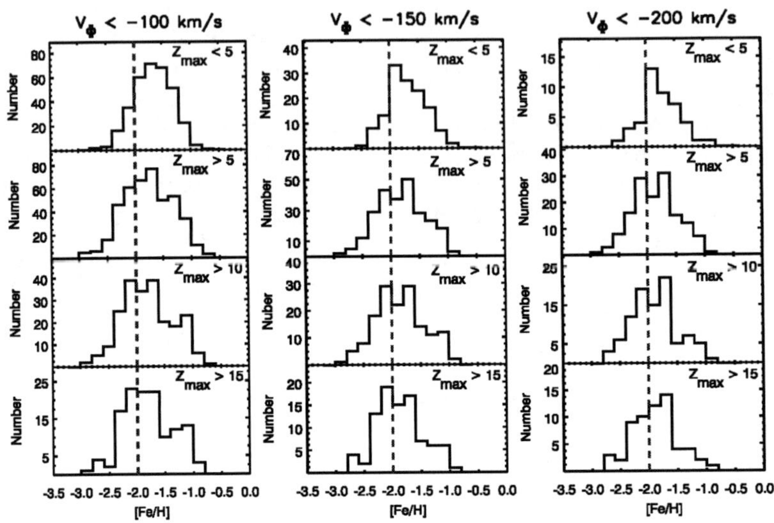

FIGURE 3. The distribution of [Fe/H] for the stars in our sample on highly retrograde orbits. The panels show various cuts in V_Φ, and for different ranges of Z_{max}. A blue dashed line at [Fe/H] $= -2.0$ is added for reference. The left-hand column applies for stars with $V_\Phi < -100$ km/s; the clearly skewed distribution of [Fe/H] exhibits an increased contribution from lower metallicity stars as one progresses from the low ($Z_{max} < 5$ kpc) to the high ($Z_{max} > 15$ kpc) subsamples. Simultaneously, the predominance of stars from the inner-halo population, with peak metallicity at [Fe/H] ~ -1.6, decreases in relative strength, and shifts to lower [Fe/H]. Similar behaviors are seen in the middle and right-hand columns, which correspond to cuts on $V_\Phi < -150$ km/s and -200 km/s, respectively.

EVIDENCE FOR THE DICHOTOMY OF THE GALACTIC HALO

There exist multiple signatures for the dichotomy of the Galactic halo; space precludes a discussion of all of them (see [1]). The change in the distribution of stellar metallicity with the rotational velocity, V or V_Φ, and Z_{max} is one of the stronger pieces of evidence. Figure 3 shows the Metallicity Distribution Function (MDF), for different values of Z_{max}, for stars with retrograde orbits. Stars from the disk population, which possess highly prograde orbits, cannot be present in this plot. We performed a Kolmogorov-Smirnoff test of the null hypothesis that the MDFs of stars shown in the lower panels for the individual cuts on V_Φ could be drawn from the MDFs of the same parent population as those shown in the upper panels; the null hypothesis is strongly rejected at a high level of statistical significance. We thus conclude that the halo comprises stars with intrinsically different distributions in metallicity. This behavior, as well as an imbalance of the energy distributions for these stars (see [1] for details), would not be expected if "the halo" is a single entity. The local sample of SDSS calibration stars exhibits a (highly statistically significant) net retrograde rotation for the outer-halo component of $V_\Phi \sim -70$ km/s. This value is in agreement with the Frenk & White analysis ([10]),

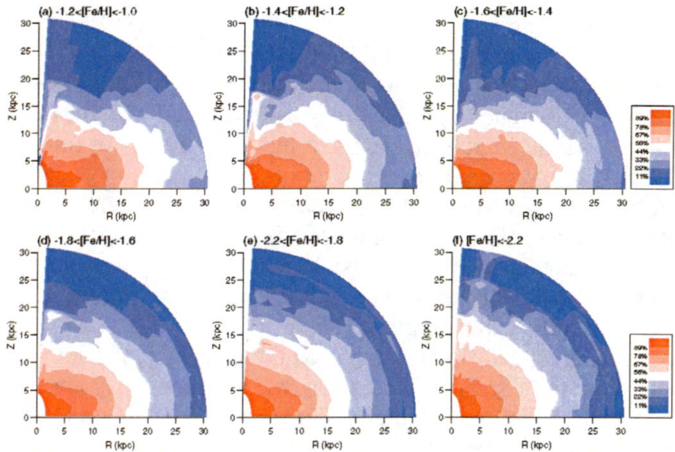

FIGURE 4. Equidensity contours of the reconstructed global density distributions for stars in our sample with various metallicities. The global density distributions are constructed from the sum of the probability density of an orbit at each location in this plane, with a weighting factor being inversely proportional to the corresponding density at the currently observed position of the star.

in which the rotational velocity is estimated on the basis of the radial velocity and the distance alone, so that there is no propagation of errors on the proper motions into the calculation of the kinematical quantities.

Other evidence for the duality of the Galactic halo can be obtained from an inversion of the orbital properties of the sample to obtain relative densities of the tracer population as a function of distance. Inspection of Figure 4 reveals that the inner regions are rather flattened at higher [Fe/H]. In particular, for the metallicity bin $-1.4 <$ [Fe/H] < -1.8 (third and fourth panel), the axis ratio is around 0.6. As [Fe/H] sweeps toward lower values, the spatial distribution becomes less flattened, and then roughly spherical at [Fe/H] < -2.2 (axis ratio ~ 0.9). This indicates that the inner-halo population dominates locally for stars with [Fe/H] > -2, whereas the outer-halo population dominates for all [Fe/H] at distances R $>$ 15-20 kpc, and also locally for stars with [Fe/H] < -2. Carollo et al. (2007) show that the the inner and outer halos also exhibit different orbital characteristics as a function of metallicity.

IMPLICATIONS OF THE DICHOTOMY OF THE HALO

The global properties of the outer-halo population clearly indicates that its formation is distinct from both the inner-halo population and the disk components of the Milky Way. Within the context of the ΛCDM paradigm, a possible scenario of the formation of the outer halo is the dissipationless chaotic merging of smaller subsystems within a pre-existing dark-matter-dominated halo. These subsystems were subjected to tidal disruption in the outer part of the dark-matter halo due to their lower masses ([11]). As candidate (surviving) counterparts for such subsystems, one might consider the currently observed low-luminosity dwarf spheroidal galaxies surrounding the Galaxy, in particular the most extreme cases recently identified from the SDSS ([12, 13]).

The difference in the MDF of the inner/outer halo components has the important consequence that the lowest metallicity stars may be associated with the outer halo. This information could be used to search for the most metal-poor stars in future surveys, by selection of stars with highly retrograde proper motions. Other important issues related to the chemical difference of the two halo components should also be explored, e.g., the determination of the primordial lithium abundance from observations of the most metal-poor stars ([14]), and the existence of a possible relationship between the numbers of carbon-enhanced metal-poor stars as a function of declining metallicity ([15]), and as a function of distance from the Galactic plane ([16]).

Very recently, Koch et al. (2007) [17] reported the existence of a strong gradient in the abundance distribution of stars located along M31's minor axis, and towards their outer halo fields, which exhibit a metallicity $[Fe/H] < -2$. This finding is consistent with the present results found for the Milky Way. Although they may differ in detail, it appears that the two dominant galaxies of the Local Group have undergone similar accretion histories.

ACKNOWLEDGMENTS

The authors are grateful to the organizers for providing assistance with travel and accomodation expenses. This work also received support from grants AST 07-07776 and PHY 02-15783; Physics Frontier Center / Joint Institute for Nuclear Astrophysics (JINA), awarded by the US National Science Foundation.

REFERENCES

1. D. Carollo et al., *Nature*, **450**, 1020–1025 (2007).
2. D. G. York et al., *Astron. J.*, **120**, 1579–1587 (2000).
3. J. A. Munn et al., *Astron. J.*, **127**, 3034–3042 (2004).
4. T. C. Beers et al., *Astron. J.*, **119**, 2866–2881 (2000)
5. Y. S. Lee et al. (a), *Astron. J.*, submitted, arXiv: 0710.5645 (2008a).
6. Y. S. Lee et al. (b), *Astrophys.J.*, submitted, arXiv: 0710.5778 (2008b).
7. C. Allende Prieto et al., *Astrophys. J.*, submitted, arXiv:0711.2886 (2008).
8. T. de Zeeuw, *Mon. Not. R. Astr. Soc.*, **216**, 599–612 (1985).
9. H. Dejonghe & T. de Zeeuw, *Astrophys. J.*, **333**, 90–129 (1988).
10. C. S. Frenk & S. D. M. White, *Mon. Not. R. Astr. Soc.*, **193**, 295–311 (1980).
11. K. Bekki & M. Chiba, *Astrophys. J.*, **558**, 666–686 (2001).
12. V. Belokurov et al., *Astrophys. J.*, **642**, L137–L140 (2006).
13. V. Belokurov et al., *Astrophys. J.*, **654**, 897–906 (2007).
14. P. Bonifacio et al., *Astrophys. J.*, **462**, 851-864 (2007).
15. S. Lucatello et al., *Astrophys. J.*, **653**, L37–L40 (2006).
16. J. Tumlinson, *Astrophys. J.*, **665**, 1361–1370 (2007).
17. A. Koch et al., *Astrophys. J.*, submitted, arXiv: 0711.4588 (2007).

5. NUCLEOSYNTHESIS IN STARS, NOVAE and SUPERNOVAE (1)

On the evolution and explosion of massive stars

Marco Limongi*,† and Alessandro Chieffi**,†

*INAF - Osservatorio Astronomico di Roma, Via Frascati 33, I-00040, Monteporzio Catone (Roma), Italy
† Center for Stellar and Planetary Astrophysics, School of Mathematical Sciences, P.O. Box, 28M, Monash University, Victoria 3800, Australia
**INAF - Istituto di Astrofisica Spaziale e Fisica Cosmica, Via Fosso del Cavaliere 100, 00133, Roma, Italy

Abstract. We review our recent progresses on the presupernova evolution of massive stars in the range $11 - 120\ M_{\odot}$ of solar metallicity. Special attention will be devoted to the effect of the mass loss rate during the Wolf-Rayet stages in determining the structure and the physical properties of the star prior the supernova explosion. We also discuss the explosive yields and the initial mass-remnant mass relation in the framework of the kinetic bomb induced explosion and hence the contribution of these stars to the global chemical enrichment of the interstellar medium

Keywords: Stellar Evolution, Nucleosynthesis, Supernovae
PACS: 97.10.Cv,97.10.Me,97.60.-s,97.60.Bw,26.20.+f,26.30.+k

INTRODUCTION

Massive stars, i.e. those stars exploding as core collapse supernovae, play a pivotal role in the chemical and dynamical evolution of the galaxies. In fact, among the other things, they provide most of the mechanical energy input into the interstellar medium via strong stellar winds and supernova explosions [1] and also synthesize most of the elements (with $4 < Z < 38$), especially those necessary to life.

Moreover, the interiors of massive stars constitute invaluable laboratories with physical conditions not seen elsewhere in the universe. Then, the understanding of the evolution and the explosion of massive stars is of paramount importance in astrophysics.

We review here our recent progresses on the presupernova evolution, the explosion and mostly the nucleosynthesis of massive stars. We mainly focus on the advanced evolutionary phases prior to the explosion, i.e. from the core He exhaustion up to the onset of the iron core collapse. However, since the advanced burning phases strongly depends on the evolutionary history of the star during the H- and He-burning stages, here we address only the aspects of these phases relevant for the further evolution of the star up to the explosion. In particular we discuss in some detail the effect of mass loss during H- and He-burning and its implications on the more advanced burning phases. The explosive nucleosynthesis as well as the contribution of massive stars to the global enrichment of the interstellar medium and their role in the chemical evolution of the galaxies is, on the contrary, described in some detail.

CP1016, *Origin of Matter and Evolution of Galaxies*,
edited by T. Suda, T. Nozawa, A. Ohnishi, K. Kato, M. Y. Fujimoto, T. Kajino, and S. Kubono

STELLAR MODELS

All the models presented in this paper have been computed by means of the latest version of the FRANEC that is described in detail in [2]. Let us just mention here that mass loss is included following [3] for the blue supergiant phase, and [4] for the red supergiant phase. For the Wolf- Rayet (WR) phase the mass loss rate provided by [5] has been adopted. The models have initial solar composition [6] and range in mass between 11 and 120 $M_\odot$.

CORE H AND HE BURNING AND THE EFFECT OF MASS LOSS

The core H burning is the first nuclear burning stage and, in these stars, is powered by the CNO cycle. The strong dependence of this cycle on the temperature implies the presence of a convective core that reaches its maximum extension just at the beginning of the central H burning and that then recedes in mass as the central H is burnt. Mass loss is rather efficient during this phase and increases substantially with the luminosity of the star (i.e. the initial mass). In stars more massive than $\sim 40\ M_\odot$ it leads to a significant reduction of the total mass. For example, at core H exhaustion the 60 $M_\odot$ model has a total mass of 38 $M_\odot$ while the 120 $M_\odot$ ends the H burning phase with a total mass of 56 $M_\odot$. The mass loss rate in this phase typically ranges from $10^{-6}\ M_\odot$/yr for the 40 $M_\odot$ to $10^{-5}\ M_\odot$/yr for the 120 $M_\odot$. The H exhausted core (He core) that forms at the central H exhaustion scales (in mass) directly with the initial mass. It is worth noting here that the presently adopted mass loss rates in the blue supergiant phase do not alter too much the initial mass-He core mass at the core H exhaustion. For example, the He core mass of a 80 $M_\odot$ model at core H exhaustion without mass loss is 26 $M_\odot$ while for the same model with mass loss it reduces to 22 $M_\odot$, i.e., only by $\sim 5\%$, although the total mass is reduced to 47 $M_\odot$, i.e., by $\sim 40\%$. On the contrary the size of the H convective core, which in turn may be affected by the overshooting and/or semiconvection, has a strong impact on the initial mass-He core mass relation and hence constitutes the greatest uncertainty in the computation of the core H burning phase.

At the core H exhaustion all the models move toward the red side of the HR diagram while the center contracts until the core He burning begins. The further fate of the models is largely driven by the competition between the efficiency of mass loss in reducing the H rich envelope during the RSG phase and the core He burning timescale. Note that, as soon as the star expands and cools, the mass loss rate switches from that provided by [3] to that given by [4]. Stars initially less massive than 30 $M_\odot$ do not loose most of their H rich mantle hence they will become and eventually explode as red supergiants. Vice versa stars initially more massive than this threshold value loose enough mass to end their life as WR stars. However, also these stars may spend a fraction of their He burning lifetime as red supergiants before becoming WR stars. The fraction of core He burning lifetime spent by one of these stars as a RSG (τ_{red}/τ_{He}) obviously depends on the efficiency of the mass loss and since it increases significantly with the luminosity (that in turn depends on the initial mass of the star), the larger the mass the higher the mass loss rate and the smaller the percentage of time spend by a star as a red supergiant. For example the 35 $M_\odot$ becomes a WNL star when the central He mass fraction is 0.52

and $\tau_{red}/\tau_{He} \sim 0.38$, while in the 80 $M_{\odot}$ the transition from RSG to BSG occurs when the central He mass fraction is 0.95 and hence τ_{red}/τ_{He} is less the 0.002. The 120 $M_{\odot}$ never moves to the red because it becomes a WNL already during the latest stages of core H burning. It is worth noting here that when a star becomes a Wolf- Rayet star, the mass loss rate typically reduces by about one order of magnitude compared to that in the RSG phase [5]. When the surface H mass fraction abundance drops below roughly 0.4 the star quickly moves to the blue side of the HR diagram and continues the core He burning as a WR star. As it is well known, central He burning occurs in a convective core whose mass size either advances in mass or remains fixed (at most). Such a behavior, typical in stars in which the He core mass does not shrink in mass during the central He burning, changes in the models in which the He core mass reduces during the central He burning (i.e. the WNE/WC/WO stars). The stars that in our set of models become at least WNE WR stars (i.e., the ones having $X_{sup} < 10^{-5}$ and $(C/N)_{sup} < 0.1$ according to the standard definition) are those with $M \geq 35$ $M_{\odot}$. Once the full H rich mantle is lost, the further evolution of the stellar model depends on the actual He core mass. In fact, as the He core is reduced by mass loss the star feels this reduction and tends to behave like a star of a smaller mass (i.e., a star having the same actual He core mass). Hence, the reduction of the He core (i.e. the total mass) during the core He burning has the following effects: 1) the He convective core reduces progressively in mass, 2) the He burning lifetime increases, 3) the luminosity progressively decreases, 3) the ^{12}C mass fraction at core He exhaustion is higher than it would be without mass loss and 4) the CO core at the core He exhaustion resemble that of other stars having similar He core masses independently on the initial mass of the star.

Stars with mass $M \geq 40$ $M_{\odot}$ experience a mass loss strong enough that the total mass is decreased down to the mass coordinate marking the maximum extension of the He convective core. Once this happens, the star becomes a WC Wolf-Rayet star (i.e., when $(C/N)_{sup} > 10$ according to the standard definition) and the products of core He burning are exposed to the surface and ejected into the interstellar medium through their stellar wind.

The relevance of the effects discussed so far, are very sensitive to the mass loss rate during the Wolf-Rayet stage. Figure 1 (left panel) shows the CO core mass at core He exhaustion as a function of the initial mass for two different prescriptions of the mass loss during the WNE/WCO WR stages, i.e., the one provided by [5] (NL00) and the one provided by [7] (LA89), the second one being higher by about 0.2-0.6 dex on average compared to the first one. From the figure it is clear that while in the case of the NL00 mass loss rate the CO core mass preserves a clear trend with the initial mass, in the case of the LA89 mass loss rate all the models show a very similar structure that resembles that of a lower mass models. It goes without saying that models with initial mass $M < 35$ $M_{\odot}$ do not differ between the two cases because they do not become WNE/WCO WR stars. Another important difference between models computed with these two different prescriptions for the mass loss is the trend of the central ^{12}C mass fraction at core He exhaustion with the initial mass. As for the CO core, also in this case, the LA89 models tend to behave like models having a similar final He core mass. Since the evolution of a massive star after core He burning is mainly driven by both the CO core mass and its chemical composition (mainly the C/O ratio), it is clear that the mass loss during the WR stage is fundamental in determining the evolutionary properties of

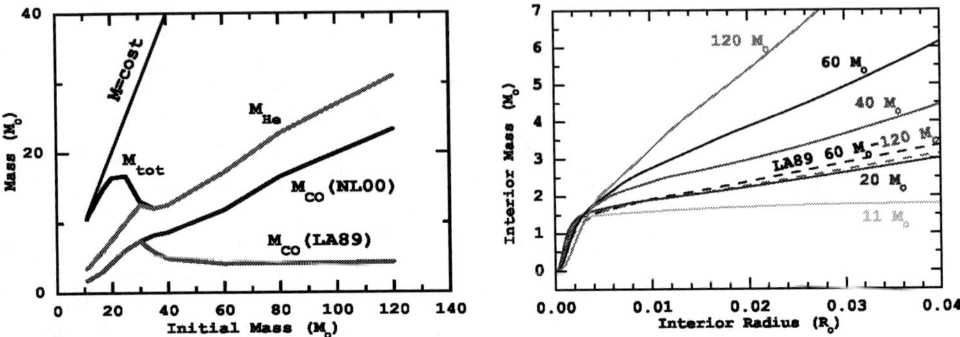

FIGURE 1. Left panel: total mass, He core mass and CO core mass at core He exhaustion as a function of the initial mass for two different prescriptions of the mass loss rate during the WNE/WCO stages (see text). Right panel: M-R relation for a subset of massive star models at the presupernova stage. The solid lines refer to the NL00 models while the dashed lines to the LA89 models respectively.

these stars during the more advanced burning stages and also their final fate (see below).

ADVANCED BURNING STAGES

The evolution of the star during the advanced burning stages is mainly driven by the size and the composition ($^{12}C/^{16}O$) of the CO core. In general each burning stage, from the C burning up to the Si burning, occurs first at the center and then, as the fuel is completely burnt, it shifts in a shell. The nuclear burning shell, then, may induce the formation of one of more successive convective zones, that can partially overlap. The general trend is that the number of convective shells scales inversely with the mass size of the CO core. For example in the 11 $M_{\odot}$ model, four consecutive C convective shells form while in the 60 $M_{\odot}$ model only one.

The complex interplay among the shell nuclear burnings and the timing of the convective zones determines in a direct way the distribution of the chemical composition and the mass-radius (M-R) relation of the star at the presupernova stage (the relevance of the M-R relation for the explosive nucleosynthesis is discussed in the next section). In general, the more efficient is the nuclear burning (i.e., the higher is the ^{12}C mass fraction left by core He burning), the higher is the number of the convective zones and the earlier is their formation, the slower is the contraction of the CO core and the shallower is the final M-R relation. This means that the higher is the mass of the CO core, the more compact is the structure at the presupernova stage. In general the mass of the CO core scales directly with the mass, but if the mass loss is so strong to significantly reduce the He core during core He burning, this scaling could not be preserved anymore (Figure 1). The different scaling of the CO core mass with the initial mass directly reflects on the scaling between the initial mass and the final M-R relation. Figure 1 (right panel) clearly shows that for the NL00 models the larger is the initial mass the more compact is the presupernova structure while, on the contrary, all the LA89 models have a very

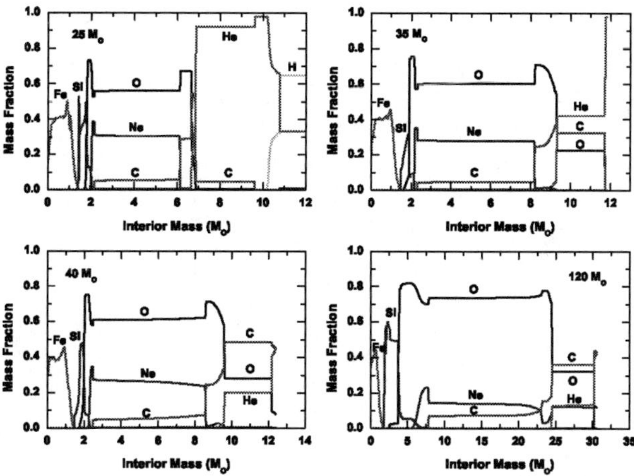

FIGURE 2. Presupernova distribution of the most abundant chemical species for a selected number of NL00 massive star models.

similar M-R relation at the presupernova stage. This is the consequence of the fact that the NL00 mass loss preserves a direct scaling between the initial mass and the CO core mass while, on the contrary, the LA89 mass loss is so strong that all the LA89 models converge toward a very similar structure. By the way, all the LA89 models have a CO core mass similar to that of the 20 $M_\odot$ computed with the NL00 mass loss. As a consequence all the LA89 models develop a final M-R relation close to that of the 20 $M_\odot$ NL00 model.

The distribution of the most abundant chemical species at the presupernova stage is shown in Figure 2 for some selected NL00 models. The 25 $M_\odot$ can be taken as representative of all the models in which the mass loss does not play a crucial role. On the contrary, the effect of mass loss is readily evident in the more massive stars. In general the presupernova star consists of an iron core of mass in the range between ~ 1.2 and ~ 1.8 $M_\odot$, the higher is the mass of the star the higher is the iron core mass, surrounded by active burning shells located at the base of zones loaded in the main products of silicon, oxygen, neon, carbon, helium and hydrogen burnings, i.e., the classical "onion structure". Thus, each zone keeps memory of the nucleosynthesis produced by the various central and/or shell burnings occurring either in a radiative environment or in a convective zone.

SIMULATED EXPLOSION AND EXPLOSIVE NUCLEOSYNTHESIS

The chemical composition left by the hydrostatic evolution is partially modified by the explosion, especially that of the more internal zones. At present there is no self consistent hydrodynamical model for core collapse supernovae and consequently we are forced to

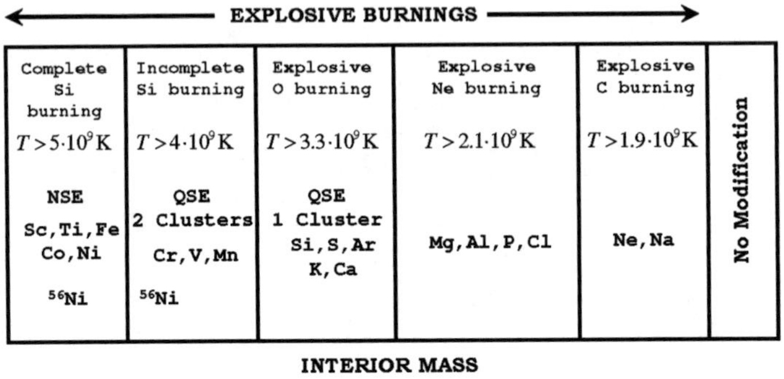

EXPLOSIVE BURNINGS

Complete Si burning	Incomplete Si burning	Explosive O burning	Explosive Ne burning	Explosive C burning	No Modification
$T > 5 \cdot 10^9$K	$T > 4 \cdot 10^9$K	$T > 3.3 \cdot 10^9$K	$T > 2.1 \cdot 10^9$K	$T > 1.9 \cdot 10^9$K	
NSE	QSE 2 Clusters	QSE 1 Cluster			
Sc, Ti, Fe Co, Ni	Cr, V, Mn	Si, S, Ar K, Ca	Mg, Al, P, Cl	Ne, Na	
^{56}Ni	^{56}Ni				

INTERIOR MASS

FIGURE 3. Schematic representation of the zones undergoing the various explosive burnings and their corresponding chemical composition .

simulate the explosion in some way in order to compute the explosive yields. The idea is to deposit a given amount of energy at the base of the exploding envelope and to follow the propagation of the shock wave that forms by means of a hydro code. The initial amount of energy is fixed by requiring a given amount of kinetic energy at the infinity (typically of the order of 10^{51} erg = 1 foe). The propagation of the shock wave into the exploding envelope induces compression and local heating and hence explosive nucleosynthesis. Zones heated up to different peak temperatures undergo different kind of explosive nucleosynthesis and hence will be characterized by different compositions (Figure 3).

Whichever is the technique adopted to deposit the energy into the presupernova model (piston, kinetic bomb or thermal bomb), in general the result is that some amount of material (the innermost one) will fall back onto the compact remnant while most of the envelope will be ejected with the desired final kinetic energy. The mass separation between remnant and ejecta is always referred to as the mass cut. This quantity strongly depends on the details of the explosive calculations, i.e., the mass location and the way in which the energy is deposited, the inner and the outer boundary conditions, and so on, hence it constitutes the most uncertain and free parameter in the explosive nucleosynthesis calculations for core collapse supernovae. It strongly affects not only the chemical yields of all those isotopes that are produced in the innermost zones of the exploding mantle, mainly ^{56}Ni and also all the iron peak elements (3), but also the relation between the initial mass and the final remnant mass, i.e., which is the mass limit or the mass interval between stars forming neutron stars and stars forming black holes after the explosion.

Figure 4 (left panel) shows the the initial-final mass relation for the NL00 models, in the assumption that the ejecta have 1 foe of kinetic energy at infinity. This choice (i.e., E_{kin} = 1 foe for all the models) implies that in stars with masses above 25-30 $M_\odot$ all the CO core, or a great fraction of it, fall back onto the compact remnant. Such a behavior is the consequence of the fact that the higher is the mass of the star the steeper is the mass-radius relation (i.e. the more compact is the structure) (Figure 1, left panel), the higher is the binding energy and hence the larger is, in general, the mass falling back

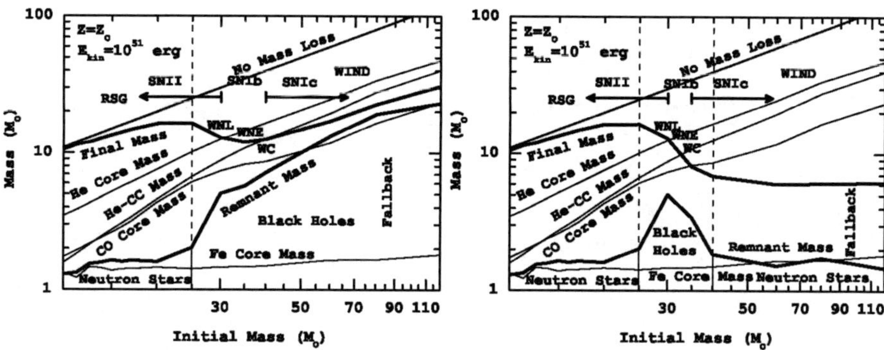

FIGURE 4. Left Panel: characteristic masses as a function of the initial mass for the NL00 models. The final kinetic energy at infinity is set to $E_{kin} = 1$ foe. Right Panel: same as left panel but for the LA89 models.

onto the compact remnant. As a consequence these stars would not eject any product of the explosive burnings, as well as those of the C convective shell, and will leave, after the explosion, black holes with masses ranging between 3 and 11 $M_\odot$. In figure 4 (left panel) the limiting masses that enter the various WR stages are also shown, i.e., WNL (30 $M_\odot$), WNE (35 $M_\odot$) and WC (40 $M_\odot$), as well as the limiting mass (30- 35 $M_\odot$) between stars exploding as Type II SNe and those exploding as Type Ib/c supernovae.

The behavior of the LA89 models is completely different because their binding energy is much smaller than their corresponding NL00 models due to the much smaller He core masses (Figure 1). In this case, the choice of a final kinetic energy of 1 foe allows, even in the more massive stars, the ejection of a substantial amount of the CO core, and hence heavy elements, leaving neutron stars as remnants (Figure 4, right panel).

The first products of the calculations described above are the yields of the various isotopes, i.e., the amount of mass of each isotope ejected by each star in the interstellar medium. The integration of these yields over an initial mass function (IMF) provide the chemical composition of the ejecta of a generation of massive stars. Figure 5 shows the production factors of all the elements obtained by assuming a Salpeter IMF ($dn/dm = km^{-2.35}$) provided by a generation of massive stars in the range 11-120 $M_\odot$ for the two cases, i.e., NL00 and LA89. This figure clearly shows that, in both cases, these stars are responsible for producing all the elements with $4<Z<38$. Moreover, the majority of the elements preserve a scaled solar distribution relative to O. Other sources, like, e.g. AGB stars and Type Ia SNe, must contribute to all the elements that are under produced by massive stars (e.g., the iron peak elements and the s-process elements above Ge). In addition, since the main difference between the two set of yields is that in the NL00 case the contribution of stars with $M > 35$ $M_\odot$ is severely suppressed, the similarity between the two set of production factors implies that these stars do not produce any specific signature on the relative distribution of the integrated yields.

The contribution of stars with $M \geq 35$ $M_\odot$, to the total yield of each element is completely different between the NL00 and LA89 cases. In the NL00 case, these more massive stars produce roughly $\sim 60\%$ of the total yields of C and N and about $\sim 40\%$

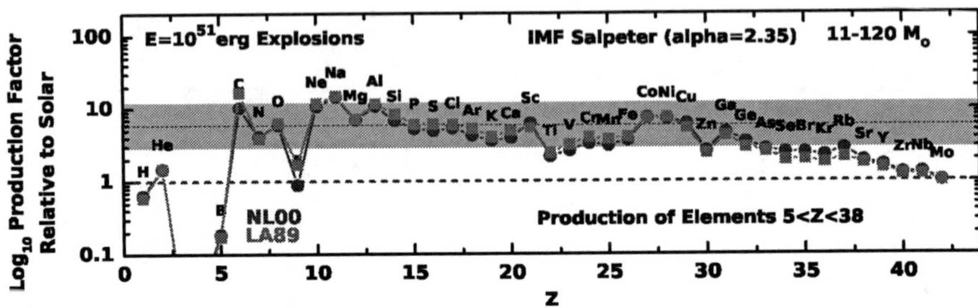

FIGURE 5. Production factor of the elements from H to Mo for a generation of massive stars in the range 11-120 $M_\odot$ (NL00 blue filled circles, LA89 red filled squares) averaged over a Salpeter IMF. The horizontal black dashed line refers to the production factor of Oxygen, PF(O), and the shaded area corresponds to a difference less than factor of 2 with respect to PF(O).

of Sc and s-process elements. This is the result of the strong mass loss experienced by these stars that allows the ejection of these elements, synthesized during H (N) and He burning (C, Sc and s-process elements), before their destruction during the more advanced burning stages and/or during the explosion. The contribution of these stars to the intermediate mass and iron peak elements is negligible because of their large remnant masses. The adoption of the LA89 modifies substantially this result. In particular, in this case stars with $M \geq 35\ M_\odot$ contribute for $\sim 30\%$ to the production of most of the elements and for $\sim 80\%$, $\sim 55\%$ and more than $\sim 100\%$ (this means that stars with $M < 35\ M_\odot$ destroy this element) to the production of C, N and F, respectively. The non negligible contribution to the majority of the elements is due to the rather small amount of fall back that allows the ejection of a substantial amount of elements produced in the innermost zones of the exploding mantle. The large contribution to C, N and F is due to the very efficient mass loss that preserves them from further destruction.

ACKNOWLEDGMENTS

I (ML) warmly thank the organizing committee for being invited to this meeting. I (ML) am also grateful to the organization for their financial support.

REFERENCES

1. Abbott, D. C. 1982, ApJ, 263, 723
2. Limongi, M., & Chieffi, A. 2006, ApJ, 647, 483
3. J.S. Vink, A. de Koter and H.J.G.L.M. Lamers, *AA*, **362**, 295 (2000)
4. C. de Jager, H. Nieuwenhuijzen and K.A. van der Hucht, *AAS*, **72**, 259 (1989)
5. T. Nugis, and H.J.G.L.M. Lamers, *AA*, **360**, 227 (2000)
6. E. Anders, and N. Grevesse, *Geochim. Cosmochim. Acta*, **53**, 197 (1989)
7. Langer, N. 1989, A&A, 220, 135

Nucleosynthesis in Jets from Collapsars

Shin-ichiro Fujimoto*, Nobuya Nishimura† and Masa-aki Hashimoto†

*Department of Electronic Control, Kumamoto National College of Technology, Kumamoto 861-1102, Japan
† Department of Physics, School of Sciences, Kyushu University, Fukuoka 810-8560, Japan

Abstract.
We investigate nucleosynthesis inside magnetically driven jets ejected from collapsars, or rotating magnetized stars collapsing to a black hole, based on two-dimensional magnetohydrodynamic simulation of the collapsars during the core collapse. We follow the evolution of the abundances of about 4000 nuclides from the collapse phase to the ejection phase using a large nuclear reaction network. We find that the *r*-process successfully operates only in the energetic jets ($> 10^{51}$ erg), so that U and Th are synthesized abundantly, even when the collapsars have a relatively small magnetic field (10^{10} G) and a moderately rotating core before the collapse. The abundance patterns inside the jets are similar to that of the *r*-elements in the solar system. The higher energy jets have larger amounts of ^{56}Ni. Less energetic jets, which have small amounts of ^{56}Ni, could induce GRB without supernova, such as GRB060505 and GRB060614.

Keywords: Accretion, accretion disks — nuclear reactions, nucleosynthesis, abundances — stars: supernovae: general — MHD — methods: numerical — gamma rays: bursts
PACS: 26.30

INTRODUCTION

During the collapse of a star more massive than 35-40$M_\odot$, the stellar core is considered to promptly collapse to a black hole. When the star has sufficiently high angular momentum before the collapse, an accretion disk is formed around the hole and jets are shown to be launched from the inner region of the disk through magnetic processes [1, 2] and neutrino heating. Gamma-ray bursts (GRBs) are expected to be driven by the jets. This scenario of GRBs is referred to as the collapsar model [4]. Assisted by the accumulating observations that imply an association between GRBs and the deaths of massive stars, this model seems to be most promising.

For accretion rates greater than $0.1M_\odot\,\mathrm{s}^{-1}$, the accretion disk is so dense and hot that nuclear burning proceeds efficiently. In fact, an innermost region of the disk related to GRBs becomes neutron-rich through electron capture on protons [5, 6, 7]. Nucleosynthesis inside the outflows from the neutron-enriched disk has been investigated with steady, one-dimensional models of the disk and the outflows [6, 8, 9]. Not only neutron-rich nuclei [6] but also *p*-nuclei [8, 9] are shown to be produced inside the outflows. This co-production of *r*- and *p*-elements in the outflows is confirmed with more elaborate calculation of the nucleosynthesis in jets from a collapsar [10], based on a two-dimensional magnetohydrodynamic (MHD) simulation [2]. The calculation, however, is performed only for the collapsar with strong magnetic fields (10^{12} G) and a rapidly rotating core whose angular velocity is $10\,\mathrm{rad\,s}^{-1}$, or R12 model in Fujimoto et al. [2, 10]. In the present study, we calculate the chemical composition of jets from six collapsars, which

CP1016, *Origin of Matter and Evolution of Galaxies*,
edited by T. Suda, T. Nozawa, A. Ohnishi, K. Kato, M. Y. Fujimoto, T. Kajino, and S. Kubono

are shown to eject the magnetically driven jets [2], to investigate the dependence of jet composition on the rotation and magnetic field of the collapsars.

MHD SIMULATIONS OF COLLAPSARS

We have carried out Newtonian MHD calculations of the collapse of rotating, magnetized massive stars of $40M_{\odot}$ [2]. We briefly present the feature of our numerical code (see [2] in detail). The numerical code for the MHD calculations employed in this paper is based on that used in the study on aspherical core collapse of a star [12]. Although the code is Newtonian, the relativistic effects of the BH are mimiced in terms of the pseudo-Newtonian potential [16]. In addition to the BH gravity, the self-gravity of the star is taken into account. We use a realistic equation of state (EOS) based on the relativistic mean field theory [13] and an another EOS [14] for lower density regime ($\rho < 10^5$g/cc), where no data is available in the EOS table with the Shen EOS. The use of the realistic EOSs is important for nucleosynthesis, because its enable us to evaluate temperatures of fluid preciously. We consider neutrino cooling through electron-positron pair capture on nuclei, electron-positron pair annihilation, and nucleon-nucleon bremsstrahlung.

The core of a star is assumed to be collapsed to a black hole promptly. Calculations are performed over the region other than the central region ($\leq$ 50km) of the stars. Spherical coordinates, (r,θ,ϕ) are used in our simulations and the computational domain is extended over 50km $\leq r \leq$ 10000km and $0 \leq \theta \leq \pi/2$ and covered with $200(r) \times 24(\theta)$ meshes. We assume that the fluid is axisymmetric and the mirror symmetry on the equatorial plane. Fluid is freely absorbed through the inner boundary of 50km, which mimics a surface of the black hole (BH) with the mass of M. The mass is initially set to be that of the central region of the progenitor $\leq$ 50km ($0.001M_{\odot}$) and is continuously increased by the mass of infalling gas through the inner boundary during the calculation.

We set the initial profiles of the density, temperature and electron fraction to those of the spherical model of a $40M_{\odot}$ massive star before the collapse [17]. The radial and azimuthal velocities are set to be zero initially, and increase due to the collapse induced by the central hole and self-gravity. The computational domain is extended from the iron core to an inner oxygen layer. The boundaries of the silicon layers between the iron core and the oxygen layers are located at about 1800 km ($1.88M_{\odot}$) and 3900km ($2.4M_{\odot}$), respectively. We adopt an analytical form of the angular velocity $\Omega(r)$ of the star before the collapse:

$$\Omega(r) = \Omega_0 \frac{R_0^2}{r^2 + R_0^2}, \tag{1}$$

as in previous work on Supernovae (SNe) [12]. Here Ω_0 and R_0 are parameters of our model. We consider three sets of (Ω_0, R_0) = (10 rad/s, 1000km) (cases with rapidly rotating core, or a R model) as in [10], (2.5 rad/s, 1000km) (cases with moderately rotating core, or a M model), and, (0.5 rad/s, 5000km) (cases with slowly rotating core, or a S model).

For the sets of Ω_0 and R_0, the maximum specific angular momentum is about 10^{17} cm^2/s, which is comparable to that of the Keplerian motion at 50km around a black hole of $3M_\odot$. Therefore the centrifugal force can be larger than the gravitational force of the central black hole and the formation of disk like structure is expected near the hole.

Initial magnetic field is assumed to be uniform, parallel to the rotational axis, and weak elsewhere ($\beta = 8\pi P/B^2 \gg 1$). In the present paper, we consider three distributions with an initial magnetic field, $B_0 = 10^8, 10^{10}$ and 10^{12} G. It should be noted that the magnetic pressure is much smaller than the other pressure before the collapse even if B_0 is equal to 10^{12} G. In brief, we have performed MHD simulations of nine collapsars in the present study, or R8, M8, S8, R10, M10, S10, R12, M12, and S12.

We briefly describe results of our MHD calculation [2]. We find that the jets can be magnetically driven from the central region of the star along the rotational axis; After material reaches to the black hole with high angular momentum of $\sim 10^{17}\text{cm}^2\text{s}^{-1}$, a disk is formed inside a surface of weak shock, which is appeared near the hole due to the centrifugal force and propagates outward slowly. The magnetic fields, which are dominated by the toroidal component, are chiefly amplified due to the wrapping of the field inside the disk. Eventually, the jets can be driven by the tangled-up magnetic fields at the polar region near the hole. The velocities of the jets are at most 10% of the velocity of light. We find that jets are ejected from six collapsars that have relatively large magnetic fields ($\geq 10^{10}$ G), or R10, M10, S10, R12, M12, and S12.

NUCLEOSYNTHESIS IN COLLAPSARS

In order to calculate chemical composition of material inside jets, we need Lagrangian evolution of physical quantities, such as density, temperature, and, velocity of the material. We adopt a tracer particle method [18] to calculate the Lagrangian evolution of the physical quantities from the Eulerian evolution obtained from our MHD calculations of the collapsars. Particles are initially placed from an iron core to an inner O-rich layer. The numbers of the particles in a layer are weighted to the mass in the layer. The total numbers of the particles, N_p, are set to be 50000, 1000, 2000, 1000, 2000, and 5000 for R10, M10, S10, R12, M12, and S12, respectively, with which we can follow ejecta through the jets appropriately. The numbers of the particles in a layer are weighted to the mass of the layer. We find that 21, 105, 18, 59, 214, and 113 particles can be ejected via the jets for R10, M10, S10, R12, M12, and S12, respectively. Hereafter, we refer to the particles through the jets as *jet particles*.

During infall to a black hole, some jet particles can attain to high temperatures greater than 9×10^9K near the black hole, where material is in nuclear statistical equilibrium (NSE). The abundances of the material in NSE can be expressed with simple analytical expressions [19], which are specified by ρ, T, and the electron fraction. The electron fraction changes through electron and positron capture on nuclei. It should be emphasized that changes in the electron fraction through neutrino interactions can be ignored because neutrino luminosity ($< 10^{52}\,\text{erg}\,\text{s}^{-1}$) is much lower than that from a proto neutron star in core collapse SNe [2, 10].

In the relatively cool regime of $T < 9 \times 10^9$K, NSE breaks and the chemical composition is calculated with a nuclear reaction network. The network includes about 4000 nuclei from neutron and proton up to Fermium, whose atomic number, Z, is 100 [see NETWORK B of Table 1 in 20]. The networks contain reactions such as two and three body reactions, various decay channels, and electron-positron capture. The experimental masses and reaction rates are adopted if available or theoretical masses and rates are taken into account otherwise [see 2, 20, for detailed nuclear data]. Moreover, spontaneous and β-delayed fission is taken into accounts in the network. We note that empirical formula [eq.(5) in 21], in which the distribution of fission yields is asymmetric, is adopted about decay products through the fission.

The jets have relatively low densities and temperatures for R10, S10, and S12. For $T_{max} < 5 \times 10^9$K, the material in the jets are abundant in α-elements, such as ^{28}Si, ^{32}S, ^{36}Ar and ^{40}Ca due to incomplete Si-burning or explosive O-burning. On the other hand, for $T_{max} > 5 \times 10^9$K, they undergo complete Si burning, which leads the synthesis of a large fraction of ^{56}Ni. The compositions of jets for R10 and S10 are therefore similar to those of ejecta from Fe and Si-rich layers of core collapse SNe. The jets for S12 is abundant in iron-group elements. This is due to slightly neutron-rich ejecta, whose electron fractions are 0.46 -0.49.

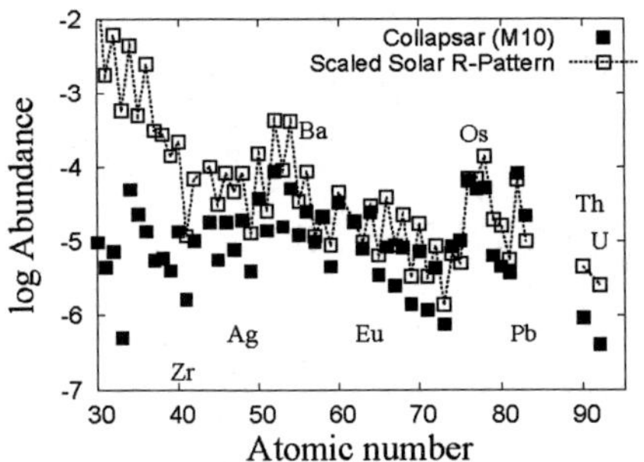

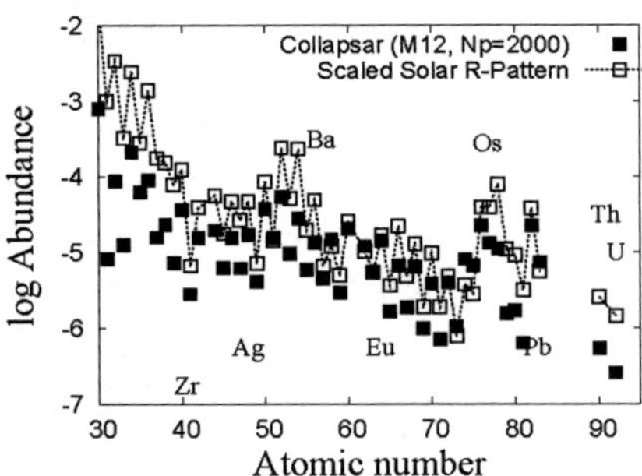

FIGURE 1. Composition of jets from collapsars with a moderately rotating core, or M10 (left panel) and M12 (right panel) as well as Eu-scaled r-nuclei in the solar system [22].

On the other hand, jets from M10 and M12 have higher densities ($> 10^9 \mathrm{g/cm^3}$) and temperatures ($> 10^{10}$ K), so that material in the jets become neutron-rich via electron capture on protons. Consequently, neutron capture reactions proceed in the jets to synthesize U and Th abundantly (Figure 1), as in the jets from R12 [10]. The composition of the jets is similar to Eu-scaled r-nuclei in the solar system for atomic numbers, $Z > 40 - 50$. It should be emphasized that r-process successfully operates even in the jets from M10, in which magnitude of magnetic fields is relatively small (10^{10} G) and rotation of the core is moderate ($2.5\,\mathrm{rad\,s^{-1}}$) before the core collapse.

Moreover, the jets are abundant in ^{56}Ni due to the complete Si burning, especially in energetic jets. Figure 2 shows masses of ^{56}Ni with respect to energies of the jets, where the energies are the sum of the kinetic energy $(E_k)_{\rm ej}$ and the internal energy $(E_i)_{\rm ej}$ of all the jet particles. We find that the ^{56}Ni masses roughly proportional to the energies of the jets. High energy jets (for M10, R12, M12, and S12) are possibly observed as GRBs with normal SNe if the collapsars are placed near the Earth. If On the other hand, low energy jets (for R10 and S10) are possibly observed to GRBs without SNe as recently observed nearby GRBs, GRB060505 and GRB060614 [23, 24], even if the jets are accompanied with GRBs that locate near the Earth.

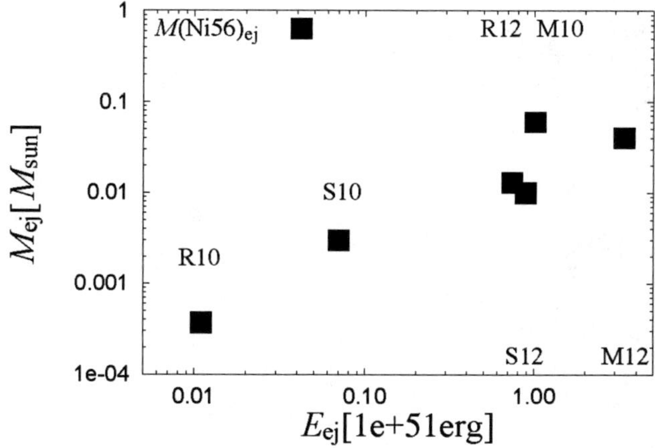

FIGURE 2. Masses of ^{56}Ni vs. energies of the jets for R10, S10, R12, S12, M10 and M12 (from left to right).

ACKNOWLEDGMENTS

This work was supported in part by Grant-in-Aid for the Scientific Research from the Ministry of Education, Science and Culture of Japan (No.17540267)

REFERENCES

1. Proga, D., MacFadyen, A. I., Armitage, P. J., & Begelman, M.,C. 2003, ApJL, 599, 5
2. Fujimoto, S., Kotake, K., Yamada, S., Hashimoto, M., & Sato, K. 2006, ApJ, 644, 1040
3. Fujimoto, S., Nishimura, N., Hashimoto, M. 2008, ApJappeared
4. Woosley, S. E. 1993, ApJ, 405, 273
5. Pruet, J., Woosley, S. E., & Hoffman, R. D. 2003a, ApJ, 586, 1254
6. Fujimoto, S., Hashimoto, M., Arai, K., & Matsuba, R. 2004, ApJ, 614, 817
7. Fujimoto, S., Hashimoto, M., Arai, K., & Matsuba, R. 2005a, in Origin of Matter and Evolution of Galaxies, ed. M. Terasawa et al. (Singapore: World Sci.), 344
8. Pruet, J., Thompson, T. A., & Hoffman, R. D. 2004, ApJ, 606, 1006
9. Fujimoto, S., Hashimoto, M., Arai, K., & Matsuba, R. 2005b, Nucl. Phys. A758, 47
10. Fujimoto, Hashimoto, M., S., Kotake, K., & Yamada, S. 2007, ApJ, 656, 382
11. Stone, J. M., & Norman, M. L. 1992, ApJS, 80, 791
12. Kotake, K., Sawai, H., Yamada, S., & Sato, K. 2004, ApJ, 608, 391
13. Shen, H., Toki, H., Oyamatsu, K., & Sumiyoshi, K. 1998, Nucl. Phys. A., 637, 435
14. Blinnikov, S. I., Dunina-Barkovskaya, N. V., & Nadyozhin, D. K. 1996, ApJS, 106, 171
15. Di Matteo, T., Perna, R., & Narayan, R. 2002, ApJ, 579, 706
16. Paczyńsky, B., & Wiita, P. J. 1980, A&A, 88, 23
17. Hashimoto, M. 1995, Prog. Theor. Phys. 94 663.
18. Nagataki, S., Hashimoto, M., Sato, K., & Yamada, S. 1997, ApJ, 486, 1026
19. Clayton, D. D. 1968, *Principles of Stellar Evolution and Nucleosynthesis* (Newyork: MacGraw-Hill).
20. Nishimura, S., Kotake, K., Hashimoto, M., Yamada, S., Nishimura, N., Fujimoto, S., & Sato, K. 2006, ApJ, 642, 410
21. Kodama, T., & Takahashi, K. 1975, Nucl. Phys. A239 489.

22. Anders, E., & Grevesse, N. 1989, Geochim. Cosmochim. Acta 53, 197
23. Fynbo, J. P. U., et al. 2006, Nature, 444, 1047
24. Gehrels, N., et al. 2006, Nature, 444, 1044

Cold r-Process in Supernovae

Shinya Wanajo

Department of Astronomy, School of Science, University of Tokyo, Bunkyo-ku, Tokyo, 113-8654, Japan; wanajo@astron.s.u-tokyo.ac.jp

Abstract. The r-process in a low temperature environment is explored, in which the neutron emission by photodisintegration does not play a role. A semi-analytic neutrino-driven wind model is utilized for this purpose. The temperature in a supersonically expanding outflow can quickly drop to a few 10^8 K, where the (n,γ)-(γ,n) equilibrium is never achieved during the heavy r-nuclei synthesis. In addition, the neutron capture competes with the β-decay owing to the low matter density. Despite such non-standard physical conditions, a solar-like r-process abundance curve can be reproduced. The cold r-process predicts, however, the low lead production compared to that expected in the traditional r-process conditions, which can be a possible explanation for the low lead abundances found in a couple of r-process-rich Galactic halo stars.

Keywords: nuclear reactions, nucleosynthesis, abundances — supernovae: general
PACS: 26.30.+k; 26.50.+x; 97.10.Tk; 97.60.Bw; 98.35.Bd

INTRODUCTION

It has been generally believed that the high entropy supernova ejecta (neutrino-driven wind) is the likely site for the rapid neutron-capture (r-) process [1, 2]. In the previous nucleosynthesis calculations related to the neutrino-driven wind scenario, the r-process has been examined in sufficiently high temperature conditions, such that $T_9 \geq 1.0$ (where T_9 is the temperature in units of 10^9 K). This physical condition, as well as the high neutron number density, has long been considered as the physical requirements to account for the solar r-process abundance curve (e.g., [3]). However, if the neutrino-driven outflow becomes supersonic, the temperature can quickly fall below $T_9 = 1$ [4, 5].

Wanajo et al. [6] have examined the r-process in such supersonically expanding outflows by introducing the "freezeout temperature" T_f (or T_{9f} in units of 10^9 K) that mimics the abrupt wind deceleration by the preceding supernova ejecta. In their calculations, the solar r-process curve around the third abundance peak ($A = 195$) was best reproduced with $T_{9f} \approx 1.0$, and the lower T_{9f} (down to 0.6) resulted in forming a lower shifted peak. As shown in [4, 5], however, it is quite possible that the wind temperature drops to a few 10^8 K before deceleration.

In this paper, the r-process in such a low temperature environment (hereafter "cold r-process") is explored. A semi-analytic neutrino-driven wind model [7] is used to obtain the thermodynamic histories of outflows. Nucleosynthesis calculations in the winds are performed with various T_{9f} including a rather low value that has not been considered in previous works. A more detailed discussion is presented in [8].

CP1016, *Origin of Matter and Evolution of Galaxies,*
edited by T. Suda, T. Nozawa, A. Ohnishi, K. Kato, M. Y. Fujimoto, T. Kajino, and S. Kubono

NUCLEOSYNTHESIS IN WINDS

The wind trajectories are obtained using the semi-analytic model of neutrino-driven winds [7, 6]. The rms average neutrino energies are taken to be 10, 20, and 30 MeV, for electron, anti-electron, and the other flavors of neutrinos, respectively. The mass ejection rate at the neutrino sphere $\dot{M}$ is determined so that the wind becomes supersonic through the sonic point. The neutron star mass is taken to be $1.4M_\odot$. The radius of neutrino sphere is assumed to be $R_\nu(L_\nu) = (R_{\nu 0} - R_{\nu 1})(L_\nu/L_{\nu 0}) + R_{\nu 1}$ as a function of neutrino luminosity L_ν, where $R_{\nu 0} = 15$km, $R_{\nu 1} = 10$km, and $L_{\nu 0} = 1 \times 10^{52}$ ergs s^{-1}. The wind trajectories are calculated for 39 constant $L_\nu = (10 - 0.5) \times 10^{51}$ erg s^{-1} with the interval of 0.25×10^{51} erg s^{-1}. In order to link the series of constant L_ν trajectories to realistic time-evolving winds, L_ν is defined as $L_\nu(t_{pb}) = L_{\nu 0}(t_{pb}/t_0)^{-1}$, where t_{pb} is the post bounce time and $t_0 = 1.0$ s. The temperature and density in each wind are set to be constant when T_9 decreases to a given freezeout temperature T_{9f}, in order to mimic the wind deceleration by the preceding ejecta.

The nucleosynthetic yields in each wind are obtained by solving an extensive nuclear reaction network code. The network consists of 6300 species between the proton and neutron drip lines, all the way from single neutrons and protons up to the $Z = 110$ isotopes [9]. Each nucleosynthesis calculation is initiated when the temperature decreases to $T_9 = 9$, at which only free nucleons exist. The initial compositions are then given by the initial electron fraction Y_e (number of proton per nucleon). In order to mimic the hydrodynamic results [1, 4], Y_e is assumed to be $Y_e(L_\nu) = (Y_{e0} - Y_{e1})(L_\nu/L_{\nu 0}) + Y_{e1}$. Y_{e0} and Y_{e1} are taken to be 0.50 and 0.15, respectively.

HOT r-PROCESS VS. COLD r-PROCESS

The nucleosynthetic yields in each T_{9f} case are mass-averaged over the 39 wind trajectories weighted by $\dot{M}(L_\nu)\Delta t_{pb}$. Figure 1 (*left*) compares the mass-averaged yields with the solar r-process abundances as functions of atomic mass number. We find that, for $T_{9f} = 1.3$, the third-peak abundances shift to the high side from $A = 195$. In addition, the abundance curve is unacceptably jagged compared to the solar r-process pattern. For $T_{9f} = 1.0$, the third-peak abundances fit the solar r-process pattern quite well. For $T_{9f} = 0.7$, however, the third-peak abundances slightly shift to the low side from $A = 195$.

This can be understood within the classical picture of r-process, in which the (n,γ)-(γ,n) equilibrium is justified. That is, for lower T_{9f}, the r-process path locates at the more neutron-rich side on the nuclide chart. As a result, the freezeout at $N = 126$ takes place at lower Z (i.e., lower A). Among the above three cases, $T_{9f} = 1.0$ can be regarded as a reasonable choice to reproduce the heavy r-process nuclei in the solar system. We find, however, that this interpretation does not hold in the lower T_{9f} cases. The $T_{9f} = 0.4$ case reproduces the third-peak abundances even better than the $T_{9f} = 0.7$ case. Moreover, for $T_{9f} = 0.1$, the third-peak abundances nicely fit the solar r-process curve.

Figure 1 (*right*) shows some key physical quantities as functions of time ($t = 0$ at $T_9 = 9$) in the $L_\nu = 2 \times 10^{51}$ erg s^{-1} case. The asymptotic entropy per nucleon in this wind is only 92 (in units of the Boltzmann constant). However, the (artificial) low initial Y_e ($= 0.22$) leads to a high neutron-to-seed ratio Y_n/Y_h ($= 109$) at the beginning of the

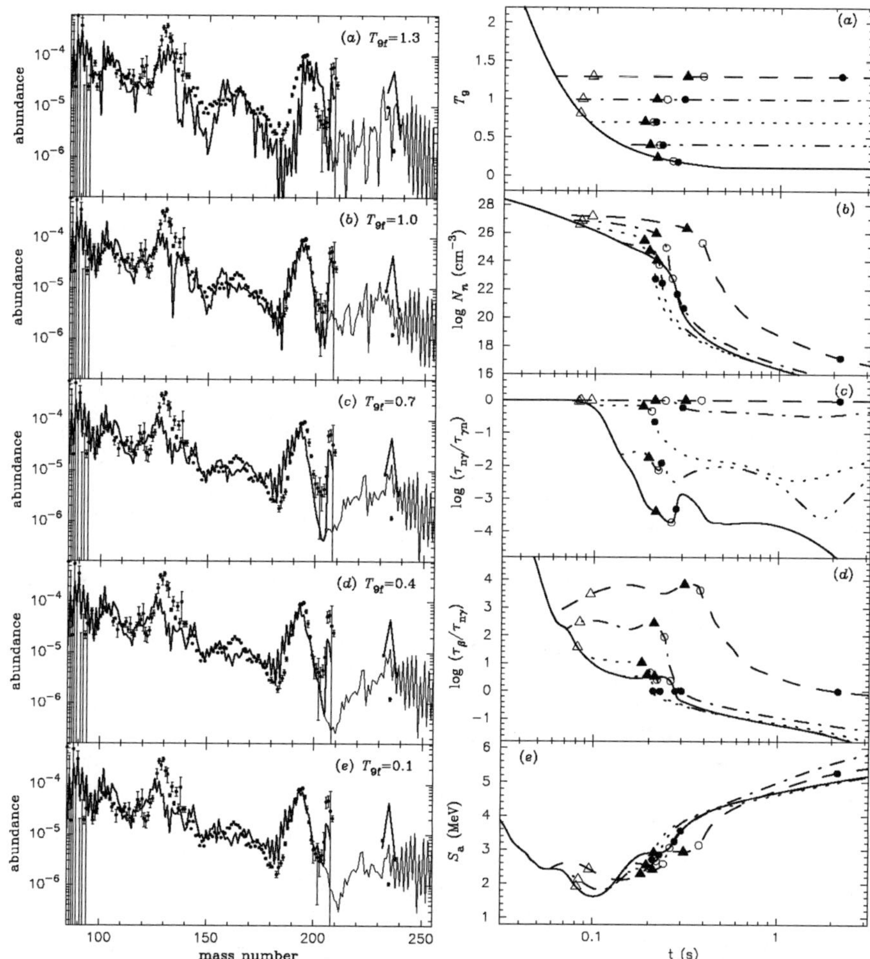

FIGURE 1. *Left:* Comparison of the mass-averaged yields (*line*) for T_{9f} = (a) 1.3, (b) 1.0, (c) 0.7, (d) 0.4, and (e) 0.1 with the solar r-process abundances (*dots*) scaled at the height of the third r-process peak ($A \approx 195$), as functions of mass number. Thick and thin lines are the yields before and after α-decay, respectively. *Right:* Time variations of (a) T_9, (b) N_n, (c) $\tau_{n\gamma}/\tau_{\gamma n}$, (d) $\tau_\beta/\tau_{n\gamma}$, and (e) S_a for $L_\nu = 2 \times 10^{51}$ erg s^{-1} as functions of time. The assigned line style in each T_{9f} case is as the same as in the top panel. The times corresponding to the second ($A = 130$) and third ($A = 195$) peak formation, n-exhaustion, and freezeout are marked by *open and filled triangles*, and *open and filled circles*, respectively.

r-process ($T_9 = 2.5$). Here, Y_n and Y_h are the abundances of free neutrons and the heavy nuclei ($Z > 2$), respectively. At $T_9 = 2.5$, the abundance-averaged mass number of the heavy nuclei is 103. This is thus taken to be representative of the third-peak forming winds. On each line, open and filled triangles denote the times when Y_n/Y_h decreases to 80 and 10, respectively. The former and latter approximately correspond to the epochs at which the second ($A = 130$) and third ($A = 195$) peaks form, respectively. Open circles denote the time when Y_n/Y_h decreases to unity, which is referred to as "n-exhaustion". No heavier nuclei are synthesized beyond this stage, but the rearrangement of the local abundance distribution continues.

The ratios $\tau_{n\gamma}/\tau_{\gamma n}$ and $\tau_\beta/\tau_{n\gamma}$ in each case are shown in Figure 1 (*right*), where $\tau_{n\gamma}$, $\tau_{\gamma n}$, and τ_β are the abundance-averaged mean lifetimes of (n,γ), (γ,n), and β-decay for $Z > 2$ nuclei, respectively (see eqs. (3) and (4) in [10]). These represent the lifetimes of the dominant species at a given time. The condition $\tau_\beta/\tau_{n\gamma} = 1$ is referred to as "freezeout" (*filled circles*). The abundance curve fixes at this time. Note that the classical r-process is characterized by the conditions $\tau_{n\gamma}/\tau_{\gamma n} \approx 1$ and $\tau_\beta/\tau_n \gg 1$.

As can be seen, the second peak forms at $t = 0.08 - 0.1$ s in all the cases, where $\tau_{n\gamma}/\tau_{\gamma n} \approx 1$. This means that the second peak forms under the condition of the (n,γ)-(γ,n) equilibrium. For $T_{9\mathrm{f}} = 1.3$, this condition holds during the whole stage of the r-process. For $T_{9\mathrm{f}} = 1.0$, $\tau_{n\gamma}/\tau_{\gamma n}$ slightly decreases after n-exhaustion and a quasi (n,γ)-(γ,n) equilibrium continues until freezeout. For $T_{9\mathrm{f}} = 0.7$, $\tau_{n\gamma}/\tau_{\gamma n}$ slightly decreases after the second-peak formation and a quasi equilibrium continues by the third-peak formation. In these cases, therefore, the temperature is enough high to maintain the (quasi) (n,γ)-(γ,n) equilibrium during the major r-process phase. Hereafter, the $T_{9\mathrm{f}} = 1.0$ case is taken to be representative of the r-process in a high temperature environment (hereafter "hot r-process").

In contrast, in lower $T_{9\mathrm{f}}$ cases, $\tau_{n\gamma}/\tau_{\gamma n}$ quickly drops below unity and the (n,γ)-(γ,n) equilibrium is never achieved after the second-peak formation. This is due to the quickly decreasing T_9 and N_n, in which the N_n-T_9 condition abruptly falls off the "waiting point validity boundary" [3, 11]. The temperature does not play any roles in such an environment, which is thus designated as the cold r-process. Hereafter, the $T_{9\mathrm{f}} = 0.1$ case is taken to be representative of the cold r-process. One may consider that, without photodisintegration, the r-process path approaches the neutron-drip line and the third peak shifts to the rather low side from $A = 195$.

The reason for this misunderstanding can be found in Figure 1d (*right*), where the lifetimes of β-decay and of neutron capture are compared. The hot r-process satisfies the classical r-process condition $\tau_\beta/\tau_n \gg 1$. In the cold r-process, however, τ_β is only a few times larger than τ_n, in which the neutron capture competes with the β-decay. This is due to the quickly decreasing N_n in the supersonically expanding wind. Consequently, the nucleosynthetic path of the cold r-process is pushed back and locates at a similar position to that of the hot r-process.

This is clearly seen in Figure 1e (*right*) that shows the r-process path in each case in terms of the neutron separation energy. Here, $S_{2n}/2$ (two-neutron separation energy divided by two) is abundance-averaged for the $Z > 2$ nuclei (S_a), which approximately represents the nucleosynthetic path at a given time. That is, $S_\mathrm{a} = 0$ is the neutron-drip line, while a higher S_a locates at closer to the β-stability. Both the cold and hot r-processes trace similar S_a histories. Notably, the freezeout in the $T_{9\mathrm{f}} = 1.0$ and 0.1 cases takes place at similar S_a values. As a consequence, similar r-process curves appear as can be seen in Figures 1b and 1e (*left*).

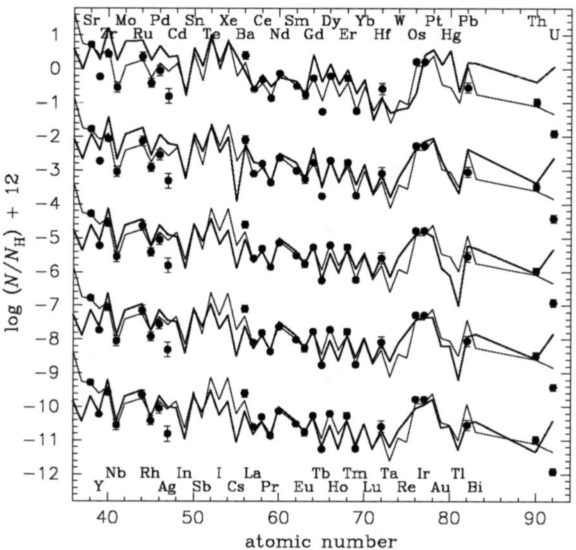

FIGURE 2. Comparison of the mass-averaged yields (*colored lines*) with abundances for CS 31082-001 (*circles*) [12, 13] and the solar r-process curve (*black lines*) scaled at the Eu ($Z = 63$) value, as functions of atomic number. The vertical scale for the uppermost set is true ($T_{9f} = 1.3$), and the others are scaled downward ($T_{9f} = 1.0, 0.7, 0.4$, and 0.1) for display purposes.

ABUNDANCE FEATURE IN THE COLD r-PROCESS

Figure 2 compares the mass-averaged abundances with the solar r-process curve and the measured elements in the r-process-rich Galactic halo star CS 31082-001. All the cases, except for $T_{9f} = 1.3$, are in reasonable agreement with the solar r-process and measured patterns. This may support the robustness of the abundance curves found in the r-process-rich stars. Note that the total mass of the r-process nuclei ($Z \geq 90$) is the same for all the cases ($\approx 2.9 \times 10^{-4} M_{\odot}$), since the seed abundances ($A \sim 100$) are already produced at $T_9 \sim 2.5$ ($> T_{9f}$).

We can see, however, a distinctive feature in the abundance curves around $A = 200-230$ (before α-decay; *thick lines* in Fig. 1, *left*) between the hot and cold r-processes. This is due to their different freezeout histories. The hot r-process follows a low S_a (≈ 2 MeV) path at n-exhaustion. At this S_a path, a trough at $A \approx 200$ appears in the abundance curve (before α-decay, Fig. 1b, *left*). This is a consequence that the nuclei at $N \approx 130-140$ are predicted to be unstable against neutron capture (see the iso-$S_{2n}/2$ curves, Fig. 6 in [10]). On the other hand, the cold r-process takes a high S_a (≈ 3 MeV) path at n-exhaustion. Therefore, the trough shifts to $A \approx 210$. The trough is further eroded by neutron capture without push-back by photodisintegration. As a consequence, the parent nuclei of Pb ($A = 210-231$ and 234) result in being significantly deficient. This leads to a visible difference in the Pb production (by a factor of 2), as can be seen in Figure 2.

IMPLICATIONS

In this paper, the r-process in a low temperature environment (cold r-process) was investigated, which is characterized by the conditions $\tau_{n\gamma}/\tau_{\gamma n} \ll 1$ and $\tau_\beta/\tau_{n\gamma} \geq 1$. Despite the fundamental difference of the cold r-process from the traditional one, a solar-like r-process curve can be reproduced. These conditions can be realized in the neutrino-driven winds from a low-mass progenitor [4]. Therefore, the low-mass end of progenitors (e.g., $8-12M_\odot$) may be the most likely site of the cold r-process. The low-mass supernovae have been also suggested to be the r-process site from Galactic chemical evolution studies [3, 14].

Among the six r-process-rich Galactic halo stars currently reported, CS 31082-001 [13] and HE 1523-0901 [15] appear to have the low Pb abundances compared to the scaled solar r-process curve. The low Pb abundance in CS 31082-001 has been in fact a most worrisome aspect, when the U/Th value is applied for the age dating [13]. This is hardly explained from the classical (or hot) r-process view. The cold r-process might enables us to consistently treat the Pb, Th, and U abundances for the cosmochronology.

Finally, it is noted that the cold r-process brings us an additional challenge to the r-process study. In most of previous calculations, the uncertainties in neutron-capture rates have not been considered seriously. This is reasonable for the hot r-process, in which uncertainties even by a factor of ten would not cause a problem. In the cold r-process, however, the neutron capture competes with the β-decay, and thus accurate neutron-capture (as well as β-decay) rates will be required to predict r-process abundances.

ACKNOWLEDGMENTS

This work was supported in part by a Grant-in-Aid for Scientific Research (17740108) from the Ministry of Education, Culture, Sports, Science, and Technology of Japan.

REFERENCES

1. Woosley, S. E., Wilson, J. R., Mathews, G. J., Hoffman, R. D., & Meyer, B. S. 1994, *Astrophys. J.*, 433, 229
2. Takahashi, K., Witti, J., & Janka, H. -T. 1994, *Astron. Astrophys.*, 286, 857
3. Mathews, G. J. & Cowan, J. J 1990, *Nature*, 345, 491
4. Arcones, A., Janka, H. -Th., & Scheck, L. 2007, *Astron. Astrophys.*, 467, 1227
5. Kuroda, T., Wanajo, S., & Nomoto, K. 2008, *Astrophys. J.*, 672, 1068
6. Wanajo, S., Itoh, N., Ishimaru, Y., Nozawa, S., & Beers, T. C. 2002, *Astrophys. J.*, 577, 853
7. Wanajo, S., Kajino, T., Mathews, G. J., & Otsuki, K. 2001, *Astrophys. J.*, 554, 578
8. Wanajo, S. 2007, *Astrophys. J.*, 666, L77
9. Wanajo, S. 2006, *Astrophys. J.*, 647, 1323
10. Wanajo, S., Goriely, S., Samyn, M., & Itoh, N. 2004, *Astrophys. J.*, 606, 1057
11. Goriely, S. & Arnould, M. 1996, *Astron. Astrophys.*, 312, 327
12. Hill, V., et al. 2002, *Astron. Astrophys.*, 387, 560
13. Plez, B., et al. 2004, *Astron. Astrophys.*, 428, L9
14. Ishimaru, Y. & Wanajo, S. 1999, *Astrophys. J.*, 511, L33
15. Frebel, A., Christlieb, N., Norris, J. E., Thom, C., Beers, T. C., & Rhee, J. 2007, *Astrophys. J.*, 660, L117

Universality of supernova gamma-process

TAKEHITO HAYAKAWA*, NOBUYUKI IWAMOTO†, TOSHITAKA KAJINO**, TOSHIYUKI SHIZUMA*, HIDEYUKI UMEDA‡ and KEN'ICHI NOMOTO‡

**Kansai Photon Science Institute, Japan Atomic Energy Agency, Kizu, Kyoto 619-0215, Japan.*
†Nuclear Data Center, Japan Atomic Energy Agency, Tokai, Ibaraki 319-1195, Japan.
***National Astronomical Observatory, Osawa, Mitaka, Tokyo 181-8588, Japan.*
‡Department of Astronomy, Graduate School of Science, University of Tokyo, Bunkyo-ku, Tokyo 113-0033, Japan.

Abstract. The solar abundances show empirical scaling laws for *p*- and *s*-nuclei with the same atomic number, which are a piece of evidence that most *p*-nuclei are dominantly synthesized by supernova γ-process. The scalings led to a novel concept of the universality of the γ-process that scalings should hold for individual supernova nucleosyntheses. Supernova γ-process calculations under various astrophysical conditions show the universality originated from three mechanism, shift of γ-process layers, change of abundance distribution by weak *s*-process, independence of nuclear reaction flows. Scaling show some deviations which may originate from contribution of other nucleosynthesis processes. The corrected abundance ratios show clear correlation, which means that the scaling is robust.

Keywords: nuclear reactions, nucleosynthesis, solar abundances, supernovae
PACS: 26.30.+k; 98.80.Ft; 91.65.Dt

INTRODUCTION

The solar abundances have provided us with hints for understanding of origin of nuclides. Pairs of two abundance peaks near neutron magic numbers in the solar abundances gave basic idea that most heavy elements are synthesized by two neutron capture reaction processes (*s*- and *r*-processes), which are the first evidence for the *r*- and *s*-processes [1]. As an other example, it has been well known a relationship, $\sigma{\cdot}N_s \sim$ constant, where σ and N_s are the neutron capture cross-section and the solar abundance for the pure *s*-nuclei [2]. Most heavy isotopes are synthesized by the *s*- and *r*-processes but origin of thirty-five neutron-deficient rare isotopes, *p*-nuclei, have been discussed over the last 50 years. Woosley and Howard found the anti-correlation between the photodisintegration reaction rates and the solar abundances of the *p*-nuclei [3], which is the first evidence that the *p*-nuclei are dominantly synthesized by photodisintegration reactions in supernova explosions (γ-process or *p*-process) [3, 4, 5]. Recently, we found two empirical scalings for the p- and s-nuclei with the same atomic number [6], which are a second evidence that most probable origin of most p-nuclei is supernova γ-process, and a novel concept of "universality of the γ-process" that the scalings hold for nuclides synthesized by individual supernova γ-processes independent of their astrophysical conditions. In this paper, we discuss universality of empirical scalings for γ-process and the reason of large deviations in the scalings originated from contaminations of other nucle-

CP1016, *Origin of Matter and Evolution of Galaxies,*
edited by T. Suda, T. Nozawa, A. Ohnishi, K. Kato, M. Y. Fujimoto, T. Kajino, and S. Kubono

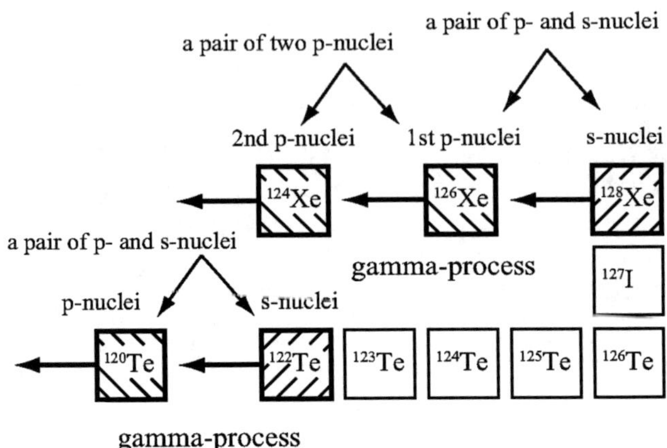

FIGURE 1. A partial nuclear chart around Te and Xe.

osyntheses.

ANALYSES OF THE SOLAR ABUNDANCES

The 1st and 2nd *p*-nuclei are lighter than the *s*-nucleus by two and four neutrons, respectively (see. Fig. 1). For example, as shown in Fig. 1, ^{126}Xe and ^{128}Xe are the first *p*-nucleus and the *s*-nucleus, respectively. Taking the abundance ratios of the *s*-nucleus to the *p*-nucleus, $N_{\odot}(s)/N_{\odot}(p)$, where $N_{\odot}(s)$ and $N_{\odot}(p)$ are the solar isotope abundances of the *s*- and *p*-nuclei, respectively. We reported the first scaling that $N_{\odot}(s)/N_{\odot}(p)$ ratios are almost constant over a wide region of atomic number (see Fig. 2). We found another empirical scaling between two *p*-nuclei with the same atomic number [6]. Nine elements have two *p*-nuclei. The ratios are almost constant over a wide range of the atomic number. The fact that the scalings appear in the solar abundances leads to a concept of the universality of the γ-process that the scalings hold for materials produced by individual supernova γ-process.

MECHANISM OF THE SCALINGS

We presented the basis of the universality of the γ-process in model calculations of core-collapse supernovae under various astrophysical conditions such as, metallicity, progenitor mass, and explosion energy as shown in Fig. 4 [7]. A massive star evolves from the main sequence to the core-collapse stage and explodes [8]. The solar abundances are adopted as the initial composition in the massive star and is affected by the weak *s*-process in evolutional states. The mass distribution of the seed nuclei affected by the weak *s*-process are used as the initial composition at the SN explosions. Most of the reaction rates are taken from a common astrophysical nuclear data library of REACLIB [9]. We calcualted in four different models. They are a standard model (1) with $M = 25$

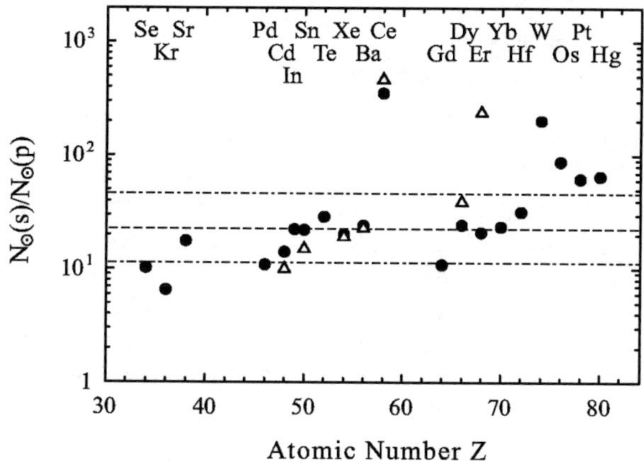

FIGURE 2. The abundance ratio of $N_{\odot}(s)/_{\odot}N(p)$ in the solar system. The circles are the ratio of the *s*-nucleus to the first *p*-nucleus. The triangles are the ratio of the *s*-nucleus to the second *p*-nucleus. The dashed line is an average value of 23. The dot-dashed lines are 11.5 and 46. Most $N_{\odot}(s)/N_{\odot}(p)$ ratios are around 23 within a factor of two.

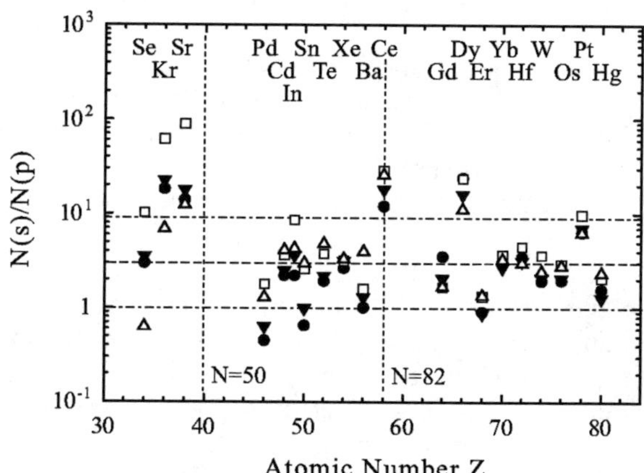

FIGURE 3. Calculated N(s)/N(p) ratios in four SN models. See text.

$M_{\odot}$, $Z = Z_{\odot}$ and $E = 10^{51}$ erg, a heavier progenitor mass model (2) with $M = 40\ M_{\odot}$, $Z = Z_{\odot}$ and $E = 10^{51}$ erg, a metal-deficient model (3) with $M = 25\ M_{\odot}$, $Z = 0.05\ Z_{\odot}$ and $E = 10^{51}$ erg, and a higher energy explosion model (4) with $M = 25\ M_{\odot}$, $Z = Z_{\odot}$ and $E = 20 \times 10^{51}$ erg (see Fig. 3). The results showed that the scalings hold for individual nucleosyntheses independent of the astrophysical conditions assumed and we found that the universality originated from three mechanisms [7].

The first mechanism is the shifts of the γ-process layers to keep their peak temperature. The γ-process layers are shifted to outer layers as the explosion energy or mass increases. Inside the γ-process layers, the seed nuclei are completely destroyed by photodisintegration reactions, whereas the p-nuclei outside the γ-process layers are not synthesized because of their low temperature. Therefore, the physical conditions of the γ-process layers are almost identical. The second mechanism is the weak s-process. It is well known that light elements of $A < 90$ are produced by successive neutron capture reactions from iron seeds in the weak s-process. However, pre-exiting heavy elements of $A > 90$ are also irradiated by neutrons and their mass distribution is changed locally to that of the main s-process in asymptotic giant branch stars. Therefore, the mass distribution patterns of the seed nuclei at the supernova explosions are almost identical. The third mechanism is independence from two nucleosynthesis flows. The p-nuclei are produced by the two flows: direct (γ,n) reactions from heavier isotopes, and β-decay after downflow by (γ,p) and (γ,α) reactions from heavier elements which may break the scalings. We calculated the contributions of both the flows for each p-nucleus. The (γ,n) reactions are dominant in the light atomic mass region of $N < 82$. The β-decay after the downflow contributes to the p-nucleus abundances in the heavy atomic mass region of $N > 82$ but does not drastically change the abundance ratios. In this way, the scalings hold for individual γ-processes.

CORRECTION OF THE SCALINGS

Here we discuss the reason of the deviations observed in the $N_\odot(s)/N_\odot(p)$ ratios. The small deviation for Gd can be explained by a contribution of a weak branch of the s-process. A unstable nucleus ^{151}Sm is known as a branching point of the s-process, where the neutron capture reaction and β-decay compete. There are large deviations for Ce, Er, and W. These three deviations can be explained by an exceptional contribution from the r-process because ^{140}Ce, ^{166}Er, and ^{182}W are not shielded against β^--decay after the freezeout of the r-process, whereas most s-nuclei are pure s-nuclei. The r-process contributions to ^{166}Er and ^{182}W are larger than the s-process contributions in theoretical calculations [10]. In addition the effect of the neutron magic number N=82 should contribute to the deviation for Ce. Figure 5 shows the corrected $N_\odot(s)/N_\odot(p)$ ratios. The circles and triangles are corrected $N_\odot(s)/N_\odot(1\text{st } p)$ and $N_\odot(s)/N_\odot(2\text{nd } p)$ ratios, respectively. The s-process abundances of ^{140}Ce, ^{168}Er, and ^{182}W are taken from calculated results in a stellar s-process model [10]. The s-process abundance for ^{152}Gd is taken from a calculated result based on neutron capture reaction rates measured at a n_TOF facility [11]. For Kr, we take N(^{82}Kr)/N(^{78}Kr) instead of N(^{80}Kr)/N(^{78}Kr), since ^{80}Kr is located on a weak branch of the s-process but ^{82}Kr is located on the main path of the s-process. It should be noted that most corrected ratios are centered around 23 within a factor of 2 (see Fig. 5).

DISTRIBUTION PATTERN OF THE SOLAR ABUNDANCES

Figure 6 shows the solar abundances of the p-nuclei and s-nuclei that are members of the first scaling. The correlation between the p- and s-nuclei can be observed in this figure. Seeger et al. pointed out that the empirical relation, $\sigma \cdot N_s \sim$ constant for the pure s-nuclei [2]. We would like to stress that the abundances of the s-nuclei that are

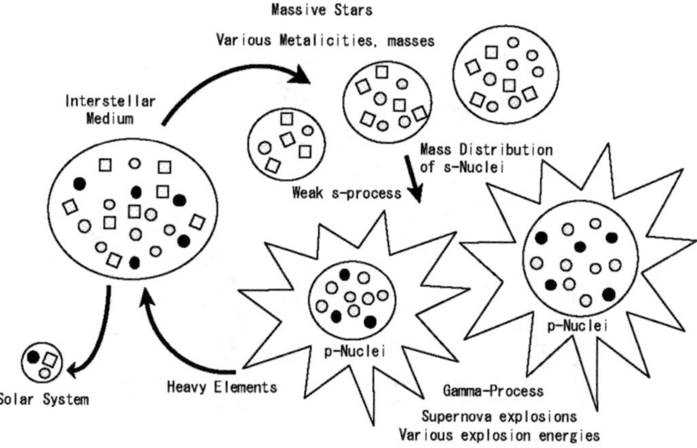

FIGURE 4. A schematic picture of a concept of universality of the γ-process in supernova explosions.

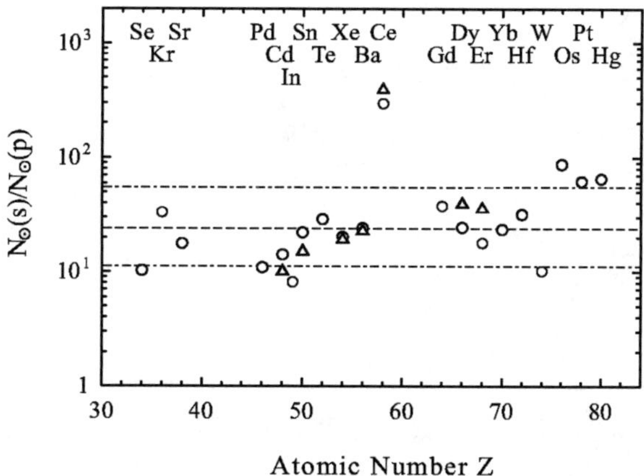

FIGURE 5. The corrected $N_\odot(s)/N_\odot(p)$ ratios. The circles are the ratio of the *s*-nucleus to the first *p*-nucleus. The triangles are the ratio of the *s*-nucleus to the second *p*-nucleus.

the members of the first scaling, N_s, are almost constant in the two mass regions of $50 < N < 82$ and $82 < N$, respectively. The abundance pattern of the *p*-nuclei shows a tendency similar to that of the *s*-nuclei. The abundances of the *p*-nuclei are almost constant in the two mass regions of $50 < N < 82$, and $82 <$ N, respectively. There are some enhancements of the *p*-nuclei: for example ^{112}Sn, ^{196}Hg. Their partner *s*-nuclei, ^{116}Sn and ^{198}Hg, also show enhancements. These enhancements show the abundances of the *p*-nuclei originated from seed *s*-nuclei are proportional to those of the *s*-nuclei with the same atomic number, respectively. These are consistent with the first scaling that the $N_\odot(s)/N_\odot(p)$ ratios are almost constant over a wide range of the atomic number. The corrected $N_\odot(s)/N_\odot(p)$ ratios and the analysis of the solar abundances show the scalings are robust.

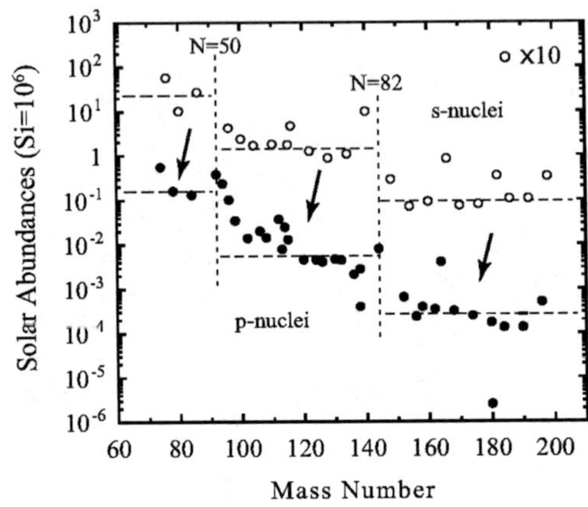

FIGURE 6. A corrected N(s)/N(p) ratios.

SUMMARY

We presented the empirical abundance scaling law that s/p ratios with the same atomic number are almost constant over a wide mass region, which is a piece of evidence that most probable origin of the p-nuclei is supernova γ-process. The scalings led to "concept of γ -process" that the scaling holds for individual γ-processes. The s/p raitos in the solar abundances show deviations originated from other nucleosyntheses as the s- and r-processes. We presented that most corrected s/p ratios are centered around 23 within a factor of 2.

REFERENCES

1. E.M. Burbidge, G.R. Burbidge, W.A. Fowler, F. Hoyle, *Rev. Mod. Phys.* **29**, 548 (1957).
2. P.A. Seeger, W.A. Fowler, D.D. Clayton, Astrophys. J. **11**, 121 (1965).
3. S.E. Woosley, W.M. Howard, Astrophys. J. Suppl. **36**, 285 (1978).
4. M. Arnould, Astron. Astrophys. **46**, 117 (1976).
5. M. Rayet *et al.*, Astron. Astrophys. 298, 517 (1995).
6. T. Hayakawa *et al.*, Phys. Rev. Lett. **93**, 161102 (2004).
7. T. Hayakawa *et al.*, Astrophys. J. **648**, L47 (2006).
8. K. Nomoto, *et al.*, in *Stellar Collapse*, ed. C.L. Fryer (Astrophysics and Space Science: Kluwer), 277 (astro-ph/0308136) (2004).
9. Thielemann, F., 2000, http://nucastro.org/reaclib.html
10. C. Arlandini, *et la.* Astrophys. J., 525, 886 (1999).
11. n_TOF collaboration, Phys. Rev. Lett. 93, 161103 (2004).

6. NUCLEOSYNTHESIS IN STARS, NOVAE and SUPERNOVAE (2)

Nucleosynthesis in novae: experimental progress in the determination of nuclear reaction rates

Alain Coc

CSNSM, CNRS/IN2P3, Université Paris Sud, UMR 8609, Bâtiment 104, F–91405 Orsay Campus, France

Abstract. The sources of nuclear uncertainties in nova nucleosynthesis have been identified using hydrodynamical nova models. Experimental efforts have followed and significantly reduced those uncertainties. This is important for the evaluation of nova contribution to galactic chemical evolution, gamma–ray astronomy and possibly presolar grain studies. In particular, estimations of expected gamma–ray fluxes are essential for the planning of observations by means of existing or future satellites.

Keywords: Nuclear Astrophysics, Novae
PACS: 26.30.Ca,26.50.+x

INTRODUCTION

Novae are thermonuclear runaways occurring at the surface of a white dwarf by the accretion of hydrogen rich matter from its companion in a close binary system[1, 2, 3, 4]. Material from the white dwarf (^{12}C and ^{16}O in the case of a CO nova, or ^{16}O, ^{20}Ne plus some Na, Mg and Al isotopes in the case of a ONe nova) provide the seeds for the operation of the CNO cycle and further nucleosynthesis. Novae are supposed to be the source of galactic ^{15}N and ^{17}O and to contribute to the galactic chemical evolution of ^{7}Li and ^{13}C. In addition they produce radioactive isotopes that could be detected by their gamma–ray emission: ^{7}Be (478 keV), ^{18}F ($\leq$511 keV), ^{22}Na (1.275 MeV) and ^{26}Al (1.809 MeV). The yields of these isotopes depend strongly on the hydrodynamics of the explosion but also on nuclear reaction rates involving stable and radioactive nuclei. Tests of the sensitivity to reaction rate uncertainties have been done using parametrized[5], semi–analytic[6], post–processed[7] nova models but also with a 1-D hydrodynamical model. Indeed, the impact of nuclear uncertainties in the hot-pp chain[8], the hot-CNO cycle[9], the Na–Mg–Al[10] and Si–Ar[11] regions have been investigated with the Barcelona (SHIVA) hydrocode. In this way, the temperature and density profiles, their time evolution, and the effect of convection time scale were fully taken into account. The nuclear reaction rates whose uncertainties strongly affect the nova nucleosynthesis having been identified, nuclear physics experiments were conducted to reduce these uncertainties. In this review, we summarize the experimental progresses made in this domain.

CP1016, *Origin of Matter and Evolution of Galaxies*,
edited by T. Suda, T. Nozawa, A. Ohnishi, K. Kato, M. Y. Fujimoto, T. Kajino, and S. Kubono

HOT CNO CYCLE

The hot–CNO cycle deserves special attention as it is the main source of energy for both CO and ONe novae and is the source for the production of ^{13}C, ^{15}N, ^{17}O (galactic chemical evolution) and ^{18}F (gamma–ray astronomy). The positrons produced in the β^+ decay of ^{18}F annihilate and are the dominant source of gamma rays during the first hours of a nova explosion[12]. Through a series of hydrodynamical calculations, major nuclear uncertainties on the production of ^{17}O and ^{18}F were pointed out in Ref. [9] (hereafter CJHT).

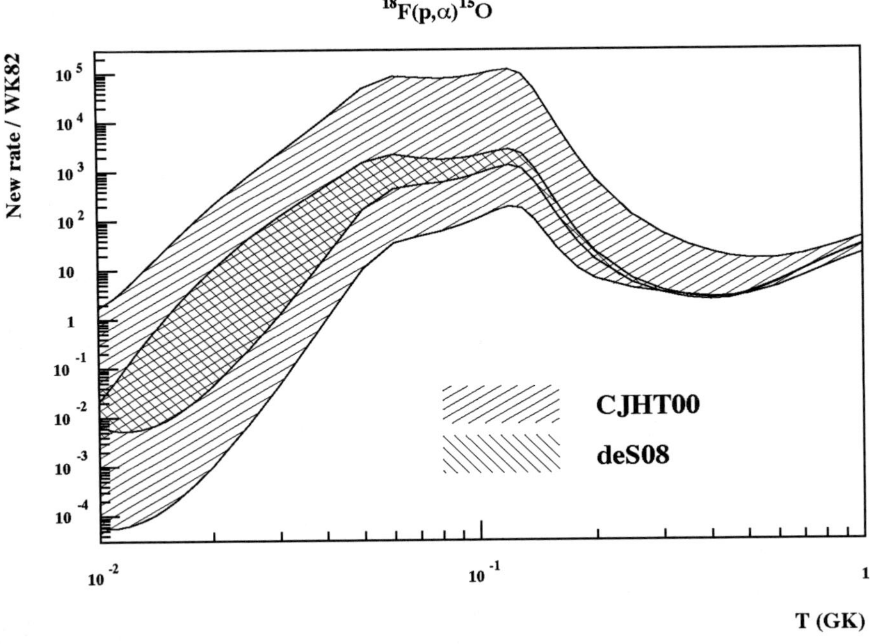

FIGURE 1. Reduction of the uncertainty[18] in the ^{18}F(p,α)^{15}O reaction rate. (An update of Fig. 6 in CJHT[9].)

In particular, due to the unknown contributions of two low energy resonances, the ^{18}F(p,α)^{15}O reaction was recognized as the main source of uncertainty for the production of ^{18}F. These resonances are postulated at 8 and 38 keV following the spectroscopic work of Utku et al.[13] who identified two levels at an excitation energy of $E_X = 6.419$ and 6.449 MeV in ^{19}Ne and assumed that they are the analogs of the $E_X = 6.497$ and 6.528 MeV 3/2$^+$ levels in ^{19}F. Two (d,p) transfer reaction experiments, in inverse kinematics using a ^{18}F beam, were performed at Louvain–la–Neuve (LLN)[14] and Oak–Ridge (ORNL)[15]. They enabled to extract the neutron spectroscopic factor(s) of the (experimentally unresolved) ^{19}F analog levels. Since the assignments of these analog levels separated only by 30 keV is not settled, even assuming the equality of spectroscopic factors between analog levels, it is not yet possible to determine the relative contributions of these two resonances. However, the extracted spectroscopic factors imply that they must be included in the calculation of the reaction rate.

In addition, if the spin and parity assignments are correct, they should interfere with the $3/2^+$ broad resonance located at the resonance energy of E_R = 665 keV. This resonance and the $3/2^-$ one at 330 keV have been precisely measured directly[16, 17] at ORNL and LLN. Interferences among the E_R = 8, 38 and 665 keV resonances are expected to have a maximum effect right in the energy range of interest for nova. It is hence extremely important to determine their constructive or destructive nature. Until recently, the only constraint in this energy region was provided by an *off–resonance* measurement[17] at 380 keV with a large error bar. A new direct measurement of the $^{18}F(p,\alpha)^{15}O$ cross section was recently performed at LLN[18] at center of mass energies of E_{CM} = 726, 666, 485 and 400 keV. The two higher energy points are located at the top of the 665 keV resonance, for comparison with previous experiments, while the lower ones are close to the limit of the interference region. With these new results, R-matrix calculations including up to four $3/2^+$ levels were performed to help constrain the astrophysical S–factor. Even though more experiments are needed in the interference region, the reduction of the uncertainty on the $^{18}F(p,\alpha)^{15}O$ reaction rate is important when one takes into account those recent measurements[14, 15, 17, 18], as shown in Figure 1. In CJHT, the uncertainty on the production of ^{18}F due to this reaction was a factor of $\sim$300; it can now be estimated to a factor of $\sim$10.

We have up to now assumed that the reaction rate is dominated by the three $3/2^+$ and the $3/2^-$ resonances but the comparison with the spectrum in ^{19}F suggests that several levels are missing in the ^{19}Ne spectrum. In particular two $1/2^+$ (ℓ=0) broad levels, one at $\approx$1 MeV above and the other below threshold, have been predicted by microscopic[19] calculations. If they exist they would lead to a significant contribution in the relevant energy range. Data analysis of an inelastic scattering experiment performed at LLN would provide information on this possible $\approx$1 MeV level[20].

The $^{17}O(p,\alpha)^{14}N$ and $^{17}O(p,\gamma)^{18}F$ reactions were also identified as sources of uncertainties for the production of ^{18}F. The latter leads to the formation of ^{18}F from the ^{16}O seed nuclei trough the $^{16}O(p,\gamma)^{17}F(\beta^+)^{17}O(p,\gamma)^{18}F$ chain while the former diverts the flow reducing both the ^{18}F and ^{17}O yields. According to the NACRE compilation[21], the uncertainty on these rates used to come from the, at that time unobserved, resonance around 190 keV, resulting in an additional factor of $\sim$10 uncertainty on the production of ^{18}F [9]. The NACRE rates were based on experimental data which were later found to be inaccurate. After several measurements performed first at LENA[23] and in Orsay[22], there is now a good agreement on the resonance energy (e.g. 183.2$\pm$0.6 keV[22]) and (p,α) strength (e.g. 1.6$\pm$0.2 meV[22]) but a small discrepancy exists for the (p,γ) strength (1.2$\pm$0.2 μeV[23] and 2.2$\pm$0.4 μeV[22]).

Figures 2 and 3 display the evolution of the ^{17}O+p rates since the NACRE compilation. They show that both rates are now known with sufficiently good accuracy for nova applications. At nova temperatures, the $^{17}O(p,\gamma)^{18}F$ rate is lower and the $^{17}O(p,\alpha)^{14}N$ rate is higher than the corresponding NACRE recommended rates. As a result, the ^{18}F and ^{17}O nova production is smaller. (See Fox et al.[23] for reanalysis of the 66 keV and subthreshold resonance contributions.)

The reaction rates involved in the production of ^{18}F and ^{17}O are now much better known and new hydrodynamical calculations are underway to update their yields and to

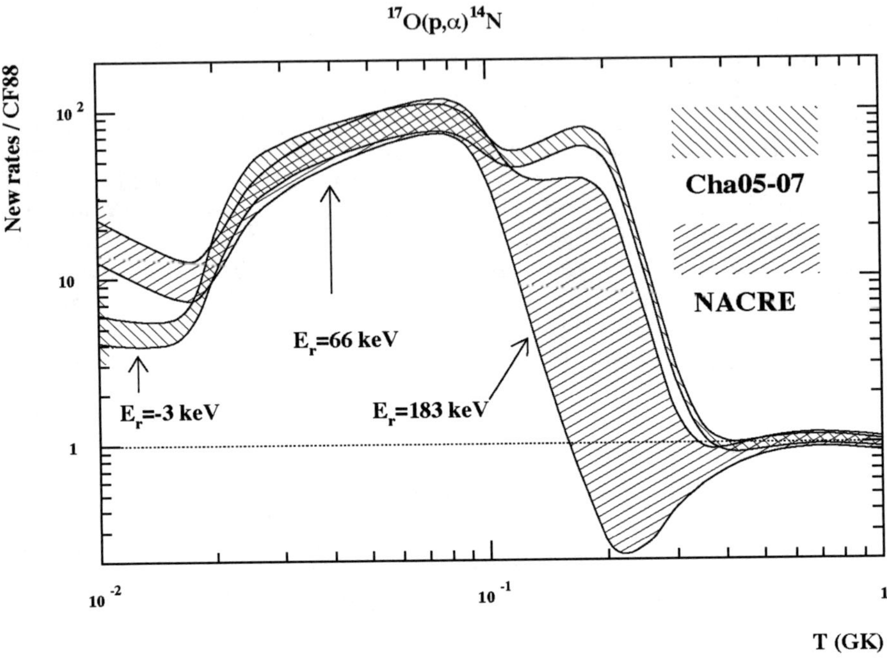

FIGURE 2. Reduction of the uncertainty in the $^{17}O(p,\alpha)^{14}N$ reaction rate[22]. (An update of Fig. 8 in CJHT[9].)

better understand their nucleosynthesis.

OTHER REGIONS OR REACTIONS

Nuclear uncertainties on the production of ^{7}Li and ^{7}Be are negligible compared with the hydrodynamics (rise time in temperature)[8]. Even though some nuclear reaction rates are still uncertain, leaks from the CNO cycle are negligible at novae temperatures. In particular, experimental data on the $^{15}O(\alpha,\gamma)^{19}Ne$[24] and $^{19}Ne(p,\gamma)^{20}Ne$ reactions[25] are now sufficient to rule out any significant nuclear flow out of the CNO cycle. Production of heavier elements depends on the presence of $^{20-22}Ne$, ^{23}Na, $^{24-26}Mg$ and ^{27}Al in ONe white dwarfs.

^{22}Na production

The decay of ^{22}Na ($\tau_{1/2}$ = 2.6 y) is followed by the emission of a 1.275 MeV γ–ray. Observations of this gamma ray emission have up to now only provided upper limits that

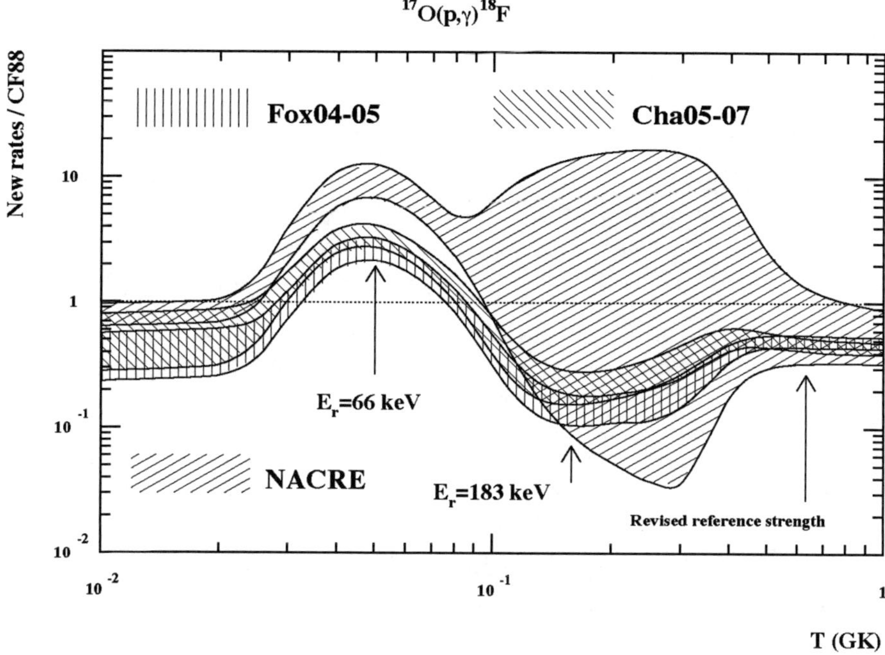

FIGURE 3. Reduction of the uncertainty in the $^{17}O(p,\gamma)^{18}F$ reaction rate according to Chafa et al.[22] or Fox et al.[23]. (An update of Fig. 9 in CJHT[9].)

are compatible with model predictions. Its detection remains a goal for the present (i.e. the INTEGRAL satellite) and future gamma-ray observatories. Calculating the expected ^{22}Na yields used to be hampered by nuclear uncertainties in the $^{21}Na(p,\gamma)^{22}Mg$ and $^{22}Na(p,\gamma)^{23}Mg$ reaction rates[10].

Destruction of ^{22}Na in nova proceeds through the $^{22}Na(p,\gamma)^{23}Mg$ reaction. Since the NACRE compilation a Gammasphere experiment[26] and a β–decay experiment[27] have improved the spectroscopy of ^{23}Mg which has reduced the uncertainty below 10^8 K by a factor of $\sim$10. However, the authors also discovered a new level which could lead to a yet unobserved resonance at 190 keV. The rate uncertainty at nova temperatures still remains large: a factor of $\sim$10.

Photodisintegration of ^{22}Mg which is important at nova temperatures, prevents further processing but the $^{21}Na(p,\gamma)^{22}Mg$ reaction remains important for ^{22}Na production. The existence of two branches, $^{21}Na(p,\gamma)^{22}Mg(\beta^+)^{22}Na$ and $^{21}Na(\beta^+)^{21}Ne(p,\gamma)^{22}Na$, with different timescales affects the production of ^{22}Na[10]. The uncertainty on the $^{21}Na(p,\gamma)^{22}Mg$ rate used to come from the unknown contributions of three unobserved resonances associated with the ^{22}Mg levels at E_X = 5.714, 5.837 and 5.962 MeV. Thanks to the experiments conducted at the TRIUMF-ISAC facility using a ^{21}Na beam, *i)* the E_R = 206 keV (E_X = 5.714 MeV) resonance strength has been precisely measured[28] and *ii*) the contribution of the other two levels was found to be negligible[29].

^{26}Al production

Due to the long lifetime of $^{26g.s.}$Al ($\tau_{1/2}$ = 0.717 My), a single nova gamma ray emission (1.809 MeV) is far too faint to be observed. However novae can contribute to the accumulation of this isotope in the Galaxy. The major nuclear uncertainties affecting its production were identified to be the ^{25}Al(p,γ)^{26}Si and $^{26g.s.}$Al(p,γ)^{27}Si reactions[10].

The $^{26g.s.}$Al(p,γ)^{27}Si reaction governs the ^{26}Al ground state destruction in novae. For nucleosynthesis calculations, its rate was often adopted from the *unpublished* work of Vogelaar. The NACRE compilation excluding unpublished results assigns a large uncertainty to this rate at nova temperatures. The strength of the resonance at E_R = 188 keV was the source of this uncertainty and influenced directly the ^{26}Al production [10]. It has now been measured directly with a ^{26}Al beam at the TRIUMF-ISAC facility and found[30] to be within a factor of 1.6 from the unpublished Vogelaar's value. Nevertheless, the uncertainty remains large (orders of magnitudes) below 10^8 K because of the lack of information on lower energy resonances.

Depending on the ^{24}Mg/^{25}Mg initial abundance ratio, the ^{25}Al(p,γ)^{26}Si reaction can have a crucial role in the formation of $^{26g.s.}$Al as it provides a diversion from the ^{24}Mg(p,γ)^{25}Al(β^+)^{25}Mg(p,γ)$^{26g.s.}$Al flow. Following the ^{25}Al(p,γ) reaction, ^{26}Si can either decay to the short lived 228 keV isomeric level of ^{26}Al or be destroyed by subsequent proton capture[31]. In either case, it bypasses the long lived ground state of ^{26}Al. (At nova temperature, the isomer and ground state in ^{26}Al have to be considered as separate species[32].) Orders of magnitude uncertainties arose from missing levels in ^{26}Si, in particular a 3^+ (ℓ=0) level. Spectroscopic studies[33, 34, 35, 36, 37] of ^{26}Si have localized a 4^+ and a 1^+ level, a probable 3^+ level at E_X = 5.912 MeV and a 0^+ level at E_X = 5.946 MeV. The corresponding resonance strengths have not been measured, but the uncertainty on this rate has been considerably reduced.

Heavy elements production

No significant amount of elements beyond aluminum are normally found in the composition of white dwarfs. The production of "heavy elements", i.e. from silicon to argon, rely on the nuclear flow out of the Mg-Al region through ^{28}Si and subsequently through ^{30}P. Its relatively long lifetime ($\tau_{1/2}$= 2.5 m) can halt the flow unless the ^{30}P(p,γ)^{31}S reaction is fast enough. This reaction is also important to calculate the silicon isotopic ratios. The results can be compared to the values measured for some presolar grains that may have a nova origin[38]. Due to the limited spectroscopic data available for the ^{31}S nucleus, a Hauser–Feshbach rate has been used, up to now, in nova nucleosynthesis calculations. This statistical model which assumes a high level density, is certainly not appropriate for such a low mass nucleus and temperature domain. The uncertainty was difficult to determine but two orders of magnitude was the usual estimate. The spectroscopy of ^{31}S is not yet completed in the range of excitation energy important for nova but about ten levels have been observed[39, 40, 41, 42, 43] in that region with spins and parities generally assigned. The relatively high level density prevented all these levels to be experimentally resolved and confirmations are still needed. The resonance strengths used to calculate the thermonuclear reaction rate are also obtained by *assuming* typical values for spectroscopic factors. The resulting reaction rate, even though still uncertain, present a significant improvement and seems[43] close to the Hauser–Feshbach one used in previous nucleosynthesis calculations.

CONCLUSIONS

Detailed calculations performed with the SHIVA hydrodynamical code have enable the identification of nuclear uncertainties affecting nova nucleosynthesis. We see that a great progress has been made in these last ten years thanks to experimental efforts, in particular for the ^{17}O(p,γ)^{18}F, ^{17}O(p,α)^{14}N, ^{18}F(p,γ)^{19}Ne, ^{18}F(p,α)^{15}O, ^{21}Na(p,γ)^{22}Mg, ^{22}Na(p,γ)^{23}Mg ^{25}Al(p,γ)^{26}Si, $^{26g.s.}$Al(p,γ)^{27}Si and ^{30}P(p,γ)^{31}S reactions which were identified as the most influential. However, further efforts are required for the ^{22}Na(p,γ)^{23}Mg, ^{25}Al(p,γ)^{26}Si, ^{30}P(p,γ)^{31}S reactions and especially for the ^{18}F(p,α)^{15}O reaction. For this last reaction where contribution of interfering broad resonance tails are essential, progress should come from direct measurements with intense ^{18}F beam (TRIUMF) or from indirect (THM[44]) measurement planned at the CRIB of the Center for Nuclear Studies (Wako).

ACKNOWLEDGMENTS

I am indebted to Margarita Hernanz and Jordi José for a now more than twelve years collaboration on nova nucleosynthesis and to Nicolas de Séréville for frequent discussions. Many thanks also to Carmen Angulo, Christian Iliadis, Faïrouz Hammache, François de Oliveira Santos and Claudio Spitaleri for long time collaborations.

REFERENCES

1. S. Starrfield, J.W. Truran, W.M. Sparks, and G.S. Kutter, Astrophys. J., **176**, 169 (1972).
2. R.D. Gehrz, J.W. Truran, R.E. Williams, and S. Starrfield, Publ. Astron. Soc. Pacific, **743**, 3–26 (1998).
3. J. José and M. Hernanz, Astrophys. J., **494**, 680 (1998).
4. J. José and M. Hernanz, J. Phys., **G34**, R431–R458 (2007).
5. L. van Wormer, J. Goerres, C. Iliadis, M. Wiescher and F.-K. Thielemann, Astrophys. J., **432**, 326 (1994).
6. A. Coc, R. Mochkovitch, Y. Oberto, J.-P. Thibaud and E. Vangioni-Flam, Astron. Astrophys., **299**, 479-492 (1995).
7. C. Iliadis, A. Champagne, J. José, S. Starrfield, and P. Tupper, Astrophys. J. S., **142**, 105 (2002).
8. M. Hernanz, J. José, A. Coc, and J. Isern, Astrophys. J., **465**, L27-L30 (1996) [astro-ph/9604102].
9. A. Coc, M. Hernanz, J. José, and J.P. Thibaud (CJHT), Astron. Astrophys., **357**, 561–571 (2000) [astro-ph/0003166].
10. J. José, A. Coc and M. Hernanz, Astrophys. J., **520**, 347-360 (1999) [astro-ph/9902357].
11. J. José, A. Coc, and M. Hernanz Astrophys. J., **560**, 897–906 (2001) [astro-ph/0106418].
12. J. Gomez–Gomar, M. Hernanz, J. José and J. Isern, Mon. Not. R. Astron. Soc., **296**, 913 (1998).

13. S. Utku, J.G. Ross, N.P.T. Bateman, D.W. Bardayan, A.A. Chen et al., Phys. Rev., **C57**, 2731 (1998).
14. N. de Séréville, A. Coc, C. Angulo, M. Assunção, D. Beaumel et al., Phys. Rev.,**C67** 052801(2003)[nucl-ex/0304014] and Nucl. Phys., **A791**, 251–266 (2007) [nucl-ex/0702034].
15. R.L.Kozub, D.W.Bardayan, J.C.Batchelder, J.C.Blackmon, C.R.Brune et al., Phys. Rev., **C71**, 032801 (2005) and Phys. Rev., **C73**, 044307 (2006).
16. R.Coszach, M.Cogneau, C.R.Bain, F.Binon, T.Davinson et al. Phys. Lett., **353B**, 184 (1995); *see also references in Ref. [14].*
17. D.W.Bardayan, J.C.Batchelder, J.C.Blackmon, A.E.Champagne, T.Davinson et al., Phys. Rev. Lett., **89**, 262501 (20002) *and references therein.*
18. N. de Séréville, C. Angulo, A. Coc, N. L. Achouri, A. Casarejos et al., *submitted.*
19. M.Dufour & P.Descouvemont, Nucl. Phys., **A785**, 381 (2007).
20. F. de Oliveira–Santos and J.C. Dalouzy, *private communication.*
21. C. Angulo, M. Arnould, M. Rayet, P. Descouvemont, D. Baye et al. (NACRE), Nucl. Phys., **A656**, 3-183 (1999) and http://pntpm.ulb.ac.be/nacre.htm.
22. A. Chafa, V. Tatischeff, P. Aguer, S. Barhoumi, A. Coc et al., Phys. Rev. Lett., **95**, 031101 (2005) and Phys. Rev., **C75**, 035810 (2007).
23. C. Fox, C. Iliadis, A.E. Champagne et al., Phys. Rev. Lett., **93**, (2004) 081102 and Phys. Rev., **C71**, 055801 (2005).
24. W.P. Tan, J.L. Fisker, J. Görres, M. Couder, and M. Wiescher, Phys. Rev. Lett., **98**, 242503 (2007).
25. M. Couder, C. Angulo, E. Casarejos, P. Demaret, P. Leleux, and F. Vanderbist, Phys. Rev., **C69**, 022801 (2004).
26. D.G. Jenkins, C.J. Lister, R.V.F. Janssens, T.L. Khoo, E.F. Moore et al., Phys. Rev. Lett., **92**, 031101 (2004)
27. V.E. Iacob, Y. Zhai, T. Al-Abdullah, C. Fu, J.C. Hardy et al., Phys. Rev., **C74**, 045810 (2006).
28. S. Bishop, R.E. Azuma, L. Buchmann, A.A. Chen, M.L. Chatterjee et al., Phys. Rev. Lett., **90**, 162501 (2003).
29. J.M. D'Auria, R.E. Azuma, S. Bishop, L. Buchmann, M.L. Chatterjee et al., Phys. Rev., **C69**, 065803 (2004).
30. C. Ruiz, A. Parikh, J. José, L. Buchmann, J.A. Caggiano et al., Phys. Rev. Lett., **96**, 252501 (2006).
31. Y. Togano et al., *these proceedings.*
32. A. Coc, M.-G. Porquet and F. Nowacki Phys. Rev., **C61**, 015801 (2000) [astro-ph/9910186].
33. D.W. Bardayan, J.C. Blackmon, A.E. Champagne, A.K. Dummer, T. Davinson et al., Phys. Rev., **C65**, 032801 (2002).
34. J.A. Caggiano, W. Bradfield-Smith, R. Lewis, P.D. Parker, D.W. Visser et al., Phys. Rev., **65**, 055801 (2002).
35. Y. Parpottas, S.M. Grimes, S. Al-Quraishi, C.R. Brune, T.N. Massey et al., Phys. Rev., **C70**, 065805 (2004).
36. D.W. Bardayan, J.A. Howard, J.C. Blackmon, C.R. Brune, K.Y. Chae et al., Phys. Rev., **C74**, 045804 (2006).
37. D. Seweryniak, P.J. Woods, M.P. Carpenter, T. Davinson, R.V.F. Janssens et al., Phys. Rev., **C75**, 062801 (2007).
38. S. Amari, X. Gao, L.R. Nittler, E. Zinner, J. José et al., Astrophys. J., **551**, 1065 (2001).
39. D.G. Jenkins, C.J. Lister, M.P. Carpenter, P. Chowdhury, N.J. Hammond et al., Phys. Rev., **C72**, 031303 (2005)
40. D.G. Jenkins, A. Meadowcroft, C.J. Lister, M.P. Carpenter, P. Chowdhury et al., Phys. Rev., **C73**, 065802 (2006).
41. A. Kankainen, T. Eronen, S.P. Fox, H.O.U. Fynbo, U. Hager et al., European Physical Journal, **A27**, 67 (2006).
42. Z. Ma, D.W. Bardayan, J.C. Blackmon, R.P. Fitzgerald, M.W. Guidry et al., Phys. Rev., **C76**, 015803 (2007).
43. C. Wrede, J.A. Caggiano, J.A. Clark, C. Deibel, A. Parikh, and P.D. Parker, Phys. Rev., **C76**, 052802(R) (2007).
44. C. Spitaleri, *these proceedings*; L. Sergi et al., *these proceedings.*

Lithium synthesis in low metallicity AGB stars

Nobuyuki Iwamoto

Nuclear Data Center, Japan Atomic Energy Agency, Tokai, Ibaraki 319-1195, Japan

Abstract. We evolve thermally pulsing AGB star models in the mass range of $1-8\ M_\odot$. The metallicity of the models is assumed to be [Fe/H] $\simeq -3$. Mass loss is taken into account to investigate the abundance patterns of the yields ejected from the AGB models. In the 1 and 2 $M_\odot$ AGB models hot bottom burning (HBB) does not take place at the base of the convective envelope during interpulse phases, but the low-mass models produce ^{7}Li in an H-flash event. The occurrence of this event is associated with the ingestion of protons from the overlying H-rich envelope into the He-flash driven convective shell during thermal pulse phase. The resulting ^{7}Li abundances are higher than the primordial one based on the analysis of the WMAP data. The present investigation also confirms the efficient production of ^{7}Li by the HBB in the intermediate-mass ($4-8\ M_\odot$) AGB stars. If these AGB stars belong to a binary system with a low-mass companion, mass loss from the primary AGB star transfers the materials enriched in ^{7}Li to the surface of the secondary star and makes the surface composition Li-rich. The formation of the Li-rich stars, however, strongly depends on the mass loss history and binary separation. The nucleosynthesis of the other light elements up to the phosphorus is also followed until the end of the AGB phase. We find that the yields of the low metallicity AGB stars well reproduce the abundance patterns of extremely metal-poor stars.

Keywords: stellar evolution, AGB stars, nucleosynthesis, lithium, metal-poor stars
PACS: 26.20.Np,97.10.Cv,97.10.Me,97.10.Tk,97.30.Hk

INTRODUCTION

An asymptotic giant branch (AGB) star is an evolved giant after the end of the core He-burning phase, and therefore, is composed of carbon and oxygen core, thin He-rich layer and H-rich convective envelope (see e.g., [1]). The stars with the initial mass of about 8 to 10 $M_\odot$ experience off-center carbon-ignition, by which the composition of the core in massive AGB stars becomes oxygen and neon-rich. A variety of chemical elements such as light elements of Li, C, N, Na and heavy elements of Zr, Ba and Pb is produced in the evolution of AGB stars. The He shell burning is thermally unstable and thermal runaway occurs repeatedly (i.e., thermal pulses). This event develops the He convective shell in the He-rich layer, and distributes produced nuclei over the He intershell. After the thermal pulse ceases, the convective envelope penetrates into the He-rich layer and carries the produced nuclei (^{12}C and *s*-process elements) into the surface (i.e., the third dredge-up). The other important nucleosynthetic site in intermediate-mass AGB stars is the H-burning shell at the bottom of the H-rich convective envelope (hot bottom burning [HBB]) during interpulse phases. The HBB converts the dredged-up ^{12}C and ^{16}O into ^{14}N via CNO cycle and makes the surface N-rich.

In the metallicity of [Fe/H] $\lesssim -2.5$ AGB stars with mass of $M \lesssim 3\ M_\odot$ experience an H-ingestion event (HIE) [2, 3]. The mass-metallicity range was referred to as Case II and II′ in Fujimoto et al. [2]. For [Fe/H] > -2.5, the He convective shell which develops

CP1016, *Origin of Matter and Evolution of Galaxies,*
edited by T. Suda, T. Nozawa, A. Ohnishi, K. Kato, M. Y. Fujimoto, T. Kajino, and S. Kubono

during thermal pulse phases does not get into touch with the H-rich outer layer due to the existence of a large entropy barrier at the boundary between the envelope and He-rich layer. However, since low metallicity AGB models have a lower entropy barrier at the boundary, the penetration of the He convective shell into H-rich envelope is allowed. The convective shell brings protons into hotter He-burning region. The ingested protons are captured by the ^{12}C produced by the triple α reaction. This event releases a large amount of energy and results in an H-flash. The released energy splits the He convective shell into two shells, one is driven by the H-flash and another is driven by the He-flash. Soon after the H-flash ceases, the envelope convection penetrates into the region where the H-flash driven convective shell is developed, and carries the produced ^{14}N and ^{13}C as well as ^{12}C into the envelope.

Lithium is one of the few elements which are primordially produced by the Big Bang nucleosynthesis. The abundance is determined from the power spectrum of the fluctuations in the cosmic microwave background observed by the WMAP [4] and is $A(\mathrm{Li}) = 2.64$ [5], where $A(\mathrm{Li}) = \log N(\mathrm{Li})/N(\mathrm{H}) + 12$, and N is the number abundance of the element. This value is slightly higher than the Spite plateau value of $A(\mathrm{Li}) = 2.1 - 2.2$, which is observed in metal-poor warm dwarfs [6].

Lithium-7 (^{7}Li) is one of the fragile nuclei. The temperature, at which $^7\mathrm{Li}(p,\alpha)^4\mathrm{He}$ becomes active, is about 2.5×10^6 K. However, in the convective envelope of intermediate-mass AGB stars, HBB plays an important role in producing ^{7}Li when the temperature at the base of the convective envelope exceeds 5×10^7 K. The $^3\mathrm{He}(\alpha,\gamma)$ reaction firstly produces ^{7}Be at the base. Part of the ^{7}Be is convectively transported outward, and then ^{7}Be decays into ^{7}Li via the $^7\mathrm{Be}(\mathrm{e}^-,\nu_e)$ reaction at the location far from the hot H-burning region. The produced ^{7}Li is also brought into the region where H-burning occurs, and then is destroyed via $^7\mathrm{Li}(p,\alpha)$ reaction. This process (i.e., Cameron-Fowler mechanism [7]) makes the surface of AGB stars Li-rich [8]. The Li-rich phase lasts until ^{3}He contained in the envelope is used up.

Tumlinson [9] argued that the depletion of surface ^{7}Li abundance in HE1327–2326 [10], which is one of the hyper metal-poor stars ever known, favor the binary mass-transfer scenario of low-mass pairs because no Li is included in the transfered gases which is mixed with the envelope of the secondary star with the primordial Li abundance. The binary scenario to explain the surface abundance patterns of hyper metal-poor stars are developed by Suda et al. [11] in which they investigated the origin of HE0107–5240 [12]. They concluded that the abundance pattern of the light elements is attributed to the yields ejected during an AGB phase of the primary star with mass less than 2.5 $M_\odot$, without considering Li evolution.

In this work we evolve thermally pulsing AGB stars and investigate the production of ^{7}Li and the other light elements in the low- and intermediate-mass models. The effects of binary mass transfer from the primary low-metallicity AGB star, which is now a white dwarf, on the surface chemical composition of the secondary star that is now observed are explored.

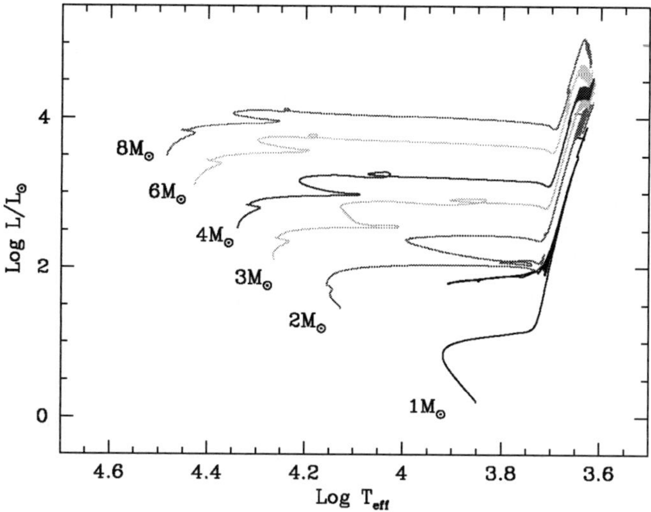

FIGURE 1. Evolutionary tracks of $1-8\,M_{\odot}$ models with $Z=2\times10^{-5}$ are shown from zero-age main sequence to the end of AGB phase. The small wiggles in the most luminous stage of each stellar model represent the evolutionary changes caused by thermal pulses and mass loss during the AGB phase.

STELLAR MODEL

We evolve stellar models without rotation and overshooting from zero-age main-sequence (ZAMS) up to the end of AGB phase and calculate the nucleosynthesis through the evolution. The ZAMS masses of the models are adopted in the range of $1-8\,M_{\odot}$. The heavy element abundance Z is assumed to be 2×10^{-5} which corresponds to the metallicity of [Fe/H] $\simeq -3$. For the initial composition we adopt the scaled solar abundances without α-element enhancements. We assume no initial ^{7}Li abundance. We use the mixing length theory for convective mixing and energy transport. For the convective stability we take the Schwarzschild criterion. Mass loss is an important factor to determine the evolution of stars and the compositions of the yields. However, since the mass loss rate is poorly known for extremely metal-poor AGB stars as well as for solar-like metallicity AGB stars, we may assume an empirical mass loss rate. We adopt the mass loss rate formulated by Blöcker [13]. The rate is based on the numerical simulation of dust-driven mass loss from the atmosphere of a low-mass star with long-period pulsation and is described by the following form:

$$\dot{M}_B = \eta_B\, 4.83\times10^{-9} M_{\rm ZAMS}^{-2.1} L^{2.7} \dot{M}_R, \tag{1}$$

where $M_{\rm ZAMS}$ and L are the ZAMS mass and surface luminosity in solar units, respectively, and $\dot{M}_R$ is the Reimers' mass loss rate which is multiplied by a factor of $\eta_R=0.5$. The value of η_R might be too large for an extremely metal-poor 1 $M_{\odot}$ model on the first ascending red giant branch (RGB) because of the low Z in the envelope. We use

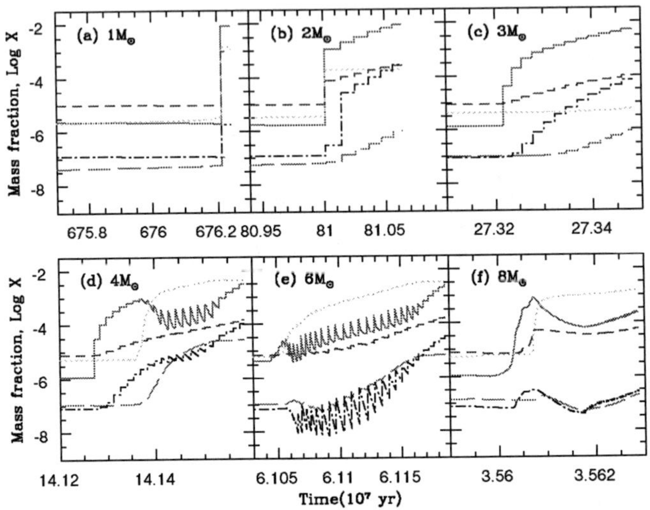

FIGURE 2. Evolution of surface ^{12}C (solid lines), ^{14}N (dotted lines), ^{16}O (dashed lines), ^{22}Ne (dash-dotted lines) and ^{23}Na (long dashed lines) abundances (by mass) for $1-8\ M_\odot$ AGB models. The HIE synthesizes large amounts of ^{14}N as well as ^{12}C and ^{16}O in 1 and 2 $M_\odot$ models.

$\eta_R = 0.1$. The factor η_B is further introduced to take into account the effect of the low-metal contents. In this work we set $\eta_B = 0.1$. The resulting rate is by a factor of 0.05 (0.01 for the 1 $M_\odot$ model) smaller than the original Blöcker's one. In calculating the stellar evolution with mass loss, the Reimers' rate is firstly used until the Blöcker's rate exceeds the Reimers' one. The transition from Reimers' to Blöcker's rates occurs during the early AGB phase where surface luminosity increases rapidly. The other details on the stellar evolution code can be seen in Iwamoto et al. [3].

RESULTS

The evolutionary tracks of $1-8\ M_\odot$ models are shown from ZAMS up to the end of the AGB phase in Fig. 1. The models with mass higher than 2 $M_\odot$ experience the central He burning phase at the blue side of the HR diagram before ascending the RGB. The 1 $M_\odot$ model undergoes an off-center He-ignition at the beginning of the core He burning. In the 8 $M_\odot$ model C-burning ignites at the off-center. After the termination of the central C-burning this model evolves to the AGB phase. Small wiggles of the tracks at the most luminous AGB stages are seen in Fig. 1. Such effective temperature-luminosity changes are attributed to the thermal pulse and mass loss.

Figure 2 shows the evolution of surface ^{12}C, ^{14}N, ^{16}O, ^{22}Ne and ^{23}Na abundances (by mass) for $1-8\ M_\odot$ models. The 1 and 2 $M_\odot$ models experience the HIE only once and they create large amounts of ^{12}C and ^{14}N, which are promptly brought into the surface. After the HIE the mass loss rate becomes high owing to the high CNO abundances in

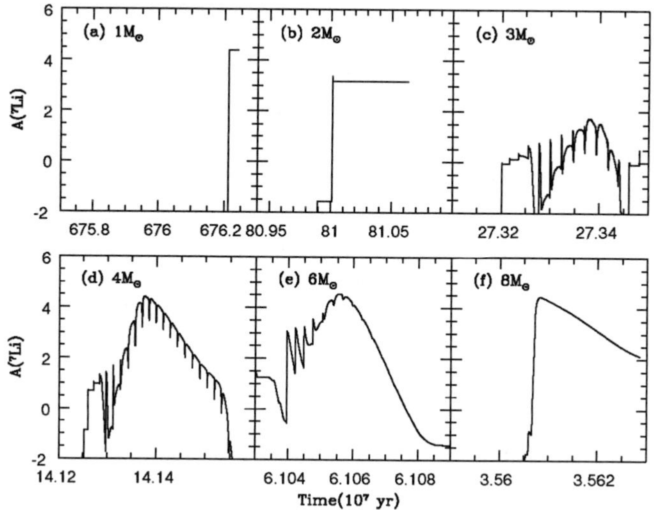

FIGURE 3. Evolution of surface Li abundances for $1-8\,M_\odot$ AGB stars.

the envelope. The 1 $M_\odot$ model loses the envelope, and thus ends the AGB evolution. In contrast, the 2 $M_\odot$ model has a large envelope mass and experiences the successive thermal pulses and third dredge-up. This results in the further increase of ^{12}C, while the ^{14}N remains unchanged. In the models with $M \geq 3\,M_\odot$ HIE does not take place. The third dredge-up brings a large amount of ^{12}C into the convective envelope. Since the HBB ignites in the models with $M \geq 4\,M_\odot$, the dredged-up ^{12}C is converted into ^{14}N during interpulse phases. As their envelope masses are lost by the mass loss, the efficiency of HBB becomes low. As a result, the dredged-up ^{12}C remains unburnt and the ^{12}C abundances increase rapidly at the later phase of the AGB evolution. Sodium enhancement is also found in 4 and 6 $M_\odot$ models. Carbon is first transformed into ^{14}N in H-burning and later into ^{22}Ne by captures of α-particles in the He layer during thermal pulse phases. The synthesized ^{22}Ne is carried into the envelope through the third dredge-up. Finally the HBB transmutes ^{22}Ne to ^{23}Na via the NeNa cycle of H-burning. Figure 2 implies that the production efficiency of ^{23}Na decreases as the HBB does not work effectively. In the lower mass models the ^{22}Ne(n,γ) reaction in the He convective shell is important to produce ^{23}Na, which is dredged-up into the envelope together with ^{22}Ne [14].

Figure 3 displays the evolution of the surface ^{7}Li abundances. As explained above, the production of ^{7}Li occurs via the Cameron-Fowler mechanism in the AGB models with $M \geq 3\,M_\odot$. The $4-8\,M_\odot$ models show larger ^{7}Li abundances than the solar system abundances $A(^7\mathrm{Li}) = 3.31$ [15]. The decrease in ^{7}Li abundances after the maximum value, is attributed to the consumption of ^{3}He contained in the convective envelope because ^{3}He is the source for the ^{7}Li synthesis. The maximum ^{7}Li abundance for 3 $M_\odot$ model is smaller than that for the $4-8\,M_\odot$ models by two orders of magnitude, since

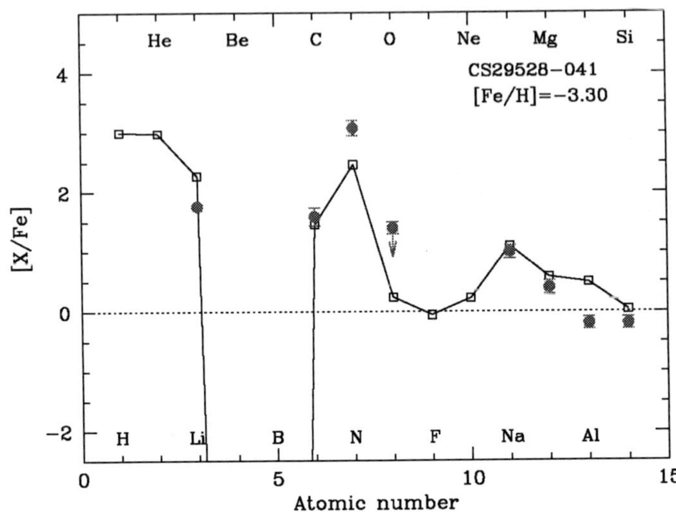

FIGURE 4. Comparison of the yields from the 6 $M_\odot$ AGB model ([Fe/H] $\simeq -3$) with the abundance pattern of CS 29528–041 (filled circles). The metallicity of CS 29528–041 is [Fe/H] $= -3.3$. For the O abundance the upper limit is given from the observation.

the convective base temperature is lower than those attained in the higher mass models. In contrast to the higher mass models, the reduction of ^{7}Li from the maximum value is responsible for the inefficient HBB (i.e., lower convective base temperature) due to the mass loss. The low metallicity AGB stars with masses lower than about 3 $M_\odot$ can also create ^{7}Li. The production is related with the occurrence of the HIE. At the onset of the thermal pulse the He convective shell extends upward and finally penetrates into the base of the H-rich envelope. The convection brings protons and ^{3}He into the convective shell. In an H-flash driven convective shell formed after the proton-ingestion, the ^{3}He(α,γ) reaction creates ^{7}Be, which is accumulated in the shell. Soon after the H-flash terminates, the envelope convection dredges-up ^{7}Be and distributes synthesized ^{7}Li over the envelope. Since the HIE occurs only once, the surface ^{7}Li abundances of the 1 and 2 $M_\odot$ models are not affected by the later AGB evolution as seen in Fig. 3a and 3b. The ejected amounts of ^{7}Li are large enough to influence the surface Li abundance of the binary companion even in the low-mass models.

In order to compare the yield ejected from the AGB models with the observed abundance pattern of the extremely metal-poor stars, we take into account a dilution in the surface convective zone of the low-mass binary companion which is now observed. The dilution factor is defined as $f_{\rm dil} = \log M_{\rm mix}/M_{\rm acc}$, where $M_{\rm acc}$ is the mass accreted from an AGB star to the secondary, and $M_{\rm mix}$ is the secondary envelope mass including the accreted gas with AGB compositions. We first attempt to explain the abundance pattern of CS 29528–041 which has the metallicity of [Fe/H] $= -3.3$ [16]. From the observed ratio of [C/N] $= -1.48$, an intermediate-mass AGB star can be a progenitor. In Figure 4 the yield of the 6 $M_\odot$ AGB model is compared with the abundance pattern of CS 29528–041. The value of $f_{\rm dil} = 0.66$ dex is adopted in this case. The observed pattern is well reproduced by this model, although there are slight underproduction of N and overproduction of Al. The carbon isotopic ratio is not observed, but the present investigation predicts ^{12}C/^{13}C $= 11.8$. The Li abundance in the transfered material is close to the Spite plateau value of $A(\rm Li) \simeq 2.2$. As a result, the diluted abundance of Li is large relative to the observation, but some processes might decrease to the observed value during the long lifetime of the low-mass secondary star.

The Li abundances in the yields of the 1, 2, 4, and 8 $M_\odot$ AGB models are larger than the Spite plateau and primordial values. If the materials transfered to the binary companion is dominant, the Li-enriched low-metallicity dwarfs might be found out. We note that the Li abundance is very sensitive to the binary separation and mass loss history of AGB stars.

The other characteristic feature is the enhancement of the fluorine (^{19}F), which is produced in the He convective shell during thermal pulses [17]. The abundances in the yields are [F/Fe] $= 3.07$, 2.20, and 1.92 for the 2, 3, and 4 $M_\odot$ models, respectively, assuming [Fe/H] $= -3$ [18]. In the AGB models with $M > 4\ M_\odot$, fluorine is efficiently destroyed by the HBB. The resulting abundances are [F/Fe] $\lesssim 0$.

ACKNOWLEDGMENTS

This work has been supported by the Grant-in-Aid for Young Scientists (B) (17740163).

REFERENCES

1. Herwig, F., *Ann. Rev. Astron. Astrophys.*, **43**, 435–479 (2005).
2. Fujimoto, M. Y., Ikeda, Y., & Iben, I. J., *Astrophys. J. Lett.*, **529**, L25–L28 (2000).
3. Iwamoto, N., Kajino, T., Mathews, G. J., Fujimoto, M. Y., & Aoki, W., *Astrophys. J.*, **602**, 377–388 (2004).
4. Spergel, D. N., et al., *Astrophys. J. Suppl.*, **148**, 175–194 (2003).
5. Spergel, D. N., et al., *Astrophys. J. Suppl.*, **170**, 377–408 (2007).
6. Spite, F., & Spite, M., *Astron. Astrophys.*, **115**, 357–366 (1982).
7. Cameron, A. G. W., & Fowler, W. A., *Astrophys. J.*, **164**, 111–114 (1971).
8. Sackmann, I.-J., & Boothroyd, A. I. *Astrophys. J. Lett.*, **392**, L71–L74 (1992).
9. Tumlinson, J., *Astrophys. J.*, **665**, 1361–1370 (2007).
10. Frebel, A., et al., *Nature*, **434**, 871–873 (2005).
11. Suda, T., Aikawa, M., Machida, M. N., Fujimoto, M. Y., & Iben, I. J., *Astrophys. J.*, **611**, 476–493 (2004).
12. Christlieb, N., et al., *Nature*, **419**, 904–906 (2002).
13. Blöcker, T., *Astron. Astrophys.*, **297**, 727–738 (1995).
14. Gallino, R., Bisterzo, S., Husti, L., Käppeler, F., Cristallo, S., & Straniero, O., *International Symposium on Nuclear Astrophysics - Nuclei in the Cosmos*, Proc. of Science, 2006, pp. 100.1.
15. Anders, E., & Grevesse, N., *Geochim. Cosmochim. Acta*, **53**, 197–214 (1989).
16. Sivarani, T., et al., *Astron. Astrophys.*, **459**, 125–135 (2006).
17. Lugaro, M., Ugalde, C., Karakas, A. I., Görres, J., Wiescher, M., Lattanzio, J. C., & Cannon, R. C., *Astrophys. J.*, **615**, 934–946 (2004).
18. Straniero, O., Cristallo, S., Gallino, R., & Dominguez, I., Memorie della Societa Astronomica Italiana, **75**, 665–669 (2004).

^{80}Se(γ,n)^{79}Se cross section and s-process branching at ^{79}Se

A. Makinaga*, H. Utsunomiya*, T. Kaihori*, T. Yamagata*, H. Akimune*, S. Goriely†, H. Toyokawa**, T. Matsumoto**, H. Harano**, H. Harada‡, S. Goko‡, F. Kitatani‡, K. Y. Hara§, S. Hohara¶ and Y.-W. Lui‖

*Department of Physics, Konan University, Okamoto 8-9-1, Higashinada, Kobe 658-8501, Japan
†Institut d'Astrophysique, Université Libre de Bruxell, Campus de la Plane, CP-226, 1050, Brussels, Belgium
**National Institute of Advanced Industrial Science and Technology, Tsukuba 305-8568, Japan
‡Japan Atomic Energy Agency, Tokai-mura, Naka, Ibaraki 319-1195, Japan
§Tandem Accelerator Complex, University of Tsukuba, Tsukuba, Ibaraki 305-8577, Japan
¶Atomic Energy Reseach Institute, Kinki University, Kowakae 3-4-1, Osaka 577-8502, Japan
‖Cyclotron Institute, Texas A & M University, College Station, Texas 77843, USA

Abstract.
Photoneutron cross sections were measured for ^{80}Se near the neutron separation energy with laser Compton-backscattered γ-ray beams at AIST. Neutron capture rates are evaluated for ^{79}Se with the photoreaction data as experimental constraints on the E1 γ-strength function, a key parameter in the Hauser-Feshbach statistical model calculation. Solving the set of differential equations under a single neutron exposure, we analyzed the solar abundance ratio of the weak components of ^{80}Kr and ^{82}Kr in terms of the s-process branching at ^{79}Se. We discuss the region of temperature and neutron density allowed for the weak s-process nucleosynthesis.

Keywords: photoneutron cross section, weak s-process, neutron capture
PACS: 25.20.-x, 24.60.Dr, 26.50.+x

INTRODUCTION

The s-process nucleosynthesis is characterized by two distinct processes producing the main and the weak components. The main component reproduces the solar s-process abundances of heavy nuclei with $94 < A < 204$ under an exponential distribution of neutron exposures [1, 2]. In contrast, the weak component is called for to explain the s-process production of nuclei with $A < 90$ and is better characterized by a single neutron exposure [3]. The weak s-process nucleosynthesis is believed to take place mainly during the core He-burning phase of massive stars at neutron densities $n_n = 10^6$-10^7cm^{-3} and temperatures around T=3 $\times 10^8$K [4]. A possible reprocessing of the s-process matter may take place during advanced burning phases of massive stars, in particular at the bottom of the shell C-burning at a temperature $T \simeq 10^9$K and neutron density bursting from 10^9 to 10^{11}cm^{-3} on time scale of a few years [4].

The β decay rate of ^{79}Se exhibits a strong temperature dependence due to thermal population of the isomeric state at 95.7 keV, potentially making ^{79}Se an s-process thermometer. While experimental information on the β^- decay half-life for the isomer (logft = 4.70) [8] and the ground state ($T_{1/2}$ = 295 ky) [9] in ^{79}Se has been improved greatly,

CP1016, *Origin of Matter and Evolution of Galaxies*,
edited by T. Suda, T. Nozawa, A. Ohnishi, K. Kato, M. Y. Fujimoto, T. Kajino, and S. Kubono

neutron capture cross sections have remained unknown experimentally, hampering a proper functioning of the thermometer.

A direct measurement of (n,γ) cross sections for ^{79}Se is currently not feasible for the unavailability of a target sample. The photodisintegration of ^{80}Se is a good probe of the E1 γ strength function near neutron threshold, a key nuclear parameter of the Hauser-Feshbach statistical model. In this paper, we use (γ,n) cross sections for ^{80}Se as experimental constraints on the E1 γ strength function to improve the statistical-model predictions of (n,γ) cross sections for ^{79}Se. Using the predicted neutron capture cross sections for ^{79}Se, we analyze the abundance ratio of s-only nuclei of ^{80}Kr and ^{82}Kr with a focus on the weak s-process component in terms of the s-process branching at ^{79}Se. The analysis is made by solving the set of differential equations under a single neutron exposure. The analysis shows that the weak abundance ratio is reproduced under the typical temperature and neutron density for core helium-burning phase (kT = 26 keV, n_n = 10^7 cm^{-3}) and a reprocessing in the shell carbon-burning phase (kT = 91 keV, n_n = 10^{10} cm^{-3}) of massive stars.

EXPERIMENT

Beams of quasi-monochromatic γ rays were produced from laser Compton scattering (LCS) in the electron storage ring TERAS at the National Institute of Advanced Industrial Science and Technology. The LCS γ-ray beams were used to irradiate a sample of 1996.0 mg/cm^2 ^{80}Se enriched to 99.95% that is encapsulated in an aluminum container in the range of the average γ-ray energy 9.98 - 11.80 MeV. A Nd:YVO_4 Q-switch laser was operated at 20 kHz in the second harmonics (λ = 532 nm). The γ-ray beams had the same macroscopic time structure of 80 ms beam-on and 20 ms beam-off as that of the laser. The ^{80}Se sample was mounted at the center of a 4π-type neutron detector consisting of 20 ^{3}He counters embedded in a polyethylene moderator in a triple-ring configuration.The neutron detection efficiency is more than 56% in the neutron energy range below 1 MeV. The so-called ring ratio technique [10] was used to determine the average neutron energies. The LCS γ beam was measured with a 120% high-purity germanium detector (HPGe). The LCS beam was monitored with a large volume (8" in diameter × 12" in length) NaI(Tl) detector. Photoneutron cross sections were determined at the average γ-ray energies with the Taylor expansion method [11].

RESULT

Results of the present photoneutron cross section measurement for ^{80}Se are shown in Fig. 1. For comparison, the data previously obtained with continuous bremsstrahlung [12] and quasi-monochromatic γ rays produced in the positron annihilation in flight [13] are also shown. We found significantly smaller cross sections near the neutron separation energy in comparison with previous measurements [12, 13]. The experimental data were analyzed with the Hauser-Feshbach model code TALYS [14]. Two different sets of input parameters are found to reproduce rather well the experimental photoreaction cross section. The first one (called Comb-QRPA) includes microscopic models, namely

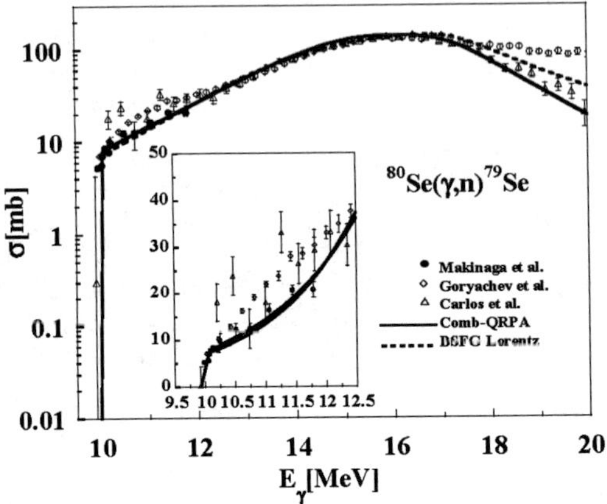

FIGURE 1. Photoneutron cross sections for ^{80}Se.

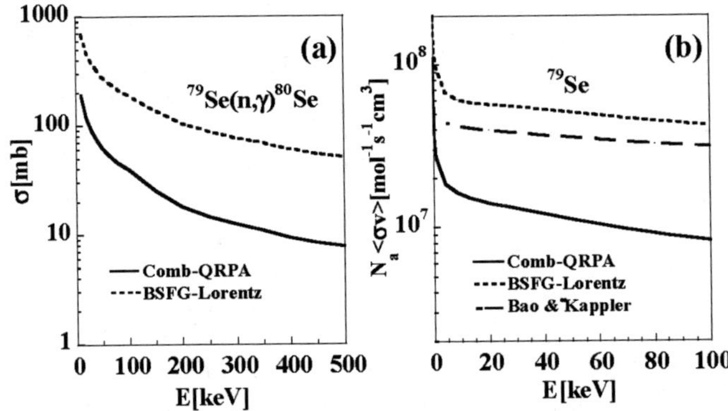

FIGURE 2. Neutron capture cross sections for ^{79}Se (*a*) and neutron capture rate for ^{79}Se (*b*) predicted with the Hauser-Feshbach model with the microscopic and macroscopic parameters of the nuclear level density and the γ strength function.

the combinatorial model of nuclear level densities (NLD) [15] and the Hartree-Fock BCS+ quasi-particle random phase approximation model [16] for the E1 γ-strength function (γSF). The second one (called BSFG-Lorentz) uses more phenomenological (macroscopic) models, namely the back-shifted Fermi gas (BSFG) NLD model and the Lorentzian-type E1 strength of [17].

Neutron capture cross sections for ^{79}Se were predicted within the framework of the Hauser-Feshbach model with the microscopic and the macroscopic parameters of the E1 γSF and the NLD that best reproduce the present photoneutron cross section. Figs. 2(a) and 2(b) show the (n,γ) cross sections and the neutron capture rate, respectively. While the present capture rate with the macroscopic parameters is 1.4 times larger than that of [18] in the temperature range $T_9 = 0.1 - 1.0$, the rate with the microscopic parameters is 3.7 - 5.0 times smaller than the macroscopic rate.

S-PROCESS BRANCHING AT ^{79}SE

The best way to analyze the ratio of the weak s-process abundances of ^{80}Kr and ^{82}Kr, N^w_{80}/N^w_{82}, may be solving the set of differential equations for the s-process reaction network under a single neutron exposure [3]. The differential equation is expressed by

$$\frac{dN_i}{dt} = -T_i N_i + \sum_j \Lambda_{j\to i} N_j. \tag{1}$$

Here, the first term on the right-hand side represents the destruction rate, where T_i is the total decay rate for nucleus i which is equal either to the neutron capture rate or to a sum of the neutron capture and the β-decay rates at a branch-point nucleus. The second term represents the production rate, where Λ is either the neutron capture rate or the β-decay rate; the sum is taken over all nuclei j that contribute to the production of nucleus i. We treated individual s-process paths separately and summed up all contributions.

For a given s-process path, Eq. 1 can be written by

$$\frac{dN_i}{dt} = -\lambda_i N_i + \lambda_{j\to i} N_j, \tag{2}$$

or equivalently, in terms of neutron exposure $\tau = n_n\, \sigma\, \Delta\, t$ with the duration time Δ t of neutron irradiation,

$$\frac{dN_i}{d\tau} = -\sigma_i N_i + \sigma_{j\to i} N_j. \tag{3}$$

In writing Eq. 3, the rate λ is related to the cross section σ by $\lambda_n = n_n \langle\sigma\, v\rangle = n_n\, \sigma\, v_T$ for neutron capture, which is applied to the β-decay as well. The abundance ratio, N^w_{80}/N^w_{82}, was taken from [19], where for the main component the result of the classical model was subtracted from the solar abundance.

We solved the set of differential equations (Eq. 3) with the initial condition, $N(^{56}\text{Fe})=1$ and N(other)=0. The exact solution of Eq. 3 under a single neutron exposure is found in [22] (Eq. (7-48) and Eq. (7-49)). Fig. 3 shows some ranges of temperature and neutron density that are consistent with the ratio of the weak s-process krypton abundances, N^w_{80}/N^w_{82}, for the microscopic and macroscopic neutron capture rates, respectively. The lines in Fig. 3 represent the neutron densities and temperatures corresponding to the central value of N^w_{80}/N^w_{82}. As shown in Fig. 3, the allowed range is very sensitive to the neutron exposure. One can see that the typical temperature and neutron density for both the core He-burning phase (kT=26 keV and n_n=10^7 cm^{-3}) and the shell C-burning phase (kT=91 keV and n_n=10^{10} cm^{-3}) of massive stars are included in the allowed range.

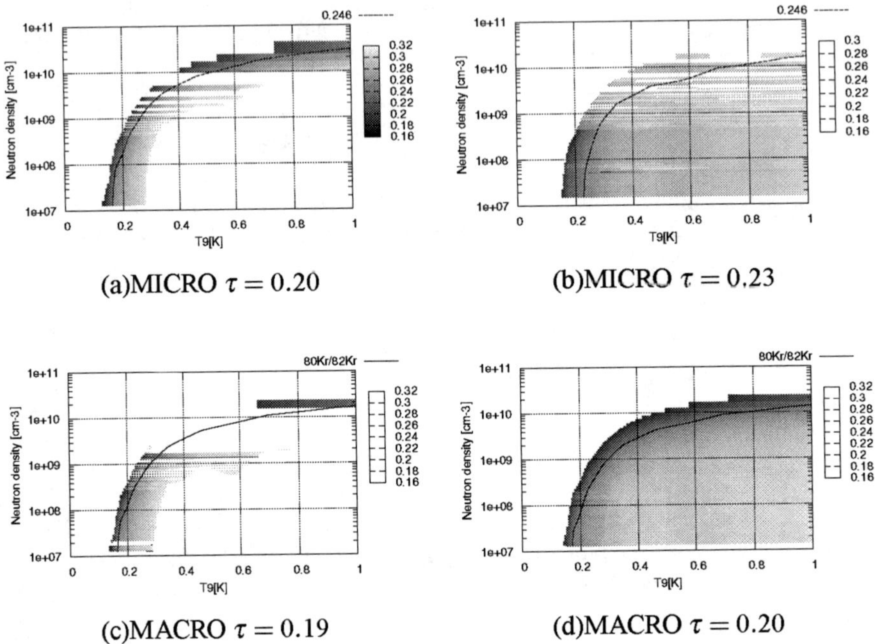

(a)MICRO $\tau = 0.20$ (b)MICRO $\tau = 0.23$

(c)MACRO $\tau = 0.19$ (d)MACRO $\tau = 0.20$

FIGURE 3. The range of neutron density and temperature which is consistent with the abundance ratio, N^w_{80}/N^w_{82}, for the microscopic nuclear parameters (Comb-QRPA) at $\tau = 0.20$ (a) and $\tau = 0.23$ (b) and for the macroscopic parameters (BSFG-Lorentz) at $\tau = 0.19$ (c) and $\tau = 0.20$ (d). See text for details.

CONCLUSION

We measured photoneutron cross sections for ^{80}Se near the neutron separation energy. The cross sections were used as experimental constraints on the E1 γ strength function for ^{80}Se, a key nuclear parameter in the Hauser-Feshbach statistical-model calculation. Using the experimentally constrained E1 γ strength function, we model-predicted neutron capture rates for ^{79}Se, a missing quantity for the s-process branching at ^{79}Se. The neutron capture rate predicted with the microscopic nuclear parameters (Comb-QRPA) in the statistical model is 4 - 5 times smaller than that predicted with the macroscopic parameters (BSFG-Lorentz). With the microscopic and the macroscopic neutron capture cross sections for ^{79}Se, the ratio of the abundances of ^{80}Kr and ^{82}Kr resulting from the branching at the s-process thermometer nucleus ^{79}Se was analyzed by solving the set of differential equations under a single neutron exposure. Regardless of the two different neutron capture rates for ^{79}Se, the resultant range of temperature and neutron density includes the typical stellar condition of both the core He-burning and the shell C-burning phases of massive stars. Thus, we conclude that the weak components of ^{80}Kr and ^{82}Kr can be processed in the core He-burning phase of massive stars and reprocessed during the advanced burning phase at the bottom of the shell C-burning.

ACKNOWLEDGMENTS

We thank K. Takahashi for his comments on the β-decay rate at high temperatures and electron number densities. This work is supported by the Japan Private School Promotion Foundation and the Konan- ULB convention. S.G. acknowledges the FNRS support.

REFERENCES

1. P.A. Seeger, W.A. Fowler, D.D. Clayton, Astrophys. J. Suppl. **11**, 121 (1965).
2. R.A. Ward, M.J. Newman, D.D. Clayton, Astrophys. J. Suppl. **31**, 33 (1976).
3. H. Beer, R.L. Macklin, Astrophys. J. **339**, 962 (1989).
4. L.-S. The, M.F. El Eid, B.S. Meyer, Astrophys. J. **655**, 1058 (2007).
5. G. Walter, H. Beer, F. Käppeler, R.-D. Penzhorn, Astron. Astrophys. **155**, 247 (1986).
6. G. Walter, H. Beer, F. Käppeler, G. Reflo, F. Fabbri, Astron. Astrophys. **167**, 186 (1986).
7. F. Käppeler, H. Beer, K. Wisshak, Rep. Prog. Phys. **52**. 945 (1989).
8. N. Klay, F. Käppeler, Phys. rev. **C38**, 295 (1988).
9. B. Singh, Nucl. Data Sheet **70**, 437 (1993).
10. B.L. Berman and S.C. Fultz, Rev. Mod. Phys. **47**, 713(1975).
11. H. Utsunomiya *et al.*, Phys. Rev. C **74**, 025806 (2006).
12. A.M. Goryachev, G.N. Zalensnyi, B.A. Tulupov, Izvestiya Akademii Nauk SSSR. Seriya Fizicheskaya, **39**, 134 (1975).
13. P. Carlos *et. al.*, Nucl. Phys. **A258**, 365 (1976).
14. A.J. Koning, S. Hilaire and M.C. Duijvestijn, Proc. Int. Conf. on Nuclear Data for Science and Technology (eds. C. Haight et al.) AIP Conference Vol. 769, p. 1154, 2005.
15. S. Hilaire, S. Goriely, Nucl. Phys. **A779**, 63 (2006).
16. S. Goriely and E. Khan, Nucl. Phys. **A706**, 217 (2002).
17. C.M. McCullagb, M.L. Stelts and R.E. Chrien, Phys. Rev. C**23**, 1394 (1981)
18. Z.Y. Bao, H. Beer, F. Käppeler, F. Voss, K. Wisshak, T. Rauscher, At. Data Nucl. Data Tables **76**, 70 (2000).
19. C. Arlandini *et. al.*, Astrophys. J. **525**, 886 (1999).
20. K. Takahashi, K. Yokoi, At. Data Nucl. Data Tables **36**, 375 (1987).
21. K. Takahashi, private communication.
22. D.D. Clayton, *Principles of Stellar Evolution and Nucleosynthesis* (The University of Chicago Press, Chicago, 1989).

7. NUCLEAR DATA FOR ASTROPHYSICS

Recent Efforts in Data Compilations for Nuclear Astrophysics

Iris Dillmann*,† and the JINA-CARINA Collaboration**

*Institut für Kernphysik, Forschungszentrum Karlsruhe, D-76021 Karlsruhe, Germany
†Physik Department E12, Technische Universtät München, D-85748 Garching, Germany
**www.jinaweb.org/events/ect07/talks/list.htm

Abstract. Some recent efforts in compiling data for astrophysical purposes are introduced, which were discussed during a JINA-CARINA Collaboration meeting on *"Nuclear Physics Data Compilation for Nucleosynthesis Modeling"* held at the ECT* in Trento/ Italy from May 29th- June 3rd, 2007. The main goal of this collaboration is to develop an updated and unified nuclear reaction database for modeling a wide variety of stellar nucleosynthesis scenarios. Presently a large number of different reaction libraries (REACLIB) are used by the astrophysics community. The "JINA Reaclib Database" on *http://www.nscl.msu.edu/~nero/db/* aims to merge and fit the latest experimental stellar cross sections and reaction rate data of various compilations, e.g. NACRE and its extension for Big Bang nucleosynthesis, Caughlan and Fowler, Iliadis et al., and KADoNiS.

The KADoNiS (Karlsruhe Astrophysical Database of Nucleosynthesis in Stars, *http://nuclear-astrophysics.fzk.de/kadonis*) project is an online database for neutron capture cross sections relevant to the *s* process. The present version v0.2 is already included in a REACLIB file from Basel university (*http://download.nucastro.org/astro/reaclib*). The present status of experimental stellar (n,γ) cross sections in KADoNiS is shown. It contains recommended cross sections for 355 isotopes between ^{1}H and ^{210}Bi, over 80% of them deduced from experimental data.

A "high priority list" for measurements and evaluations for light charged-particle reactions set up by the JINA-CARINA collaboration is presented. The central web access point to submit and evaluate new data is provided by the Oak Ridge group via the *http://www.nucastrodata.org* homepage. "Workflow tools" aim to make the evaluation process transparent and allow users to follow the progress.

Keywords: reaction rate libraries, compilations, data evaluation, stellar neutron cross sections, *s* process, *p* process
PACS: 25.40.Lw, 26.30.+k, 26.30.Ef, 29.87.+g, 97.10.Cv

STELLAR NEUTRON CAPTURE COMPILATIONS

Status of stellar (n,γ) cross sections

The pioneering work for stellar neutron capture cross sections was published in 1971 by Allen and co-workers [1]. In this paper the role of neutron capture reactions in the nucleosynthesis of heavy elements was reviewed and a list of recommended (experimental or semi-empirical) Maxwellian averaged cross sections at kT= 30 keV (MACS30) presented for nuclei between Carbon and Plutonium.

The idea of an experimental and theoretical stellar neutron cross section database was picked up again by Bao and Käppeler [2] for *s*-process studies. This compilation published in 1987 included cross sections for (n,γ) reactions (between ^{12}C and ^{209}Bi), some (n,p) and (n,α) reactions (for ^{33}S to ^{59}Ni), and also (n,γ) and (n,f) reactions for

CP1016, *Origin of Matter and Evolution of Galaxies,*
edited by T. Suda, T. Nozawa, A. Ohnishi, K. Kato, M. Y. Fujimoto, T. Kajino, and S. Kubono

long-lived actinides. A follow-up compilation was published by Beer et al. in 1992 [3].

In the update of 2000 the Bao compilation [4] was extended down to ^{1}H and – like the original Allen paper – semi-empirical recommended values for nuclides without experimental cross section information were added. These estimated values are normalized cross sections derived with the Hauser-Feshbach code NON-SMOKER [5], which account for known systematic deficiencies in the nuclear input of the calculation. Additionally, the database provided stellar enhancement factors and energy-dependent MACS for energies between kT= 5 keV and 100 keV.

The KADoNiS (Karlsruhe Astrophysical Database of Nucleosynthesis in Stars) project [6] is based on these previous compilations and aims to be a regularly updated database. The current version KADoNiS v0.2 (January 2007) is already the second update and includes – compared to the previous Bao et al. compilation [4] – 38 updated and 14 new recommended cross sections. A paper version of KADoNiS ("v1.0") is planned for 2008, which also will – like the first Bao compilation from 1987 [2] – include (n,p) and (n,α) reactions for light isotopes and (n,γ) and (n,f) reactions for long-lived actinides at kT= 30 keV.

In total, data sets are available for 355 isotopes, including 76 radioactive nuclei (22%) on or close to the s-process path. For 12 of these radioactive nuclei experimental data is available: ^{14}C, ^{93}Zr, ^{99}Tc, ^{107}Pd, ^{129}I, ^{135}Cs, ^{147}Pm, ^{151}Sm, ^{154}Eu, ^{163}Ho, ^{182}Hf, and ^{185}W. The remaining 64 radioactive nuclei are not yet measured in the stellar energy range and are represented only by semi-empirical cross section estimates with typical uncertainties of 25 to 30%. Almost all (n,γ) cross sections of the 277 stable isotopes have been measured. The few exceptions are ^{17}O, $^{36,38}Ar$, ^{40}K, ^{50}V, $^{72,73}Ge$, ^{77}Se, $^{98,99}Ru$, ^{131}Xe, ^{138}La, ^{158}Dy and ^{195}Pt. Most of these cross sections are difficult to determine because they are not accessible by activation measurements or samples are not available in sufficient amounts and/or enrichment for time-of-flight measurements.

Fig. 1 shows the status of KADoNiS v0.2 for the mass region above iron ($Z\geq 26$) at kT=30 keV. The even-N nuclei are now almost completely measured, but partially not yet included in the latest KADoNiS version from January 2007. The odd-N nuclei exhibit predominantly only semi-empirical cross sections because most of them are radioactive.

Fig. 2 gives the uncertainties at kT=30 keV of all experimental (n,γ) cross sections measured beyond iron. The average uncertainty in this region is $\pm 6\%$ (indicated by the vertical dashed lines), but may be considerably larger at lower and higher temperatures. Mavericks with uncertainties $\geq 20\%$ belong to ^{62}Ni (where two recent, but contradictory measurements were not yet included [7, 8]) and the light Pt isotopes $^{190,192,194,195}Pt$.

Extensions of KADoNiS

Several extensions are planned for KADoNiS in 2008. The s-process library will be complemented in the near future by some (n,p) and (n,α) cross sections measured at kT=30 keV, as it was already done in [2]. Additionally some "gaps" in the list of stable isotopes will be filled. For different reasons in the previous compilations the datasets for ^{2}H, ^{6}Li, ^{9}Be, $^{10,11}B$, ^{17}O, and ^{138}La were missing.

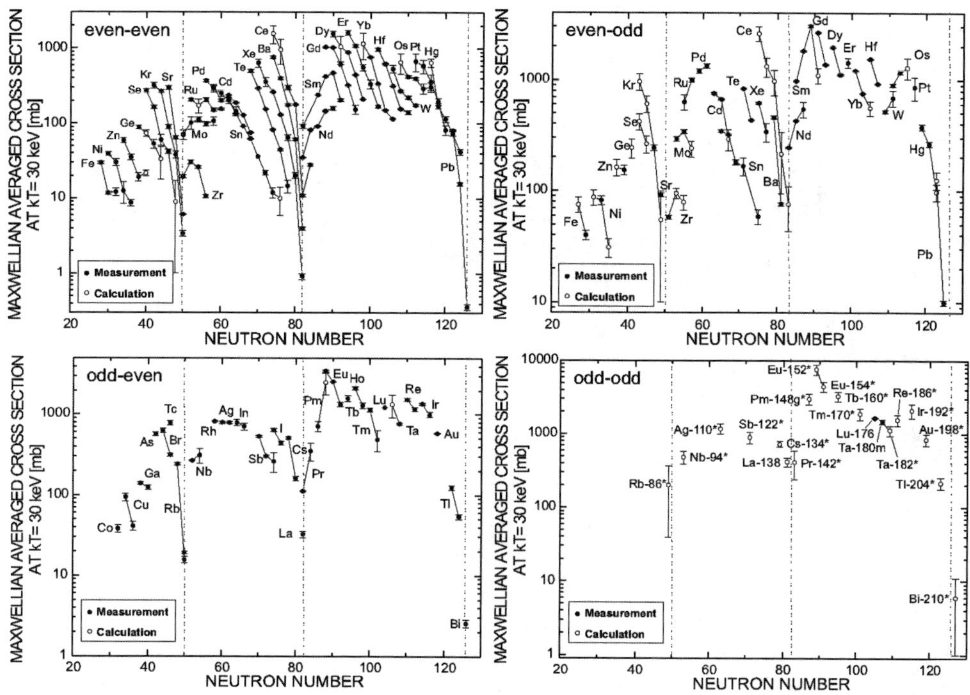

FIGURE 1. Maxwellian averaged cross sections at kT=30 keV for even-even, even-odd, odd-even, and odd-odd isotopes beyond Iron ($Z\geq 26$) listed in KADoNiS. Filled circles show measured cross sections, open circles indicate semi-empirical estimates.

Furthermore it is planned to include more radioactive isotopes, which are relevant for s-process nucleosynthesis at higher neutron densities (up to 10^{11} cm^{-3}). These isotopes are more than one unit away from the "classical" s-process path on the neutron-rich side of stability, and their stellar (n,γ) cross sections have to be extrapolated from known cross sections with the statistical Hauser-Feshbach model. The present list covers 73 new isotopes and is available on the KADoNiS homepage at *nuclear-astrophysics.fzk.de/kadonis/new_iso_list.txt*. However, only a few of these radioactive isotopes can be measured with present techniques (e.g. ^{60}Fe), the largest fraction has to wait for future radioactive ion beam (RIB) facilities.

Once all available datasets are implemented in the s-process database, a re-calculation of semi-empirical estimates based on the latest experimental results of neighboring nuclides will be performed and KADoNiS v1.0 can be published.

The second big part of the KADoNiS project is an experimental p-process database. A test version is already online at *nuclear-astrophysics.fzk.de/kadonis/pprocess/*. This p-process database aims to be a collection of the available experimental data close to or within the Gamow window of the p process (T_9= 2-3 GK, corresponding to $E_p\approx$ 1.5-6 MeV or $E_\alpha\approx$ 4-14 MeV). However, the largest fraction of p-process reactions concerns

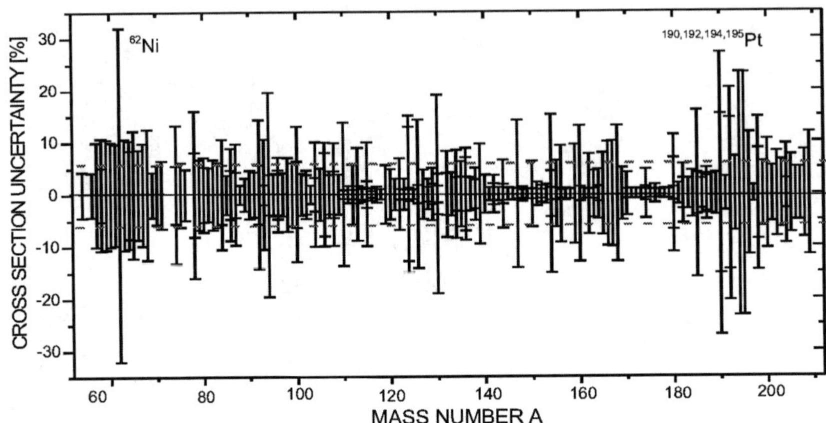

FIGURE 2. Present experimental uncertainties of stellar (n,γ) cross sections beyond iron ($Z\geq26$) in KADoNIS v0.2 (January 2007). The average uncertainty in this region is $\pm6\%$ (without semi-empirical estimates). ^{62}Ni and some light Pt isotopes exhibit much larger uncertainties.

short-lived radioactive nuclei, which have to be inferred from theoretical work, e.g. the Hauser-Feshbach statistical model [9, 5].

Some experimental information is available for charged-particle reactions, but the largest amount of data concerns (n,γ) data, which is connected via detailed balance with the respective (γ,n) reactions needed for the "γ process" mechanism of the p process. Most of this neutron capture data was measured with the activation technique at one single energy (kT=25 keV), and thus has to be extrapolated to the respective γ-process energies (kT=170-260 keV) with the help of energy-dependencies from the Hauser-Feshbach theory.

Of utmost importance in this respect are photodisintegration and neutron-induced particle-exchange reactions. (γ,n) and (n,γ) reactions influence the reaction flow strongly in the whole p-process mass region between A=70-208. On the other hand (γ,p) and (n,p) are only important for the production of light p-process isotopes up to the N=82 shell (^{144}Sm), whereas (γ,α) and (n,α) exhibit a very strong influence only for heavy p isotopes [10, 11].

OTHER JINA-CARINA COLLABORATION EFFORTS

Light charged-particle reactions

The JINA-CARINA collaboration has set up a priority list for measurements and evaluations of light charged-particle reactions. Several reactions were grouped together according to their priority, and the evaluation and experimental status was discussed. The 20 reactions classified with "high priority" are listed in Table 1. Among them are the 3α and ^{12}C(α,γ) reactions, as well as the neutron source reactions for the s process, ^{13}C(α,n) and ^{22}Ne(α,n).

TABLE 1. "High priority" list for measurements and evaluations of light charged-particle reactions published by the JINA-CARINA Collaboration.

Reaction			
(p,γ)	(p,α)	(α,x)	(α,n)
$^{7}Be(p,\gamma)^{8}B$	$^{17}O(p,\alpha)^{14}N$	$D(\alpha,\gamma)^{6}Li$	$\alpha(\alpha,n)^{7}Be$
$^{14}N(p,\gamma)^{15}O$	$^{18}F(p,\alpha)^{15}O$	3α	$^{15}O(\alpha,n)^{18}Ne$
$^{21}Na(p,\gamma)^{22}Mg$		$^{12}C(\alpha,\gamma)^{16}O$	$^{17}O(\alpha,n)^{20}Ne$
$^{24}Mg(p,\gamma)^{25}Al$		$^{14}O(\alpha,p)^{17}F$	$^{13}C(\alpha,n)^{16}O$
$^{25}Mg(p,\gamma)^{26}Al$		$^{18}Ne(\alpha,p)^{21}Na$	$^{22}Ne(\alpha,n)^{25}Mg$
$^{25}Al(p,\gamma)^{26}Si$			$^{25}Mg(\alpha,n)^{28}Si$
$^{26}Al(p,\gamma)^{27}Si$			

Unified reaction rate library

The main problem of nucleosynthesis modeling in astrophysics is the diversity of available reaction rate libraries (REACLIBs) depending on the respective requirements. More or less every group uses its "own" library, which makes inter-comparisons of different modeling results impossible. The most widely used libraries are based on SMOKER or NON-SMOKER predictions [5] with some experimental information of light isotopes (*http://download.nucastro.org/astro/reaclib*). Additionally the reaction libraries include (partially) experimental results or fitting parameters from NACRE [12] (*http://pntpm.ulb.ac.be/Nacre/ nacre_d.htm*) and its extension for Big Bang nucleosynthesis [13] (*http://pntpm.ulb.ac.be/bigbang*), Caughlan and Fowler [14], Iliadis et al. [15], or very recently KADoNiS [6] (*http://nuclear-astrophysics.fzk.de/kadonis*).

The aim of the collaboration is to merge all of these different libraries and include them in the unified JINA Reaclib Database (*http://www.nscl.msu.edu/~nero/db/*). In this context the latest available experimental results (which also have been evaluated) will be fitted and provided in the most recent REACLIB version, but also previous libraries can be downloaded from this website.

A reaction rate library from Basel university (*http://download.nucastro.org/astro /reaclib*) has been recently updated with cross sections from KADoNiS v0.2 [11]. In a first approach the existing fitting parameters (based on NON-SMOKER cross sections [5]) for the (n,γ) and (γ,n) rates were only normalized to the KADoNiS cross sections at kT=30 keV. These 355 isotopes have to be checked again for cross sections measured with the time-of-flight technique, which exhibit a proper energy-dependence. In these cases a careful refitting is necessary, and the new fit parameters will be added into the respective updated REACLIB.

Workflow tools

The central access point for submission and evaluation of new experimental data is provided via *http://nucastrodata.org/infrastructure.html*. By launching the interface, the display for "Computational Infrastructure for Nuclear Astrophysics" is opened, where users can browse the status of evaluations. On the same website "Scientific contributors" can send their data to evaluators/ referees/ editors, which will comment on this. Flowcharts will show the progress to ensure the transparency of the whole evaluation process.

SUMMARY

Presently a lot of effort is ongoing concerning a transparent evaluation progress of reaction rates and the allocation of a unified reaction library for nucleosynthesis modeling. Central web access points provided by the JINA-CARINA collaboration are *http://nucastrodata.org/* and *http://www.nscl.msu.edu/~nero/db/*. A priority list for measurements and evaluations of light charged-particle reactions has been set-up by the collaboration.

ACKNOWLEDGMENTS

I.D. was supported by the Swiss National Science Foundation Grants 2024-067428.01 and 2000-105328. The Joint Institute for Nuclear Astrophysics (JINA) is supported by the National Science Foundation through the Physics Frontier Center Program. The CARINA network (Challenges and Advanced Research In Nuclear Astrophysics) is supported by the EURONS through the 6th EC Research Framework Program (FP6).

REFERENCES

1. B. Allen, J. Gibbons, and R. Macklin, *Adv. Nucl. Phys.* **4**, 205 (1971).
2. Z. Bao, and Käppeler, *At. Data Nucl. Data Tables* **36**, 411 (1987).
3. H. Beer, F. Voss, and R. Winters, *Astrophys. J. Suppl.* **80**, 403 (1992).
4. Z. Bao, H. Beer, F. Käppeler, F. Voss, K. Wisshak, and T. Rauscher, *At. Data Nucl. Data Tables* **76**, 70 (2000).
5. T. Rauscher, and F.-K. Thielemann, *At. Data Nucl. Data Tables* **75**, 1 (2000).
6. I. Dillmann, M. Heil, F. Käppeler, R. Plag, T. Rauscher, and F.-K. Thielemann, *Proc. 12th International Symposium on Capture Gamma-Ray Spectroscopy and Related Topics (CGS 12), Notre Dame, IN/USA, AIP Conference Proceedings* **819**, 123 (2006).
7. H. Nassar, M. Paul, I. Ahmad, D. Berkovits, M. Bettan, P. Collon, S. Dababneh, S. Ghelberg, J. Greene, A. Heger, M. Heil, D. Henderson, C. Jiang, F. Käppeler, H. Koivisto, S. OŠBrien, R. Pardo, N. Patronis, T. Pennington, R. Plag, K. Rehm, R. Reifarth, R. Scott, S. Sinha, X. Tang, and R. Vondrasek, *Phys. Rev. Lett.* **94**, 092504 (2005).
8. A. Tomyo, Y. Temma, M. Segawa, Y. Nagai, H. Makii, T. Shima, T. Ohsaki, and M. Igashira, *Ap. J.* **623**, L153 (2005).
9. W. Hauser, and H. Feshbach, *Phys. Rev.* **87**, 366 (1952).
10. W. Rapp, J. Görres, M. Wiescher, H. Schatz, and F. Käppeler, *Ap. J.* **653**, 474 (2006).
11. I. Dillmann, *Ph.D. thesis, University of Basel/ Switzerland* (2006).
12. C. Angulo, M. Arnould, M. Rayet, P. Descouvemont, D. Baye, C. Leclercq-Willain, A. Coc, S. Barhoumi, P. Aguer, C. Rolfs, R. Kunz, J. W. Hammer, A. Mayer, T. Paradellis, S. Kossionides, C. Chronidou, K. Spyrou, S. Degl'Innocenti, G. Fiorentini, B. Ricci, S. Zavatarelli, C. Providencia, H. Wolters, J. Soares, C. Grama, J. Rahighi, A. Shotter, and M. Lamehi Rachti, *Nucl. Phys. A* **656**, 3 (1999).

13. P. Descouvemont, A. Adahchoura, C. Angulo, A. Coc, and E. Vangioni-Flam, *At. Data Nucl. Data Tables* **88**, 203 (2005).
14. G. Caughlan, and W. Fowler, *At. Data Nucl. Data Tables* **40**, 283 (1988).
15. C. Iliadis, J. M. D'Auria, S. Starrfield, W. Thompson, and M. Wiescher, *Ap. J. Suppl.* **134**, 151 (2001).

New Global Calculation of Nuclear Masses and Fission Barriers for Astrophysical Applications

P. Möller*, A. J. Sierk*, R. Bengtsson†, T. Ichikawa** and A. Iwamoto‡

*Theoretical Division, Los Alamos National Laboratory,
Los Alamos, New Mexico 87545, USA
†Department of Mathematical Physics, Lund Institute of Technology,
Box 118, SE - 22100 Lund, Sweden
**RIKEN Nishina Center, RIKEN, Wako, Saitama, 351-0198, Japan
‡Japan Atomic Energy Agency (JAEA), Tokai-mura, Naka-gun,
Ibaraki, 319-1195, Japan

Abstract. The FRDM(1992) mass model [1] has an accuracy of 0.669 MeV in the region where its parameters were determined. For the 529 masses that have been measured since, its accuracy is 0.46 MeV, which is encouraging for applications far from stability in astrophysics. We are developing an improved mass model, the FRDM(2008). The improvements in the calculations with respect to the FRDM(1992) are in two main areas. (1) The macroscopic model parameters are better optimized. By simulation (adjusting to a limited set of now known nuclei) we can show that this actually makes the results *more* reliable in new regions of nuclei. (2) The ground-state deformation parameters are more accurately calculated. We minimize the energy in a four-dimensional deformation space ($\varepsilon_2, \varepsilon_3, \varepsilon_4, \varepsilon_6$) using a grid interval of 0.01 in all 4 deformation variables. The (non-finalized) FRDM (2008-a) has an accuracy of 0.596 MeV with respect to the 2003 Audi mass evaluation before triaxial shape degrees of freedom are included (in progress). When triaxiality effects are incorporated preliminary results indicate that the model accuracy will improve further, to about 0.586 MeV.

We also discuss very large-scale fission-barrier calculations in the related FRLDM (2002) model, which has been shown to reproduce very satisfactorily known fission properties, for example barrier heights from ^{70}Se to the heaviest elements, multiple fission modes in the Ra region, asymmetry of mass division in fission and the triple-humped structure found in light actinides. In the superheavy region we find barriers consistent with the observed half-lives. We have completed production calculations and obtain barrier heights for 5254 nuclei heavier than $A = 170$ for all nuclei between the proton and neutron drip lines. The energy is calculated for 5009325 different shapes for each nucleus and the optimum barrier between ground state and separated fragments is determined by use of an "immersion" technique.

Keywords: Nuclear Masses, Fission Barriers
PACS: 21.10,24.75.+i

INTRODUCTION

We discuss a new global calculation of ground-state (gs) and fission-barrier nuclear-structure data. The gs calculation includes nuclear masses, deformation parameters, and spins of odd-even nuclei. We also give some details of a very large-scale calculation of fission barriers for 5254 heavy nuclei. Some aspects of the calculations were reported at the *International Conference on Nuclear Data and Technology, April 22–27, 2007, Nice* [2]. We present here additional results that are new since that meeting. We use the macroscopic-microscopic method; full details of the model are found in Refs. [1, 3, 4].

CP1016, *Origin of Matter and Evolution of Galaxies,*
edited by T. Suda, T. Nozawa, A. Ohnishi, K. Kato, M. Y. Fujimoto, T. Kajino, and S. Kubono

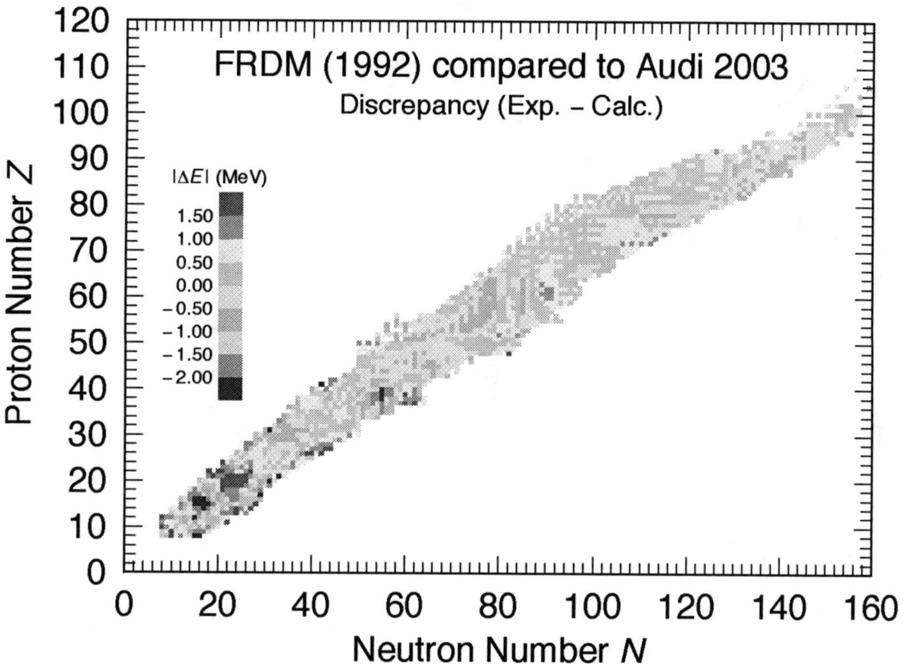

FIGURE 1. Difference between the Audi 2003 mass evaluations and the FRDM (1992) mass table. No systematic divergence far from stability is observed.

MASS AND FISSION-BARRIER CALCULATIONS

A goal for a theory of nuclear masses is that it be able to accurately predict masses of nuclei for which no measured values are available. Data for unknown nuclei are needed in many simulations; one example is simulations of the r-process. Our FRDM (1992) which was finalized in 1992 and published in 1995 [1], was adjusted to a 1989 data base of nuclear masses [5]. We have since compared it to nuclear masses measured after 1989. There are 529 such masses in the Audi 2003 mass evaluation [6]. The model accuracy for these *predicted* nuclear masses is 0.46 MeV; much better than in the region where the model parameters were adjusted, see discussion in [2]. In Fig. 1 we compare calculated masses to the *entire* Audi 2003 evaluation, that is the 529 nuclei measured since 1989 are also included. They are mainly located along the upper and lower edges of the plotted region, that is towards the neutron and proton drip lines. In Fig. 2 we show a similar plot for the HFB-8 mass model [7]. This model exhibits larger staggering between odd and even nuclei (a problem that may have been improved in later model versions) and, in the heavy regions, a more systematic variation in the error between the proton and neutron drip lines.

We have now improved the FRDM (1992) mass model. Successive improvements are

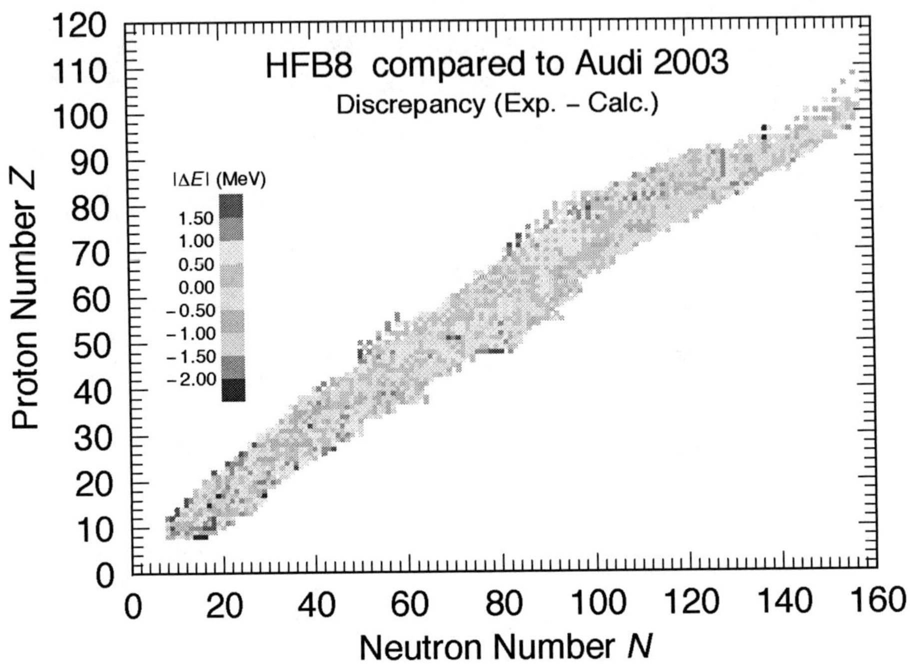

FIGURE 2. Difference between the Audi 2003 mass evaluations and the HFB-8 mass table.

listed in table 1. The first three lines show the published FRDM (1992) compared to the data set to which it was adjusted (Audi 1989, labeled "1"), the totality of masses in the most recent mass evaluation (Audi 2003, labeled "2") and new masses in A2003 relative to A1989 (labeled "3"). Because of a 100 000 fold increase in computer power since the FRDM (1992) calculation was carried out we can now considerably refine the calculation. On line 4 (92-a) we show the result of a better optimization of the constants to the 1989 data set. Line 5 compares this better optimized model to the 529 new masses. It is interesting to note that the better optimized model has better predictive power (0.42 MeV versus 0.46 MeV)! Line 6 compares to the entire 2003 data set. The next three lines (92-b) differ only in that fission barriers are not included in the adjustment. We have earlier [8, 3, 4] observed that the FRDM should not be applied to fission barriers. The next line (06-a) shows the effect of adjusting the model parameters to the 2003 data set. The model is extraordinarily stable, its accuracy only changing by 0.0017 MeV.

In our 1992 mass calculation [1] the determination of the ground-state deformation was carried out by interpolation in a coarsely spaced grid in the two deformation coordinates ε_2 and ε_4; ε_3 and ε_6 were studied only approximately (see [1]). Here we considerably refine this calculation, by studying the stability of *all* minima found in the limited-space potential-energy-surface calculations in a full 4D deformation space in the coordinates ε_2, ε_3, ε_4 and ε_6. We use a grid with a grid interval 0.01 in all 4

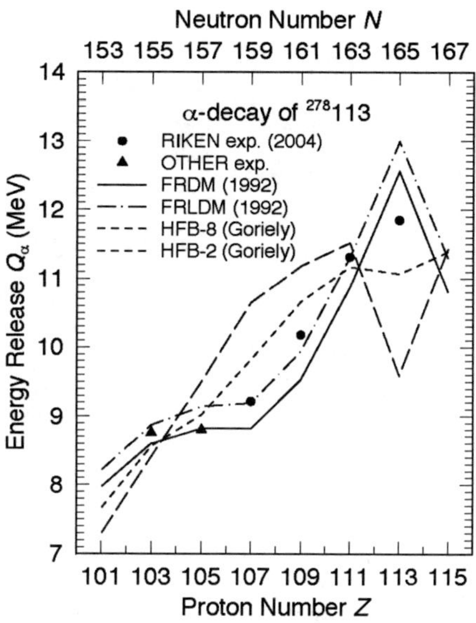

FIGURE 3. Calculated and observed Q_α values in the α-decay of element $^{278}113$. See the discussion in the text.

deformation parameters. We start by calculating a 4D "cube" around a minimum found in the previous mass calculation in 2D ε_2-ε_4 space [1]. Such a cube consists of 81 grid points, 80 of them on the surface of the cube. The lowest energy will be for a point on the surface of this cube, unless the initial point determined from the limited 2D calculation

TABLE 1. FRDM (1992) and successive enhancements, compared to different data sets. The first column indicates a model designation, the second which data set the model was Adjusted/Compared to, and the last two columns the mean deviation and the model accuracy. In column 2 "1", "2", and "3" stand for the Audi 1989 mass evaluation [5], the Audi 2003 mass evaluation [6], and masses that are in the 2003 evaluation but not in the 1989 evaluation ("new" masses), respectively. The model constants are given in the middle section. The top line gives the original model constants [1].

Model	A/C	a_1	a_2	J	Q	a_0	c_a	C	γ	μ_{th}	$\sigma_{th;\mu=0}$
(92)	1/1	16.247	22.92	32.73	29.21	0.00	0.436	60	0.831	0.0156	0.6688
(92)	1/3									0.1755	0.4617
(92)	1/2									0.0607	0.6314
(92)-a	1/1	16.245	23.02	32.22	30.73	−2.24	0.465	104	0.927	0.0000	0.6614
(92)-a	1/3									0.0174	0.4208
(92)-a	1/2									0.0114	0.6180
(92)-b	1/1	16.286	23.37	32.34	30.51	−5.21	0.468	179	1.027	0.0000	0.6591
(92)-b	1/3									0.0031	0.4174
(92)-b	1/2									0.0076	0.6157
(06)-a	2/2	16.274	23.27	32.19	30.64	−5.00	0.450	169	1.000	0.0000	0.6140
(07)-a	2/2	16.231	22.96	32.11	30.83	−3.33	0.460	119	0.907	0.0000	0.5964

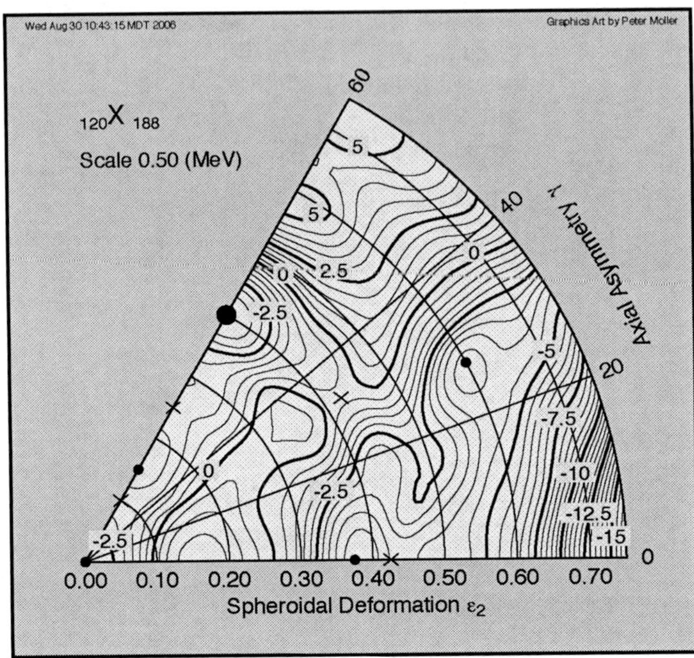

FIGURE 4. Calculated potential-energy surface for $^{308}120$. The filled dots indicate local minima, the x symbols significant saddles. The large filled dot designates the ground state. Although this minimum is not the lowest minimum it has the highest saddle with respect to fission and is therefore considered to be the ground-state minimum.

accidentally turns out to be the location of the local minimum. If not, we construct a new 4D cube around the point corresponding to the lowest energy on the surface of the initial cube, taking care not to recalculate energies that are already calculated. We continue in this fashion until the lowest-energy point is *not* on the surface of the last cube investigated. It is then the interior point in this last cube that is the minimum in the full 4D space. We investigate all the minima found in the 2D space in this fashion. However, we found that sometimes there exists one minimum for $\varepsilon_3 = 0$ and another at $\varepsilon_3 \approx 0.1$. separated by a low ridge. We therefore repeat the search for minima with the above starting points, except $\varepsilon_3 = 0.1$ in all the starting points. The lowest of the 4D minima is the optimum choice for the ground state, with the exception that for heavy nuclei the stability with respect to fission needs to be investigated as discussed below.

This refined calculation improves the accuracy of our original study to 0.596 MeV, (07)-a in Table 1. We can anticipate that including axial asymmetry effects [9] will further improve the accuracy to about 0.586 MeV. In our work we include essential refinements that in practice are not possible to consider in self-consistent Hartree-Fock calculations. To illustrate one issue, we show in Fig. 3 a comparison of calculated and observed Q_α values for the first $^{278}113$ decay chain that was observed at RIKEN. The cusp at $N = 165$ in the FRDM and FRLDM models occurs because of a change in gs deformation from deformed to spherical. In the HFB-8 calculation the cusp occurs because of a deformation change from $\beta_2 = 0.21$ to $\beta_2 = 0.42$ However, although the more deformed minimum is the lowest found in the HFB-8 calculation, our experience tells us that it will have a very low barrier with respect to fission. Therefore the higher-energy, less deformed minimum should be tabulated as the "ground state". We illustrate further this issue in Fig. 4 in a potential-energy plot for $^{308}120$. The deepest axially symmetric minimum is at $\varepsilon_2 = 0.375$ with $E \approx -4.7$ MeV. However, the barrier is only about 1 MeV high; the saddle is nearby at $\varepsilon_2 = 0.425$. Instead we need to designate the minimum with the highest barrier with respect to fission as the ground-state (an even more sophisticated approach would be to calculate the half-life with respect to all decay modes for all minima, but we do not take this step here). This model leads us to assign the minimum at $\varepsilon_2 = 0.40$ and $\gamma = 60$ and $E \approx -4.1$ MeV as the ground state. Equivalently, the ground state is oblate with $\varepsilon_2 = -0.40$. The saddle is at $\varepsilon_2 = 0.425$ and $\gamma = 32.5$ and $E \approx -0.25$ MeV. The barrier with respect to fission is about 3.9 MeV. We have for all nuclei in our mass studies (and fission-barrier studies) used a water-immersion technique to assign the ground-state to the correct minimum. So far HFB mass calculations do not use such techniques and must therefore be considered unreliable for heavy nuclei.

We have calculated fission potential-energy surfaces for 5254 nuclei from $A = 170$ to $A = 330$. For each nucleus the energy is calculated for 5009325 different shapes in a 5-dimensional deformation space. From these calculations we can determine all minima, saddle points between minima, valleys leading to different scission configurations (that is, different fission *modes*), ridge heights between valleys, and the distinctly different saddle points that provide the entry doorways to the different fission modes. Triaxial shapes were studied in the vicinity of the first barrier peak in a 3D calculation. We are in the process of tabulating these results. Our calculations agree well with various observed fission properties in the heavy- and superheavy-element regions. For example, we obtain 4 to 6 MeV barriers for the new elements between $Z = 107$ and $Z = 113$ that were observed in cold-fusion reactions. Barriers of this magnitude are required so that the fission half-lives are sufficiently long to be compatible with the observed half-lives, which are in the range μs to ms. This work was carried out under the auspices of the National Nuclear Security Administration of the U.S. Department of Energy at Los Alamos National Laboratory under Contract No. DE-AC52-06NA25396.

REFERENCES

1. P. Möller, J. R. Nix, W. D. Myers, and W. J. Swiatecki, Atomic Data Nucl. Data Tables **59** (1995) 185.
2. P. Möller, R. Bengtsson, K.-L. Kratz, and H. Sagawa, Proc. International Conference on Nuclear Data and Technology, April 22–27, 2007, Nice, France, to be published, and `http://t16web.lanl.gov/Moller/publications/nd2007.html`.
3. P. Möller, D. G. Madland, A. J. Sierk, and A. Iwamoto, Nature **409** (2001) 785.
4. P. Möller, A. J. Sierk, and A. Iwamoto, Phys. Rev. Lett. **92** (2004) 072501.
5. G. Audi, Midstream atomic mass evaluation, private communication (1989), with four revisions.
6. G. Audi, A. H. Wapstra, and C. Thibault, Nucl. Phys. **A729** (2003) 337.
7. M. Samyn, S. Goriely, M. Bender, J.M. Pearson, Phys Rev. C **70** (2004) 044309.
8. P. Möller and A. Iwamoto, Phys. Rev. C **61** (2000) 047602.
9. P. Möller, R. Bengtsson, P. Olivius, B. G. Carlsson, and T. Ichikawa, Phys. Rev. Lett. **97** (2006) 162502.

Development of Experimental Nuclear Reaction Data File for Astrophysics

N .Otuka [1*,†], M. Aikawa**, Y. Hirabayashi[‡,*], A. Ohnishi[§,*] and K. Katō[§,*]

*Nuclear Reaction Data Centre (JCPRG), Faculty of Science, Hokkaido University, Sapporo 060-0810, Japan
†Nuclear Data Center, Japan Atomic Energy Agency, Tokai 319-1195, Japan
**OpenCourseWare, Hokkaido University, Sapporo 060-0811, Japan
‡Information Initiative Centre, Hokkaido University, Sapporo 060-0811, Japan
§Department of Physics, Faculty of Science, Hokkaido University, Sapporo 060-0810, Japan

Abstract. Experimental nuclear reaction data compilation activity in Japan for the NRDF database and international collaboration of experimental nuclear reaction data exchange by the EXFOR format are presented. Recent development of an experimental nuclear reaction data file (NRDF/A) for astrophysics, which is based on the EXFOR compilation, is also introduced. All services are available at http://www.jcprg.org/.

Keywords: NRDF, EXFOR, NRDF/A, nuclear data, cross section, reaction rate
PACS: 25.00.00, 26.00.00, 29.87.+g, 95.30.-k

INTRODUCTION

The nuclear reaction rate is a key input parameter both in astronuclear physics and in reactor physics. Nucleosynthesis in the early universe and in stars has been studied in astronuclear physics using the nuclear network calculation, and it is similar to the burn-up calculation in reactor physics. The nuclear reaction rate is obtained from the excitation function of the cross section value, therefore cross section values in a wide energy range must be known. It is still difficult for us to prepare the reaction rates and cross sections *ab initio* from nuclear theory precisely. On the contrary, there are various data libraries of reaction rates and cross sections. For example, Maxwellian-averaged cross sections (MACS) and reaction rates at kT = 1 keV to 1 MeV were calculated from the Japanese Evaluated Nuclear Evaluated Library Ver. 3.3 (JENDL-3.3) [1], which is evaluated on the basis of experimental cross sections. The MACS is tabulated for the use of astrophysics [2]. Another example is the compilation of MACS by Z. Y. Bao et al. [3] mainly based upon experimental MACS near kT = 30 keV as well as theoretical calculations by NON-SMOKER [4]. In the current situation, therefore, experimental cross sections play a crucial role in the MACS and reaction rate evaluation for the study of nucleosynthesis. Here we report recent activity of experimental nuclear reaction data compilation in Japan and the world. Then we also show a recent attempt of data compilation for astrophysical uses.

---

[1] Present address: Nuclear Data Section, International Atomic Energy Agency (IAEA), Wagramerstraße 5, P.O.Box 100, A-1400 Wien, Austria

CP1016, *Origin of Matter and Evolution of Galaxies*,
edited by T. Suda, T. Nozawa, A. Ohnishi, K. Kato, M. Y. Fujimoto, T. Kajino, and S. Kubono

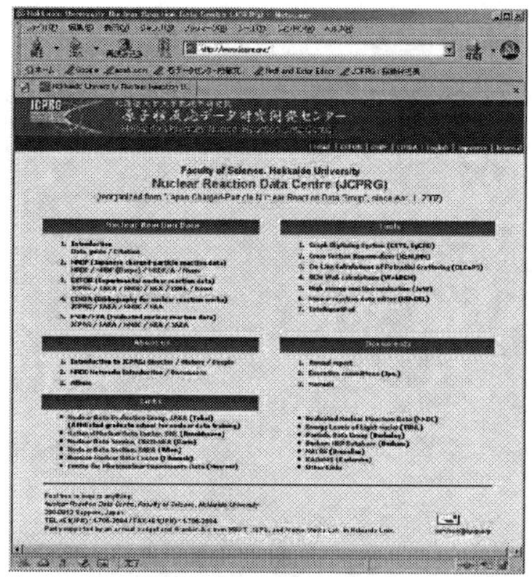

FIGURE 1. JCPRG home page (http://www.jcprg.org/)

HOKKAIDO UNIVERSITY NUCLEAR REACTION DATA CENTRE (JCPRG)

JCPRG is a nuclear data centre to compile experimental data for charged-particle (proton, light- and heavy-ions, charged-meson) induced nuclear reactions and was established in 2007 as the successor of Japan Charged-Particle Nuclear Reaction Data Group. Data has been compiled in "Nuclear Reaction Data File" (NRDF) format, and can be retrieved via the internet. As a member center of the Network of Nuclear Reaction Data Centres (NRDC) coordinated by International Atomic Energy Agency (IAEA), a part of our data files are also translated to "EXchange FORmat" (EXFOR) which is designed to allow exchange of nuclear reaction data between the centers in the world. In addition to experimental reaction data compilation, we have developed various software products such as web-based data retrieval engine and graph point reader (digitizer) [2]. We also periodically scan major journals in nuclear physics published in one year to grasp the latest experimental developments in nuclear physics. The JCPRG home page (http://www.jcprg.org/) provides our data services (database, software etc.) as shown in **Fig. 1**.

---

[2] Our digitizers (GSYS and SyGRD) are free and available at http://www.jcprg.org/gsys/gsys-e.html.

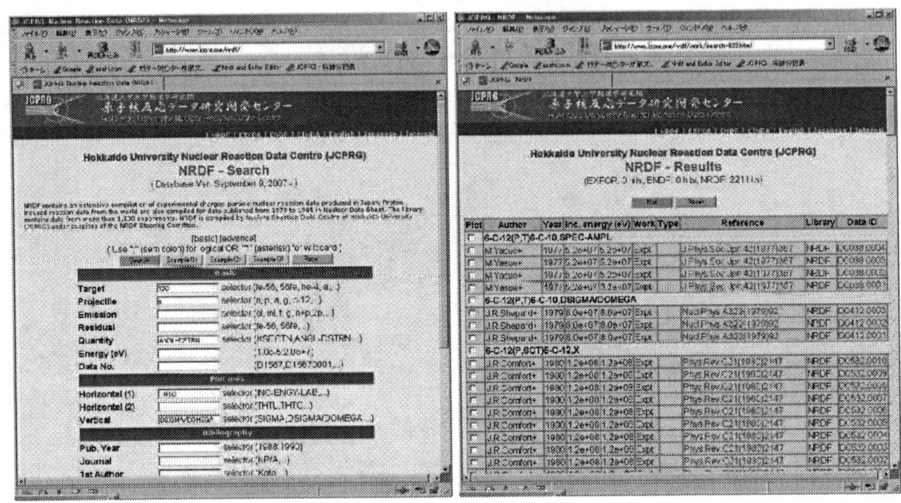

FIGURE 2. An example of NRDF search (http://www.jcprg.org/nrdf/)

GENERAL PURPOSE EXPERIMENTAL DATA COMPILATION

Nuclear Reaction Data File — NRDF

Nuclear Reaction Data File (NRDF) is a format and database system for charged-particle nuclear reaction data. When we started our activity in 1970s, there was no definite way to accumulate charged-particle nuclear reaction data in the world because of great diversity in incident particle (proton, deuteron, α, ^{12}C etc.), reaction type (elastic, inelastic, break-up, spallation etc.), physical quantity (cross section, differential cross section, thick target yield, fission yield etc.) and parameter (incident energy, angle, emission energy, excitation energy etc.). This situation is different in neutron induced reaction data (especially in application), where the spectrum is much narrower than that of charged-particle nuclear reaction data. Through intense discussions in the Japanese Study Group (SG), the data format and system have been developed, and compilation of charged-particle nuclear reaction data started in 1978 [5]. NRDF has accumulated charged-particle nuclear reaction data from Japanese facilities published in 1978 and later [3]. As of February 2008, data in 1770 published articles are compiled in NRDF. The database managing system and retrieval system were developed on medium scale computers at Computer Center of Hokkaido University at the beginning stage, and they are now set up on a Linux workstation. The query form and an example of search result of NRDF are shown in **Fig. 2**.

---

[3] Proton-induced reaction data from the world listed in "recent reference" of Nuclear Data Sheets from 1979 to 1985 are also compiled.

TABLE 1. Nuclear Reaction Data Centres (NRDC). Italicized web sites provide EXFOR on-line services.

Data Centre	Internet
National Nuclear Data Center (NNDC)	*www.nndc.bnl.gov*
OECD Nuclear Energy Agency Data Bank (NEA-DB)	*www.nea.fr/databank*
IAEA Nuclear Data Section (NDS)	*www-nds.iaea.org*
Russian Nuclear Data Centre (CJD)	*www.ippe.obninsk.ru/podr/cjd*
ATOMKI Charged-Particle Nuclear Reaction Data Group	www.atomki.hu/atomki/CyclAppl
Nuclear Structure and Reaction Data Centre (CAJaD)	
Center for Photonuclear Experimental Data (CDFE)	*cdfe.sinp.msu.ru*
Chinese Nuclear Data Center (CNDC)	
Center of Nuclear Physics Data (CNPD)	cnpd.vniief.ru
JAEA Nuclear Data Center (JAEA-NDC)	wwwndc.jaea.go.jp
Hokkaido University Nuclear Reaction Data Centre (JCPRG)	*www.jcprg.org*
KAERI Nuclear Data Evaluation Laboratory (KAERI-NDEL)	atom.kaeri.re.kr
Ukrainian Nuclear Data Center (UkrNDC)	ukrndc.kinr.kiev.ua

International collaboration on experimental reaction data exchange

In contrast to charged-particle nuclear reaction data, neutron induced reaction data has a longer history of compilation because of the application to reactor design. It is however impossible for one centre to cover all data from the world even if we set up a specific scope (e.g. neutron induced reaction data). In 1960s, the four data centres at Brookhaven, Saclay, Obninsk and Vienna compiled neutron induced reaction data and exchanged their compilation to cover the data from the world. In order to exchange data files efficiently, the four centres agreed to implement a common system of the structure and content of exchanged data in 1969 [6], which is referred to as the Exchange Format (EXFOR). In order to extend the EXFOR format to charged-particle nuclear reaction data, the Karlsruhe Charged Particle Nuclear Data Group (KaChaPaG) prepared a coding sample of charged-particle nuclear reaction data with a format close to the EXFOR format. After several meetings in 1975 and 1976, the extended EXFOR was approved as the exchange format of both neutron and charged-particle induced reaction data. Further, photonuclear reaction data compilation joined to the exchange in 1980. JCPRG developed a translator from the NRDF format to the EXFOR format (NTX) [8], and submitted the first transmission in 1982. Now 13 data centres shown in **Table 1** join the data exchange activity as the member centres of the Network of Nuclear Reaction Data Centres (NRDC) [9].

As of February 2008, more than 17500 works of neutron, charged-particle and photon induced reaction data are accumulated in the format, and they are disseminated by both on-line (web) and off-line (e.g. CD, DVD) services [4] .

[4] Two centres, Charged Particle Nuclear Data Group of Kernforschungszentrum Karlsruhe (KaChaPaG, Germany), and Nuclear Data Group of Institute of Physical and Chemical Research (RIKEN, Japan) have also contributed to the compilation activity in the past.

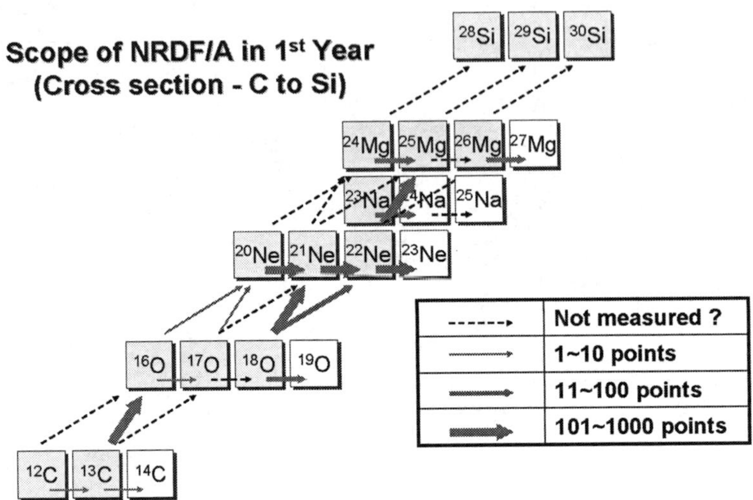

FIGURE 3. Reactions compiled in the NRDF/A files

COMPILATION FOR ASTRONUCLEAR PHYSICS — NRDF/A

Both NRDF and EXFOR are experimental reaction data compilations for general purpose. Their formats are very flexible to cover a wide range of reactions and physical quantities. For a specific usage (e.g. astrophysics, reactor physics), however, data files compiled in a particular format may be convenient. The Computational Format (C4) [7] is a data format for ENDF (Evaluated Nuclear Data Format) evaluators who mainly evaluate neutron-induced reaction data (cross sections, angular distributions, double differential cross sections etc.) in the incident energy range from 10^{-5} to 20 MeV. C4 is designed to tabulate the revelant data derived from the original EXFOR files. This format consists of several columns, and defines content and character length of each column, and consequently it can be read by various computer languages easily.

In astronuclear physics, cross sections of neutron, proton and light ion induced reactions up to keV energies are of interest to obtain reaction rates of these reactions. JCPRG is developing a new format and database (Nuclear Reaction Data File for Astrophysics, NRDF/A) for the reaction rate evaluation for the astrophysics. We chose 31 key reactions of (n, γ), (α, γ) and (α, n) for target nuclides from ^{12}C to ^{26}Mg (**Fig. 3**) for the first test stage of NRDF/A, and compiled cross section data for these reactions by translation from the EXFOR format to the NRDF/A format. An example of the NRDF/A compilation is shown in **Fig.4** for $^{13}C(\alpha, n)^{16}O$.

SUMMARY

We have compiled charged-particle nuclear reaction data measured in Japanese facilities in the NRDF database and translated a part of NRDF files to the EXFOR format for

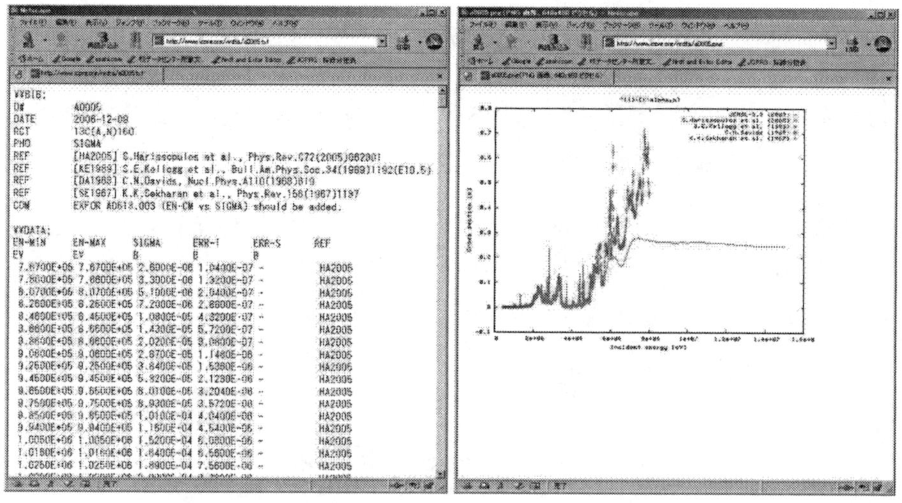

FIGURE 4. Data table and graph in NRDF/A for $^{13}C(\alpha, n)^{16}O$

international exchange of experimental nuclear reaction data. In addition to the experimental nuclear reaction data compilation for general purposes, we started developing an experimental database specialized for astrophysics based on the EXFOR compilation. We are planning to extend this database to a wider range of reactions, including resonance parameters.

ACKNOWLEDGMENTS

This work was partially supported by Grant-in-Aid for Publication of Scientific Research Results (No. 188064) from Japan Society for the Promotion of Science (JSPS).

REFERENCES

1. K. Shibata, T. Kawano, T. Nakagawa, O. Iwamoto, J. Katakura, T. Fukahori, S. Chiba, A. Hasegawa, T. Murata, H. Matsunobu, T. Ohsawa, Y. Nakajima, T. Yoshida, A. Zukeran, M. Kawai, M. Baba, M. Ishikawa, T. Asami, T. Watanabe, Y. Watanabe, M. Igashira, N. Yamamuro, H. Kitazawa, N. Yamano, H. Takano, J. Nucl. Sci. Technol., **39**, 1125 (2002).
2. T. Nakagawa, S. Chiba, T. Hayakawa, T. Kajino, At. Data Nucl Data Table, **91**, 77, 2005.
3. Z. Y. Bao, H. Beer, F. Käppeler, F. Voss, K. Wisshak, T. Rauscher, At. Data Nucl Data Table, **76**, 70, 2000.
4. T. Rauscher, F. K. Thielemann, At. Data Nucl Data Table, **75**, 1, 2000.
5. M. Togasi, H. Tanaka, J. Info. Sci., **4**, 213 (1982).
6. A. Lorenz, International Nuclear Data Committee Report INDC(NDU)-16, 1969.
7. D. E. Cullen, A. Trkov, International Atomic Energy Agency Report IAEA-NDS-80 Rev.1, 2001.
8. M. Chiba, T. Katayama, H. Tanaka, J. Info. Sci., **12**, 153, 1986.
9. V. Pronyaev, O. Schwerer (ed.), International Nuclear Data Committee Report INDC(NDS)-401 Rev.4, 2003.

The r-process element abundance with a realistic fission fragment mass distribution

S. Chiba*, H. Koura*, T. Maruyama*, M. Ohta†, S. Tatsuda†, T. Wada**, T. Tachibana‡, K. Sumiyoshi§, K. Otsuki¶ and T. Kajino||,††

*Advanced Science Research Center, JAEA Tokai, Naka, Ibaraki 319-1195, Japan
†Department of Physics, Konan University, 8-9-1 Okamoto, Kobe 658-8501, Japan
**Department of Pure and Applied Physics, Kansai University, Suita, Osaka 564-8680, Japan
‡Senior High School of Waseda University, Tokyo, Japan
§Numazu College of Technology, Ooka 3600, Numazu 410-8501, Japan
¶GSI Theory Department, Planckstrasse1 64291 Darmstadt Germany
||National Astronomical Observatory, 2-21-1 Osawa, Mitaka, Tokyo 181-8588, Japan
††University of Tokyo, Japan

Abstract. Effect of the β-delayed fission in r-process abundance is investigated with a realistic model for the fission fragment mass distribution (FFMD). The data base for the FFMD is constructed based on the two-center shell model and multi-dimensional Langevin calculation. The β-decay rates including neutron emission and β-delayed fission are also newly calculated with 2nd version of the the gross theory. The differences appeared in the final element abundance calculated with and without fission process, with different β-delayed fission rates are demonstrated.

Keywords: r-process, nucleosynthesis, fission, FFMD, β-delayed fission, abundance
PACS: 26.00.00,26.30.Hj,26.30.-k

INTRODUCTION

Fission plays an important role in the r-process, which determines not only the yields of actinides and above, but those of medium to heavy nuclei are also influenced by the presence of fission[1, 2, 3]. It is especially interesting since the so-called region of universality at $56 \leq Z \leq 75$ elements just falls in the range of fission fragments. In this sense, the fission fragment mass distribution is a clue to understand quantitatively the effects of fission in the r-process.

In this work, we emphasize our efforts to generate a nuclear-data set for an r-process network calculation including fission. As our first attempt, we consider only the β-delayed fission. Necessary data are calculated for the fission fragment mass distribution that reproduces the measured values and β-delayed fission probabilities, which are then used for an r-process network calculation to see how they will influence the final abundance distributions.

CP1016, *Origin of Matter and Evolution of Galaxies*,
edited by T. Suda, T. Nozawa, A. Ohnishi, K. Kato, M. Y. Fujimoto, T. Kajino, and S. Kubono

COMPUTATIONAL METHODS

Reaction Network

We follow the basic streamline of network calculation presented by Terasawa *et al.*[4] where the reaction network of Meyer *et al.*[5] was extended to include a full network covering light nuclei below $Z \leq 10$. Our calculation includes about 5000 nuclei, which starts from a mixture of neutrons, protons and electrons. Evolutaion of temperature, density and initial electron fraction were based on the numerical simulation by a general-relativistic hydrodynamics for a neutrino-driven wind in type II SNe[4], and the result was fitted with a simple exponential model[1] expressed as

$$T_9(t) = T_0 \exp(-t/t_{exp}) + T_{out}; \text{ and } \rho(t) = 3.3 \times 10^5 S/T_9(t)^3. \quad (1)$$

Here, T_9 denotes temperature in unit of 10^9 K and ρ density in the unit of (g/cm^3). We used $T_0 = 8.4$, $T_{out} = 0.7$, $t_{exp} = 7.0$ ms, and varied the entropy S to simulate different astrophysical conditions. The initial electron fraction was set to 0.427[4].We start the network calculation at $T_9 = 9$.

The reaction rates were taken from many sources (such as Orito *et al.*[6], Kajino-Boyd[7] and NACRE[8]). The nuclear masses are taken from the work by Koura *et al.* (KTUY05)[9]. The β-decay rates were calculated by 2nd version of the gross theory (GT2) where the Q-values were taken from KTUY05. The β-delayed neutron emission and β-delayed fission rates were also calculated. The spontaneous fission rates and α-decay rates were also calculated in terms of corresponding Q-values and fission barriers of KTUY05[10].

Figure 1 shows various regions where different modes of fission, namely the β-delayed fission, spontaneous fission and neutron-induced fission, are dominant. The thick line designates a typical r-process path under circumstances we consider in this work. We notice that the r-process path runs into the region where β-delayed fission is dominant. Therefore only β-delayed fission is taken into account here as our first attempt to include fission.

Fission Fragment Mass Distribution

The fission fragment mass distribution (FFMD) is expressed as a superposition of 3 Gaussian functions; one of them corresponds to the symmetric fission, and the rest to the asymmetric ones. What are needed here are the mass asymmetry α of asymmetric fission mode, ratio of symmetric to asymmetric fission modes and width of the mass distribution. These parameters were determined by considering potential energy surface (PES) and results of multi-dimensional Langevin calculation in a semi-phenomenological way[11, 12].

The PES's near the saddle and scission points of the fissioning nuclei are calculated by the two-center shell model[13, 14]. The fragment separation z, fragment deformation δ and mass asymmetry α are selected as collective coordinates to specify nuclear shape. We took an average of mass asymmetry corresponding to the standard 1 and standard 2

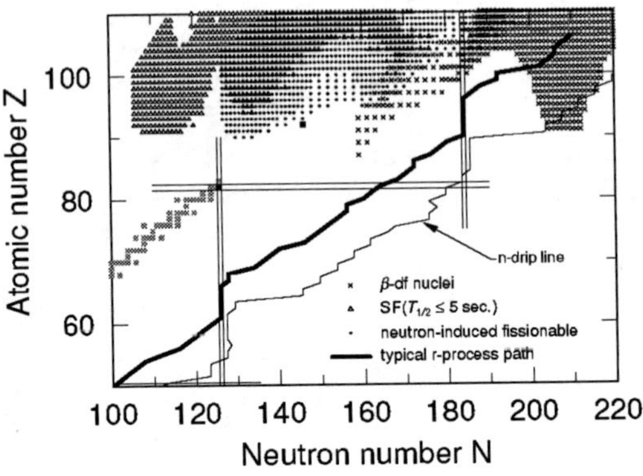

FIGURE 1. Regions predicted by KTUY05 where different modes of fission (β-delayed : x, spontaneous : triangles and neutron-induced : dots) are dominant

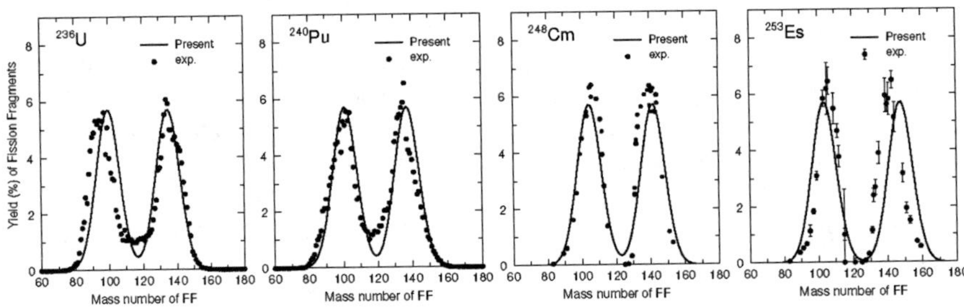

FIGURE 2. FFMD of fissioning ^{236}U, ^{240}Pu, ^{248}Cm and ^{253}Es. The present results (solid lines) are compared with experimental data (circles).

located as valleys between δ=0.2 and 0.3 near the scission point as that of asymmetric fission component[11].

The ratio of symmetric to asymmetric fission is determined firstly for Fm isotopes. We assume that the ratio is maximum for ^{264}Fm and to be 0.71 based on multi-dimensional Langevin calculation[15]. Then, the potential depths for the symmetric and asymmetric fission near scission points are compared for Fm isotopes, and we use the experimental fact that the symmetric mode is prominent for isotopes between ^{257}Fm and ^{266}Fm. The ratios of the symmetric/asymmetric fission are determined phenomenologically to reproduce the gross feature of experimental data. For other elements, the symmetric/asymmetric ratios are determined by comparing the difference of the potential depths for the symmetric and asymmetric modes to those of Fm isotopes. The width of the fragment distribution was determined according to the results of Langevin calculation, and was set to 7 (in mass) for simplicity. The number of pre-fission neutrons is assumed to be 2.

The FFMD was calculated for $98 \leq Z \leq 120$ and $120 \leq N \leq 200$, and data tables containing the mass asymmetry α and symmetric/asymmetric ratio were constructed. The present FFMD are compared with measured values in Fig. 2.

β-decay Rates and β-delayed Fission Probabilities

The calculation of delayed fission probabilities P_f and delayed neutron emission probabilities P_n are performed by using 2nd version of the gross theory (GT2) of nuclear beta-decay[16]. They are expressed as,

$$P_k = \frac{\int_{-Q}^{0} S_\beta(E) f(-E) \frac{\Gamma_k}{\Gamma_n+\Gamma_\gamma+\Gamma_f} dE}{\int_{-Q}^{0} S_\beta(E) f(-E) dE}, \quad k = f, n, \gamma \qquad (2)$$

where subscript k indicates f for fission, n for neutron emission and γ for γ emission. Denominator of the above equation corresponds to the total β-decay rate (except for a normalization constant). The symbol $S_\beta(E)$ denotes the β-strength function, f the integrated Fermi function, Q the β-decay Q-value (we adopted from KTUY05), and λ is the decay constant of the β decay. In the gross theory, the β-strength to all the final nuclear levels are treated as an averaged function based on sum rules of the β-strength functions. This theory takes account of the Fermi, Gamow-Teller, and the first forbidden transitions of rank 0, 1 and 2.

The decay widths Γ_n and Γ_f are expressed as

$$\begin{aligned} \Gamma_n(E) &= \frac{1}{2\pi} \frac{1}{\rho(E)} \frac{2MR^2}{\hbar^2} g \int_0^{E-S_n} \rho^*(E - S_n - \varepsilon) \varepsilon d\varepsilon, \text{ and} \\ \Gamma_f(E) &= \frac{1}{2\pi} \frac{1}{\rho_f(E)} \int_0^{E-B_f} \rho_f^*(E - B_f - \varepsilon) \end{aligned} \qquad (3)$$

where $\rho(E)$ is the level density in the compound nucleus, $\rho^*(E)$ that of the residual nucleus, S_n the neutron separation energy and g the single-particle state density. The symbol $\rho_f(E)$ denotes the level density of the nucleus before fission, $\rho_f^*(E)$ is that on the saddle point and B_f the fission barrier height.

In the calculation of the level density, we follow the method in ref.[17] which is based on Gilbert-Cameron type composite formula, namely the Fermi gas and constant temperature models with correction of shell-effects by Ignatyuk's method. The Γ_γ was taken from the empirical estimation by Malecky *et. al.*[18]

Initially, we took the B_f values calculated by KTUY05 model. However, we noticed that they were higher than experimental values systematically. Therefore, for the moment, we reduced the B_f values calculated with the KTUY05 model by 3 MeV and calculated the fission decay width. The calculated β-delayed fission probabilities are shown in Fig. 3, where those with the present fission barrier model are given in the left panel, while those with the fission barriers of Myers and Swiatecki[19] are given in the right panel. We notice that there is a considerable difference in the results calculated with different fission barrier models.

RESULTS AND DISCUSSION

The calculated r-process abundances (Y) are shown in Fig. 4 where the results with the KTUY05 fission barrier (reduced by 3 MeV) (left) are compared with those with

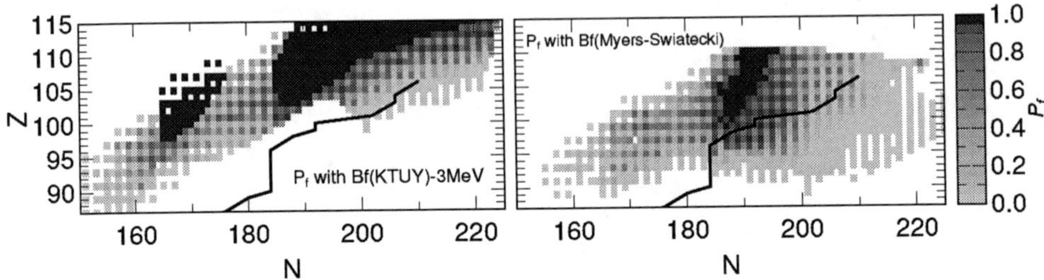

FIGURE 3. β-delayed fission probabilities calculated with fission barriers of KTUY05, but reduced by 3MeV (left) and those with Myers-Swiatecki 1999 (right). Thick solid lines denote a typical r-process path.

Myers-Swiatecki (MS) 1999 fission barrier (right). We notice that the global feature is similar for the both models and the observed solar r-abundance is well reproduced above A of about 150 . We recognize, however, a difference in the distribution of fission fragments, namely, the distribution of FF with the MS barriers is more asymmetric than those with KTUY05 barriers. It gives a significant difference in the sub-peak abundance around 160, where the yield is almost doubled if the MS barrier is used compared to the calculation in which the fission process is ignored (captioned as "without fission" in the figure).

Entropy (S) dependence of ratios of various isotopes and (group of) elements are calculated and shown in Fig. 5. In our setup, the reaction flow does not go enough into the region of actinides at S=200, so the trend is quite different at this entropy. However, when the entropy becomes higher, more and more nuclei at and above actinides are produced, and then destroyed by the β-delayed fission. Therefore the abundance of the heaviest elements tend to be almost stationary (only slowly increasing), whereas more and more FF are accumulated as S gets higher. The S-dependence of $Y(56 \leq Z \leq 75)/Y(90 \leq Z)$ above $300 \leq S$ is understood in such a way. Anyway, these ratios indicate that the abundance ratios are rather robust if the entropy is high enough, which might be an origin of the universality. On the contrary, the trend is wildly different if the entropy is low. There is therefore a possibility that the r-process abundance is sensitive to the astrophysical conditions if the entropy is not high. However, there seems to be a rather large uncertainty in the fission barriers, so the quantitative conclusions are to be awaited. From the nuclear-data point of view, we definetly need neutron-induced fission rates in terms of the KTUY05 mass and fission barriers.

A part of this work was carried out under auspices of Grant-in-Aid for Scientific Research of JSPS under contract number 18560805, 19740151, 18340071 and 18540295, and a grant for cooperative reserach from National Astronomical Observatory of Japan.

REFERENCES

1. K. Otsuki *et al.*, *New Astronomy* **8**, 767(2003) and references therein.
2. I.V. Panov *et al.*, *Nucl. Phys.* **A747**, 633(2005) and references therein.
3. G.M.-Pinedo *et al.*, *Progress in Part. and Nucl. Phys.* **59**, 199(2007) and references therein.

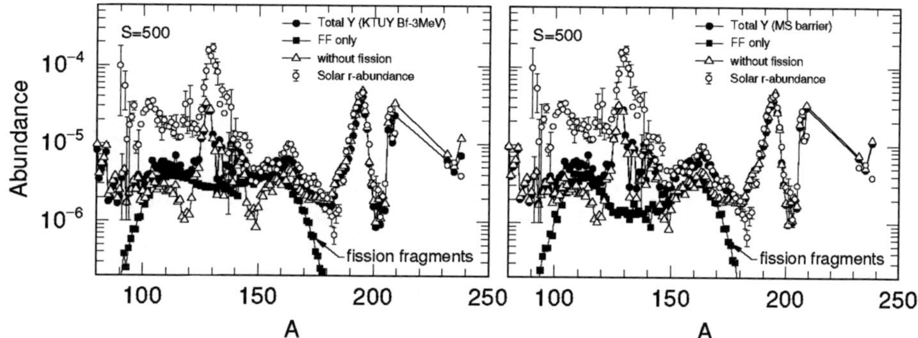

FIGURE 4. r-process mass abundance calculated with various assumptions on β-delayed fission. Left : full calculation including β-delayed fission rates calculated with fission barriers of KTUY05 reduced by 3 MeV (full circles) is compared with a calculation without β-delayed fission (triangles), those for fission fragments generated by the β-delayed fission (squares) and observed solar r-abundance (open circles). Right : same as left but the fission barriers were taken from the work of Myers-Swiatecki 1999.

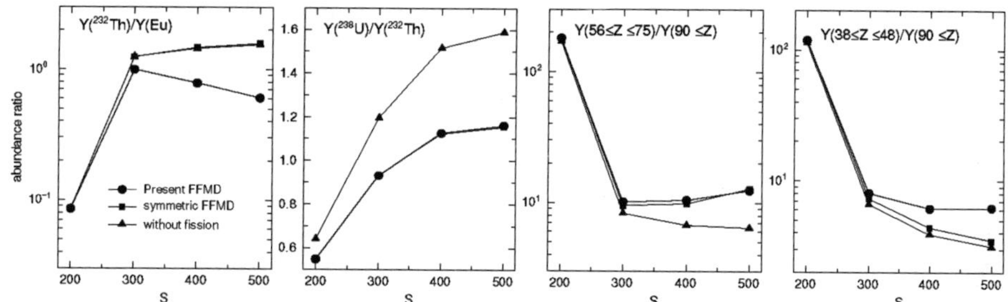

FIGURE 5. Entropy-dependences of ratios of various isotopes and elements in r-process abundance. The β-delayed fission rates calculated with KTUY05 fission barrier (reduced by 3 MeV) are used.

4. M. Terasawa *et al.*, *ApJ* **562**, 470(2001).
5. B.S. Meyer et al., *ApJ* **399**, 656(1992).
6. M. Orito et al., *ApJ* **488**, 515(1997).
7. T. Kajino and R.N. Boyd, *ApJ* **359**, 267(1990).
8. C. Angulo *et al.*, *Nucl. Phys.* **A656**, 3(1999).
9. H. Koura *et al.*, *Prog. Theor. Phys.* **113**, 305(2005).
10. H. Koura, *AIP Conf. Proc.* **704**, 60(2004).
11. M. Ohta *et al.*, to appear in *Proc. Int. Conf. on Nucl. Data for Science and Technology*, Nice, France, April 2007; M. Ohta *et al.*, presented at *Int. Nucl. Phys. Conf.*, Tokyo, Japan, May 2007.
12. S. Tatsuda *et al.*, this symposium.
13. S. Suekane *et al.*, JAERI-memo, 5918(1974), unpublished.
14. A. Iwamoto *et al.*, *Prog. Theor. Phys.*, **55**, 115(1976).
15. T. Asano *et al.*, *J. Nucl. Radiochemical. Sci.* **5**, 1(2004).
16. T. Tachibana, M. Yamada and Y .Yoshida, Prog. Theor. Phys. 84 (1990) 641; T. Tachibana and M. Yamada, in Proc. Int. Conf. on Exotic Nuclei and Atomic Masses, ENAM95 (1995), p.763.
17. T. Kawano, S. Chiba and H. Koura, *J. Nucl. Sci. Tech.* **43**, 1(2006).
18. H. Malecky *et. al.*, *Yad. Phys.* **13**, 240(1971).
19. W.D. Myers and W.J. Swiatecki, *Phys. Rev. C* **60**, 014606(1999).

8. NUCLEAR STRUCTURE and REACTIONS FOR ASTROPHYSICS (1)

Structure and Reaction Experiments for Nuclear Astrophysics

K. E. Rehm

Physics Division
Argonne National Laboratory
9700 South Cass Av.
Argonne, IL, 60439, USA

Abstract. Astrophysics is a rapidly growing field, driven by observations from ground and space-based telescopes. Nuclear physics provides critical input parameters, such as cross sections, masses and half-lives, which are important for the interpretation of the observational data. In this contribution recent accomplishments from structure and reaction experiments relevant to studies in nuclear astrophysics will be discussed.

Keywords: Nuclear Astrophysics
PACS: 26.20.-7, 26.30.-k, 26.50.+x

INTRODUCTION

When Galileo Galilei in 1609 pointed his new telescope towards Jupiter's moons he looked at objects located about 30 light-minutes away. Since then our capability to look towards the edge of the universe has grown enormously. With the biggest telescopes we are now able to go back in time to events, which occurred about 10 billion years ago, an improvement by about 14 orders of magnitude.

The study of nuclear reactions, which occur in the stars and produce the elements we observe in our world today, is a much younger science. It started in 1919, when Ernest Rutherford discovered [1] events, which were later explained as originating from the $^{14}N(\alpha,p)^{17}O$ reaction, i.e. converting nitrogen into oxygen, a process with a cross section of about 0.1 b. About 80 years later experiments performed in the underground laboratory at Gran Sasso [2] have pushed these limits down to the fb region, again an improvement by almost 14 orders of magnitude.

There are more recent examples how astronomy has influenced physics. The search for dark matter, weakly interacting material, which is required to describe the movement of stars in galaxies, has contributed considerably to the development of new detectors. The observation of an accelerated expansion of the universe, which started with studies of type Ia supernovae, brought us the concept of dark energy. All this has led to a renaissance in nuclear astrophysics.

Depending on the temperature and the astrophysical environment nature produces the elements in the universe through a variety of reaction chains which are schematically

CP1016, *Origin of Matter and Evolution of Galaxies,*
edited by T. Suda, T. Nozawa, A. Ohnishi, K. Kato, M. Y. Fujimoto, T. Kajino, and S. Kubono

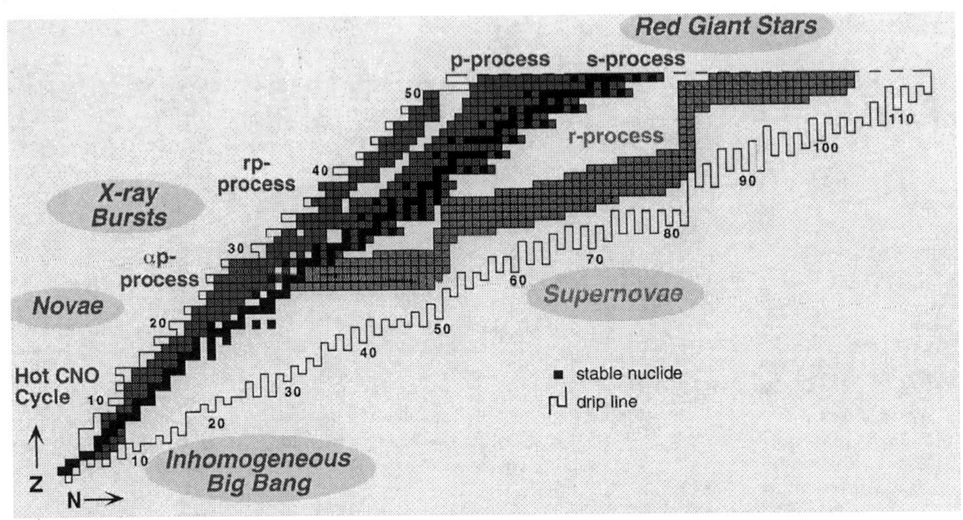

FIGURE 1. Examples of reaction paths in nucleosynthesis (taken from Ref. 3).

shown in Fig. 1. There has been considerable progress during the last few years in our attempts to repeat in terrestrial laboratories what stars do on a grand scale. We have started to measure reactions at energies which are typical of the conditions encountered in stellar environments [2]. The development of radioactive beams during the last 15 years [3] has allowed us to measure processes that occur during explosive scenarios, such as novae or supernovae explosions. This synergy between observations and other fields of physics has made nuclear astrophysics an exciting area of research. In this contribution I will give a brief overview about some of the recent results of nuclear structure and reaction studies, which are of importance to astrophysics.

REACTIONS IN THE PP-CHAIN

This reaction chain, which converts four protons into ^{4}He, is the main energy source of stars similar in weight to the sun, and is therefore of considerable interest. An earlier compilation by experimentalists and theorists has outlined several reactions that need to be known with improved accuracies [4]. The first step in the pp-chain, the reaction $p+p \rightarrow d+e^+ + \nu$, is still outside of today's experimental capabilities, but good theoretical predictions exist [5] for this process. The second step, $d+p \rightarrow {}^3He+\gamma$ has been measured at the LUNA facility down to solar energies [6], as has the subsequent reaction $^3He+{}^3He \rightarrow {}^4He+2p$ [2]. The reactions producing heavier isotopes (e.g. ^{7}Be, ^{8}B) are typically weak branches in the pp-chain, but are important in connection with recent solar neutrino experiments, which are sensitive to the ^{8}B neutrino spectrum. For these reactions several new S-factor measurements have been performed.

The long-standing discrepancy between experiments measuring either the direct gammas from the ^{4}He + ^{3}He $\rightarrow ^7$Be reaction or the ^{7}Be activity has been resolved in a recent LUNA experiment giving a S_{34} value of 0.56±0.017 keVb [7]. This value agrees within the uncertainties with the recent measurement from the University of Washington, which gives S_{34}=0.595±0.018 keVb [8]. The cross section for the subsequent reaction, ^{7}Be(p,γ)^{8}B, which determines the neutrino spectrum of the sun, has been the subject of a long debate. In addition to direct measurements [9], Coulomb dissociation and the Asymptotic Normalization Coefficient techniques have been used to determine the relevant S-factor S_{17}. After earlier disagreements among the different measurements, newer results seem to converge to values that agree within their respective uncertainties: S_{17}=22.1±0.6±0.6 eVb (direct measurement [10]) and S_{17}=20.6±0.8±1.2 eVb (Coulomb dissociation [11]).

The solar neutrino spectrum measured in the underground detectors is distorted due to the MSW effect. In order to determine the effect of neutrino oscillations the shape of the 'standard' ^{8}B neutrino spectrum has to be known with high precision. As pointed out in Ref. [12] earlier measurements show a bias of $\sim \pm 100$ keV in the spectra. There are two new measurements of the ^{8}B neutrino spectrum, one using a beam of ^{8}B particles [13], which are implanted into a Si detector, the other producing ^{8}B via the ^{6}Li(^{3}He,n)^{8}B reaction [14]. These two measurements performed with different techniques are in excellent agreement. With all these new measurements the reactions in the pp-chain are now known to better than $\sim$5%.

THE CNO CYCLE

The conversion of four protons into ^{4}He in the presence of the catalyst ^{12}C was first proposed independently by von Weizäcker [15] and Bethe [16] to explain the energy production in the sun. The higher Coulomb barriers in the CNO material, however, require higher temperatures, which occur only in stars heavier than the sun. Cross sections for the (p,γ) proton captures as well as for the final ^{15}N(p,α)^{12}C reaction have been measured in the past by various groups and it was well known that the cycle time of the CNO cycle is determined by its slowest reaction: ^{14}N(p,γ)^{15}O. It came therefore as a surprise when two recent independent experiments gave an S-Factor about a factor of two smaller than found in the earlier experiments. One experiment, done at the underground LUNA facility extended, for the first time, these measurements into the Gamow window [17]. The other method, performed above ground, used an active shield in order to reduce the background in the gamma detector [18]. As discussed in Ref. [19] the smaller cross sections for the ^{14}N(p,γ)^{15}O reaction increases the CNO cycle time and, thus, the lifetimes of globular clusters, which are now in better agreement with the values obtained by other techniques.

THE HOT CNO CYCLE AND ITS BREAKOUT REACTIONS

The reactions discussed in the two previous sections occur, with few exceptions on such a slow pace that there is always enough time for a β-unstable nucleus, produced in a reaction to decay back to its stable isobar. With increasing temperatures the reaction rates increase, until the β-decays of the unstable nuclei produced in the reactions limit the cycle time and, thus, the energy production. In some cases the β-decay can be bypassed with proton or α-induced reactions on the unstable reaction products, leading to a reaction path on the proton-rich side of the mass valley. For the CNO cycle the β-decay of ^{13}N ($T_{1/2}$= 9.97 m) can be bypassed with a (p,γ) reaction leading to the nucleus ^{14}O with a shorter, 70.6 s half-life. At high enough temperatures this β-decay can again be bypassed via the ^{14}O(α,p) reaction, leading to ^{17}F which is then converted into ^{18}Ne ($T_{1/2}$=1.67 s) via the ^{17}F(p,γ) reaction. This reaction network, called the hot CNO cycle, is the subject of extensive research at many radioactive ion beam facilities worldwide, with sometimes conflicting results. It is impossible to give credit to all experiments performed in this mass region. One example concerns the ^{14}O(α,p)^{17}F reaction. Because of the need of a He gas target and a radioactive ^{14}O beam a direct measurement is very difficult. Its cross section was studied first by the time-inverse ^{17}F(p,α)^{14}O reaction, which gives a lower limit for the astrophysical reaction rate [20, 21]. Direct measurements have been performed by two groups [22, 23], which, however, differ among themselves and also with the results of the time-inversed experiments. This reaction requires further studies with higher intensity beams and better targets.

Another example concerns ^{18}F, where the interest is driven in part by the hope to detect with orbiting γ-ray satellites the 511 keV γ radiation from the decay of ^{18}F ($T_{1/2}$= 1.83 h) produced in a Nova explosion. For this reason the energy dependence of reactions producing and destroying ^{18}F are being investigated worldwide. The beam intensities available again complicate these measurements, especially at very low energies, where the cross sections are in the 1-10 μb/sr region [24, 25, 26].

Another series of reactions, which is also hotly debated, concerns the breakout reactions from the hot CNO cycle into the rapid proton capture (rp) process. The Q-value in the mass 18-20 region are such, that once nuclei heavier than ^{19}Ne have been formed, the isotopes can not be recycled back into CNO material, and the reaction chain continues via (p,γ) reactions towards heavier nuclei. There are three reactions that cross the ^{19}Ne 'border': ^{18}F(p,γ)^{19}Ne, ^{15}O(α,γ)^{19}Ne and ^{18}Ne(α,p)^{21}Na. The breakout via the first reaction is hampered by the competing, strong ^{18}F(p,α)^{15}O reaction and so far only an upper limit for the (p,γ) reaction rate has been obtained [27, 28]. A direct measurement of the second reaction is beyond our present experimental capabilities. A large number of indirect attempts have been made to measure half-lives and α/γ branching ratios for levels above the (α,γ) threshold in ^{19}Ne. At the 2σ level, however, only upper limits for the reaction rates have been obtained so far [29, 30, 31, 32, 33].

The large number of publications as well as the numerous open questions in the hot

CNO cycle indicate that this is an active, contentious area of research.

γ-RAY ASTRONOMY AND THE RP-PROCESS

Long-lived radio-isotopes are of special interest in nuclear astrophysics. If their γ-lines can be detected by orbiting γ-ray telescopes one can determine the amount of material produced at this particular site. These radio-isotopes provide, therefore, important markers for comparisons with theoretical calculations. Besides the two isotopes ^{7}Be and ^{18}F, already mentioned above, there are several other long-lived radio-isotopes (^{22}Na, ^{26}Al, ^{44}Ti, ^{56,57}Co, ..), which have been the topic of intense research in recent years. While the latter four radio-isotopes have all been seen experimentally, an observation of ^{22}Na ($T_{1/2}$= 2.6 y) has not yet been reported. Although predicted to be observable in the Novae Her1991 and Cyg1992 [34], the search for the 1.275 MeV γ-ray from ^{22}Na with the COMPTEL satellite was unsuccessful [35]. Since then several laboratory experiments have been performed, which put the production rate of ^{22}Na in Novae on a much better footing. The reaction ^{21}Na(p,γ)^{22}Mg, important for the production of ^{22}Na, was measured directly for the first time with a radioactive ^{21}Na beam [36]. A high-resolution study of ^{23}Mg, populated in the fusion-evaporation reaction ^{12}C(^{12}C,n)^{23}Mg reaction, gave spins and excitation energies of several states of ^{23}Mg in the vicinity of the ^{22}Na(p,γ) threshold [37]. With these new results the detection limit of the INTEGRAL satellite was determined to be $\sim$1 kpc, which is smaller than the distance of the two Novae ($\sim$3 kpc), mentioned above.

There is a similar, but less complete story for the neighboring radio-isotope ^{26}Al ($T_{1/2}$=7.17x10^{5}y, E_γ=1.809 MeV). The situation for this isotope is more complicated, due to the presence of an excited 0^+ isomer, which decays without emitting a 1.8 MeV γ- ray. The reaction ^{26}Al(p,γ)^{27}Si, destroying the radio-isotope has recently been measured with a radioactive ^{26}Al beam [38], as has the reaction ^{25}Mg(p,γ)^{26}Al, which produces ^{26}Al [39]. The level structure of ^{26}Si, the nucleus that feeds ^{26}Al, is, however, still under intense debate, and experiments, trying to answer some of these questions, are being pursued at many facilities.

Direct reaction rate measurements for nuclei of the rp-process above A$\sim$30 have not yet been performed, because in most cases radioactive beams with good beam qualities and sufficient intensities are not yet available. There are, however, a few measurements employing indirect techniques to determine spins, excitation energies or spectroscopic factors for some of the critical reactions [40, 41, 42]. Most of the reaction rates in the rp-process, however, still rely only on theoretical (e.g. Hauser-Feshbach) estimates.

Another experimental observable that can be used to give us a better understanding of the processes going on during the rp-process are the light-curves from X-ray bursts [43, 44]. Especially the falloff of the light-curve was found to be sensitive to the half-lives of several critical waiting points above ^{56}Ni. Although the β-decay life times of these nuclei (^{64}Ge, ^{68}Se, ^{72}Kr) are quite well known, the hydrogen-rich environment allows for a bypass of the slow β decay via two-proton capture reactions. The rates for these processes depend critically on the Q-values (i.e. the masses) of the nuclei involved. For this reason a series of high-precision mass measurements with Penning traps have recently been performed [45, 46, 47], which give reliable estimates for the effective half-lives of these waiting point nuclei.

THE R-PROCESS

The stellar abundance curves of the elements tells us that about half of the heavy nuclei are produced in a neutron-rich, explosive environment, and type II supernovae are the prime candidates for the site of the so-called rapid neutron capture process (r-process), where starting from certain seed nuclei, through a series of neutron-capture reactions followed by β decays, heavier nuclei up to ^{208}Pb and beyond are being formed. Because all nuclei in the r-process path have to be produced during the time scale of the explosion, the sum of their β-decay half-lives has to be commensurate with the explosion time. For this reason the r-process follows a path through short-lived, exotic, neutron-rich nuclei, which for the most part has not been accessible to experiments in the past. In order to pin down the path of the r-process the half-lives of nuclei, especially along the closed neutron shells at N =50, 82 and 126 as well as their mass values are needed. Furthermore a small neutron capture rate for nuclei in the vicinity of the closed-neutron shells can lead to an increased impedance of the reaction flow. For this reason neutron capture rates of neutron-rich nuclei are needed.

Depending on the temperature and the neutron density during the explosion, the r-process can follow different routes. It is thought to proceed through the closed shell nuclei ^{78}Ni and ^{82}Ge along the N=50 shell before reaching N=82 at ^{124}Mo. It leaves the N=82 shell in the vicinity of ^{130}Cd and reaches the neutron number N=126 around ^{188}Sm. While most of these nuclei, especially the heaviest ones are outside the range of present experiments, considerable progress has been made for nuclei along the N =50 and 82 shell closures. The isotopes ^{125,126}Pd have recently been produced in a first experiment at the RIKEN RIBF facility [48] and at the N=82 shell a series of experiments at ISOLDE has pushed the half-life measurements down to ^{129}Ag [49]. Life-time measurements along the N=50 line are now complete through a recent experiment determining the half-life of the doubly-magic nucleus ^{78}Ni [50].

While mass measurements of most of the nuclei located directly in the r-process path are presently outside our experimental capabilities, extensions of mass measurements to nuclei located between the valley of stability and the r-process path can already provide valuable information. Recent experiments on neutron-rich Zr and Mo [51] and Ba and La [52] isotopes have shown that neutron-rich nuclei such as ^{110}Mo are less bound than predicted by the 2003 Mass Compilation [53]. If this trend continues, the r-process path might be closer to the valley of stability than assumed in the present calculations.

The short half-lives of the r-process nuclei make it impossible to use them as targets in (n,γ) experiments. For this reason these neutron capture experiments have to be performed in inverse kinematics with the short-lived nuclei as a beam. However, since neutron targets do not exist, surrogate reactions have to be used, with (d,p) reactions being the prime candidates. These reactions, performed by bombarding deuterium targets (e.g. CD_2) with heavy ion beams and detecting the outgoing protons in highly pixilated Si detector arrays, provide many challenges, especially with regard to the Q-value resolution that can be achieved. Among the reactions studied so far, the ^{82}Ge(d,p)^{83}Ge reaction was the first to involve a nucleus located in the r-process path [54]. More recently similar studies of the ^{132}Sn(d,p)^{133}Sn [55] and ^{130}Sn(d,p)^{131}Sn [56] reactions have been performed. Although these nuclei are not directly in the r-process path, the experiment nevertheless provide important information about the experimental difficulties encountered in similar experiments at a future RIB facility. Another important question that needs to be answered concerns the relation between (n, γ) capture and (d,p) reactions. Some related experiments have been performed [57, 58] with stable beams in the mass 130 region, but a full understanding of this relation has not yet been achieved.

SUMMARY

There has been a strong revival of nuclear astrophysics in recent years. Triggered by advances in observational astronomy, as well as by the availability of underground accelerators and radioactive ion beams, we have started to unravel some of the processes going on in quiescent and explosive stellar burning in terrestrial laboratories. Low-background environments, high intensity beams and detector arrays with high efficiencies are required to perform these difficult experiments, which have provided us with a first glimpse into this exciting area of nuclear physics. Next generation facilities and detectors are needed to give us a quantitative understanding of these phenomena.

ACKNOWLEDGMENTS

This work was supported by the US Department of Energy, Office of Nuclear Physics, under contract No. DE-AC02-06CH11357 and by the NSF Grant No. PHY -02-16783 (Joint Institute for Nuclear Astrophysics).

REFERENCES

1. E. Rutherford, *Phil. Mag.* **37**, 581 (1919).
2. R. Bonetti et al., *Phys. Rev. Lett.* **82**, 5205 (1999).
3. M. S. Smith and K. E. Rehm, *Annu. Rev. Nucl. Part. Sci.*, **51**, 91 (2001).
4. E. Adelberger et al., *Rev. Mod. Phys.* **70**, 1265 (1998).
5. R. Schiavilla al., *Phys. Rev. C* **58**, 1263 (1998).
6. C. Casella et al., *Nucl. Phys. A* **706**, 202 (2003).
7. F. Confortola et al., *Phys. Rev. C* **75**, 065803 (2007).

8. T. A. D. Brown et al., *Phys. Rev. C* **76**, 055801 (2007).
9. B. Filippone et al., *Phys. Rev. Lett.* **50**, 412 (1983).
10. A. Junghans et al., *Phys. Rev. C* **68**, 065803 (2003).
11. F. Schümann et al., *Phys. Rev. C* **73**, 015806 (2006).
12. J.N. Bahcall et al., *Phys. Rev. C* **54**, 411 (1996).
13. W. Winter et al., *Phys. Rev. C* **73**, 025503 (2006).
14. M. Bhattarcharya et al., *Phys. Rev. C* **73**, 055802 (2006).
15. F. von Weizsäcker, *Physik. Zeitsch.* **39**, 633 (1938).
16. H. Bethe, *Phys. Rev.* **55**, 436 (1939).
17. A. Formicola et al., *Phys. Lett. B* **591**, 61 (2004).
18. R. C. Runkle et al., *Phys. Rev. Lett.* **94**, 082503 (2005).
19. G. Imbriani et al., *Astron. Astrophys.* **420**, 625 (2004).
20. B. Harss et al., *Phys. Rev. C* **65**, 035803 (2002).
21. J. Blackmon et al., *Phys. Rev. C* **72**, 034606 (2005).
22. M. Notani et al., *Nucl. Phys. A* **738**, 411c (2004).
23. C. B. Fu et al., *Phys. Rev. C* **76**, 021603 (2007).
24. R. Kozub et al., *Phys. Rev. C* **73**, 044307 (2006).
25. K. Y. Chae et al., *Phys. Rev. C* **74**, 12801 (2006).
26. N. de Sereville et al., *Nucl. Phys. A* **791**, 251 (2007).
27. K. E. Rehm et al., *Phys. Rev. C* **55**, R566 (1997).
28. S. Utku et al., *Phys. Rev. C* **57**, 2731 (1998); Erratum *Phys. Rev. C* **58**, 1354(1998).
29. B. Davids et al., *Phys. Rev. C* **67**, 065808 (2003).
30. K. E. Rehm et al., *Phys. Rev. C* **67**, 065809 (2003).
31. W. Tan et al., *Phys. Rev. C* **72**, 041302 (2005).
32. R. Kanungo et al., *Phys. Rev. C* **74**, 045803 (2006).
33. W. Tan et al., *Phys. Rev. Lett.* **98**, 242503 (2007).
34. S. Starrfield et al., *Astrop. J* **391**, L7 (1992).
35. A. F. Iyudin et al., *Astron. Astrophys.* **300**, 422 (1995).
36. S. Bishop et al., *Phys. Rev. Lett.* **90**, 162501 (2003).
37. D. Jenkins et al., *Phys. Rev. Lett.* **92**, 031101 (2004).
38. C. Ruiz et al., *Phys. Rev. Lett.* **96** 202501 (2006).
39. A. Arazi et al., *Phys. Rev. C* **74**, 025802 (2006).
40. R. R. C. Clement et al., *Phys. Rev. Lett.* **92**, 172502 (2004).
41. A. M. Amthor et al., *Bull. Am. Phys. Soc.* **43**, 103 (2007).
42. K. E. Rehm et al., *Phys. Rev. Lett.* **80**, 676 (1998).
43. J. Fisker et al., arXiv 0703311 (2007).
44. S. Wooseley et al., *Astrop J. Suppl.* **151**, 75 (2004).
45. J. Clark et al., *Phys. Rev. Lett.* **92**, 192501 (2004).
46. J. Clark et al., *Phys. Rev. C* **75**, 032801 (2007).
47. D. Rodriguez et al., *Phys. Rev. Lett.* **93**, 161104 (2004).
48. T. Motobayashi, contribution to this conference.
49. K. L. Kratz et al., Proc. Intern. Conf. on Fission and Properties of Neutron-rich Nuclei, World Scientific 1998, p,586.
50. P. T. Hosmer et al., *Phys. Rev. Lett.* **94**, 112501 (2005).
51. U. Hager et al., *Phys. Rev. Lett.* **96**, 042504 (2006).
52. G. Savard et al., *Intern. J. Mass Spectr.* **251**, 252 (2006).
53. A. H. Wapstra, G. Audi and C. Thibault, *Nucl. Phys. A* **729**, 129 (2003).
54. J. S. Thomas et al., *Phys. Rev. C* **71**, 021302 (2005).
55. K. Jones et al., *Bull. Am. Phys. Soc.* **52**, 43 (2007).
56. R. Kozub et al., *Bull. Am. Phys. Soc* **52**, 92 (2007).
57. I. Tomandl et al., *Nucl. Phys. A* **717**, 149 (2003).
58. J. Honzatko et al., *Nucl. Phys. A* **756**, 249 (2005).

RECENT ASTROPHYSICAL APPLICATIONS OF THE TROJAN HORSE METHOD TO NUCLEAR ASTROPHYSICS

C.Spitaleri [1], S.Cherubini[1], V.Crucillá[1], M.Gulino[1], M.La Cognata[1],L.Lamia[1], R.G. Pizzone[1], S.M.R. Puglia[1], G.G.Rapisarda[1], S.Romano[1], M.L.Sergi[1], A.Tumino[1], R.Tribble[2], A.Banu[2], T.Al-Abdullah[2], C.Fu[1], V.Goldberg[2], A.Mukhamedzhanov[2], G.Tabacaru[2], L.Trache[2], Y.Zhai[2], V.Kroha[3], V.Burjan[3], Z.Hons[3], J.Mrazek[3], E.Somorjai[4], G. Kiss[4], Chengbo Li[5]

[1] *Dipartimento di Metodologie Fisiche e Chimiche per l'Ingegneria, Universitá di Catania, Catania, Italy and INFN- Laboratori Nazionali del Sud, via Santa Sofia 62, Catania, 195100, Italy ,*
[2] *Cyclotron Institute, Texas A&M University, College Station,Texas, Usa,*
[3] *Nuclear Physics Institute of ASCR, p.r.i. Rez near Prague, Czech Republic*
[4] *Atomki, Debrecen, Hungary*
[5] *China Institute of Atomic Energy, Beijing, P.R.China*

Abstract. The Trojan Horse Method (THM) is an unique indirect technique allowing to measure astrophysical rearrangement reactions down to astrophysical relevant energies. The basic principle and a review of the recent applications of the Trojan Horse Method are presented. The applications aiming to the extraction of the bare astrophysical $S_b(E)$ for some two-body processes are discussed.

Keywords: indirect methods, cross section measurements, nuclear astrophysics
PACS: 24.50.+g,26.20.Np

INTRODUCTION

The thermonuclear fusion reactions are the origin of many chemical elements, and their isotopes, that can be found in the universe [1, 2, 3]. The present picture is that all elements from carbon to uranium have been produced entirely within stars during their fiery lifetimes and explosive deaths. A few of the lightest elements were formed before the stars even existed, during the birth of the universe itself. In addition, a few of the most reactive light elements appear to have been synthesized in intergalactic space by cosmic rays. Thus theories of nucleosynthesis have identified the most important sites of element formation and also the diverse nuclear processes involved in their production. The detailed understanding of the origin of the chemical elements and their isotopes combines astrophysics and nuclear physics, and it forms what is called nuclear astrophysics. A good knowledge of the rates of these fusion reactions is essential for understanding this broad picture [4].

The cross section $\sigma(E)$ of nuclear fusion reaction A(x,c)C is of course governed by the laws of quantum mechanics where, in most cases, the Coulomb and centrifugal barriers arising from nuclear charges and angular momenta in the entrance channel of

CP1016, *Origin of Matter and Evolution of Galaxies,*
edited by T. Suda, T. Nozawa, A. Ohnishi, K. Kato, M. Y. Fujimoto, T. Kajino, and S. Kubono

the reaction strongly inhibit the penetration of one nucleus into another. This barrier penetration leads to a steep energy dependence of the cross section. It is the challenge to the experimentalist to make precise $\sigma(E)$ measurements over a wide range of energies, as our fragmented knowledge of nuclear physics prevents us from predicting $\sigma(E)$ on purely theoretical grounds. For these reasons bare nucleus cross section $\sigma_{pl}(E)$ (for stellar plasma) of the (p,α) reaction at the Gamow energy (E_G) should be known with an accuracy better than 10% [3, 5] because of their crucial role in understanding the first phases of the Universe history and the subsequent stellar evolution. Unfortunately the presence of Coulomb barrier, in the reactions with charged particles, is a limit, often insuperable, to perform cross section measurements at ultra-low energies. Indeed the Coulomb barrier of height E_C in charged-particle induced reactions cause an exponential decrease of the cross section $\sigma_b(E)$ at $E < E_C$, $\sigma_b(E) \sim exp(-2\pi\eta)$, leading to a low-energy limit of direct $\sigma_b(E)$ measurements, which is typically much larger than E_G. Owing to the strong Coulomb suppression, the behavior of the cross section at E_G is usually extrapolated from the higher energies by using the definition of the smoother astrophysical factor S(E):

$$S_b(E) = E\sigma_b(E)exp(2\pi\eta) \tag{1}$$

where η is the Sommerfeld parameter, $exp(2\pi\eta)$ is the inverse of the Gamow factor, which removes the dominant energy dependence of $\sigma(E)$ due to the barrier penetrability.

Although the $S_b(E)$-factor allows for an easier extrapolation, large uncertainties to $\sigma_b(E_G)$ may be introduced, for instance, due to the presence of unexpected resonances, or high energy tails of sub-threshold resonances. In order to avoid the extrapolation procedure, a number of experimental solutions were proposed in direct measurements.

In recent years the availability of high-current low-energy accelerators, such as those at underground laboratories together improved target and detection techniques has allowed us to perform $\sigma_b(E)$ measurements in some cases down to E_G or at least close to E_G [6]. Then in principle no $\sigma_b(E)$ extrapolation would be needed anymore for these reactions. However, the measurements in laboratory at ultra-low energies suffer from the complication due to electron screening effect [4, 7]. This leads to an exponential increase of the laboratory cross section measured $\sigma_s(E)$ [or equivalently of the astrophysical factor $S_s(E)$] with decreasing energy relative to the case of bare nuclei. This can be described by an enhancement factor defined by the relation

$$f_{lab}(E) = \sigma_s(E)/\sigma_b(E) = exp(\pi\eta U_e/E) \tag{2}$$

In this equation U_e is the electron screening potential in the laboratory which is different from the U_{pl} present in the stellar environment. Clearly, a good understanding of U_e is needed in order to calculate σ_b from the experimental data σ_s using equation (2). The effective cross section $\sigma_{pl}(E)$ in the stellar plasma is connected to the bare nucleus cross section $\sigma_b(E)$ and to the stellar electron screening enhancement factor f_{pl} by the relation

$$\sigma_{pl}(E) = \sigma_b(E)f_{pl}(E) = \sigma_b(E)\cdot exp(\pi\eta U_{pl}/E) \tag{3}$$

with U_{pl} is the plasma potential energy.

If $\sigma_b(E)$ is measured at the ultra-low energies E_G and f_{pl} is estimated within the framework of the Debye-Hückel theory, it is possible to estimate from Eq.(3) the effective cross section $\sigma_{pl}(E)$ in the stellar plasma. In turn, the understanding of U_e may help to better understand U_{pl}, needed to calculate σ_{pl}.

Then, although it is possible to measure cross sections in the Gamow energy range, the bare cross section σ_b is extracted by extrapolating the direct data behavior at higher energies where negligible electron screening contribution is expected.

BASIC THEORY OF THE TROJAN HORSE METHOD

Alternative methods for determining bare nucleus cross sections of astrophysical interest are needed. In this context a number of indirect methods, e.g. the Coulomb dissociation (CD) [8, 9], the Asymptotic Normalization Coefficient method (ANC)[10, 11, 12, 13, 14] and the THM were developed (Table 1).The THM has already been applied several times to reactions connected with fundamental astrophysical problems. Some of them make use of direct reaction mechanisms.

In particular, the THM is a powerful tool which selects the quasi-free (QF) contribution of an appropriate three-body reaction performed at energies well above the Coulomb barrier to extract a charged particle two-body cross section. In the framework of the extrapolation problems linked to the presence of electron screening effects a number of experimental measurements were carried out in order to measure the bare nucleus cross section in reactions of astrophysical interest (Table 1).

The idea of the THM [15] is to extract the cross section of an astrophysically relevant two-body reaction

$$A + x \rightarrow c + C \tag{4}$$

at low energies from a suitable chosen three-body quasi-free reaction

$$A + a \rightarrow c + C + S \tag{5}$$

This is done with the help of direct reactions theory assuming that the nucleus a having a strong $x \oplus S$ cluster structure. In many applications (Table 1) this assumption is trivially fulfilled: a = deuteron, x = proton, S = neutron. This three-body reaction (5) can be described by a Pseudo-Feynman diagram, where only the first term of the Feynmann series is retained. The upper pole describes the virtual break-up of the target nucleus a into the clusters x and S; S acts as spectator to the $A + x \rightarrow c + C$ reaction which takes place in the lower pole.

This description is called Impulse Approximation (IA) [16] and the cross section of the three body reaction can be factorized into two terms corresponding to the two poles. In Plane Wave Impulse Approximation (PWIA) it is given by [17, 18]:

$$\frac{d^3\sigma}{dE_c d\Omega_c d\Omega_C} \propto KF \left(\frac{d\sigma}{d\Omega_{cm}}\right)^{off} \cdot |\Phi(\vec{p}_s)|^2 \tag{6}$$

TABLE 1. The two-body reactions studied via Trojan Horse Method

	Direct reaction	Indirect reaction	E_{inc} (MeV)	Q (MeV)	q_t (MeV/c)	TH_{nucl}	ref
[1]	$^7Li(p,\alpha)^4He$	$^7Li(d,\alpha\ \alpha)n$	19,22	15.122	391	$d = (p\otimes n)$	[21]
[2]	$^6Li(d,\alpha\ )^4He$	$^6Li(^6Li,\alpha\ \alpha)^4He$	5	22.372	136	$^6Li = (\alpha\otimes d)$	[22]
[3]	$^6Li(p,\alpha)^3He$	$^6Li(d,\alpha\ ^3He)n$	14, 25	1.795	263	$d = (p\otimes n)$	[23]
[4]	$^{11}B(p\ ,\alpha)^8Be$	$^{11}B(d\ ,^8Be\ \alpha)n$	27	6.36	370	$d = (p\otimes n)$	[24]
[5]	$^{10}B(p,\alpha)^7Be$	$^{10}B(d,^7Be\ \alpha)n$	27	-1.079	360	$d = (p\otimes n)$	[25]
[6]	$^9Be(p,\alpha)^6Li$	$^9Be(d,^6Li\ \alpha)n$	22	-0.099	360	$d = (p\otimes n)$	[26]
[7]	$^2H(^3He,p)^4He$	$^6Li(^3He,p\ \alpha)^4He$	5, 6	16.88	83-95	$^6Li = (\alpha\otimes d)$	[27]
[8]	$^2H(d,p)^3H$	$^2H(^6Li,t\ p)^4He$	14	2.59	112	$^6Li = (\alpha\otimes d)$	[28]
[9]	$^{12}C(\alpha,\alpha)^{12}C$	$^{12}C(^6Li,\alpha\ ^{12}C)^2H$	20	1.47	130	$^6Li = (\alpha\otimes d)$	[29]
[10]	$^{15}N(p,\alpha)^{12}C$	$^{15}N(d,\alpha^{12}C)n$	60	2.74	645	$d = (p\otimes n)$	[30, 31]
[11]	$^{18}O(p,\alpha)^{15}N$	$^{18}O(d,\alpha^{15}N)n$	54	1.76	668	$d = (p\otimes n)$	[32]
[12]	$^{17}O(p,\alpha)^{14}N$	$^{17}O(d,\alpha^{14}N)n$	41	-1.032	560	$d = (p\otimes n)$	[33]

where:

- $[(d\sigma/d\Omega)_{cm}]^{off}$ is the half-off-energy-shell differential cross section for the two body A(x,c)C reaction at the center of mass energy E_{cm} given in post collision prescription by:

$$E_{cm} = E_{c-C} - Q_{2b} \tag{7}$$

  where Q_{2b} is the two body Q-value of the $A + x \rightarrow c + C$ reaction and E_{c-C} is the relative energy between the outgoing particles c and C;
- KF is a kinematical factor containing the final state phase-space factor and it is a function of the masses, momenta and angles of the outgoing particles:

$$KF = \frac{\mu_{Aa} m_c}{(2\pi)^5 \hbar^7} \frac{p_C p_c^3}{p_{Aa}} \left[\left(\frac{\vec{p}_{Bx}}{\mu_{Bx}} - \frac{\vec{p}_{Cc}}{m_c} \right) \cdot \frac{\vec{p}_c}{p_c} \right]^{-1} \tag{8}$$

- $\Phi(\vec{p}_s)$ is the Fourier transform of the radial wave function $\chi(\vec{r})$ for the x-S inter-cluster motion, usually described in terms of Hänkel, Eckart and Hulthén functions depending on the x-S system properties.

In the experimental works [Table1] the validity conditions of the IA appear to be fulfilled.

We stress that one cannot extract the absolute value of the two-body cross section. However, the absolute value can be deduced through normalization to the direct data available at energies above and/or below the Coulomb barrier. Thanks to this, since we select the region of low momentum p_s for the spectator ($p_s \leq 40$ MeV/c) the PWIA approach can be used for the further analysis of the experimental results. If $|\Phi(\vec{p}_s)|^2$ is known and KF is calculated, it is possible to derive $[(d\sigma/d\Omega)_{cm}]^{exp}$ from a measurement of $d^3\sigma/dE_c d\Omega_c d\Omega_C$ by using Eq.(6). Under appropriate kinematical conditions, the

TABLE 2. Nuclei with cluster structure which can have been used as Trojan Horse nuclei and their principal properties.

	THM -nucleus	binding energy MeV	Clusters	Intercluster motion
1	d	2.225	p - n	l=0
2	t	6.257	n - d	l=0
3	^{3}He	5.494	p - d	l=0
4	^{6}Li	1.475	α - d	l=0
6	^{7}Li	2.468	α- t	l=1
7	^{7}Be	1.587	α- ^{3}He	l=1
8	^{9}Be	2.467	α- ^{5}He	l=0

three-body reaction is considered as the decay of the "Trojan Horse" a into the cluster x and S and the interaction of A with x inside the nuclear field, whereby the nucleus S can be considered as a spectator during the reaction. Table 2 shows the Trojan horse nuclei used in experiments.

If the bombarding energy E_A is chosen high enough to overcome the Coulomb barrier in the entrance channel of the three-body reaction, both Coulomb barrier and electron screening effects are negligible.

We stress that in our approach (Table 2) the initial projectile velocity is compensated for by the binding energy of particle x inside a. Thus the two-body reaction can be induced at very low (even vanishing) $A-x$ relative energy. In this way it is possible to extract the two-body cross section from Eq.(9) after inserting the appropriate penetration function G_l in order to account for the penetrability affecting direct data below the Coulomb barrier [19, 20]. The complete formula is given by:

$$\left(\frac{d\sigma}{d\Omega}\right) \propto \left[\frac{d^3\sigma}{dE_c d\Omega_c d\Omega_C}\right] \cdot [KF|\Phi(\vec{p}_s)|^2]^{-1} \cdot G_l. \tag{9}$$

As shown above, since in the experimental works the IA validity conditions are fulfilled, the PWIA was applied for the extractions of the two-body cross-section. In this approximation the differential two-body cross-section of Eq.(9) is expressed by:

$$\left(\frac{d\sigma}{d\Omega}\right) = \sum_l C_l P_l \left(\frac{d\sigma_l}{d\Omega}\right). \tag{10}$$

DATA ANALYSIS AND RESULTS

In order to select the region where the QF mechanism is dominant, coincidence events for spectator momenta close to zero momentum were considered in all analysis. Monte Carlo calculations were then performed to extract the $(KF \cdot |\Phi(\vec{p}_s)|^2)$ product. The momentum distribution entering the calculation was that given in [21, 22]. The geometrical efficiency of the experimental setup as well as the detection thresholds of PSD's were taken into account.

Following the PWIA prescription of Eq.(6), the two-body cross-section $d\sigma/d\Omega_{cm}$ was derived dividing the selected three-body coincidence yield by the result of the Monte Carlo calculation.

The experimental way of testing the basic assumptions in the polar description of the QF reaction was applied [34, 35, 36] :

- the behavior of the angular distribution $\sigma(\theta)_{THM}$ indirectly extracted must be the same of that of the direct $\sigma(\theta_{cm})_{DIR}$ angular distribution;
- and/or the behavior of the excitation function $\sigma(E)_{THM}$ must be the same of the direct one at energies higher than those of astrophysical interest.

A good agreement of the two trends (direct and indirect cross sections) is a necessary condition (test of applicability of the approximation) to be fulfilled before extracting the astrophysical S(E) factor by means of the THM.

As already mentioned the THM applies to a suitable three-body reaction, which is performed in a kinematically complete experiment. The experimental set-up is optimized in order to cover the angular regions where the quasi-free process is expected to be favored. The two-body cross section is then extracted from the three-body coincidence yield within a spectator momentum window usually ranging from -30 to + 30 MeV/c. Note that the deduced two-body cross section is the nuclear part alone, this being the main feature of the THM. In order to deduce the experimental S(E) factor from standard definition, the nuclear cross section is multiplied by the proper transmission coefficient G_l(E).

In all the cases considered hereby the behavior of the indirectly extracted excitation function, after normalization, is similar to the direct one in the whole investigated range except at ultra-low energies where electron screening effect is no longer negligible in the direct data.

The corresponding three body reactions investigated with the relevant references are reported in Table 1.

ASTROPHYSICAL APPLICATIONS AND CONCLUSIONS

The present paper presents the basic features of the THM and a review of recent applications to several reactions of importance in astrophysics.These results show the possibility of extracting the bare nucleus two-body cross section via THM. In particular among these reactions some were performed in order to study the main reactions connected to the Li, Be and B depletion [21-25]. While the study in the case of the ^{6}Li$(p,\alpha)^{3}$He and ^{7}Li$(p,\alpha)^{4}$He, shows quite conclusive results, the problems connected to the ^{6}Li$(d,\alpha)^{4}$He reaction require further studies.

The study of the ^{9}Be$(p,\alpha)^{6}$Li reaction is still in progress since a clear J^{π} assignment for a resonance in the three body cross section is not yet available. The ^{10}B$(p,\alpha)^{7}$Be reaction has been recently measured again in order to provide better energy resolution. The available results arising from this experiment are still preliminary.

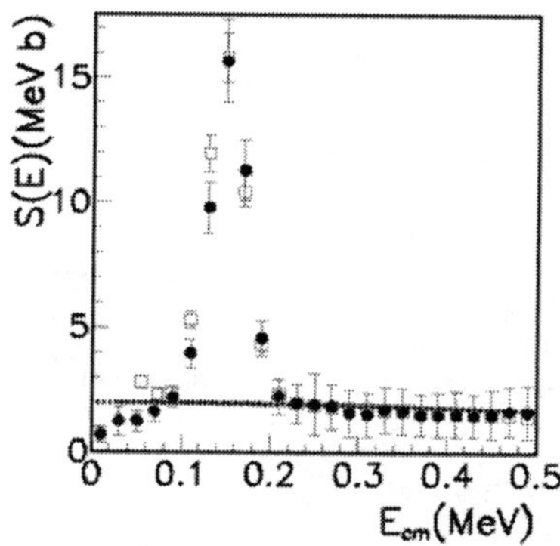

FIGURE 1. New THM data for the $^{11}B(d,\alpha^{8}Be)n$ (S0)=1,98 $\pm$0,20)compared with direct data

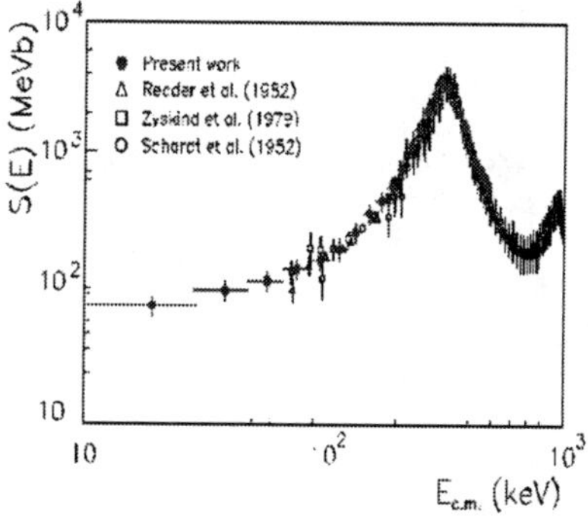

FIGURE 2. THM data for the $^{15}N(p,\alpha)^{12}C$ (S0)=62 $\pm$10) compared with direct data

The same should be done for AGB stars nucleosynthesis where the $^{15}N(p,\alpha)^{12}C$ plays a key role for ^{19}F production.

Recently, p-capture reactions on ^{11}B and ^{15}N were also investigated by selecting the quasi-free contribution to the $^{15}N(d,\alpha^{12}C)n$, and $^{11}B(d,\alpha^{8}Be)n$ three-body processes. The extracted S(E) factors for these reactions are shown in Figs 1 and 2. Their importance is indeed strongly related to the stellar structure and evolution as well as to nucleosynthesis. However, a lot remains to do in the future to achieve reliable information for many key reactions and processes. New theoretical developments are strongly needed especially for the study of electron screening effects in fusion reactions in order to meet progress in the application field (fusion reactors).

REFERENCES

1. E.M.Burbige,G.R.Burbige,W.A.Fowler,F.Hoyle, Rev.Mod.Phys.**29**, 547 (1957)
2. W.A.Fowler, Rev. Mod. Phys.**56**, 149 (1984)
3. C.Rolfs, W.S. Rodney, Cauldrons in the cosmos, University of Chicago Press, Chicago (1988)
4. F. Strieder, C.Rolfs, C.Spitaleri, P.Corvisiero, Naturwissenschaften **88**, 461, (2001)
5. C.Rolfs, Prog. Part. Nucl. Phys., **153** , 23 (2001)
6. R.Bonetti et al., Phys. Rev.Lett., **82**, 5205 (1999)
7. H.J Assenbaum, K.Langanke, C.Rolfs, Z.Phys.,**327** 461 (1987)
8. G. Baur and H. Rebel, J. Phys. G **20**, 1 (1994) and referen ces therein
9. G. Baur and H. Rebel, Annu. Rev. Nucl. Part. Sci. **46**, 321 (1996)
10. X.D.Tang et al., Phys. Rev. C**67**, art.n.015804 (2003)
11. A.Azhari et al., Phys.Rev., **C63** , 055803 (2001)
12. A.M.Mukhamedzhanov, C.A.Gagliardi, R.E.Tribble Phys. Rev. C**63**, art.n.024612 (2001)
13. A.M.Mukhamedzhanov et al., Phys. Rev. C **56**, 1302 (1997)
14. A.M.Mukhamedzhanov, R.E.Tribble Phys. Rev. C**59**, 3418(1999)
15. G. Baur, Phys. Lett. B **178**, 135 (1986)
16. G.F. Chew, Phys.Rev. **80**, 196(1950)
17. U.G.Neudatchin, Y.F.Smirnov, At. Energy Rev., **3** , 157 (1965)
18. G.Jacob, Th. A. Maris, Rev. Mod. Phys. **38** , 121 (1966)
19. S. Cherubini et al., Ap. J. **457**, 855(1996).
20. C. Spitaleri et al. Phys. Rev. C **60**, 055802 (1999)
21. M. Lattuada et al., Ap.J. **562**, 1076 (2001)
22. C. Spitaleri et al., Phys. Rev. C **63**, 005801 (2001)
23. A. Tumino et al., Phys. Rev. **C 67), 065803 (2003)**.
24. C.Spitaleri et al. Phys. Rev.C . **69** 055806 (2004)
25. L.Lamia et al., Nucl. Phys. A.**787** 309 (2007)
26. S.Romano et al., Eur. Phys. J. A**27**, 221 (2006)
27. M.La Cognata et al., Phys. Rev. C **72** 065802(2005)
28. A.Rinollo et al., Nucl. Phys. A **758**146c (2005)
29. C. Spitaleri et al., Eur. Phys. J **A 7**, 181 (2000)
30. M.La Cognata et al., Eur. Phys.Journ. A, Hadrons and Nuclei,**27**,249 (2006).
31. M.La Cognata et al., Phys. Rev C **76**,065804 (2007).
32. M.La Cognata et al. J.Phys. G Nucl. Part. Phys. **35**, 014014 (2008).
33. M.L.Sergi et al. Proceedings of Omeg07 Conference(2007).
34. A.K.Jain et al., Nucl.Phys.A,**153**,49,(1970)
35. A.K.Jain et al., Nucl. Phys., **A216**, 519 (1973)
36. A. Guichard et al., Phys. Rev. C **4**, 700 (1971)

Breakup of radioactive nuclear beams at intermediate energies as indirect method for nuclear astrophysics

L. Trache[a], F. Carstoiu[b], C.A. Gagliardi[a], R.E. Tribble[a], A. Banu[a]

[a]Cyclotron Institute, Texas A&M University, College Station, TX 77843, USA
[b]Institute for Physics and Nuclear Engineering "H. Hulubei", Bucharest, Romania

Abstract. We discuss the use of one-nucleon removal reactions of loosely bound nuclei at intermediate energies as an indirect method in nuclear astrophysics. These breakup reactions are good spectroscopic tools and can be used to study a large number of loosely bound proton- or neutron-rich nuclei over a wide range of beam energies. They are peripheral processes that can be used to extract asymptotic normalization coefficients (ANC) from which direct capture proton reaction rates of astrophysical interest can be calculated parameter free. We emphasize the importance of reaction model calculations and of exclusive measurements to check them. We review several cases: the breakup of ^{8}B, ^{9}C, ^{15}C and ^{23}Al. Firrst we review how we have used the data for the breakup of ^{8}B at energies from 30 to 1000 MeV/nucleon on light and heavy targets to extract the astrophysical factor $S_{17}(0)=18.7\pm1.9$ eV·b for the key reaction for solar neutrino production. Glauber model calculations in the eikonal approximation and in the optical limit using different effective NN interactions were found to give consistent, though slightly different results, for both ^{8}B and ^{9}C cases, enabling also us to evaluate the precision of the method. The third case is that of a neutron-rich nucleus and the ANC alone does not lead to unambiguous estimates of the associated (n,γ) capture reaction. The ^{23}Al case is that of an sd-shell nucleus, suspected of halo properties, for which our recent GANIL experiment revealed a complex configuration mixing in its ground state. The method outlined here has the big advantage that it can be used for beams of low quality, such as cocktail beams, and intensities as low as a few pps. As such it is very appropriate for the existing and future radioactive beam facilities. The breakup reactions are therefore complementary or a very good alternative to the use of proton and neutron transfer reactions (the ANC method) which require radioactive nuclear beams of much better purity and intensity.

Keywords: nuclear astrophysics; indirect methods; breakup at intermediate energies.
PACS: 25.60-t, 26.30.+k, 25.60.Ge, 23.40.-s, 26.65.+t

INDIRECT METHODS IN NUCLEAR ASTROPHYSICS

An important challenge of direct nuclear astrophysics measurements at low energies stems from the very low cross sections when reactions between charged particles are involved, due to Coulomb repulsion [1]. This is, for example, the case with radiative proton capture (p,γ) reactions occurring in H-burning. Another important challenge in the study of stellar reactions in general is that in most cases the targets involved are unstable nuclei. These two reasons combined lead to the use of indirect methods. Another non-negligible aspect is that direct and indirect methods have different systematic errors, and an agreement of results, particularly in crucial

CP1016, *Origin of Matter and Evolution of Galaxies*,
edited by T. Suda, T. Nozawa, A. Ohnishi, K. Kato, M. Y. Fujimoto, T. Kajino, and S. Kubono

cases, is extremely important, while disagreements lead to deepening our search. The combination of different methods can also give useful complementary information (see e.g. [2]).

A number of indirect methods have been proposed and used for nuclear astrophysics, with some of them adopted from general nuclear physics studies and some tailored specifically for this purpose. The list, however, is not very long: Coulomb dissociation [3,4], transfer reactions (the ANC method) [5], breakup at intermediate energies [6,7], the Trojan horse method [8], and other spectroscopic methods, in particular the location and study of resonances.

We will concentrate below in describing the breakup method, its characteristics and advantages for radioactive ion beams (RIB), rather than on reporting nuclear astrophysics data.

BREAKUP REACTIONS = SPECTROSCOPIC TOOL

Work done in the last decade in several laboratories has demonstrated that one-nucleon removal reactions (or breakup reactions) can be a good and reliable spectroscopic tool. In a typical experiment a loosely bound projectile at energies above

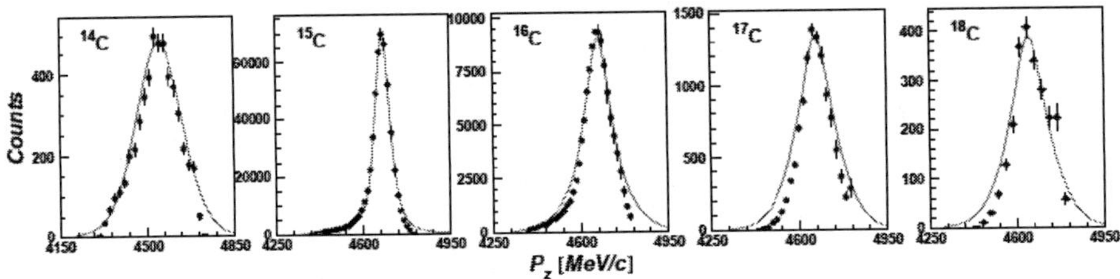

FIGURE 1. Examples of core momentum distributions measured for isotopes of C. The curves represent calculated distributions. Selected from Sauvan et al. [9].

the Fermi energy impinges on a target and loses one nucleon. The momentum distributions (parallel and/or transversal) of the remaining core measured after reaction give information about the momentum distribution of the removed nucleon in the wave function of the ground state of the projectile. The shape of the momentum distributions allow to determine the quantum numbers *nlj* of the s.p. wave function. Figure 1, selected from Ref. [9], shows a number of such transversal momentum distributions measured for $^{14\text{-}18}$C isotopes at energies around 50 MeV/u. A "normal" momentum distribution in ^{14}C, a very narrow one in ^{15}C, pointing to a large spatial extension (halo nucleus) and an $s_{1/2}$ neutron state, and one distribution reflecting configuration mixing in ^{17}C, are all seen in neighboring isotopes.

It was shown in Ref. [6] that breakup reactions are very peripheral and was demonstrated that the integrated breakup cross sections can be used to extract the asymptotic normalization coefficients (ANC) of the s.p. wave function. For this to be true, we need careful reaction model calculations. They need to reproduce all available data from such measurements if they are to be believed. We have investigated this aspect in detail in Ref. [7].

It was shown earlier that the s.p. quantum numbers and the ANCs are sufficient to determine the astrophysical S-factors for radiative proton capture reactions, due their peripheral character [10, 11]. We will briefly discuss that the relation is not that simple for neutron ANCs and (n,γ) reactions.

Breakup of ^{8}B: an independent determination of $S_{17}(0)$

We have used all available ^{8}B one-proton removal data existent in the literature to extract the ANC of its g.s. wave function. The data were taken in different laboratories, with various methods and apparatus (therefore, with different sources of errors), on targets ranging from light (^{9}Be) to heavy (^{208}Pb) and at energies between 28 and 1000 MeV/u. The data were analyzed using Glauber model type of calculations, in a potential approach and in the optical limit, and using various effective nucleon-nucleon interactions. We have verified that all momentum distributions measured are reproduced by calculations before using the calculations to extract absolute values for the ANC. No new parameters were introduced and no fits were made for anything but the ANC. Details are presented in Ref. [7], we sum up the conclusions here. The first was that a consistent ANC value can be extracted from all experiments and all types of calculations. The second was that a certain dependence on the effective interactions used is observed, and this sets a limit of about 10% on the accuracy of the values we can extract with this method.

The average ANC value obtained was translated into a value of the astrophysical S-factor for the ^{7}Be(p,γ)^{8}B reaction, the crucial reaction for the production of solar neutrinos. We found $S_{17}(0)=18.7\pm1.9$ eV$\cdot$b, a value slightly lower than other measurements made by direct and indirect methods, but consistent, within the error bars, with their average. It is beyond the scope of this paper to go into the details of a comparison.

Breakup of ^{9}C: a case where breakup has no competition

Relatively scarce (and not very precise) data exist on the breakup of ^{9}C. From them we have determined the ANC of the one proton in ^{9}C and evaluated the $S_{18}(0)$ astrophysical factor for the ^{8}B(p,γ)^{9}C reaction [12, 7]. The reaction is important in the hot pp chain, as a way of bypassing the A=8 mass gap. The method used is the one sketched above, except that here no momentum distributions were available to further check the calculations and the experimental errors are larger that the differences found between the calculations made with various effective NN interactions. Rather, the remarkable fact here is that virtually no other types of measurements were possible to date and probably even in the future. The only data competing here are from nuclear and Coulomb breakup [13].

Breakup of ^{23}Al: to determine configuration mixing

Space-based gamma-ray telescopes have had for some time the ability to detect gamma rays of cosmic origin, proving that nucleosynthesis is an ongoing process in our own Galaxy. Gamma rays following the β-decay of long-lived ^{26}Al, ^{44}Ti, ^{56}Ni, etc., have been detected. It was also predicted that ^{22}Na ($t_{1/2}$=2.6 y) is going to be produced in the thermonuclear runaway of the so-called ONe (Oxygen-Neon) novae [1,14]. The novae models using the currently available nuclear reaction rates led to the conclusion that enough ^{22}Na is produced and survives for the gamma-ray line E_γ=1275 keV from its decay to be observed at the sensitivity level of the current satellite based telescopes. However, the line was not observed and the question arising is why? The nuclear data came under scrutiny. Two reactions were proposed as mechanisms for the depletion of ^{22}Na or its precursor ^{22}Mg: ^{22}Na(p,γ)^{23}Mg [15] and ^{22}Mg(p, γ)^{23}Al [16].

The ground state spin and parity for ^{23}Al was uncertain (before our experiments), with assignments that included $1/2^+$, $3/2^+$ and $5/2^+$. The mirror nucleus ^{23}Ne has J^π =$5/2^+$ for its ground state and it was natural to assume the same spin for ^{23}Al. However, recently it was claimed that proton-rich ^{23}Al is a halo nucleus and that this can only be explained if the last proton is in the $2s_{1/2}$ orbital (i.e. J^π=$1/2^+$ for ^{23}Al g.s.), not in $1d_{5/2}$ (level inversion). Using $1/2^+$ instead of $5/2^+$, we calculated the astrophysical S-factor and stellar reaction rate for ^{22}Mg(p,γ)^{23}Al and found an increase of 30-50 times over the current estimate for the temperature range T_9=0.1-0.3. This would result in a significant depletion of ^{22}Mg before it β decays into ^{22}Na and, if confirmed, could explain the non-observation of the 1.275 MeV γ-ray. It was important for both nuclear structure and for nuclear astrophysics to determine unambiguously the spin and parity of the ground state of ^{23}Al. We proposed to do it in two ways. One was by measuring the core momentum distribution from ^{23}Al breakup at intermediate energies. The other way was to use β-decay measurements [17,18].

We proposed at GANIL an experiment that could determine the spin and parity, the ANC and the configuration mixing in the ground state of ^{23}Al from the measurement of the momentum distribution of the core after the proton breakup on a light target. Calculations show that the momentum distribution is about 2 times narrower for a $2s_{1/2}$ orbital than for a $1d_{5/2}$ orbital. The experiment was carried out using a special experimental setup. The Ge clover detectors of EXOGAM were arranged in a new configuration at the target position of the magnetic spectrograph SPEG. Also 12 NaI detectors were included in the system. The primary beam 95 MeV/u ^{32}S was delivered on a 450 mg/cm^2 carbon target in SISSI. SISSI was tuned for a ^{23}Al secondary beam at magnetic rigidity Bρ=1.95 Tm. That gave about 15 kHz of secondary beam on the second target (C at 180 mg/cm^2), a cocktail made mostly of ^{24}Si, ^{23}Al, ^{22}Mg and ^{21}Na, with ^{23}Al @ 50 MeV/u being about 1.4% of the total.

With SPEG tuned at 1.75 Tm we could see that we measured the momentum distributions for the cores of the beam nuclei after the removal of one proton: ^{23}Al, ^{22}Mg, ^{21}Na and ^{20}Ne. We have inclusive momentum distributions and in coincidence with gammas. The data analysis is still in progress. The conclusion to date are: (1) the method works; (2) there is a large amount of configuration mixing in the g.s. of ^{23}Al; (3) ANCs can be determined for a number of other nuclei present in the beam cocktail. These ANCs will be used to determine direct capture astrophysical S-factors and rates relevant for reactions in H-burning, all impossible to determine from any other measurements. This is the main advantage of breakup reactions at intermediate energies: it can be used for cocktail beams of low quality.

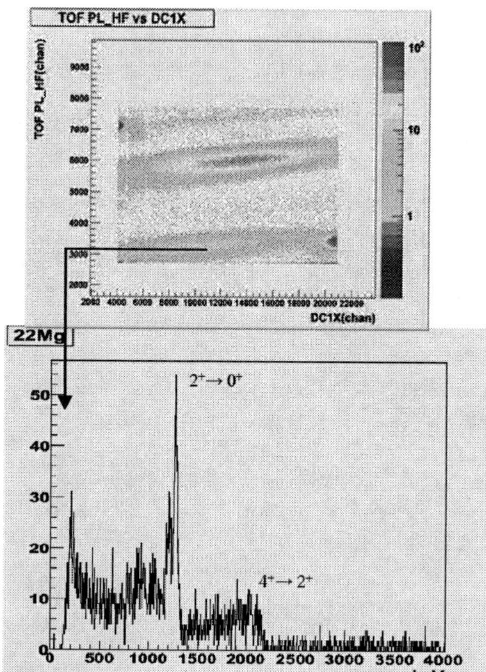

FIGURE 2. Considerable configuration mixing is found in the ground state of ^{23}Al. Top part shows that mass A=22 separated in the focal plane of SPEG splits into ^{22}Mg and ^{22}Na (two different mechanisms). Bottom part shows the gamma-rays measured at target position in coincidence with ^{22}Mg core in SPEG.

Breakup of ^{15}C and (n,γ) cross sections

Neutron breakup reactions are also peripheral. Indeed, they may be more so than those for protons, due to the lack of the proton-core Coulomb barrier that restricts spatially the protons inside the nucleus. Therefore, neutron ANCs can be easily and reliably extracted from breakup. We found $C^2(s_{1/2})=1.48(18)\ fm^{-1}$ for ^{15}C. However, when we want to transform these ANCs into (n,γ) reaction cross sections, we face the problem that the neutron capture reactions are not necessarily peripheral. Neutron capture can also happen inside the nucleus. To have a correct cross section evaluation we need to know the neutron-nucleus interaction at small distances too. No clear solution exists for this problem so far.

CONCLUSIONS

We have illustrated an indirect method to be used in nuclear astrophysics with radioactive nuclear beams. One-proton (or neutron) removal (breakup) reactions at intermediate energies are peripheral and can be used to determine quantum numbers, configuration mixings and ANCs. In the proton case they can be used to evaluate the direct component of the astrophysical S-factors for radiative proton capture reactions. In the neutron case, further knowledge of the neutron-nucleus potential is needed.

The method presents a number of experimental advantages: the reaction cross sections are large (~ 100 mb) and beams of low intensities can be used (as low as 1 pps); at intermediate energies beam tracking can be done and low quality cocktail beams from fragmentation give information on multiple cases in the same experiment. The use of RNBs and indirect methods allows us to extend our reach to regions of unstable nuclei of crucial importance for our understanding of cosmic nucleosynthesis.

ACKNOWLEDGEMENTS

This paper is based on work done along the years with our colleagues. They are acknowledged in the references quoted here. For the GANIL experiment, work was done with the participation of several people from GANIL and from the "Exotic nuclei" group of Nigel Orr at LPC Caen, the groups of M. Chartier at the University of Liverpool, of W. Catford at the University of Surrey.

This work was supported by the U.S. Department of Energy under Grant No. DE-FG03-93ER40773, the Welsh Foundation, by the Romanian National Authority for Scientific Research under grant CEX M3-165 and by the EURONS program.

REFERENCES

1. C. Rolfs and W.S. Rodney, *Cauldrons in the Cosmos*, University of Chicago Press, 1988.
2. L. Trache, in *"Exotic Nuclei and Nuclear/Particle Astrophysics (II)" Proceedings of the Carpathian Summer School of Physics 2007*, edited by L. Trache and S. Stoica, API Conference Proceedings 972, Melville, NY, 2008.
3. G. Baur, CA Bertulani and H. Rebel, *Nucl. Phys.* **A458**, 188 (1986).
4. T. Motobayashi et al., *Phys. Rev. Lett.* **73**, 2680 (1994).
5. see e.g. A.M. Mukhamedzanov, C.A. Gagliardi and RE Tribble, *Phys. Rev. C* **63**, 024612 (2001).
6. L. Trache, F. Carstoiu, C.A. Gagliardi and RE Tribble, *Phys. Rev. Lett.* **87**, 271102 (2001).
7. L. Trache, F. Carstoiu, C.A. Gagliardi and RE Tribble, *Phys.Rev. C 69, 032802* (2004).
8. see e.g. G.R. Pizzone et al., in *"Exotic Nuclei and Nuclear/Particle Astrophysics (II)" Proceedings of the Carpathian Summer School of Physics 2007*, edited by L. Trache and S. Stoica, API Conference Proceedings 972, Melville, NY, 2008, and references therein.
9. E. Sauvan et al., *Phys. Rev. C* **69**, 044603 (2004).
10. R.F. Cristy and I. Duck, *Nucl. Phys.* **24**, 89 (1961).
11. X.M. Xu et al, *Phys. Rev. Lett.* **73**, 2027 (1994).
12. L. Trache et al., *Phys. Rev. C* **66**, 035801 (2002).
13. T. Motobayashi, *Nucl. Phys.* **A718**, 101c (2003).
14. J. Jose, A. Coc, M. Hernanz, *Astroph. J.* **520**, 347(1999); **560,** 897 (2001).
15. M. Wiescher et al., *Nucl. Phys.* **A484**, 90 (1988).
16. See e.g., F. Stegmuller et al., *Nucl. Phys.* **A601**, 168 (1996) and references therein.
17. V.E. Iacob et al, *Phys. Rev. C* **74**, 045810 (2006).
18. Y. Zhai, thesis, Texas A&M University, 2007.

Investigation of Stellar $^{26}Si(p,\gamma)^{27}P$ Reaction via Coulomb Dissociation

Y. Togano*, T. Gomi†, T. Motobayashi**, Y. Ando*, N. Aoi**, H. Baba**, K. Demichi*, Z. Elekes‡, N. Fukuda**, Zs. Fülöp‡, U. Futakami*, H. Hasegawa*, Y. Higurashi**, K. Ieki*, N. Imai§, M. Ishihara**, K. Ishikawa¶, N. Iwasa||, H. Iwasaki††, S. Kanno**, Y. Kondo¶, T. Kubo**, S. Kubono‡‡, M. Kunibu*, K. Kurita*, Y. U. Matsuyama*, S. Michimasa‡‡, T. Minemura§§, M. Miura¶, H. Murakami*, T. Nakamura¶, M. Notani¶¶, S. Ota‡‡, A. Saito‡‡, H. Sakurai**, M. Serata*, S. Shimoura‡‡, T. Sugimoto**, E. Takeshita**, S. Takeuchi**, K. Ue††, K. Yamada**, Y. Yanagisawa**, K. Yoneda** and A. Yoshida**

*Department of Physics, Rikkyo University, Tokyo 171-8501, Japan
†National Institute of Radiological Sciences, Chiba 263-8555, Japan
**RIKEN Nishina Center, RIKEN, Saitama 351-0198, Japan
‡ATOMKI, 4001 Debrecen, Hungary
§High Energy Accelerator Research Organization (KEK), Ibaraki 305-0801, Japan
¶Department of Physics, Tokyo Institute of Technology, Tokyo 152-8551, Japan
||Department of Physics, Tohoku University, Miyagi 980-8578, Japan
††Department of Physics, University of Tokyo, Tokyo 113-0033, Japan
‡‡Center for Nuclear Study, University of Tokyo, Saitama 351-0198, Japan
§§National Cancer Center, Chiba 277-8577, Japan
¶¶Physics Division, Argonne National Laboratory, IL 60439, USA

Abstract. The Coulomb dissociation of the proton-rich nuclei ^{27}P was studied experimentally using ^{27}P beams at 57 MeV/nucleon with a lead target. The radiative widths of the low-lying excited state in ^{27}P were deduced. The resonant capture reaction rate of stellar $^{26}Si(p,\gamma)^{27}P$ through these states was estimated using the measured radiative widths. The astrophysical implications obtained from the extracted reaction rate will be discussed.

Keywords: Reaction induced by unstable nuclei, Coulomb excitation, and Nucleosynthesis in novae, supernovae and other explosive environments, $20 \leq A \leq 38$
PACS: 25.60.-t, 25.70.De, 26.30.+k,27.30.+t

INTRODUCTION

The nucleosynthesis process of ^{26}Al has attracted much attention in connection with the cosmic evolutions. This was triggered by the cosmic γ ray observations, where the characteristic γ rays of ^{26}Al were found in the galactic plane [1]. The information obtained from the observations is quite useful for investigating ongoing nucleosynthesis processes, since the lifetime of ^{26}Al (10^6 yr) is much shorter than the timescale of the cosmic evolution (10^{10} yr). For this purpose a lot of theoretical studies have attempted to reproduce the ^{26}Al γ-ray yield, assuming a variety of stellar sites as the origin [2, 3]. Nevertheless, they have large deviation between each model. Particularly the

CP1016, *Origin of Matter and Evolution of Galaxies*,
edited by T. Suda, T. Nozawa, A. Ohnishi, K. Kato, M. Y. Fujimoto, T. Kajino, and S. Kubono

results of explosive events, such as novae, X-ray bursts, and supernovae, have very large uncertainties caused by the insufficiency of nuclear database, especially for unstable nuclei including ^{26}Al and around.

In most of the stellar sites, the isotope ^{26}Al is produced via the reaction sequence ^{24}Mg(p,γ)^{25}Al(β^{+} ν)^{25}Mg(p,γ)^{26}Al. In the sites which have high stellar density and temperature, the explosive conditions, the proton capture on ^{25}Al becomes faster than its β decay and produces the unstable nucleus ^{26}Si which β decays to ^{26}Al. The ground state of ^{26}Al β decays to the first excited state in ^{26}Mg, giving rise to a 1.8 MeV γ ray from the deexcitation. The isomeric level ($T_{1/2}$ = 6.3 s) at E_x = 228 keV β decays predominantly to the ^{26}Mg ground state and therefore, is thought to be of no relevance to the astrophysical observations. However, in the condition whose temperature is higher than 0.4 GK, the thermal equilibrium is achieved among low lying states in ^{26}Al [4]. The β decay of ^{26}Si feeds the only isomeric state and thus affects the synthesis of ^{26}Al through the equilibration. Therefore, ^{26}Si destruction by proton capture is important to determine the amount of the ground state in ^{26}Al produced on the basis of the equilibrium.

Under this high temperature the capture reaction rate is expected to be dominated by the resonant capture via the first excited state at 1.2 MeV in ^{27}P [5]. Through the Breit-Wigner expression for resonant capture, the measurement of the radiative width Γ_γ and the proton decay width Γ_p of the state gives the relevant (p,γ) cross section. Guo *et al.* measured the asymptotic normalization coefficients (ANC) of the mirror nucleus ^{27}Mg [6]. The Γ_p of the first excited state in ^{27}P was deduced using the relationship of the ANCs for the mirror nucleus. The calculated Γ_γ value was reported by Herndl *et al.* [7] and Caggiano *et al.* [5] based on shell models. The value obtained by the calculations have very large uncertainty, thus the experimental information concerning the Γ_γ value is highly desired.

In order to extract Γ_γ values of the low-lying excited state in ^{27}P, the Coulomb dissociation method is applied. The Coulomb dissociation at intermediate energies is an alternative method to study the radioactive capture reactions of astrophysical interest at low energies [8, 9, 10]. The process can be regarded as photodisintegration by virtual photons, which is essentially the inverse of the radiative capture process [11]. The cross section is enhanced by the phase space factors as compared to its inverse reaction. One can extract the electromagnetic transition probabilities $B(\lambda,\downarrow)$ of the unbound excited state which Γ_γ values are directly obtained as

$$\Gamma_\gamma = \frac{8\pi(\lambda+1)}{\lambda\,[(2\lambda+1)!!]^2}\left(\frac{E_x}{\hbar c}\right)^{2\lambda+1} B(\lambda,\downarrow), \qquad (1)$$

where E_x is the excitation energy and λ is the multipolarity. The intermediate energy beams give us an experimental advantage that much thicker targets can be used compared to the direct measurements.

In this paper we aim to determine the Γ_γ values of resonant states in ^{26}Si(p,γ)^{27}P reaction through the Coulomb dissociation experiment and deduce the astrophysical implications from the present result.

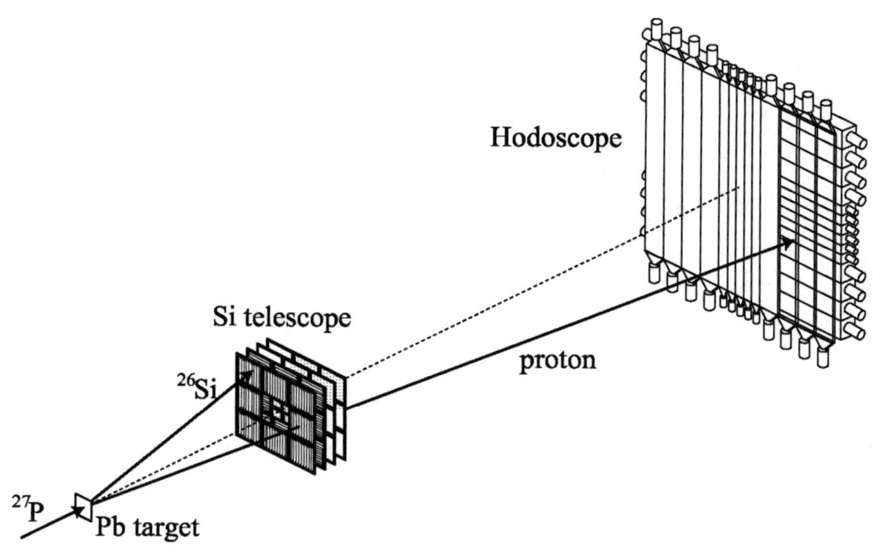

FIGURE 1. A schematic view of the experimental setup. The entire system is in vacuum.

EXPERIMENT

The experiment was performed at the RIKEN Nishina Center. A ^{27}P beam was produced via the projectile fragmentation of a 115-A-MeV ^{36}Ar beam incident on a 460 mg/cm^2 thick Be target, and separated by the RIKEN Projectile-fragment Separator (RIPS) [12]. The typical ^{27}P intensity was about 2800 count per second and the energy was 57 A MeV with an energy spread of about 1.5%. It bombarded a 125 mg/cm^2 thick lead target. The isotopic purity of ^{27}P in the secondary beam is about 1%. The major contributions are ^{26}Si, ^{25}Al, and ^{24}Mg. The particle identification for secondary beams was performed event-by-event by means of the time-of-flight (TOF) -ΔE method using a 0.5 mm-thick plastic scintillator (F2PL) located at the second focal plane of the RIPS. Two set of parallel plate avalanche counters (PPACs) were also placed at the final focal plane of the RIPS to extrapolate the position and angle of the beam at the target.

The isotope ^{27}P were excited by the lead target and disintegrated to ^{26}Si and proton. Figure 1 shows the detector system for the measurement of breakup products. The emission angles of these products were measured at a position sensitive silicon telescope located at 48 cm downstream of the target. The silicon telescope consists of four layers of detectors with 0.5 mm thickness. Each layer was of eight silicon detectors with 50 $\times$ 50 mm^2 effective areas on 56 $\times$ 56 mm^2 frames. The eight detectors in a layer formed a 3 $\times$ 3 matrix with a hole in the center. The count rate in the telescope was suppressed to 3000 count per second by introducing the hole whereas the total beam rate was about 3 $\times$ 10^5 count per second. The detectors in the first and second layers have 5 mm wide strip electrodes, which enables to measure the hit positions of the products. The energy of the ^{26}Si was also measured by the silicon telescope.

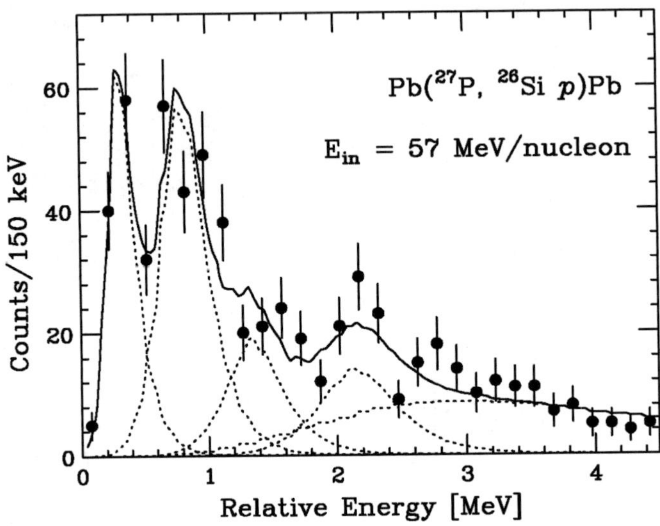

FIGURE 2. Relative energy spectrum of the ^{27}P breakup on Pb (closed circle). The data was fitted by the detector responses obtained by a Monte-Carlo simulation. The dashed curves and solid curve represent the each components and sum of the elements.

The energy of the proton, which penetrated the silicon telescope, was determined with a plastic scintillator hodoscope placed at the 2.8 m downstream of the target by measuring the TOF. The hodoscope with an active area of 1×1 m^2, consists of thirteen 5 mm thick ΔE- and sixteen 60 mm thick E-plastic scintillators. The outgoing proton was stopped in the E counters after passing through the ΔE counters.

The relative energy E_{rel} between ^{26}Si and proton was extracted by combining the positions and energies of the products. The energy E_{rel} corresponds to the center-of-mass energy, and thus the excitation energy of ^{27}P, E_x, can be determined as

$$E_x = E_{rel} + E_s, \tag{2}$$

where E_s represents the separation energy for the decay channel (0.861 MeV for ^{27}P).

RESULTS AND DISCUSSIONS

The relative energy spectrum is shown in Fig. 2. The closed circles represent the experimental data. The solid curve represents the best fit with five contributions shown by the dashed curves. The detector responses were obtained by a Monte-Carlo simulation using the GEANT4 code [13]. The peaks at 0.31 MeV and 0.8 MeV corresponds to the known first and second excited state in ^{27}P. The peaks at 1.3 MeV and 2.2 MeV are unknown states at 2.2 MeV and 3.1 MeV, respectively. The component which distributes from 0.8 MeV to 4.5 MeV corresponds to the non-resonant proton capture process.

We assumed that the non-resonant capture is dominated by the E1 transition and the astrophysical S-factor is independent of the energies.

The Coulomb dissociation cross section for the first excited state in ^{27}P was determined to be 17.5 $\pm$ 3.2 mb. The error includes the statistical one and ambiguity of the detection efficiency. Supposing the spin and parity of the state is 3/2$^+$ from the level scheme of the mirror nucleus ^{27}Mg [14], the transition between the first excited state and ground state (1/2$^+$) is induced by the M1 and E2 multipolarities. Thus the radiative width of interest, Γ_γ, has a M1 component ($\Gamma_\gamma(M1)$) and a E2 component ($\Gamma_\gamma(E2)$). In general, the M1 width is larger than the competing E2 width in the same γ-transition. However, the Coulomb dissociation is much more sensitive to the E2 component as compared with the M1 component by almost three order of magnitude in the present incident energy region. Due to this effect, the experimental cross section is expected to be exhausted by the E2 excitation, and the width $\Gamma_\gamma(E2)$ is determined to be (8.4 $\pm$ 1.5) $\times$ 10^{-5} eV from the measured cross section. To deduce the major M1 width, the E2/M1 mixing ratio δ for the first excited state in ^{27}P was estimated by coupling the known mixing ratio for the mirror nucleus ^{27}Mg [15] and the double ratio R of the mixing ratios of a mirror pair defined as

$$R = \delta_{T_Z+} / \delta_{T_Z-}, \tag{3}$$

where δ_{T_Z+} and δ_{T_Z-} represent the mixing ratios for mirror nuclei. The double ratio R was evaluated by a shell model calculation with the USDB interaction [16] using the effective charges $e_p = 1.35e$ and $e_n = 0.35e$ and effective g-factors taken to be single nucleon value. The error of the R is estimated to be 20% by comparing the calculated R with the known R [17]. The radiative width Γ_γ of the first excited state was deduced using the evaluated δ value and the measured E2 width to be $(4.2 \pm 1.3) \times 10^{-3}$ eV. The error includes the experimental error and ambiguity coming from the estimation of M1 contribution. This value is consistent with the previous estimation [5].

The cross sections for the second and third excited states were obtained to be 29.8 mb and 12.6 mb, respectively. Supposing the spins and parities of the states are 5/2$^+$ from the mirror analogy [14], their Γ_γ values were determined from the experimental cross sections alone, because the transitions to the ground state are induced by almost pure E2 transitions. They were determined to be (3.2 $\pm$ 0.6) $\times$ 10^4 eV and (5.9 $\pm$ 1.5) $\times$ 10^4 eV, respectively.

The reaction rate of the ^{26}Si(p,γ)^{27}P was calculated using the obtained Γ_γ values. Fig. 3 shows the temperature dependence of the reaction rate. The solid and dotted curves represent the present result of the reaction rate and range of its error. The error includes the ambiguities of the radiative widths and resonance energies. The dashed curve shows the direct capture component of reaction rate calculated by Caggiano *et al.* [5] based on a shell model. As seen in Fig 3 the resonant capture through the first excited state in ^{27}P is the dominant process at the temperature between 0.08 and 4 GK. This region covers the novae and X-ray bursts temperature, and hence the resonant capture through the first excited state in ^{27}P is the most important process in explosive nucleosynthesis in those astrophysical events. The rates through the second and third excited states have negligible contributions in the whole temperature range up to 5 GK. Compared with the β^+ decay rate of ^{26}Si, the proton capture reaction on ^{26}Si becomes faster at the peak temperature of heavy novae [18] where the equilibrium between the ground state and

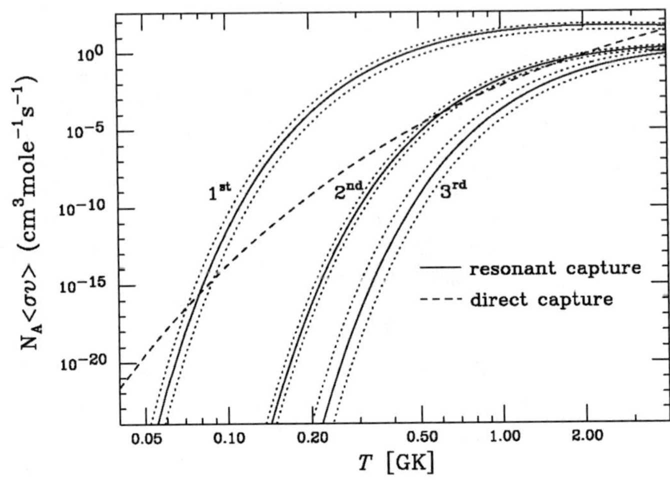

FIGURE 3. Temperature dependence of the reaction rate ^{26}Si(p,γ)^{27}P reaction. The solid and dashed curves represent the resonant and direct capture components, respectively. Dotted curve shows the margin of the errors for each resonant capture through each excited state in ^{27}P.

isomeric state in ^{26}Al is achieved. It is also indicated that the ^{26}Si(p,γ)^{27}P is faster than the β^+ decay of ^{26}Si in most cases of X-ray bursts [19].

REFERENCES

1. R. Diehl *et al., Astron. Astrophys. Suppl. Ser.* **97**, 181 (1993).
2. G. Meynet, *Astrophys. J. Suppl.* **92**, 441 (1994).
3. R. Gehrz *et al., Publ. Astron. Soc. Pac.* **110**, 3 (1998).
4. A. Coc *et al., Phys. Rev. C* **61**, 015801 (1999).
5. J. A. Caggiano *et al., Phys. Rev. C* **64**, 025802 (2001).
6. B. Guo *et al., Phys. Rev. C* **73**, 048801 (2006).
7. H. Herndl *et al., Phys. Rev. C* **52**, 1078 (1995).
8. T. Motobayashi *et al., Phys. Rev. Lett.* **73**, 2680 (1994).
9. N. Iwasa *et al., Phys. Rev. Lett.* **83**, 2910 (1999).
10. T. Gomi *et al., Nucl. Phys. A* **734**, E77 (2004).
11. G. Baur *et al., Nucl. Phys. A* **458**, 188 (1986).
12. T. Kubo *et al., Nucl. Instr. and Meth. B* **70**, 309 (1992).
13. S. Agostinelli *et al., Nucl. Inst. and Meth. A* **506**, 250 (2003).
14. P. M. Endt, *Nucl. Phys. A* **521**, 1 (1990).
15. M. J. A. de Voigt *et al., Nucl. Phys. A* **186**, 365 (1972).
16. B. A. Brown *et al., Phys. Rev. C* **74**, 034315 (2006).
17. Y. Togano *et al., in preparation* (2008).
18. C. Iliadis *et al., Astrophys. J. Suppl.* **142**, 105 (2002).
19. O. Koike *et al., Astron. & Astrophys.* **342**, 464 (1999).

α+^{6,8}He resonant scattering and exotic structures in ^{10,12}Be

Makoto Ito* and Naoyuki Itagaki†

*RIKEN, Hirosawa 2-1, Wako, Saitama 351-0198, Japan
†Department of Physics, University of Tokyo, Hongo, 113-0033, Tokyo, Japan

Abstract. The α+^{6}He low-energy reactions and the structural changes of ^{10}Be in the microscopic α+α+2N model are studied by the generalized two-center cluster model with the Kohn-Hulthén-Kato variation method. It is found that, in the inelastic scattering to the α+^{6}He(2_1^+) channel, characteristic enhancements are expected as the results of the parity-dependent non-adiabatic dynamics. The similar method is applied to the resonant scattering of α+^{8}He, and the coupling with the compound configurations of α+α+4N are discussed.

Keywords: Cluster model, Coupled channel and Distorted wave, Transfer reactions
PACS: 21.60.Gx,24.10.Eq,25.60.Je

In light unstable nuclei such as Be, B and C, the molecular orbital (MO) model well works in describing the many kinds of properties of these systems [1]. In particular, much efforts have been devoted to the Be isotopes (α+α+N+N...) in which a motion of valence neutrons couples to the two α-structure of ^{8}Be. The MO model such as π^- and σ^+-orbital around two α-cores has been shown to give a good description of the low-lying states of these isotopes [2].

In recent experiments, furthermore, the low-energy ^{6}He beam becomes available. Low-energy reaction cross-sections such as the elastic scattering with an α target [3] and the sub-barrier fusions with heavy target [4] have been accumulated. In future experiments, it will also be possible to investigate the molecular states in Be isotopes through the reactions such as α+^{6}He and α+^{8}He with low-energy ^{6,8}He-beams [5]. Therefore, it is very interesting to study theoretically the low-energy scattering of ^{6}He and ^{8}He by an α target.

In studying reaction processes exciting the molecular degrees of freedom, it is very important to construct a unified model which is capable of describing both structure and reaction on the same footing. The development of such a unified model is also useful in the study of the nuclear reaction for the astrophysical interests, because a determination of reaction rate of unstable nuclei is crucially important in the studies of explosive phenomena in the supernovae and the very early universe just after the big bang.

For this purpose, we introduce a microscopic model, the generalized two-center cluster model (GTCM) [6, 7]. In this model, it is possible to describe both molecular and atomic limit of the system of C_1+C_2+N+N+... where C_i is the i-th cluster core and N is the nucleon. In the region where two core nuclei are close, the total system is expected to form the molecular orbital structure, while in the region where two core nuclei are far apart, the molecular orbitals smoothly change into the product of the wave functions

CP1016, *Origin of Matter and Evolution of Galaxies*,
edited by T. Suda, T. Nozawa, A. Ohnishi, K. Kato, M. Y. Fujimoto, T. Kajino, and S. Kubono

consisting of the atomic orbitals. In this paper, we apply the GTCM for the ^{10,12}Be nuclei with the α+α+XN (X=2,4). We will analyze both molecular structure of ^{10,12}Be and the nuclear reaction such as the low-energy α+^{6,8}He scattering.

First, we briefly explain the framework of GTCM [6, 7] by showing the example of ^{10}Be. The basis functions are given as

$$\Phi^{J^\pi K}_{m,n}(S) \quad = \quad \hat{P}^{J^\pi}_K \cdot \mathscr{A}\{\psi_L(\alpha)\psi_R(\alpha)\varphi(m)\varphi(n)\} \ . \qquad (1)$$

The α-cluster wave function $\psi_i(\alpha)$ (i=L,R) is given by the $(0s)^4$ configuration in the harmonic oscillator (HO) potential. The position of an α-cluster is explicitly specified as the left (L) or right (R) side. The relative motion between α particles is described by a localized Gaussian function specified by the distance S [8]. A single-particle state for valence neutrons around one of α clusters is given by the atomic orbitals (AO), $\varphi(i,p_k,\tau)$ with the subscripts of a center i (=L or R), a direction p_k (k=x, y, z) of $0p$-orbitals and a neutron spin τ (=$\uparrow$ or $\downarrow$). In Eq. (1), the index $m(n)$ is an abbreviation of the AO (i,p_k,τ). The intrinsic basis functions with the full anti-symmetrization $\mathscr{A}$ are projected to the eigenstate of the total spin J, its intrinsic angular projection K and the total parity π by the projection operator $\hat{P}^{J^\pi}_K$.

The total wave function is finally given by taking a superposition over S and K as

$$\Psi^{J^\pi} \quad = \quad \int dS \sum_{iK} C^K_i(S)\, \Phi^{J^\pi K}_i(S) \qquad (2)$$

with $i \equiv (m, n)$. The coefficients $C^K_i(S)$ are determined by solving a coupled channel GCM (Generator Coordinate Method) equation [8]. If we fix the generator coordinate S and diagonalize the Hamiltonian with respect to i and K, we obtain the energy eigenvalues as a function of S, which we call the adiabatic energy surfaces (AES).

In the present calculation, we used the Volkov No.2 and the G3RS for the central and the spin-orbit part of the nucleon-nucleon (NN) interaction, respectively. The parameters in the NN interactions are modified from those in Ref. [6] so as to reproduce the threshold of α+^{6}He$_{g.s.}$ and the excitation energy of the ^{6}He(2^+_1) state [7]. The radius parameter b of HO wave functions for α clusters and valence neutrons is commonly taken as 1.46 fm. We included all the AO configurations of two neutrons that can be constructed by the $0p$-orbitals.

The AES of ^{10}Be (J^π=0^+) is shown in the Fig. 1. There appear three local minima at the short distance region of the AES. The adiabatic states (AS) at the lowest, second and third minima have the molecular orbital configurations of $(\pi^-_{3/2})^2$, $(\sigma^+_{1/2})^2$ and $(\pi^-_{1/2})^2$, respectively [2, 6, 7].

At the asymptotic region ($S \geq 8$ fm) where two α-cores are completely separated, the valence neutrons are localized around one of the α cores. The localization of the orbitals leads to the formation of the dinuclear channels such as [^{4}He+^{6}He(I)]$_L$ (white diamonds) and [^{5}He(I_1)+^{5}He(I_2)]$_{IL}$ (solid circles), in which individual channels are specified by the intrinsic spin of the clusters ($\mathbf{I}_1$, $\mathbf{I}_2$), the channel spin I ($\mathbf{I}$=$\mathbf{I}_1+\mathbf{I}_2$) and the relative angular momentum between clusters, L. The structural changes occur smoothly between the molecular orbital region and the dinuclear channels region.

To take into account the excitation of the relative motions between two α-cores, we solve the GCM equation by employing the AS from S=1 fm to S=70 fm with the mesh

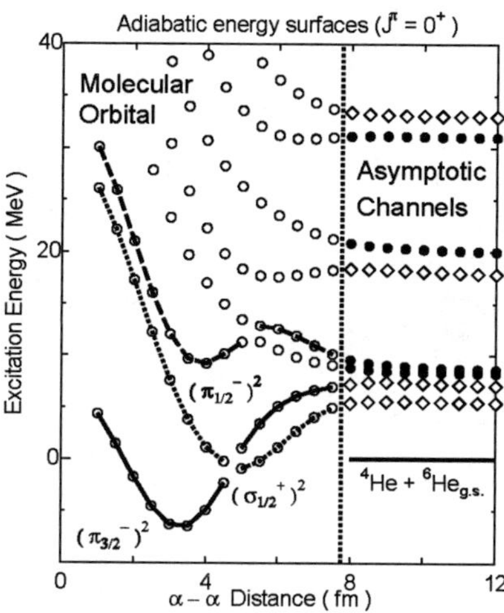

FIGURE 1. Adiabatic energy surfaces in ^{10}Be. The solid and dotted surfaces have the dominant components of $(\pi_{3/2}^-)^2$ and $(\sigma_{1/2}^+)^2$, respectively, while the dashed one has a dominant component of $(\pi_{1/2}^-)^2$.

of 0.5 fm. The calculated energy spectra are shown in the left panel in Fig. 2 (Fig. 2 A). In solving GCM, we apply the Absorbing-kernels in the Generator Coordinate Method (AGCM) in which the absorbing boundary condition is imposed outer region of the total system [9]. Due to the absorbing boundary, the resonance poles can be clearly identified in the complex energy plane.

The 0_1^+, 0_2^+ and 0_4^+ states are the poles corresponding to the respective local minima in the AES, having the molecular orbital structures. Therefore, we should call these states the "adiabatic poles", because they can be realized as the local minima in the AES. On the other hand, there is no local minimum corresponding to the 0_3^+ state. The AES of $(\pi_{3/2}^-)^2$ shown by the solid curve in Fig. 1A generates this resonance state, but only the surface AS ($S = 5 \sim 8$ fm) contribute to the formation of the 0_3^+ state. Because of the enhancement of the surface AS, the wave function in 0_3^+ has an enhanced component of $[\alpha+^6\mathrm{He}(2_1^+)]_{L=2}$ and hence, it does not have the molecular orbital configuration but the configuration similar to "the molecular resonance".

We have performed the same kind of calculation for J^{π}=1$^-$ and in left panel of Fig. 2 (Fig. 2 B), the AES for the J^{π}=1$^-$ state is shown by the dotted curves. In contrast to the results of the $J^{\pi} = 0^+$ states (Fig. 2A), there appears the avoided crossing between the lowest two AESs at S=6 fm. At the asymptotic region, the lowest and second AESs are smoothly connected to the $[\alpha+^6\mathrm{He}_{g.s.}]_{L=1}$ and $[\alpha+^6\mathrm{He}(2_1^+)]_{L=1}$ channels, respectively.

By performing the GCM calculation, we have identified two adiabatic poles corresponding to the two local minima of $(\pi_{j_z}^-\sigma_{1/2}^+)_{K=1}$. The lower pole ($j_z = 3/2$) is generated from the linear combination of the AS around the lowest local minimum, while the higher one ($j_z = 1/2$) is originated from the minimum in the third AES.

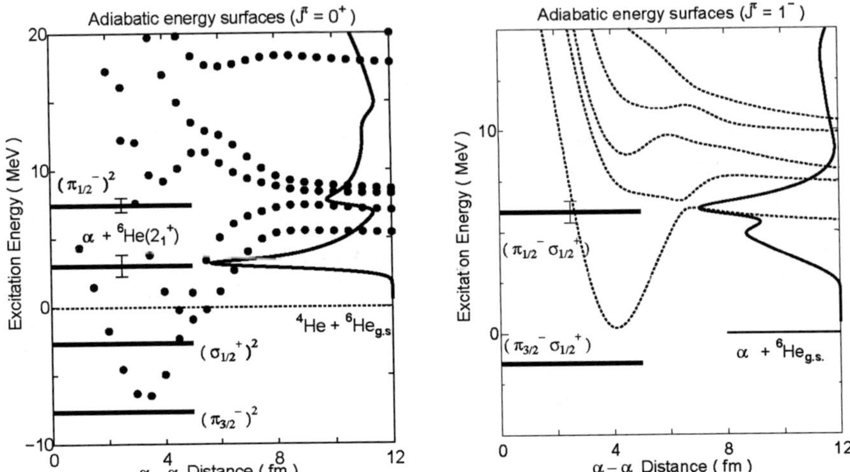

FIGURE 2. Left panel (A) : Energy spectra in ^{10}Be(0^+). The 0_1^+, 0_2^+ and 0_4^+ states have a configuration of $(\pi_{3/2}^-)^2$, $(\sigma_{1/2}^+)^2$, $(\pi_{1/2}^-)^2$, respectively, while the 0_3^+ state have the α+^{6}He(2_1^+) configuration. The solid curve shown in the right side represents the scattering matrix for the transition of α+^{6}He$_{g.s.}$ → α+^{6}He(2_1^+). Right panel (B) : The same as left panel but for J^π=1^-. In the right part, the partial cross section for the inelastic scattering of $[\alpha+^6\mathrm{He}_{g.s.}]_{L=0} \to [\alpha+^6\mathrm{He}(2_1^+)]_{L=1}$ is shown by the solid curve.

Let us consider the difference in the J^π=0^+ and 1^- in relation to the inversion doublet structure. The $(\pi_{3/2}^-)^2$ AES in J^π=0^+ and the α+^{6}He$_{g.s.}$ AES in J^π=1^- originally form the inversion doublet of α+^{6}He$_{g.s.}$ with K^π=$0^\pm$. The J^π=0^+ partner couples to the symmetric ^{5}He+^{5}He configuration and hence, it is strongly distorted to the molecular orbital as the distance S gets closer. In the J^π=1^- surface, however, the dinuclear configuration of α+^{6}He$_{g.s.}$ is well developed, which is similar situation to the ^{20}Ne=α+^{16}O system [10]. Therefore, the appearance of the avoided crossing in the negative parity has a close connection to the formation of the inversion doublet in the compound system of ^{10}Be.

We next discuss how the AES profile and the level scheme appear in the low energy reactions. We show the partial cross section for the inelastic scattering to ^{6}He(2_1^+) excitation. In solving the scattering problem, we employ the Kohn-Hulthén-Kato (KHK) method [11] where the AO basis in Eq. (1) are transformed to the asymptotic channel wave function. The present calculation is equivalent to the usual coupled-channels calculation including the channels of α+^{6}He(0_1^+,0_2^+,2_1^+,2_2^+,1^+) and ^{5}He($3/2^-$,$1/2^-$)+^{5}He($3/2^-$,$1/2^-$).

The calculated partial inelastic cross-sections are shown in the right part of Figs. 2A and 2B. In J^π=0^+, a strong peak appears at $E_{c.m.}$~ 3 MeV, although there is no definite avoided crossing in the AES. This is due to the effect of the molecular resonance, 0_3^+, which includes the large component of the exit ^{6}He(2_1^+) channel. We can also see the enhancements at $E_{c.m.}$ ~ 7 MeV, which nicely coincident to the adiabatic pole of $(\pi_{1/2}^-)^2$, 0_4^+, but it is much smaller than that generated by the molecular resonance.

In J^{π}=1^{-}, the strong enhancement can also be seen at $E_{c.m.}\sim$ 6 MeV, nevertheless there is no pole in the incident and exit channels. The energy of the enhancement is quite close to that of the avoided crossing at S=6 fm. Therefore, we can conclude that this inelastic peak is due to the Landau-Zener transition at the avoided crossing [12].

Finally, we discuss the results of the $\alpha+^{8}$He scattering. In this case, the compound nucleus is ^{12}Be=$\alpha+\alpha+4N$ and the number of the AOs which should be included in the exact KHK becomes much larger than those in ^{10}Be. Therefore, we restrict the AO configuration to the axial symmetric shape (K=0), and apply the closed channel method proposed by Kamimura [13].

In this method, the total wave function is written as

$$\Psi^{J^{\pi}(+)} = \sum_{\nu} b_{\nu} \hat{\Psi}_{\nu}^{J^{\pi}K=0} + \sum_{\beta} \varphi_{\beta}\, \chi_{\beta}^{(+)} . \tag{3}$$

The first term stands for closed channels within the finite range of S, which is the linear combination of the solutions of Eq. (2) with the subscript of the eigenvalue, ν and describes the internal compound-states before decaying into the open channels. The second term represents the binary open channels labeled by β, on which the scattering boundary condition is explicitly imposed by utilizing the KHK method [7, 13, 9]. We consider three rearrangement channels, $\alpha+^{8}\mathrm{He}_{g.s.}$, $^{6}\mathrm{He}_{g.s.}+^{6}\mathrm{He}_{g.s.}$ and $^{5}\mathrm{He}_{g.s.}+^{7}\mathrm{He}_{g.s.}$ as the open channels, in which ^{5,7}He are described by the HO wave functions.

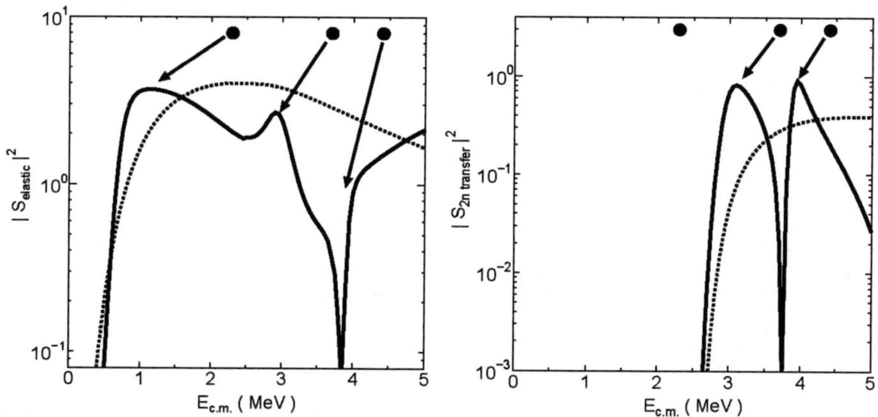

FIGURE 3. Left panel (A) : The squared magnitude of the S-matrix of the $\alpha+^{8}$He elastic scattering as a function of the collision energy (J^{π}=0^{+}). The solid and dotted curves show the S-matrix, while the circles represent the energy positions of the individual closed channels. Right panel (B) : The same as Fig. 3B but for $\alpha+^{8}\mathrm{He} \rightarrow {}^{6}\mathrm{He}+^{6}\mathrm{He}$. See text for details.

In Fig. 3, we show the S-matrix of the elastic channel (A) and the two neutron transfer one, ^{6}He+^{6}He (B). The dotted curves in the panel of A and B represent the solutions of the open channels in which only the second term of Eq. (3) is included in the calculation. In both panels, there is no sharp structures in the dotted curves, but the resonant behaviors appear in the full solution (solid curves) where the coupling to the closed compound systems (first term of Eq. (3)) is switched on. The arrows indicate the main contribution to generate the respective resonant peaks in the excitation functions. It should be noticed that the magnitude of the S-matrix is largely changed by the coupling between the open channels and the closed ones. This means that the calculation of the open channel is insufficient in estimating the cross section at the extremely low-energy collision. The closed compound system cannot decay into the asymptotic region, but their coupling is quite important in determining the yield of the reaction cross section in the slow collision.

In summary, we investigated the adiabatic properties of the ^{10,12}Be structures as well as the α+^{6,8}He$_{g.s.}$ low-energy scattering. We achieved such an unified study of the structures and the reactions by employing the generalized two-center cluster model (GTCM). In the studies of the α+^{6}He scattering, Kohn-Hulthén-Kato (KHK) method is applied. We have found the strong enhancements in the inelastic scattering cross-section to the α+^{6}He(2_1^+) in both the J^π=0^+ and J^π=1^- states. However, their origins are very different to each other. In the J^π=0^+, the enhancements is due to the molecular resonance formation, while the enhancement in J^π=1^- is originated from the Landau-Zener level-crossing between the lowest two adiabatic energy surfaces. In the scattering of α+^{8}He, the closed channel method is applied and the coupling between the open channels and the closed ones plays very important role in the extremely slow collision process. The experiment of the α+^{8}He scattering is planned at GANIL [14] and hence, it is very interesting to compare our prediction with the feature experiment.

The authors would like to thank Profs. M. Kamimura, K. Kato, S. Shimoura and Drs. A. Saito and M. Milin for useful discussions and encouragements. This work has been supported by the Grant-in-Aid for scientific Research in Japan (Nos. 18740129 and 19740124). The numerical calculations have been performed at the RSCC system, RIKEN.

REFERENCES

1. Y. Kanada-En'yo, H. Horiuchi and A. Dote, *Phys. Rev. C***60**, 064304 (1999), *ibid, Phys. Rev. C***66**, 024305 (2002), *ibid, Phys. Rev. C***68**, 014309 (2003).
2. N. Itagaki and S. Okabe, *Phys. Rev. C***61**, 044306 (2000) ; N. Itagaki, S. Okabe and K. Ikeda, *Phys. Rev. C***62**, 034301 (2000), and references therein.
3. R. Raabe *et al., Phys. Rev. C***67** 044602 (2003).
4. R. Raabe *et al., Nature* **431** 823 (2004), and references therein.
5. M. Freer *et al., Phys. Rev. Lett.* **96** 042501, N. Soić, Private communication.
6. M. Ito, K. Kato and K. Ikeda, *Phys. Lett. B***588**, 43 (2004), *ibid, in Proceedings of the international symposium on A New Era of Nuclear Structure Physics,* Y. Suzuki, S. Ohya, M. Matsuo and T. Ohtsubo (World Scientific, 2004), p. 148.
7. M. Ito, K. Yabana, K. Kato and K. Ikeda, *J. Phys. Con. Ser.* **20**, 185 (2005) ; M. Ito, *Phys. Lett. B***636** 293 (2006) ; M. Ito *J. Phys. Con. Ser.* **49**, 206 (2006).
8. H. Horiuchi *et al., Suppl. Prog. Theor. Phys.* **62**, 1 (1977) and references therein .
9. M. Ito and K. Yabana, *Prog. Theor. Phys.* **113**, 1047 (2005)
10. H. Horiuchi *et al., Suppl. Prog. Theor. Phys.* **52**, 1 (1972) ; *Suppl. Prog. Theor. Phys.* **68**, 1 (1980) and references therein.
11. M. Kamimura, *Prog. Theor. Phys. Suppl.* **62**, 236 (1977).
12. See for example, L. D. Landau and E. M. Lifshitz, *Quantum mechanics, 3rd ed.* (Elsevier, Butterworth-Heinemann, 1958). H. Nakamura, *Nonadiabatic Transition,* (World Scientific, Singapore, 2000) and references therein .
13. K. Hamaguchi et al., Phys. Lett. B **650**, 268 (2007).
14. M. Freer, N. Orr, A. W. Ashwood and M. Millin, *private communications* (2007).

9. NUCLEAR STRUCTURE and REACTIONS FOR ASTROPHYSICS (2)

Experimental Nuclear Astrophysics with Real Photon Beams

Tatsushi Shima

Research Center for Nuclear Physics, Osaka University
10-1, Mihogaoka, Ibaraki, Osaka 567-0047, Japan

Abstract. Real photon beams in the energy region up to several ten MeV provide a good opportunity for experimental studies of astrophysical nucleosynthesis caused by the electromagnetic interactions as well as the weak interactions. In this paper development of real photon sources at low energies is reviewed, and a recent experimental work on the photonuclear reactions of ^{4}He with new-generation γ-ray sources by the laser Compton-scattering method is presented.

Keywords: nucleosynthesis, photonuclear reaction, laser Compton-scattered γ-ray

PACS: 25.20.-x, 26.30.-k, 26.35.+c, 26.50.+x

INTRODUCTION

It is considered the primordial light elements were produced via the nuclear reaction chains including radiative capture reactions such as p(n,γ)d, d(p,γ)^{3}He, ^{3}H(α,γ)^{7}Li, ^{3}He(α,γ)^{7}Be, and so on [1-3]. In stellar helium-burning, the (α,γ) reactions and the (n,γ) reactions are responsible for energy generation [4-6] and heavy-element synthesis by the s-process [7,8], respectively. On the other hand, nucleosynthesis occurring in stellar explosive environments like r-, p-, and γ-processes [9-15] are supposed to be governed by the (n,γ) reactions and their inverse (γ,n) reactions. In addition, the photonuclear reactions are expected to provide useful information about analogous neutrino-induced nuclear reactions by neutral currents [16,17], which are considered to play critical roles in the dynamics [18,19] of core-collapse supernovae and nucleosynthesis in neutrino-driven wind [17,20].

For studying the above processes of nucleosynthesis, accurate data of the relevant nuclear reactions are highly demanded. Photonuclear reaction experiments by means of real photon beams are quite useful for such purposes because of the following reasons:

· The property of the electromagnetic interaction is well-known.

· The photodisintegration cross section is enhanced compared to that of its inverse radiative capture reaction due to the ratio of the phase space factors.

CP1016, *Origin of Matter and Evolution of Galaxies,*
edited by T. Suda, T. Nozawa, A. Ohnishi, K. Kato, M. Y. Fujimoto, T. Kajino, and S. Kubono

· They provide a unique opportunity to study many-body fusion processes like $\alpha(\alpha n,\gamma)^{9}Be$ and $^{4}He(2\alpha,\gamma)^{12}C$ through their inverse reactions.

So far experimental studies of the photonuclear reactions have been performed by using a variety of real photon sources. Discrete γ-rays from radioactive isotopes and nuclear reactions were used in the early days, but they were not so useful because of their fixed energies and low intensities. Then accelerator-driven photon sources have been developed to generate intense γ-rays in a wide energy range. The bremsstrahlung photons generated by Coulomb scattering of high-energy electrons in a target made of heavy element have been used for a number of photodisintegration experiments. Since their energy distributions are continuous, one needs to make differential measurements by shifting the electron energy to determine a photonuclear reaction cross section as a function of the photon energy. The photon tagging technique was also employed to determine a γ-ray energy by measuring the energy of the scattered electron in coincidence. The bremsstrahlung photons, however, have energy distributions characterized by the inverse of the photon energy, and the total flux is dominated by low energy γ-rays, which may cause serious background in experiments. To generate γ-ray beams with better quality, quasi-monochromatic γ-ray sources have been developed. In 1960's, the γ-ray sources by means of the positron-annihilation in flight were developed using positron linear accelerators [21], and were applied for systematic studies of the (γ,n) reactions [22]. Since the annihilation γ-ray beam also contains a considerable amount of bremsstrahlung photons, one has to subtract their contribution by making measurements with a positron beam and an electron beam separately at the same beam energies. The laser Compton-scattering (LCS) method was proposed to realize a more useful quasi-monochromatic γ-ray source by Milburn [23] and Arutyunian and Tumanian [24] in 1963. The LCS γ-ray is produced by the Compton backscattering of a laser light and a relativistic electron beam. The γ-ray energy E_γ is determined as a function of the energy E_L of the laser photon, the electron beam energy E_e, the incident angle θ_L and the backscattering angle θ of the photon as given by Eq. 1;

$$E_\gamma = \frac{E_L \cdot \left(1-(v_e/c)\cos\theta_L\right)}{1-(v_e/c)\cos\theta + (E_L/E_e)\cdot\left(1-\cos(\theta_L-\theta)\right)}\ , \qquad (1)$$

where v_e and c are the velocities of the electron and the light in a vacuum. By extracting the forward component of the γ-ray using a collimator, a quasi-monochromatic γ-ray beam is obtained. Fig. 1 shows an example of the LCS γ-ray pulse height spectrum. Since the LCS γ-ray contains little low-energy background photons, it enables one to measure the reaction cross section at the γ-ray energy of interest with a good signal-to-noise ratio.

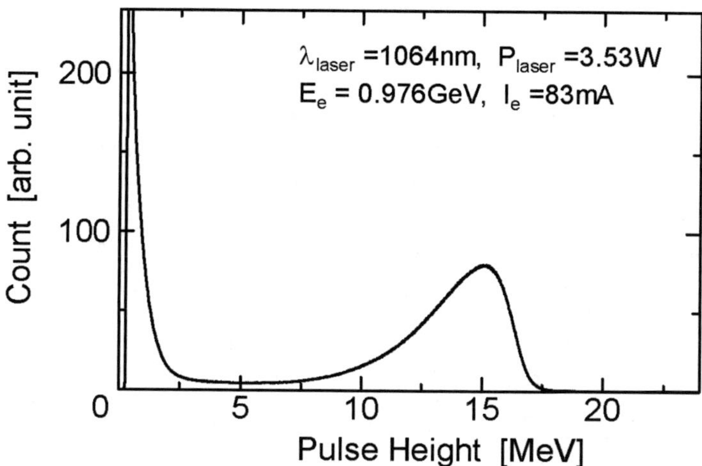

FIGURE 1. Example of the pulse height spectrum of the LCS γ-ray at NewSUBARU measured with a GSO scintillation counter. The energy and the current of the electron beam were 0.976 GeV and 83 mA, respectively. The wavelength and the output power of the Nd:YVO_4 laser were 1064 nm and 3.53 W, respectively. The LCS γ-ray was collimated by a slit made of lead with the diameter of 3 mm and the thickness of 100 mm. The dimension of the GSO detector was 76mm(W)×76mm(H)×180mm(L).

The LCS γ-ray source has the following features;

· quasi-monochromatic ($\Delta E_\gamma/E_\gamma < \sim 10\%$) and variable energy,

· little background γ-rays due to bremsstrahlung,

· small angular dispersion ($\Delta\theta < \sim 0.3$ mrad),

· considerably high intensity ($10^4 \sim 10^8$ photons/MeV/s), and

· large linear and/or circular polarization of nearly 100%.

The first realization of the LCS γ-ray source succeeded at the Lebedev institute in 1964 [25]. Since then, a number of LCS γ-ray sources have been built according to the development of new-generation synchrotron radiation facilities. At present, three LCS γ-ray sources are in operation in the energy region below a few ten MeV, which are of interest in the studies of nuclear astrophysics.

In 1985 the LCS γ-ray source at the National Institute of Advanced Industrial Science and Technology (AIST) was built using a 800 MeV electron storage ring TERAS [26]. The γ-rays in the energy range from 2 MeV to 40 MeV are produced using fundamental and harmonic lights of a Nd:YLF laser, and were applied to study the photonuclear reactions of light nuclei [27-30] as well as heavy nuclei [31-35].

The High Intensity γ-ray Source (HIγS) at the Duke FEL facility has been in operation from 1999 [36]. A unique feature of HIγS is its very high efficiency of the γ-ray production by intra-cavity scattering of the electron beam and the laser beam by means of the free electron laser, which is driven by the electron storage ring itself. At present, it covers the energy range from 2 MeV to 70 MeV, and the total γ-ray intensity of 2×10^8 photons/s and 10^9 photons/s are achieved above and below 20 MeV, respectively. Up to now the measurements of the (γ,n) reactions of light nuclei such as D [37,38] and ^{9}Be [39] have been performed. A project of the ^{16}O(γ,α)^{12}C experiment is also ongoing for precise determination of the inverse ^{12}C(α,γ)^{16}O reaction rate [39], which is relevant to the evolution of the stars in the helium-burning phase.

Most recently, a new LCS γ-ray source has become available at the NewSUBARU synchrotron radiation facility of the Laboratory of Advanced Science and Technology for Industry at the University of Hyogo, Japan [40]. It can provide γ-rays with the energy of 16~40 MeV and the total intensity of up to 2×10^6 photons/s by using a 3 W Nd:YVO_4 laser and a 1~1.5 GeV electron beam with an electric current of 200 mA. The experimental research project of nuclear astrophysics by means of the NewSUBARU LCS γ-ray beam have started in 2005. In the following, a new experimental result on the photonuclear reactions of ^{4}He is presented.

PHOTODISINTEGRATION OF ^{4}HE

Recent model calculations of core-collapse supernovae (SNe) suggest that a considerable amount of α-particles may be produced around the iron core due to the photodisintegrations [41,42], and it has been argued that the energy exchange by the neutrino inelastic scattering with the produced α-particles may assist the explosion [43]. To evaluate the contribution of the inelastic neutrino-^{4}He scattering, one needs information about nuclear responses of ^{4}He for the weak interactions by the neutral current as well as the charged current. The nuclear data of the inelastic neutrino-^{4}He scattering are also demanded to calculate nucleosynthetic yields of the r-process in the neutrino-driven wind of core-collapse SNe [16,18,20,44]. As mentioned above, the photonuclear reactions would be useful tools to probe the nuclear responses of ^{4}He to the neutral-current weak interaction. So far, the photonuclear reaction cross sections of ^{4}He have been measured in the energy region from 20 to 215 MeV using quasi-monochromatic photon beams and/or bremsstrahlung photon beams (see the references in [29]). The inverse ^{3}H(p,γ)^{4}He and ^{3}He(n,γ)^{4}He reactions were also measured in order to determine the photonuclear reaction cross sections through the principle of the detailed balance. In the energy region above E_γ~40 MeV, most of the existing data are in agreement with each other. On the other hand, the data show discrepancies as large as 50~100% below E_γ~40 MeV, and there has been arguments concerning the backgrounds caused by low-energy γ-rays, geometrical acceptance of a detector for the low-energy fragments from the photonuclear reactions, the effective thickness of the target, and so on. To determine the photonuclear reaction cross sections accurately, we performed an experiment with use of quasi-monochromatic LCS γ-ray beams and a time projection chamber (TPC) [45]. Since the TPC gas contained helium gas, it could detect the charged fragments with an efficiency of ~100% and a solid angle of 4π. The cross section data up to E_γ ~30MeV have been obtained using the LCS γ-ray source at AIST [29], and the measurement was continued in the extended energy region up to 37.2 MeV at NewSUBARU. Fig. 2 shows a schematic view of the experimental set up.

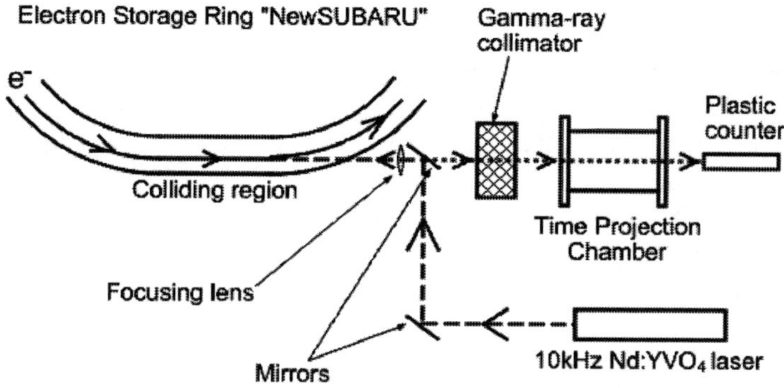

FIGURE 2. Experimental setup for the measurement of the photonuclear reactions of ^{4}He at NewSUBARU.

We generated the LCS γ-rays with the maximum energies ranging from 29.5 MeV to 37.2 MeV using a 1064 nm laser light and electron beams with the energy of 1.3 ~ 1.46 GeV. In order to study possible backgrounds contained in the incident γ-ray beam, we also made a measurement at the mean γ-ray energy of 31.6 MeV by using the LCS γ-rays produced by a 532 nm laser light and 0.974 GeV electrons.

Fig. 3 shows an example of the observed tracks of the charged fragments from the ^{4}He(γ,p)^{3}H reaction. The detected events were analyzed by checking the tracks of the charged fragments and their energy losses in the effective region of the TPC. The background was found to be dominated by the scattered electrons in the TPC, and they were easily distinguished by their meandering tracks and small energy losses. Another type of the background was due to the α-decays of natural radioactive impurities in the TPC gas, and they were rejected by requiring the existence of a reaction vertex on the γ-ray beam axis. The γ-ray intensity was measured simultaneously using a plastic scintillation counter, whose detection efficiency was calibrated with a GSO scintillation counter with the absolute efficiency of 96.0±0.5%. The energy distribution of the incident γ-ray was measured with a 6"φ×5" NaI(Tl) detector. The absolute values of the photonuclear reaction cross sections were determined from the counting rates of the reactions, the γ-ray intensity, and the number density of the target nucleus in the effective region of the TPC. The total sensitivity of the measurement was checked by measuring the d(γ,p)n reaction simultaneously by using the gas mixture of He (103 kPa) and CD_4 (25.8 kPa), where the d(γ,p)n reaction cross section has been well known experimentally in the energy region of the present interest. Further details of the present experimental method are described elsewhere [29,45].

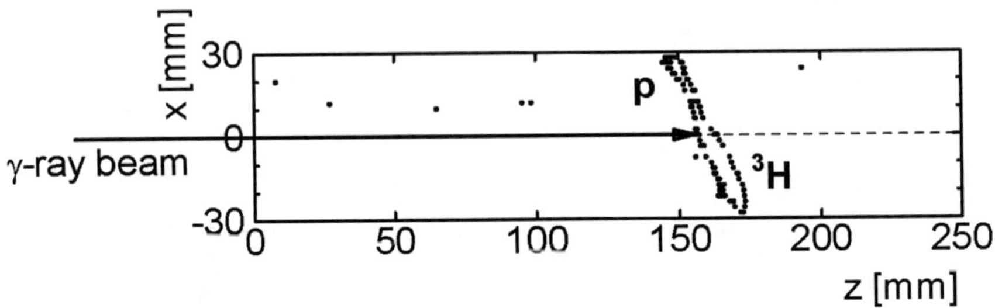

FIGURE 3. Example of the tracks of the charged fragments from the $^4He(\gamma,p)^3H$ reaction.

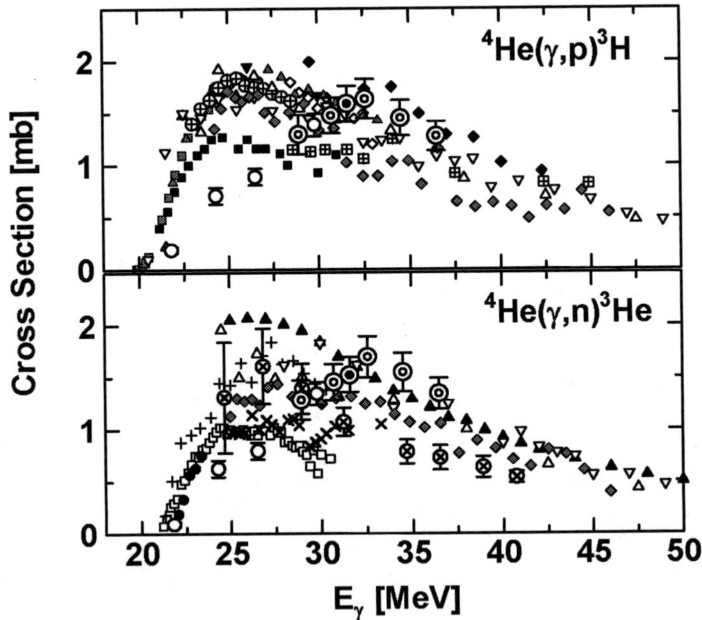

FIGURE 4. Two-body photodisintegration cross sections of ^4He. The double circles and the dotted circle stand for the preliminary data obtained from the present measurements with the laser lights with the wavelengths of 1064 nm and 532 nm, respectively. The open circles denote our previous data measured at AIST [29]. The crossed circles are the latest data of the (γ,n) reaction measured by means of a tagged photon beam by Nilsson et al [46,47]. The other symbols indicate the previous data (see the references in [29]). Upper panel; $^4He(\gamma,p)^3H$, open upward triangles; Gorbunov, open downward triangles; Arkatov *et al.*, crossed squares; Bernabei *et al.*, filled squares; Feldman *et al.*, open diamonds; Hoorebeke *et al.*, gray squares; Hahn *et al.*. Lower panel; $^4He(\gamma,n)^3He$, open upward triangles; Gorbunov, open downward triangles; Arkatov *et al.*, open squares; Berman *et al.*, diagonal crosses; Ward *et al.*, filled circles; Komar *et al.*. The error bars of the previous data are not shown for clarity.

Fig. 4 shows the preliminary result of the cross section of the $^4He(\gamma,p)^3H$ and $^4He(\gamma,n)^3He$ reactions from the present experiment together with the previous data. The present results for the laser wavelength of 1064 nm are consistent with the one for 532 nm at 31.6 MeV and also with our previous data measured at AIST [29], suggesting the cross sections were determined accurately without any systematic error concerning the incident γ-ray beams. Our previous and present results indicate that both the (γ,p) and (γ,n) cross sections monotonically increase up to the peak energy of 32.6 MeV, and decrease in the region above the peak energy. Below the peak energy, our data obtained at AIST are in disagreement with many of the previous ones, and that is to be solved by further experimental efforts. It is also interesting to compare the present data with recent theoretical calculations which are used to estimate the inelastic neutrino-^4He reaction cross sections [48-50].

SUMMARY

Because of their high performances, quasi-monochromatic LCS γ-ray beams are the promising tool for high-precision experiments on the photonuclear reactions. Recently built LCS facilities can provide intensive LCS γ-rays, covering the energy range from a few MeV up to several ten MeV, and are suitable for experimental studies of nuclear astrophysics. At the NewSUBARU LCS γ-ray source, a new experimental project for nuclear astrophysics is in progress, and several experiments have already been performed successfully [51]. The following experimental studies are currently ongoing or planned;

- photodisintegrations of light nuclei relevant to big bang nucleosynthesis and explosive nucleosynthesis,
- (γ,α) reactions relevant to α- and γ-processes,
- (γ,n) reactions of heavy nuclei including long-lived radioactive isotopes and isomers for studies of p-process and nuclear cosmochronology, and
- photo-excitations of isobaric analog states to simulate nuclear spin excitations by neutrinos.

ACKNOWLEDGMENTS

The author would like to thank Profs. Y. Nagai, H. Utsunomiya, T. Kajino, H. Akimune, S. Miyamoto and T. Mochizuki for fruitful discussions and collaborative works. This work was supported in part by Grant-in-Aid for Specially Promoted Research of the Japan Ministry of Education, Science, Sports and Culture and in part by Grant-in-Aid for Scientific Research of the Japan.

REFERENCES

1. K. M. Nollett and S. Burles, *Phys. Rev.* **D61**, 123505 (2000).
2. R. H. Cyburt, *Phys. Rev.* **D70**, 023505 (2004).
3. A. Coc, E. V-Flam, P. Descouvemont, A. Adahchour, and C. Angulo, *Astrophys. J.* **600**, 544-552 (2004).
4. F. Hoyle, *Astrophys. J., Suppl.* **1**, 121 (1954).
5. T. Rauscher, A. Heger, R. D. Hoffman, and S. E. Woosley, *Astrophys. J.* **576**, 323 (2002).
6. L. R. Buchmann and C. A. Barnes, *Nucl. Phys.* **A777**, 254-290 (2006).
7. R. L. Macklin and J. H. Gibbons, *Rev. Mod. Phys.* **37**, 166 (1965).
8. F. Käppeller, H. Beer, and K. Wisshak, *Rep. Prog. Phys.* **52**, 945 (1989).
9. S.E. Woosley and R.D. Hoffman, *Astrophys. J.* **395**, 202 (1992).
10. K. Takahashi, J. Witti and H.-Th. Janka, *Astron. Astrophys.* **286**, 857 (1994).
11. A.G.W. Cameron, *Astrophys. J.* **562**, 456 (2001).
12. S. Rosswog *et al. Astron. Astrophys.* **341**, 499 (1999).
13. M. Arnould and S. Goriely, *Phys. Rep.* **384**, 1 (2003).
14. S.E. Woosley and W.M. Howard, *Astrophys. J. Suppl.* **36**, 285 (1978).
15. M. Rayet, N. Prantzos, and M. Arnould, *Astron. Astrophys.* **227**, 271 (1990).
16. S.E. Woosley, D.H. Hartmann, R.D. Hoffman, and W.C. Haxton, *Astrophys. J.* **356**, 272 (1990).
17. B.S. Meyer, *Astrophys. J.* **449**, L55 (1995).
18. S.E. Woosley et al., *Astrophys. J.* **433**, 229 (1994).
19. H.-T. Janka and E. Muller, *Astron. Astrophys.* **306**, 167 (1996).
20. A. Heger *et al.*, *Phys. Lett.* **B606**, 258 (2005).
21. B.L. Berman and S.C. Fultz, *University of California Lawrence Livermore Laboratory Report* No. UCRL-75383.
22. B.L. Berman and S.C. Fultz, *Rev. Mod. Phys.* **47**, 713 (1975).
23. R.H. Milburn, *Phys. Rev. Lett.* **10**, 75 (1963).
24. F.R. Arutyunian and V.A. Tumanian, *Phys. Lett.* **4**, 176 (1963).
25. O.F. Kulikov, Y.Y. Telnov, E.I. Filippov, and M.N. Yakimenko, *Phys. Lett.* **13**, 344 (1964).
26. H. Ohgaki *et al.*, *IEEE Trans. Nucl. Sci.* **38**, 386 (1991).
27. H. Utsunomiya *et al.*, *Phys. Rev.* **C63**, 018801 (2001).
28. K.Y. Hara, H. Utsunomiya, S. Goko, H. Akimune, T. Yamagata, and M. Ohta, *Phys. Rev.* **D68**, 072001 (2003).
29. T. Shima *et al.*, *Phys. Rev.* **C72**, 044004 (2005).
30. S. Naito *et al.*, *Phys. Rev.* **C73**, 034003 (2006).
31. H. Utsunomiya *et al.*, *Phys. Rev.* **C67**, 015807 (2003).
32. P. Mohr *et al.*, *Phys. Rev.* **C69**, 032801 (2004).
33. T. Shizuma *et al.*, *Phys. Rev.* **C72**, 025808 (2005).
34. H. Utsunomiya *et al.*, *Phys. Rev.* **C74**, 025806 (2006).
35. S. Goko *et al.*, *Phys. Rev. Lett.* **96**, 192501 (2006).
36. H.R. Weller and M.W. Ahmed, *Modern Physics Letters.* **A18**, 1569 (2003).
37. E.C. Schreiver *et al.*, *Phys. Rev.* **C61**, 061604 (2000).
38. W. Tornow *et al.*, *Phys. Lett.* **B574**, 8 (2003).
39. HIGS Progress Report I, 2003-2004.
40. K. Aoki *et al.*, *Nucl. Instr. Meth. in Phys. Res.* **A516**, 228 (2004).
41. K. Sumiyoshi *et al.*, *Astrophys. J.* **629**, 922 (2005).
42. N. Ohnishi, K. Kotake, and S. Yamada, *Astrophys. J.* **667**, 375 (2007).
43. W. C. Haxton, *Phys. Rev. Lett.* **60**, 1999 (1988).
44. T. Yoshida, M. Terasawa, T. Kajino, and K. Sumiyoshi, *Astrophys. J.* **600**, 204 (2004).
45. T. Kii, T. Shima, T. Baba, and Y. Nagai, *Nucl. Instr. Meth. in Phys. Res.* **A552**, 329 (2005).
46. B. Nilsson *et al.*, *Phys. Lett.* **B626**, 65 (2005).
47. B. Nilsson *et al.*, *Phys. Rev.* **C75**, 014007 (2007).
48. D. Gazit, S. Bacca, N. Barnea, W. Leidemann, and G. Orlandini, *Phys. Rev. Lett.* **96**, 112301 (2006).
49. T. Suzuki, S. Chiba, T. Yoshida, T. Kajino, and T. Ohtsuka, *Phys. Rev.* **C74**, 034307 (2006).
50. S. Bacca, *Phys. Rev.* **C75**, 044001 (2007).
51. T. Hayakawa *et al.*, *Phys. Rev.* **C74**, 065802 (2006).

$E1$ and $E2$ cross sections of the $^{12}C(\alpha,\gamma)^{16}O$ reaction at $E_{cm} \sim 1.4$ MeV using pulsed α beams

H. Makii*, Y. Nagai*,†, K. Mishima**, M. Segawa*, T. Shima†, H. Ueda*,†
and M. Igashira‡

*Japan Atomic Energy Agency, Tokai, Ibaraki 319-1195, Japan
†Research Center for Nuclear Physics, Osaka University, Ibaraki, Osaka 567-0047, Japan
**RIKEN, Wako, Saitama 351-0198, Japan
‡Research Laboratory for Nuclear Reactors, Tokyo Institute of Technology, Meguro, Tokyo
152-8550, Japan

Abstract. We have installed a new system to measure the γ-ray angular distribution of the $^{12}C(\alpha,\gamma)^{16}O$ reaction at the 3.2 MV Pelletron accelerator laboratory at Tokyo Institute of Technology to accurately determine the $E1$ and $E2$ cross sections. In this experiment, we used high efficiency anti-Compton NaI(Tl) spectrometers to detect a γ-ray from the reaction with a large S/N ratio, intense pulsed α beams to discriminate true events from neutron induced background with a time-of-flight (TOF) method, and the monitoring system of target thickness. We succeeded in removing a background due to neutrons and could clearly detect the γ-ray from the $^{12}C(\alpha,\gamma)^{16}O$ reaction with high statistics.

Keywords: α-induced reaction, Radiative capture, Angular distribution, Stellar helium burning
PACS: 25.55.-e, 25.40.Lw, 26.20.Fj, 27.20.+n

INTRODUCTION

The $^{12}C(\alpha,\gamma)^{16}O$ reaction cross section at the stellar temperature, at a center-of-mass energy of $E_{cm} \approx 0.3$ MeV, plays an important role in determining the mass fraction of ^{12}C and ^{16}O after stellar helium burning, abundance distribution of the elements between carbon and iron, and the iron core mass before supernovae explosion [1]. Therefore, it is quite important to accurately determine the cross section at $E_{cm} \approx 0.3$ MeV. The direct measurement of the cross section at $E_{cm} \approx 0.3$ MeV, however, is not possible using a current experimental technique, because the cross section is too small, around 10^{-17} barn. Hence, one has to derive the cross section by extrapolating a measured cross section at $E_{cm} \geq 1.0$ MeV into the range of stellar temperature by using theoretical calculations [2–4]. The total cross section of the $^{12}C(\alpha,\gamma)^{16}O$ reaction, $\sigma_{tot.}(E_{cm} \sim 0.3$ MeV), has been known to be dominated by electric dipole ($E1$) and electric quadrupole ($E2$) α-capture reactions into the ground state of ^{16}O [2, 3]. Extensive studies of cross-section measurements at $E_{cm} \geq 1.0$ MeV, including the cascade transitions via the excitation states of ^{16}O, have been carried out over a period of 30 years using both the direct α radiative capture reaction by ^{12}C [5–9] and the inverse ^{12}C radiative capture reaction by ^{4}He [10–15]. The former reaction was employed to determine both the $E1$ cross section, $\sigma_{E1}(E_{cm})$, and the $E2$ one, $\sigma_{E2}(E_{cm})$, by measuring the angular distribution of the characteristic γ ray from the $^{12}C(\alpha,\gamma)^{16}O$ reaction to the ground state in ^{16}O, since the $E1$ and $E2$ transition amplitudes for the γ ray contribute to

CP1016, *Origin of Matter and Evolution of Galaxies*,
edited by T. Suda, T. Nozawa, A. Ohnishi, K. Kato, M. Y. Fujimoto, T. Kajino, and S. Kubono

$\sigma_{tot.}(E_{cm} \sim 0.3$ MeV) with a different dependence on E_{cm} [2–4]. The former reaction, however, suffers from background events of high-energy neutrons from the ^{13}C(α,n)^{16}O and/or ^{9}Be(α,n)^{12}C reactions with large cross sections, about $10^{7}-10^{8}$ times larger than $\sigma_{tot.}(E_{cm})$ at $E_{\alpha,lab} \leq 2.5$ MeV (see Ref. [16] and references there in). Note that the reaction Q values are $+2.22$ and $+5.70$ MeV for ^{13}C(α,n)^{16}O and ^{9}Be(α,n)^{12}C, respectively, and that a small amount of ^{13}C and/or ^{9}Be is contained in the target.

Despite a significant progress has been made via these extensive studies, the measured $\sigma_{E1}(E_{cm})$ and $\sigma_{E2}(E_{cm})$ still have large uncertainties. It is therefore highly required to carry out a new experiment to accurately determine these $\sigma_{E1}(E_{cm})$ and $\sigma_{E2}(E_{cm})$ by measuring the γ-ray angular distribution of the ^{12}C(α,γ)^{16}O reaction. In the present experiment, we measured the γ-ray angular distribution by employing three anti-Compton NaI(Tl) spectrometers to obtain a sufficient γ-ray yield and using an intense pulsed α-beam to discriminate true γ-ray events from background events due to high energy neutrons from the ^{13}C(α,n)^{16}O and/or ^{9}Be(α,n)^{12}C reactions with a time-of-flight (TOF) method.

EXPERIMENTAL METHOD

The present experiment was carried out using an intense pulsed α-beam, which was provided from the 3.2 MV Pelletron accelerator at the Research Laboratory for Nuclear Reactors of Tokyo Institute of Technology. An average beam current was about 8 μA at a repetition rate of 4 MHz. A pulse width of the beam was about 1.9 ns (FWHM), which was sufficient to distinguish a true event from the ^{12}C(α,γ)^{16}O reaction from neutron related background events. The γ-ray angular distributions of the ^{12}C(α,γ)^{16}O reaction were measured by constructing a new measurement system. Since a detailed description of the new system is published in elsewhere [17], we briefly describe the present experimental method.

We used three high efficiency anti-Compton NaI(Tl) spectrometers as a γ-ray detector to obtain true γ-ray signals with a good S/N ratio. A schematic view of the experimental setup is shown in Figure 1. Each spectrometer consists of a central NaI(Tl) detector with a diameter of 22.9 cm and a length of 20.3 cm, and an annular NaI(Tl) detector with a size of 5.1 cm in thickness and 36.8 cm in length. The central NaI(Tl) detector was placed 33 cm away from the ^{12}C targets to separate a true γ-ray event from the ^{12}C(α,γ)^{16}O reaction from neutron-induced background events with a TOF method during off-line data analysis. In addition, a 10 cm thick boron-doped polyethylene block was placed in front of the central NaI(Tl) detector to attenuate the flux of neutrons from the ^{13}C(α,n)^{16}O and/or ^{9}Be(α,n)^{12}C reactions. These NaI(Tl) spectrometers were set at 40°, 90° and 130° with respect to the α-beam direction. The γ-ray yield at angles of 40° and 130° provided information on the $E2$ strength and that at 90° gave information on the $E1$ strength. The absolute γ-ray detection efficiency and the response function of the NaI(Tl) spectrometer were calculated using a Monte Carlo code, GEANT4 [18], by taking into account the γ-ray absorption in all materials placed between the ^{12}C target position and NaI(Tl) spectrometer. The calculated efficiency was normalized to the measured one, which was obtained by using the standard γ-ray source with a well-known intensity (^{60}Co and ^{88}Y), and 2.838-, 5.110-, 8.940-, and 10.76-MeV γ-rays from

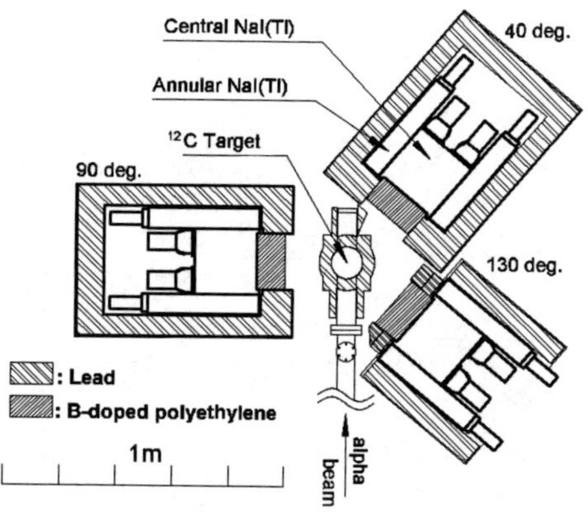

FIGURE 1. Schematic view of the NaI(Tl) spectrometers placed at 40°, 90° and 130° with respect to the α-beam direction.

^{27}Al$(p,\gamma)^{28}$Si measured at $E_p = 992$ and 2046 keV [19, 20].

An enriched (more than 99.95 %) ^{12}C target foil with a thickness of about 300 $\sim$ 400 μg/cm^2 was made by a thermal cracking of methane. A gold with a thickness of 0.1mm was used as a backing of ^{12}C target. We took the Rutherford backscattering spectrum (RBS) of α-particles from the enriched ^{12}C targets with a Si detector to determine their thickness and the incident α-particle intensity. The spectrum was used also to check any change of the target thickness during the measurement.

ANALYSIS

Typical γ-ray and TOF spectra taken by the central NaI(Tl) detector are shown in Figures 2 (a) and (b), respectively. In Figure 2 (a) we see clearly several discrete γ rays in the energy rage from 1 to 3 MeV and at 6.8 MeV, and continuum γ-rays up to about 11 MeV. They were produced by the collision of neutrons from the ^{13}C$(\alpha,n)^{16}$O reaction with various materials including the NaI(Tl) detector itself. In Figure 2 (b) a high peak at about 15 ns and a low one at about 3 ns were observed, and the events at the narrow peak at 0 ns contained events caused by α-beam bombardment on enriched ^{12}C targets at the target position (Figure 3 (a)). The events at the high and the low peaks were assigned to be caused by γ-ray events from neutron induced reactions as is the case with the γ-ray spectrum shown in Figure 2 (a). A detailed description of the neutron induced background is given in Refs. [16, 17]. Note that events in the plateau region below about -5 ns in Figure 2 (b) are due to time-independent background events caused by scattered neutrons and cosmic rays.

We could obtain true events from the ^{12}C$(\alpha,\gamma)^{16}$O reaction by subtracting both the neutron induced background and the time-independent background with use of the TOF method as shown in Figure 3 (b), where we see the discrete γ-ray from the α-capture of ^{12}C into the ground state of ^{16}O with a large S/N ratio. A dashed line is the fitted spectrum for the γ-ray using the response function, which was calculated by taking

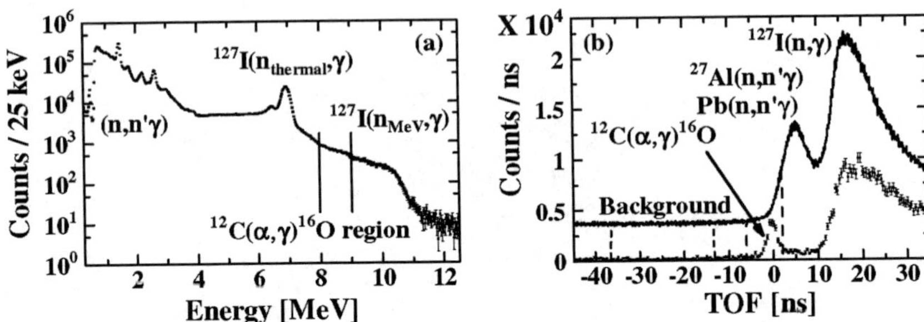

FIGURE 2. (a) γ-ray spectrum taken by the 90° central NaI(Tl) detector at E_{cm} = 1.6 MeV. The sharp peak at 6.8 MeV was due to the thermal neutron capture reaction by ^{127}I. (b) TOF spectrum taken by he 90° central NaI(Tl) detector at E_{cm} = 1.6 MeV obtained by putting the γ-ray energy gate in all regions (solid line) and in the region of between 8.0 MeV and 9.0 MeV (closed circles), respectively. A dashed line indicates the time-independent background events due to thermal neutrons and cosmic rays.

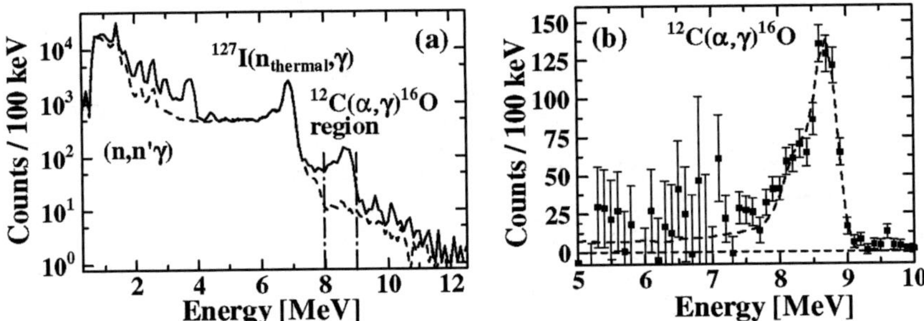

FIGURE 3. (a) Foreground and background spectra obtained by gating narrow peak at $-6 < \mathrm{TOF} < 2$ ns (solid line) and at $-36 < \mathrm{TOF} < -13$ ns (dashed line), respectively, of the TOF spectrum in Figure 2 (b). (b) Background subtracted (net) γ-ray energy spectrum. A dashed line indicates the fitted spectrum using the response function of the spectrometer ($E_{cm} = 1.6$ MeV, $\theta_\gamma = 90°$).

account of the energy loss of α-particles in the ^{12}C target, energy dependence of the ^{12}C$(\alpha,\gamma)^{16}$O cross section, Doppler shift of the γ-ray and the intrinsic energy resolution of the NaI(Tl) spectrometer. Regarding the energy dependence of the cross section we took only the Coulomb part of the cross section, since the nuclear part of the cross section, an astrophysical S-factor, can be assumed to be constant in the present α-beam energy range. It should be mentioned that the average energy of full-energy γ-ray peak obtained by the calculation is in good agreement with observed one. This indicates that the observed peak was due to the ^{12}C$(\alpha,\gamma)^{16}$O reaction. An observed γ-ray yield for each NaI(Tl) spectrometer was obtained by integrating the counts of the response function fitted to measured data.

The RBS spectrum of α-particles is shown in Figure 4. Since the thickness of the ^{12}C target was thin, about 300 $\sim$ 400 μg/cm^2, and the target backing was gold with a thickness of 0.1 mm, incident α-particles were scattered by gold while loosing their energies in passing through the ^{12}C target. The target thickness of ^{12}C was determined

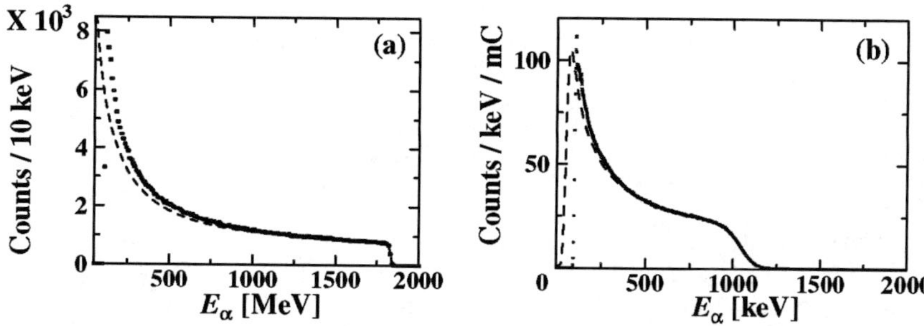

FIGURE 4. RBS spectrum obtained with a Si detector by bombarding 2.270 MeV α-particles on ^{197}Au targets (a) and enriched ^{12}C targets on a gold backing (b), respectively. A dashed line indicates the calculated spectrum using a simulation code.

accurately by fitting the RBS spectrum using the simulation code SIMNRA [21] as shown in Figure 4.

Since the cross section of the $^{12}C(\alpha,\gamma)^{16}O$ reaction changes with decreasing the energy of α-beam over the target thickness, it is necessary to define an effective cross section by considering the energy dependence of the cross section as given in Ref. [22]. Here we took the Coulomb part of the cross section as is the case with the calculation of the response function for the characteristic γ-ray from the $^{12}C(\alpha,\gamma)^{16}O$ reaction.

RESULTS

In order to obtain the σ_{E1} and σ_{E2} the resulting differential cross sections were fitted using the angular distribution formula, which is given in Ref. [5]. Here we used elastic scattering data of Tischhauser *et al.* [23] for the relative phase between $E1$ and $E2$ amplitude of the $^{12}C(\alpha,\gamma)^{16}O$ reaction. This treatment was proposed by Barker [24]. The thus obtained result for σ_{E1} agree with that by Dyer and Barnes [5], Redder *et al.* [6], Kunz *et al.* [8], Assunção *et al.* [9] and also R-matrix calculation by Azuma *et al.* [25]. For σ_{E2}, our present result agree with the data set of Redder *et al.* [6] and Assunção *et al.* [9]. The details of the comparison between our result and other data set will be published in elsewhere [26].

SUMMARY

With use of a new system we succeeded to measure the γ-ray angular distributions at $E_{cm} = 1.6$ MeV and 1.4 MeV using intense pulsed α-beams together with high efficiency anti-Compton NaI(Tl) spectrometers with a large S/N ratio and high statistics. A Rutherford backscattering spectrum of α particles from the ^{12}C targets taken on-line gave precise information on the target thickness during beam irradiation. Consequently, we could determine accurately the σ_{E1} and σ_{E2} with small statistical and systematic uncertainties. To obtain the energy dependence of both σ_{E1} and σ_{E2}, we have already measured the angular distribution at lower energy, down to $E_{cm} \sim 1.2$ MeV. The data analysis to determine the absolute cross section is now in progress.

ACKNOWLEDGMENTS

We would like to thank Prof. I. Sugai and Mr. Y. Takeda for making enriched ^{12}C targets with excellent performance, and Mr. K. Tosaka for his help in measuring the pulse width of α beam and for his nice operation of the Pelletron accelerator. The present work was supported by a Grant-in-Aid for Specially Promoted Research of the Japan Ministry of Education, Science, Sports, and Culture.

REFERENCES

1. T. A. Weaver, and S. E. Woosley, *Phys. Rep.* **227**, 65–96 (1993).
2. F. C. Barker, and T. Kajino, *Aust. J. Phys.* **44**, 369–396 (1991).
3. L. Buchmann, et al., *Phys. Rev. C* **54**, 393–410 (1996).
4. C. Angulo, and P. Descouvemont, *Phys. Rev. C* **61**, 064611 (2000).
5. P. Dyer, and C. A. Barnes, *Nucl. Phys.* **A233**, 495–520 (1974).
6. A. Redder, et al., *Nucl. Phys.* **A462**, 385–412 (1987).
7. J. M. L. Ouellet, et al., *Phys. Rev. C* **54**, 1982–1998 (1996).
8. R. Kunz, et al., *Phys. Rev. Lett.* **86**, 3244–3247 (2001).
9. M. Assunção, et al., *Phys. Rev. C* **73**, 055801 (2006).
10. K. U. Kettner, et al., *Z. Phys* **A308**, 73–94 (1982).
11. R. M. Kremer, et al., *Phys. Rev. Lett.* **60**, 1475–1478 (1988).
12. G. Roters, et al., *Eur. Phys. J.* **A6**, 451–461 (1999).
13. L. Gialanella, et al., *Eur. Phys. J.* **A11**, 357–370 (2001).
14. D. Schürmann, et al., *Eur. Phys. J.* **A26**, 301–305 (2005).
15. C. Matei, et al., *Physical Review Letters* **97**, 242503 (2006).
16. H. Makii, et al., *Phys. Rev. C* **76**, 022801 (2007).
17. H. Makii, et al., *Nucl. Instrum. Methods Phys. Res. A* **547**, 411–423 (2005).
18. S. Agostinelli, et al., *Nucl. Instrum. Methods Phys. Res. A* **506**, 250–303 (2003).
19. D. L. Kennedy, et al., *Nucl. Instrum. Methods* **140**, 519–527 (1977).
20. F. Zijderhand, et al., *Nucl. Instrum. Methods Phys. Res. A* **286**, 490–496 (1990).
21. M. Mayer, Simnra user's guide, Report ipp 9/11, Max-Planck-Institut für Plasmaphysik, Garching, Germany (2006).
22. C. Rolfs, and W. S. Rodney, *Caudrons in the Cosmos*, University of Chicago Press, 1988.
23. P. Tischhauser, et al., *Phys. Rev. Lett.* **88**, 072501 (2002).
24. F. C. Barker, *Aust. J. Phys.* **40**, 25–38 (1987).
25. R. E. Azuma, et al., *Phys. Rev. C* **50**, 1194–1215 (1994).
26. H. Makii, et al., to be submitted (2008).

Experimental study of the variation of alpha elastic scattering cross sections along isotopic and isotonic chains at low energies

G. G. Kiss*, D. Galaviz†,**, Gy. Gyürky*, Z. Elekes*, Zs. Fülöp*, E. Somorjai*, K. Sonnabend**, A. Zilges**, P. Mohr‡, J. Görres§, M. Wiescher§, N. Özkan¶, T. Güray¶, C. Yalcin¶,* and M. Avrigeanu‖

*Institute of Nuclear Research (ATOMKI), H-4001 Debrecen, POB. 51, Hungary
†Instituto de Estructura de la Materia, E-28006 Madrid, Spain
**Technische Universität Darmstadt, D-64289 Darmstadt, Germany
‡Strahlenterapie, Diakonie-Klinikum, D-74523 Schwäbisch Hall, Germany
§University of Notre Dame, Notre Dame, Indiana 46556, USA
¶Kocaeli University, Department of Physics, TR-41380 Umuttepe, Kocaeli, Turkey
‖"Horia Hulubei" National Institute for Physics and Nuclear Engineering, 76900 Bucharest, Romania

Abstract. To improve the reliability of statistical model calculations in the region of heavy proton rich isotopes alpha elastic scattering experiments have been performed at ATOMKI, Debrecen, Hungary. The experiments were carried out at several energies above and below the Coulomb barrier with high precision. The measured angular distributions can be used for testing the predictions of the global and regional optical potential parameter sets. Moreover, we derived the variation of the elastic alpha scattering cross section along the Z = 50 (^{112}Sn - ^{124}Sn) isotopic and N = 50 (^{89}Y - ^{92}Mo) isotonic chains. In this paper we summarize the efforts to provide high precision experimental angular distributions for several A $\approx$ 100 nuclei to test the global optical potential parameterizations applied to p-process network calculations.

Keywords: Astrophysical p-process, Alpha elastic scattering, Optical potential
PACS: 24.10.Ht Optical and diffraction models, 25.55.-e 3H-, 3He-, and 4He-induced reactions, 25.55.Ci Elastic and inelastic scattering

MOTIVATION

The 35 stable most proton-rich nuclei between Se and Hg are called p-nuclei [1]. The natural isotopic abundance of these nuclei is 10-100 times less than the more neutron-rich isotopes. Contrary to the synthesis of the other heavy elements, the p-nuclei cannot be produced by neutron-capture.

It is generally accepted that the main stellar mechanism synthesizing these nuclei — the so-called p-process — is initiated by (γ,n) photodisintegration reactions on already existing more neutron-rich seed nuclei. As the neutron separation energy increases along this path toward more neutron deficient isotopes, (γ,p) and (γ,α) reactions become stronger and process the material toward lower masses [2, 3, 4]. In the case of the synthesis of light p-nuclei (γ,p) reactions and for the middle and heavy p-nuclei (γ,α) reactions are playing more important role.

Modeling the synthesis of the p-nuclei and calculating their abundances requires an

CP1016, *Origin of Matter and Evolution of Galaxies*,
edited by T. Suda, T. Nozawa, A. Ohnishi, K. Kato, M. Y. Fujimoto, T. Kajino, and S. Kubono

TABLE 1. Charge and mass number of the target nuclei and α beam energies for each of the measured angular distributions.

nucleus	Z	N	$E_{c.m.}$ [MeV]	Reference
^{89}Y	39	50	15.51, 18.63	[9]
^{92}Mo	42	50	13.20, 15.69, 18.62	[11]
^{106}Cd	48	58	15.55, 17.01, 18.90	[12, 16]
^{110}Cd	48	62	15.58, 18.78	[17]
^{116}Cd	48	68	15.60, 18.81	[17]
^{112}Sn	50	62	13.90, 18.84	[13]
^{124}Sn	50	74	18.89	[13]

extended reaction network calculation involving thousands of reactions on hundreds of stable and unstable nuclei. A very small fraction of these reaction cross sections has been measured in the laboratory so reaction rate calculations are based on the Hauser-Feshbach statistical model, using global optical potentials. The form of the proton-nucleus optical potential is relatively well known contrary to the alpha-nucleus optical potential where our knowledge is limited especially at energies below the Coulomb barrier mainly due to the lack of relevant experimental data.

Considerable theoretical and experimental efforts have been devoted in recent years to improve our knowledge on the alpha-nucleus optical potential at low energies of astrophysical relevance [5, 6, 7, 8, 9, 10]. In this paper we summarize the experimental efforts to investigate the variation of the alpha elastic scattering cross sections along isotonic and isotopic chains to provide data for testing the available global and regional parameterizations.

ELASTIC SCATTERING EXPERIMENTS

The branching points of the p-process at high masses are unstable proton-rich nuclei, where the branching dominantly proceeds through the (γ,α) reaction. Since no alpha elastic scattering was performed so far on these unstable nuclei, the global parameterizations used to determine these (γ,α) reaction rates are based on the analysis of angular distributions of alpha scattering on stable nuclei nearby. To prove its reliability, global optical potentials must be able to reproduce the measured angular distributions along isotopic and isotonic chains. In recent years, the variation of the alpha elastic scattering cross sections as the function of neutron and proton numbers at the $A \approx 100$ mass region has been investigated at ATOMKI, Debrecen.

To improve the low energy alpha elastic scattering database, several experiments have been performed [11, 12, 13, 14]. In Table 1. the measured isotopes and energies are listed. The uncertainties of the alpha beam energies are typically $\leq 0.5\%$.

The targets were produced with evaporating highly enriched (95+%) target material onto a thin Carbon foil. The beamspot was small and well defined in order to reduce the uncertainty of the scattering angle. In the case of each isotope complete angular distributions have been measured between 20° and 170°. The absolute cross sections are

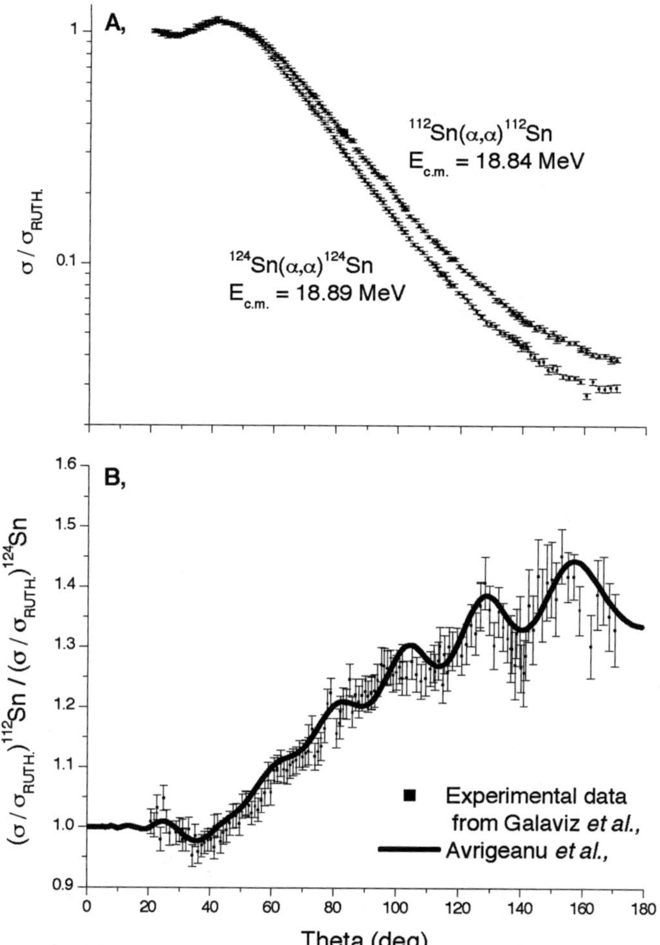

FIGURE 1. Comparison of the Rutherford-normalized alpha elastic scattering cross sections of the ^{112}Sn$(\alpha,\alpha)^{112}$Sn and ^{124}Sn$(\alpha,\alpha)^{124}$Sn reactions (A panel) measured by Galaviz *et al.*, [13]. The ratio of the measured angular distributions versus the angle in the center-of-mass frame is shown on the B panel. The predictions for the ratio of the angular distributions using the improved potential of M. Avrigeanu *et al.*, [8] is presented also on the B panel.

covering more than five orders of magnitude with almost the same accuracy ($\leq 5\%$ total uncertainties). More details about the experimental setup can be found in [9, 15].

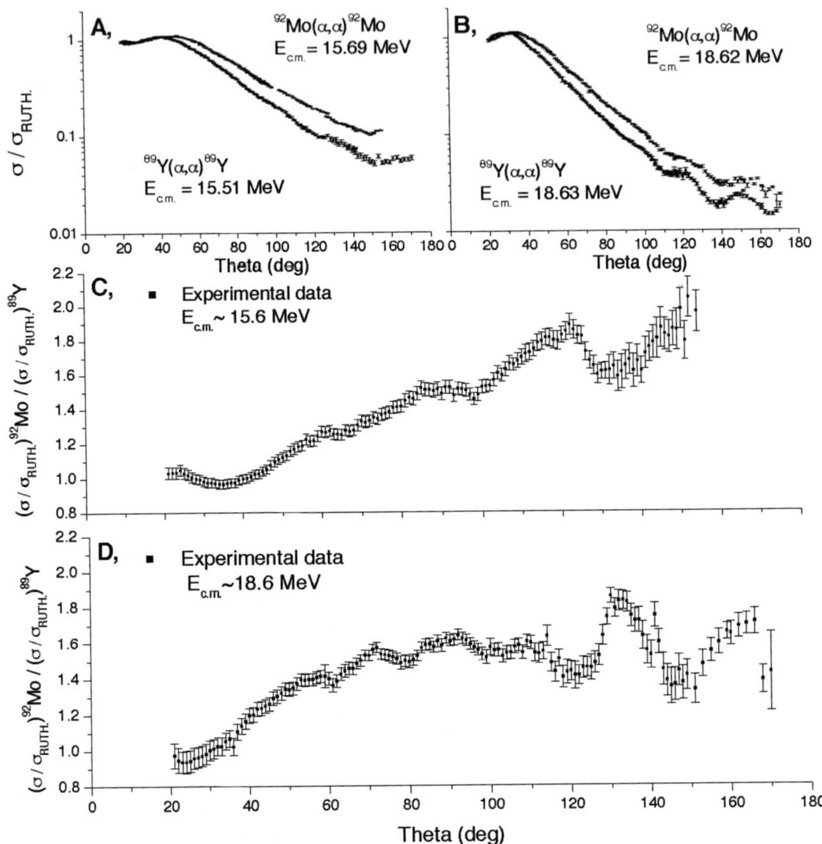

FIGURE 2. Investigation of the variation of the elastic scattering cross section along the N=50 isotonic chain. The measured angular distributions at two different alpha energies of the $^{89}Y(\alpha,\alpha)^{89}Y$ and $^{92}Mo(\alpha,\alpha)^{92}Mo$ [11] reactions are shown in the A and B parts. The ratios of the Rutherford normalized cross sections $(\sigma / \sigma_{RUTH.})^{92}$Mo / $(\sigma / \sigma_{RUTH.})^{89}$Y at $E_{c.m.} \approx$ 15.6 and 18.6 MeV versus the angle in the center-of-mass frame are shown in C and D panels.

Investigation of the variation of the alpha elastic scattering cross sections along the N=50 isotopic chain

To test the global optical model parameterizations, angular distributions of the $^{112}Sn(\alpha,\alpha)^{112}Sn$ and $^{124}Sn(\alpha,\alpha)^{124}Sn$ at $E_{c.m.}$ = 18.84 and 18.89 MeV have been measured. It was found that the elastic alpha scattering cross sections of the ^{112}Sn and ^{124}Sn differ by roughly 30-40% at backward angles.

Figure 1. shows the measured Rutherford-normalized elastic scattering cross sections (part A) and the ratio of the cross sections versus the angle (part B). It was evident that the global alpha-nucleus potentials fail to reproduce either the strength or the oscillation pattern for backward angles [13]. However, by increasing the depth of the imaginary part of the optical potential it was possible to extract a potential that is able to describe the experimental data [8] as it shown in the B part of Fig. 1.

Investigation of the variation of the scattering cross section along the Z=50 isotonic chain

In order to investigate the behavior of the optical potential parameters along the N=50 isotonic chain we derived the ratio of the elastic scattering angular distributions of two N=50 neutron-magic nuclei: ^{92}Mo and ^{89}Y. The $^{92}\mathrm{Mo}(\alpha,\alpha)^{92}\mathrm{Mo}$ reaction has been investigated by Fülöp *et al.* at $E_{c.m.}$=13.20, 15.69 and 18.62 [11]. Recently, an elastic scattering experiment on ^{89}Y has been performed. Complete angular distributions have been measured at $E_{c.m.}$= 15.51 and 18.63 MeV.

Contrary to the previously investigated even-even isotopes, the nucleus ^{89}Y is even-odd. Experimental alpha scattering data on even-odd isotopes in the A=100 mass region is scarce. In the case of ^{89}Y there is big difference between the energy as well as the spin of of the ground and the first excited state (the energy of the first excited state is 908 keV and the spins are $1/2^-$ and $9/2^+$ respectively). The A and B parts of Fig. 2. shows the measured alpha elastic angular distributions.

The ratio of the elastic alpha scattering cross section of the ^{92}Mo and ^{89}Y shows an oscillation pattern. The C and D parts of Fig. 2. shows the ratio of the Rutherford normalized cross sections of the $^{92}\mathrm{Mo}(\alpha,\alpha)^{92}\mathrm{Mo}$ and $^{89}\mathrm{Y}(\alpha,\alpha)^{89}\mathrm{Y}$ at $E_{c.m.} \approx 15.5$ and 18.6 MeV. Although there are small differences in the center-of-mass energies ($\approx$ 170 keV at the 15.5 MeV data and $\approx$ 15 keV in the case of the 18.6 MeV data), the ratio of the Rutherford normalized cross sections is well defined because the dominating $1/E^2$ dependence of the scattering cross section is taken into account by Rutherford normalization. Therefore, the ratio of Rutherford normalized cross sections is a very sensitive test for global alpha-nucleus potential parameterizations.

The comparison of the measured angular distribution and the predictions using several global optical potential parameterization is in progress.

OUTLOOK: ELASTIC SCATTERING EXPERIMENTS ON CADMIUM ISOTOPES

The ratio of the (α,α) scattering cross sections of Z=50 proton and N=50 neutron magic nuclei has been used to test the global alpha nucleus optical potentials. Recently, the angular distributions of elastic alpha scattering on ^{106,110,116}Cd isotopes were measured at ATOMKI, Debrecen [16, 17]. The new high precision experimental data allow to investigate the variation of the alpha-nucleus optical potential parameters along the Z=48 isotopic chain. Furthermore, it is also possible to compare the elastic alpha scattering cross sections of the N=62 nuclei: the ^{110}Cd and ^{112}Sn at $E_{c.m.} \approx 18.8$ MeV in order to provide a further test for optical potential parameterizations. Such studies will help to improve global alpha-nucleus potentials and thus reduce the corresponding uncertainties of modern p-process models.

ACKNOWLEDGMENTS

This work was supported by OTKA (K68801, T49245), DFG (SFB634), MTA-OTKA-NSF grant 93/049901, the Scientific and Technical Research Council of Turkey (TUBITAK) grant TBAG-U/111 (104T2467) and the Joint Institute of Nuclear Astrophysics (www.JINAweb.org) NSF-PFC grant PHY02-16783. Gy.Gy. is supported by the Bolyai grant. D.G. acknowledges financial support from the Mario Bueno grant (CSIC and Hungarian Academy of Sciences) EST001024.

REFERENCES

1. S. E. Woosley and W. M. Howard, *Astrophys. J. Suppl.* **36**, 285 (1978).
2. M. Arnould and S. Goriely, *Phys. Rep.* **384**, 1 (2003).
3. T. Rauscher, *Phys. Rev. C* **73**, 015804 (2006).
4. W. Rapp, J. Görres, M. Wiescher, H. Schatz, and F. Käppeler, *Astrophys. J.* **653**, 474 (2006).
5. V. Demetriou and M. Axiotis, *AIP conference proc.* **891**, 281 (2006).
6. P. Demetriou, C. Grama and S. Goriely *Nucl. Phys.* **A707**, 253 (2002).
7. M. Avrigeanu, W. von Oertzen and V. Avrigeanu, *Nucl. Phys. A* **764**, 246 (2006).
8. M. Avrigeanu and V. Avrigeanu, *Phys. Rev. C* **73**, 038801 (2006).
9. G. G. Kiss, Gy. Gyürky, Zs. Fülöp, E. Somorjai, D. Galaviz, A. Kretschmer, K. Sonnabend, A. Zilges, P. Mohr and M. Avrigeanu, *J. Phys. G* **35**, 014037 (2008).
10. P. Mohr, Zs. Fülöp, H. Utsunomiya *Eur. Phys. J.* **32**, 357 (2007).
11. Zs. Fülöp, Gy. Gyürky, Z. Máté, E. Somorjai, L. Zolnai, D. Galaviz, M. Babilon, P. Mohr, A. Zilges and T. Rauscher, *Phys. Rev.C* **64**, 065805 (2001).
12. G. G. Kiss, Zs. Fülöp, Gy. Gyürky, Z. Máté, E. Somorjai, D. Galaviz, A. Kretschmer, K. Sonnabend and A. Zilges, *Eur. Phys. J.* **27**, 197 (2006).
13. D. Galaviz, Zs. Fülöp, Gy. Gyürky, Z. Máté, P. Mohr, T. Rauscher, E. Somorjai, and A. Zilges, *Phys. Rev. C* **71**, 605802 (2005).
14. P. Mohr, T. Rauscher, H. Oberhummer, Z. Máté, Zs. Fülöp, E. Somorjai, M. Jaeger and G. Staudt, *Phys Rev. C* **55**, 1523 (1997).
15. D. Galaviz Ph.D. thesis TU Darmstadt (2004).
16. D. Galaviz, A. Kretschmer, K. Sonnabend, A. Zilges, G. G. Kiss, Zs. Fülöp, Gy. Gyürky and E. Somorjai, *in preparation*
17. G. G. Kiss, Gy. Gyürky, Z. Elekes, Zs. Fülöp, E. Somorjai, D. Galaviz, J. Görres, M. Wiescher, N. Özkan, T. Gürray, C. Yalcin *in preparation*

Neutron capture reaction in oxygen nuclei near threshold energy regions

K. Yamamoto*, H. Masui†, K. Katō**, T. Wada‡ and M. Ohta*

*Department of Physics, Konan University, 8-9-1 Okamoto, Kobe 658-8501, Japan
†Information Processing Center, Kitami Institute of Technology, Kitami 090-8507, Japan
**Division of Physics, Hokkaido University, Sapporo 060-0810, Japan
‡Department of Pure and Applied Physics, Kansai University, 3-3-35 Yamate-cho, Suita 564-8680, Japan

Abstract. For the key reactions $^{16}O(n,\gamma)^{17}O$ and $^{17}O(n,\gamma)^{18}O$ on nucleosyntheses in stellar interior, the theoretical estimate for the cross section and reaction rate have been made. The careful description of the nuclear structure is important in very low energy regions. The Cluster Orbital Shell Model is adopted for reproducing the structure of the nuclei. Our results for the cross section of the $^{16}O(n,\gamma)^{17}O$ reaction are consistent with that of the microscopic two cluster model. The straightforward application becames possible to the case of the $^{17}O(n,\gamma)^{18}O$ reaction with a substantial reliance.

Keywords: Reaction,Complex scaling method,astrophysical energy
PACS: 21.60.Gx, 23.20.-g, 28.20.Np

INTRODUCTION

We investigate the neutron capture reaction near threshold energies in light nuclei ^{16}O and ^{17}O. For the investigation of the stellar evolution in terms of the element abundance, there are two important ingredients: the scenario in the stellar evolution model and the related nuclear reaction data at astrophysical energies.

In general, the nuclear data at astrophysical energy regions involves a lot of uncertainties due to experimental errors and ambiguity originating in compilation including extrapolation processes. Sometime, the rate of a specific reaction can affect decisively to produce heavier elements in a stellar nucleosynthesis. We call it the key reaction in this synthesis process. The reliable theoretical estimation for the key reaction is necessary for the network calculation to get the element abundance, especially when we have no experimental data for that reaction. For example, in the s-process of the helium shell flash model in metal-poor stars [1], we found that the reactions $^{17}O(n,\gamma)^{18}O$ and $^{17}O(n,\alpha)^{14}C$ are the key reactions [2], and insisted the importance of the theoretical analysis. In Fig. 1, we show the element synthesis paths for the s-process in metal-poor star [1]. We see that the key nuclide ^{17}O is an important gateway to produce heavier elements.

In the present work, we discuss the cross section of the neutron capture reaction near threshold energies for the nuclides ^{16}O and ^{17}O. In particular, the reaction $^{17}O(n,\gamma)^{18}O$ should be estimated precisely since this reaction has a crucial role in the nucleosynthesis in the above mentioned helium shell flash model and we have no experimental data so far.

CP1016, *Origin of Matter and Evolution of Galaxies*,
edited by T. Suda, T. Nozawa, A. Ohnishi, K. Kato, M. Y. Fujimoto, T. Kajino, and S. Kubono

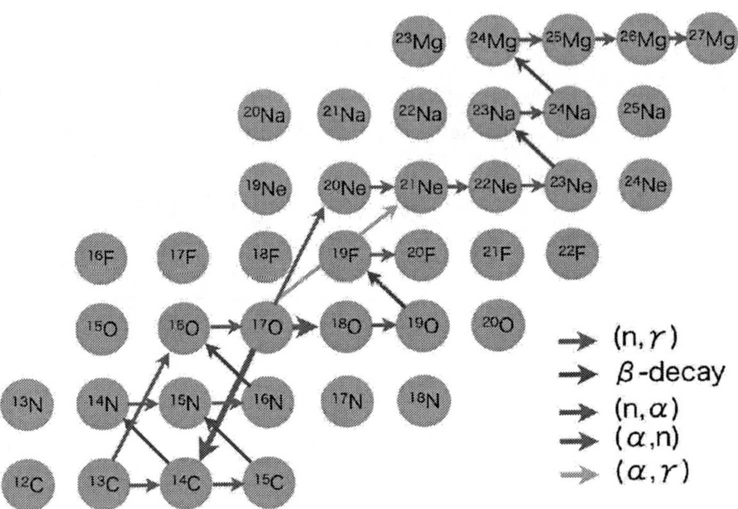

FIGURE 1. The elements synthesis diagram in the s-process. We can understand that nuclide ^{17}O becomes the key isotope or nuclide on synthesis paths.

For the calculation of the reaction cross sections, we use the wave function taken account of the nuclear structure described by the Cluster Orbital Shell model (COSM) [3]. The Complex Scaling Method (CSM) is used to treat unbound and resonance states accurately [3]. The reproduction of the energy levels of the final nucleus and the proper treatment of the long range tail of the scattering state are necessary for obtaining the accurate estimation of the reaction rate near threshold energies. To this end, the scattering wave function is described by the Lippmann-Schwinger equation. To confirm the validity whether our method is well applicable or not to the reaction $^{17}\mathrm{O}(\mathrm{n},\gamma)^{18}\mathrm{O}$, the cross section for the reaction $^{16}\mathrm{O}(\mathrm{n},\gamma)^{17}\mathrm{O}$ will be presented and will be compared with the available nuclear data compilations.

METHOD OF THEORETICAL ESTIMATION

For the calculation of the reaction cross section, we use the wave function reproducing well the nuclear structure. In this study, the wave function for the bound, resonance and continuum states are described with the Cluster Orbital Shell Model and Complex Scaling Method. In the final state and the scattering state i.e. the initial state, we estimate their wave function by the same framework.

We consider the reaction $B(b,\gamma)A$. The reaction cross section can be expressed as

$$\sigma_{E\lambda}^{b\gamma}(E) = \frac{2(2J_A+1)}{(2J_B+1)(2J_b+1)} \frac{k_\gamma^2}{k_{CM}^2} \frac{(2\pi)^3(\lambda+1)}{\lambda[(2\lambda+1)!!]^2} \left(\frac{E_\gamma}{\hbar c}\right)^{2\lambda-1} \frac{dB(E)}{dE}, \tag{1}$$

where k_γ and k_{CM} are the wave number of the photon and the center-of-mass motion between core and their valence nucleons, respectively, E_γ is the photon energy, λ is the multipolarity.

Here, $dB(E)/dE$ [$e^2fm^{2\lambda}$/MeV] is the transition strength distribution from the bound states to the scattering state, as follows:

$$\frac{dB(E)}{dE} = \frac{1}{2J_A+1}|\langle\Phi_A^\theta||(\hat{O}_\lambda^\dagger)^\theta||\Phi_{B+b}^\theta(E)\rangle|^2, \tag{2}$$

where $\hat{O}_\lambda$ is the electric transition operator, Φ is complex-scaled wave function form and a set of biorthogonal bases. The suffix θ denotes the scaling angle to rotate the cuts of the Riemann sheets. For a set of the basis function of the wave function, we use the Gaussian basis function for one radial component viz each valence nucleon has the basis function. The maximum range of the Gaussian basis function is about 30 fm and 100 fm for the bound state and continuum (scattering) state, respectively. When we describe the wave function of the bound state and the continuum state with the set of the Gaussian basis, the transition matrix of the Eq.(2) is as follows,

$$\begin{aligned} &|\langle\Phi_A^\theta||(\hat{O}_\lambda^\dagger)^\theta||\Phi_{B+b}^\theta(E)\rangle|^2 \\ &= \sum_\nu \int \langle\tilde{\Phi}_A||(\hat{O}_\lambda^\dagger)||\Phi_{E_\nu}^{(B+b)}\rangle\langle\tilde{\Phi}_{E_\nu}^{(B+b)}||\hat{O}_\lambda||\Phi_A\rangle\delta(E-E_\nu) \\ &= \frac{-1}{\pi}\mathrm{Im}\left[\sum_j \langle\tilde{\Phi}_A^\theta||(\hat{O}_\lambda^\dagger)^\theta||\Phi_j^{(B+b)}(E_j^\theta)\rangle\langle\tilde{\Phi}_j^{(B+b)}(E_j^\theta)||\hat{O}_\lambda^\theta||\Phi_A^\theta\rangle\frac{1}{E-E_j^\theta}\right]. \end{aligned} \tag{3}$$

where the summation takes an element j of the complete set. The quantities E_j^θ are the energies of the continuum states by the COSM + CSM. For the reaction at very low energy (E <a few keV), the transition matrix is affected by the long range tail of the scattering wave function. When we use the Gaussian basis function in the COSM + CSM for the scattering wave function at the very low energy, it is difficult to treat its solution precisely. Therefore, we modify the formulation for calculating the scattering wave function, that is, we apply the Lippmann-Schwinger equation for describing the scattering state precisely. The modification frees us from the troublesome treatment of the long range tail for the very low energy. We take the Lippmann-Schwinger equation by the full Hamiltonian $H(\theta)$ between core with valence nucleons,

$$|\Psi_{B+b}^{LS,\theta}(E)\rangle = |\psi_{B+b}^\theta(E)\rangle + \frac{1}{E-H(\theta)}V(\theta)|\psi_{B+b}^\theta(E)\rangle, \tag{4}$$

where the hamiltonian is $\hat{H}_i = \hat{t}_i' + \hat{V}_i^d + \hat{V}_i^{ex} + \hat{V}_i^{ls} + \lambda\hat{\Lambda}_i$, $\hat{V}^d$, $\hat{V}^{ex}$ and $\hat{V}^{ls}$ are the direct part of the folding potential, the exhange part and effective LS potential, respectively [3]. The $\hat{\Lambda}$ is the Pauli principle part that Pauli principle between nucleons in the core and a valence one is treated in term of the orthogonality condition model (OCM) [6].

Then the transition probability shown in Eq.(3) by using Eq. (4) is expressed by

$$
\begin{aligned}
&|\langle\Phi_A^\theta||(\hat{O}_\lambda^\dagger)^\theta||\Psi_{B+b}^{LS,\theta}(E)\rangle|^2 \\
&= \Big|\langle\Phi_A^\theta||(\hat{O}_\lambda^\dagger)^\theta||\psi_{B+b}^\theta(E)\rangle \\
&+\sum_j \langle\Phi_A^\theta||(\hat{O}_\lambda^\dagger)^\theta||\Phi_j^{(B+b)}(E_j^\theta)\rangle\langle\Phi_j^{(B+b)}(E_j^\theta)||V^\theta||\psi_{B+b}^\theta(E)\rangle\frac{1}{E-E_j^\theta}\Big|^2
\end{aligned} \quad (5)
$$

RESULTS AND DISCUSSION

We have performed a calculation of the capture reaction cross section in the reaction $^{16}O(n,\gamma)^{17}O$ from the thermal to the astrophysical energy regions. In our calculation, the energy eigenvalues of 5/2$^+$ state and 1/2$^+$ state for the nuclide ^{17}O are -4.184 and -3.277 (MeV) with respect to the $^{16}O+n$ threshold. The experimental values are -4.143 and -3.273 (MeV) for the 5/2$^+$ and 1/2$^+$ states, respectively.

To calculate the electric transition matrix, we use the COSM and the CSM. It is noted that the scattering wave function is calculated by using the Lippmann-Schwinger equation. We compare the present results with that of another theoretical calculation by the microscopic cluster model (GCM) [7].

In Fig. 2, we show the E1 transition partial cross sections of the reaction $^{16}O(n,\gamma)^{17}O$ for the ground state and the first excitation state. The solid line and the dotted line indicate the transition to the ground state and the first excitation state by our method at astrophysical energy regions, respectively. The open circles and the open triangles are the experimental data in Ref. [8]. The component of the experimental date are expected to be the transition from the p-wave of the continuum. The solid squares and the solid circles connected by the solid line are the results of another theoretical calculation by means of the Generator coordinate method (GCM) of the microscopic two cluster model. The results of the solid and the dotted line are calculated by our method, that is, we use the COSM + CSM and the COSM + CSM + LS eq. for the final state and the scattering state, respectively. The present results using our method well reproduce the experimental data in astrophysical energy regions. Our results also have consistency with the theoretical calculation by GCM [7].

In Fig. 3, we expand the calculation of the cross sections by our framework to very low energy region for the reaction $^{16}O(n,\gamma)^{17}O$. The solid line and the dotted line present the E1 transition cross sections to the ground state and the 1st excitation state, respectively, and the dot-dashed line is E2 transition cross section to the ground sate. The open triangles and the open circles express the experimental data at the astrophysical energy regions and the open square is the experimental data at thermal energy regions [9]. Our results show that the cross section at the astrophysical energy regions is dominated by the E1 transition from the component of the p-wave scattering state. In the thermal energy region, we infer that the experimental data can not be explained by the M1 transition.

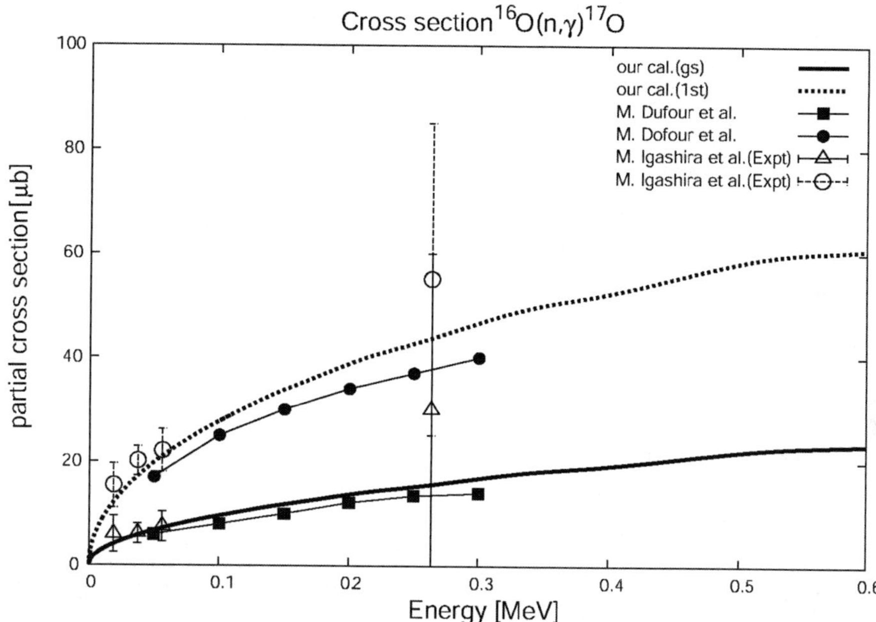

FIGURE 2. E1 transition cross section of the reaction $^{16}O(n,\gamma)^{17}O$. Solid line and dotted line indicate our calculations with COSM + CSM + Lippmann-Schwinger equation. They are corresponding to the transition from continuum to ground state and the first excitation state, respectively. The open circles and triangles are the experimental data taken from Ref. [8]. The solid circle and the solid square connected by the solid line are the capture reaction cross section to the ground state and the 1st excitation state by GCM calculations [7].

SUMMARY

By using the Cluster orbital shell model, the Complex scaling method and the Lippmann-Schwinger equation, the neutron capture reaction cross section in the reaction $^{16}O(n,\gamma)^{17}O$ well reproduces the experimental data and is consistent with another theoretical calculation at astrophysical energy regions as shown in Fig. 2. The description for the long range tail of the scattering wave function by the Lippmann-Schwinger equation is a better way to calculate the cross section from astrophysical energy to thermal energy regions. The procedure of the present model calculation can be applied straightforward to the reaction $^{17}O(n,\gamma)^{18}O$ which may have an important role in the nucleosynthesis mentioned in this report. The calculation is on the way.

REFERENCES

1. T. Nishimura N. Iwamoto, T. Suda, M. Aikawa, M. Y. Fujimoto & I. Iben Jr., Proceedings of the International Symposium "*Origin of Matter and Evolution of Galaxies*", ed. by Kobono *et al.*, AIP Conference proceedings **847**, (2006), p455

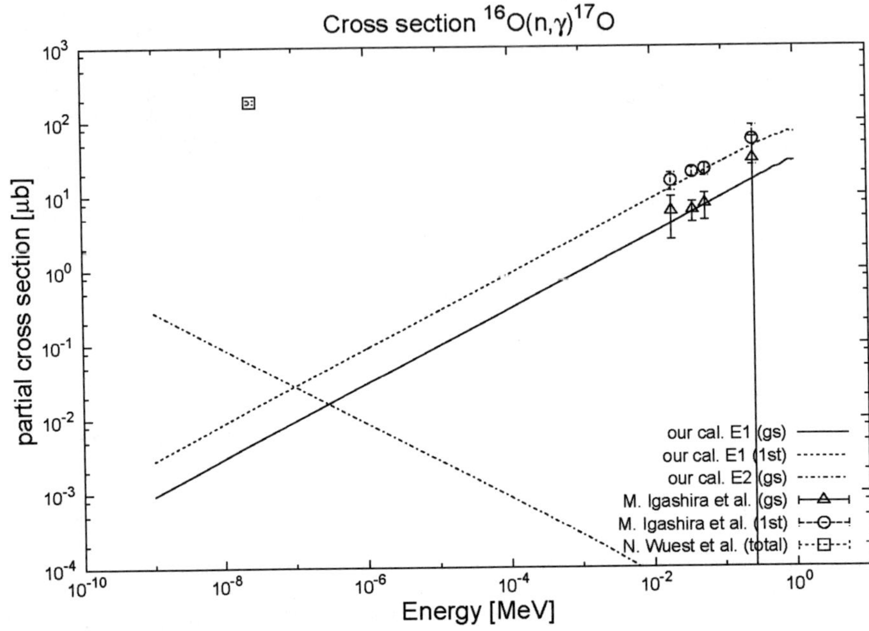

FIGURE 3. E1 and E2 transition cross section with our framework. The solid and dotted lines indicate the E1 transition cross section, dot-dashed line is the E2 transition cross section. The circles and triangles are the experimental data in E1 transition at astrophysical energy [8]. The square show the experimental data at thermal energy [9].

2. K. Yamamoto, T. Wada, M. Ohta, T. Nishimura, M. Y. Fujimoto, K. Katō, T. Suda & M. Aikawa, Proceedings of the International Symposium "*Tours symposium on Nuclear Physics VI*", AIP Conference proceedings **891**, (2007) p405.
3. H. Masui, K. Katō & K. Ikeda, *Phys. Rev.* **C 73**, 034318 (2006).
4. S. Aoyama, T. Myo, K. Katō, K. Ikeda *Prog. Theor. Phys.*, **116** (2006) 1.
5. T. Myo, K. Katō, S. Aoyama & K. Ikeda, *Phys. Rev.* **C 63**, 05431 (2001).
6. S. Saito, *Prog. Theor. Phys. Suppl.* **62**, 11 (1977).
7. M. Dufour & P. Descouvemont, *Phys. Rev.* **C 72**, 015801 (2005).
8. M. Igashira *et al.*, *Astrophys. J.* **441**, L89 (1995).
9. N. Wuest *et al.*, *Phys. Rev.* **C 19**, 1153 (1979).

Low Energy Li+p,d Reactions In Liquid Plasmas And The Effect Of Liquefied Li^+ Ions On The Screening Potential

J. Kasagi, H. Yonemura, Y. Toriyabe, A. Nakagawa and T. Sugawara

Laboratory of Nuclear Science, Tohoku University, Sendai, Japan 982-0826

Abstract. Thick target yields of α particles emitted in the $^6Li(d,\alpha)^4He$ and $^7Li(p,\alpha)^4He$ reactions were measured for Li target in the solid and liquid phase. Observed reaction rates for the liquid Li are always larger than those for the solid. This suggests that the stopping power of hydrogen ion in the liquid Li metal might be smaller than in the solid. Using the empirically obtained stopping power for the liquid Li, we have deduced the screening potentials of the Li+p and Li+d reactions in both phases. The deduced screening potential for the liquid Li is about 500 eV larger than for the solid. This difference is attributed to the effect of liquefied Li^+ ions. It is concluded that the ionic screening is much stronger than the electronic screening in a low-temperature dense plasmas.

Keywords: Low energy nuclear reaction, Li+p and Li+d reaction in liquid Li, liquid metal Li, screening energy.
PACS: 01.30.Cc

INTRODUCTION

Low energy nuclear reactions play an important role in nuclear synthesis and energy production in stars, where thermal nuclear reactions take place in various plasma conditions. In order to simulate such processes, reaction rates should be known from the nuclear reaction experiments in a laboratory. The progress of the accelerators made it possible to measure nuclear cross sections for variety of nuclear reactions, especially by using radio active beams. However, the reaction rate depends on not only the cross section but also the environments very strongly. For example, it has been well known that the screening by electrons enhances the nuclear cross section very much, even in the laboratory experiment where a target nucleus is usually in atom or molecule. Thus, the nuclear reactions in various conditions should be explored more in order to estimate the reaction rates in stars, although it is impossible to prepare for very high density plasmas, at present.

We have developed measurements of low-energy nuclear reactions in metal environments [1], where target nuclei are surrounded by conduction electrons; i.e., target nuclei in degenerated electron plasmas. The screening potentials of the D+D and Li+D reactions in such conditions were found out to be very large[2-4]: about 600 eV for D+D reaction in PdO host, for example. Subsequently, we have

CP1016, *Origin of Matter and Evolution of Galaxies,*
edited by T. Suda, T. Nozawa, A. Ohnishi, K. Kato, M. Y. Fujimoto, T. Kajino, and S. Kubono

tried to prepare another condition of the environment for low-energy nuclear reactions. In the present work, we report on nuclear reactions in liquid Li metal plasmas, for the first time, where Li^+ ions are moving freely in a conduction electron sea, and much higher density ($\rho \sim 10^{22}/cm^3$) can be realized than in laboratory gas plasmas.

PLASMA PROPERTIES OF LIQUID LI

One of the quantity which characterizes a plasma is the Wigner-Seits radius, $a_{ws}= (3/4\pi n)^{1/3}$: n is the number density of particles. The radius is 0.17 nm for both the Li^+ ions and the electrons calculated with $n_{Li} = n_e \sim 4.6\times10^{22}/cm^3$. The so-called plasma-parameter, $L= h/(2\pi MkT)^{1/2}/a_{ws}$, where M is the mass of particles, is estimated to be 0.1 for the Li^+ ions and 15 for the electrons. The particles with $L<<1$ can be considered classical ones, while those with $L>>1$ can be considered quantum ones. Therefore, the liquid Li may be regarded as plasma consisting of classical Li+ ions and quantum electrons. It should be noticed that the number density of particles is much larger than that of gas plasma realized in the laboratory. Thus, the liquid metal plasma in laboratory can realize similar plasma conditions in the core of Jupiter with slightly lower temperature and density.

The present work aims at obtaining the screening potential of the Li+p and Li+d reactions in liquid Li. Since the target Li is surrounded by the conduction electrons in addition to the bound electrons, the screening potential due to both electrons is estimated to be $U_e = 3e^2 \times (1/\lambda_{be}^2+1/\lambda_{ce}^2)^{1/2}$, where $\lambda_{be(ce)}$ is a screening length due to the bound (conduction) electrons. They are simply estimated to be 24 pm from the adiabatic approximation and 61 pm from the Thomas-Fermi approximation for the bound and conduction electrons, respectively. Thus, the screening potential of 194 eV is expected from the electrons. For the solid Li case, the screening effect is provided only by these electrons; i.e., the screening potential U_{sol} = 194 eV is predicted.

For the liquid Li, in addition to the electrons, the effect of classical Li^+ ion gas should be considered. In this case, a screening length can be estimated by the Debey model which gives λ_{Li} = 6.7 pm at T = 520 K; much shorter than those originated from quantum electrons. Thus the screening potential of the Li + p(d) reaction in the liquid Li is estimated to be U_{liq} = 673 eV; almost 500 eV difference may be expected between the solid and the liquid target.

One of the interesting question is whether the screening due to positive ions can work effectively or not. Since the mass of ions is much lager than that of electrons, positive ions cannot respond quickly to change, and, hence, the ionic screening might be reduced very much.

EXPERIMENTAL PROCEDURE

The experiments were performed by using proton and deuteron beams obtained from a low-energy ion generator at Laboratory of Nuclear Science at Tohoku University. Natural Li (92.4% ^7Li, 7.6% ^6Li) and enriched ^6Li were used for ^{7}Li+p and ^{6}Li+d reactions, respectively. A technique to generate the liquid Li metal target has been developed. A lump of natural Li or enriched ^{6}Li metal was placed horizontally on a small saucer which can be heated up to 500 °C in a vacuum chamber. The temperature of the surface of the Li target was monitored directly by a radiation thermometer. The melting point of the Li metal is about 180 °C; a phase change was easily known by watching the temperature. A beam was injected from the upper part of the chamber, with its angle of 30° with respect to the vertical line. Alpha particles emitted in the $^{6}Li(d,\alpha)^{4}He$ and $^{7}Li(p,\alpha)^{4}He$ reactions were measured with a Si detector of 300 μm in thickness. A 5-μm thick Al foil covered the detector surface to prevent electrons and scattered beam particles from hitting the detector directly.

Thick target yields of α particles from the $^{7}Li(p,\alpha)^{4}He$ and $^{6}Li(d,\alpha)^{4}He$ reactions were measured for the solid (T~60 °C) and the liquid (T~250 °C) Li target as a function of bombarding energy between 25 and 70 keV by 2.5 keV steps. The beam current was measured from the target, on which a permanent dipole magnet was placed to suppress secondary electron emissions. Its intensity was adjusted for each bombarding energy so as to keep the input beam power constant.

RESULTS AND DISCUSSION

Observed excitation functions show clear difference for the liquid and the solid target. It turned out that the reaction rates for the liquid Li are always larger than those for the solid in both the $^{7}Li(p,\alpha)^{4}He$ and $^{6}Li(d,\alpha)^{4}He$ reactions. The thick target yield at the bombarding energy E_b is described as

$$Y(E_b) = (const.) \times \rho_{Li} \int_0^{E_b} \frac{\sigma(E+U_s)}{dE/dx} dE . \quad (1)$$

Here, ρ_{Li} is the number density of Li, dE/dx is the stopping power of Li metal, and σ(E) is the cross section of the $^{7}Li(p,\alpha)^{4}He$ or $^{6}Li(d,\alpha)^{4}He$ reaction. The enhancement due to the screening potential U_s is expected only at very low bombarding energies. Thus, larger reaction rates for the liquid target observed for $E_{p,d} > 40$ keV are considered mainly due to the reduction of the stopping power in the liquid phase.

In the following preliminary analysis, the density of Li is taken from ref. [5] and the stopping power for the solid Li is from ref. [6]. For the cross sections, astrophysical S-factor is taken from ref. [7] for the $^{7}Li(p,\alpha)^{4}He$ reaction and from ref. [8] for the $^{6}Li(d,\alpha)^{4}He$ reaction.

Effective stopping power for hydrogen ions in liquid Li

Ratios of the reaction rates in the liquid target to the solid are shown in Fig. 1 as a function of bombarding energy per nucleon (E/amu). The data plotted with red squares correspond to the $^{7}Li(p,\alpha)^{4}He$ reaction for $E_p > 35$ keV, and green circles to the $^{6}Li(d,\alpha)^{4}He$ reaction for $E_d > 40$ keV. As seen in the figure, the ratio becomes larger and larger as the E/amu increases. This surprises us very much, but the phenomena were easily reproduced in the continuous measurement of the yield versus temperature. In such measurements, the yield is suddenly decreased when the target phase is changed from the liquid to the solid.

In Fig. 1, the two data sets are smoothly connected as if described by a function of E/amu or velocity of the hydrogen. This indicates that the stopping power of hydrogen ion in the liquid Li metal might be smaller than in the solid. In order to compare the screening potential for both in the liquid and solid phase, the stopping power in the liquid Li metal is indispensable. It is, then, deduced empirically so as to reproduce the data in Fig. 1. The stopping power for the liquid target used in the following analysis is $(dE/dx)_{liq} = F(E/amu) \times (dE/dx)_{sol}$; the function F is a quadratic function of E/amu and is determined to reproduce the solid curve in Fig. 1. The origin of the reduction of the stopping power in the liquid is not known at present. However, we try to deduce the screening potential in the liquid target as well as in the solid target.

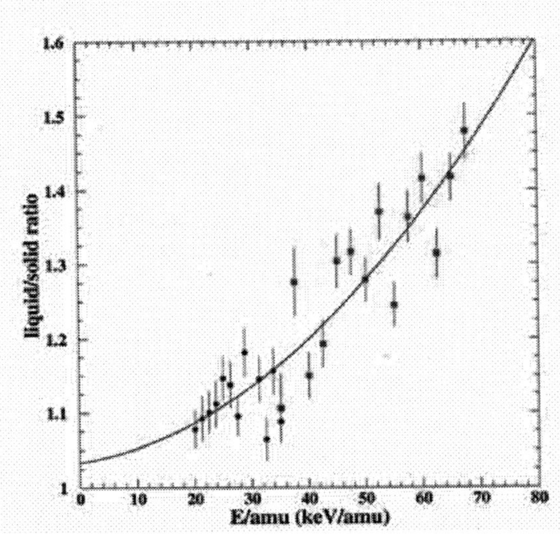

FIGURE 1. Ratio of the reaction rates in the liquid to solid phase as a function of E/amu. Data plotted with red squares are from the $^7Li(p,\alpha)^4He$ reaction and those with green circles from the $^6Li(d,\alpha)^4He$ reaction.

Screening potential for Li+p,d reactions in liquid Li

We try to deduce the screening potentials of the Li+p and Li+d reactions in the liquid and solid phase. The thick target yields of α particles measured in the ^{6}Li+d reaction are shown in Fig. 2; the left part for the solid Li and the right part for the liquid Li. In the upper part, the thick target yields normalized at 70 keV are plotted as a function of the bombarded energy. In the lower part, the enhancement factor which is the experimental yield divided by that calculated by eq. (1) with $U_s = 0$ is plotted.

It is clear that the reaction rates are more strongly enhanced with the decrease of the bombarding energy, as can be seen in the lower part of Fig. 2. Also noticed is the fact that much larger enhancement is observed in the liquid target. The screening potential U_s of the Li+d reaction is deduced by fitting the calculated yields to the experimental ones. The deduced values are $U_{sol} = 350 \pm 50$ eV and $U_{liq} = 900 \pm 50$ eV, respectively for the solid Li and the liquid Li. The difference of the screening energies is about 550 eV.

For the ^{7}Li+p reaction, similar results of the screening potential have been obtained. In this case, however, the data only for $E_p < 45$ keV are analyzed, because of large uncertainties of the stopping power for higher energy region. The screening energies from the ^{7}Li+p reaction are $U_{sol} = 360 \pm 100$ eV and $U_{liq} = 1000 \pm 200$ eV, respectively for the solid Li and the liquid Li. Again, very large difference between solid and liquid is obtained.

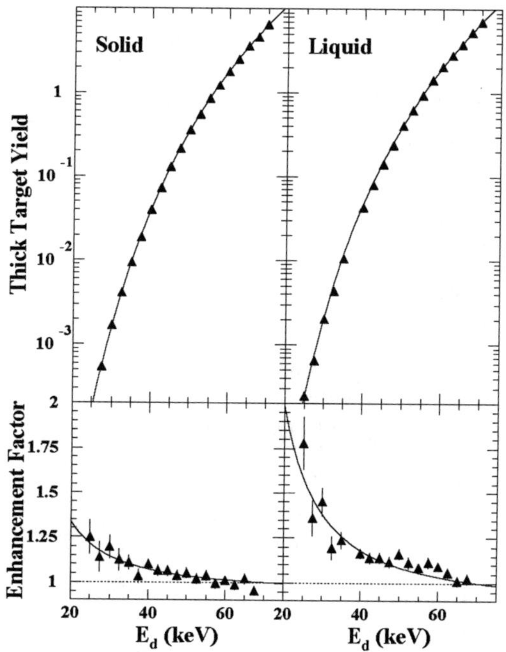

FIGURE 2. Thick target yield of a particles emitted in the ^{6}Li(d,α)^{4}He reaction for the solid and liquid target. In the upper part, the data normalized to the yield at 70 keV are plotted. In the lower part, the experimental yields divided by the yield calculated without screening energy are plotted.

As already discussed, the simple plasma picture gives the screening energies, U_{sol} = 194 eV and U_{liq} = 673 eV, respectively, for the solid and the liquid Li metal. The experimental ones deduced in the present analysis gives slightly larger values, U_{sol} =

350 ± 50 eV and $U_{liq} = 900 \pm 50$ eV. Although the simple plasma picture does not explain well the screening energy for each phase, the difference between the solid and the liquid is well explained. Therefore, we can conclude that the ionic screening mechanism affects the reaction rate very much in the liquid Li metal.

SUMMARY

We have investigated the ^{7}Li+p and ^{6}Li+d reactions for bombarding energies between 25 and 70 keV with liquid Li target, for the first time. The effects of the solid-liquid phase transition are clearly seen in the reaction rates. The reaction yield in the liquid phase is always larger than in the solid phase. This observation suggests that the stopping power in the liquid Li is smaller than in the solid. Using the data of the yield ratio between the liquid and the solid for $E_b > 40$ keV, we have made an empirical correction to the stopping power of the liquid Li.

Screening potentials for the Li+p,d reaction are successfully obtained for the liquid Li as well as the solid. It turns out that the liquid Li provides much lager screening potential than the solid: the difference is about 500 eV in the present preliminary analysis. This difference is very well explained by a simple plasma picture of the solid and the liquid Li metal. It can be concluded that the ionic screening is much stronger than the electronic screening in a low-temperature dense plasmas.

REFERENCES

1. J. Kasagi, *Prog. Theor. Phys. Supp.* **154**, 365 (2004).
2. H. Yuki et al., *JETP Lett.* **68**, 823 (1998).
3. J. Kasagi et al., *J. Phys. Soc. Jpn,* **71**, 2881 (2002).
4. J. Kasagi et al., *J. Phys. Soc. Jpn.* **73**, 608 (2004).
5. Y. Shimizu, A. Mizuno, T. Masaki and T. Itami, *Phys. Chem. Phys.* **4**, 4431 (2002).
6. J.F. Ziegler and J.P. Biersack, code SRIM, **http://www.srim.org**.
7. S. Engstler et al., *Z. Phys. A***342**, 471 (1992).
8. M. Lattuada et al., *Astro. J.* **562**, 1076 (2001).

10. EXPLOSION MECHANISM OF SUPERNOVAE

Explosions inside Ejecta and Most Luminous Supernovae

S.I.Blinnikov

Institute for Theoretical and Experimental Physics (ITEP), Moscow, 117218, Russia
RESCEU, School of Science,
Tokyo University, Tokyo 113-0033, Japan

Abstract. The extremely luminous supernova SN2006gy is explained in the same way as other SNIIn events: light is produced by a radiative shock propagating in a dense circumstellar envelope formed by a previous weak explosion. The problems in the theory and observations of multiple-explosion SNe IIn are briefly reviewed.

Keywords: supernovae, explosion mechanism
PACS: 26.30.-k, 26.30.Ef, 97.60.Bw

INTRODUCTION

The discovery of SN2006gy [1, 2] demonstrates that some supernova (SN) events produce 10 or even 100 times more visible photons than other, already powerful explosions. The anomalously high power of the emission of SN2006gy demands an explanation.

SN2006gy is of SNIIn type and it revived interest in SN models where light is produced by a long living radiative shock which propagates in a dense circumstellar envelope.

I discuss problems in the theory of SNIIn events and in the physics of supercritical radiative shocks. The powerful visible light of SN2006gy can be easily explained by a radiative shock born due to a collision of SN ejecta with a dense cloud formed by a weak explosion some years before the SN. Strong X-ray emission of SNIIn near maximum light may be absent since it is absorbed by the dense cloud or not produced at all in the radiation-dominated shock.

SUPERNOVA TYPES

The models with a long living radiative shock running in a dense circumstellar envelope were invoked earlier [3, 4] to explain the unusual properties of other powerful supernovae with narrow emission lines in their spectra, SNIIn. SN2006gy also belongs to the same SNIIn class.

A simple diagram below illustrates the relation between different astronomical types of supernovae which are classified purely on the appearance of their spectra near maximum light, irrespective of the underlying physics. For example, we believe that SNe II are born when a giant star with an H-rich atmosphere has a powerful explosion in its core. This explosion may be a result of a catastrophic collapse of the stellar core. The

CP1016, *Origin of Matter and Evolution of Galaxies*,
edited by T. Suda, T. Nozawa, A. Ohnishi, K. Kato, M. Y. Fujimoto, T. Kajino, and S. Kubono

details of the explosion of core-collapsing SNe are still not clear, in spite of many important results obtained by workers in this field.

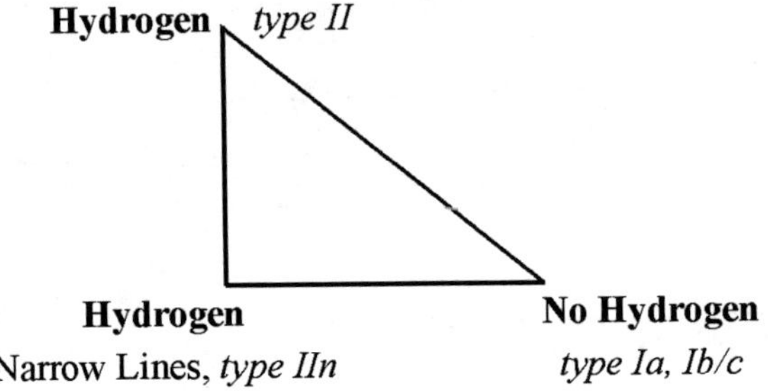

Taking into account the physics of SN explosion we can produce a couple of other, a bit different, diagrams for SN types. I want to subdivide the SN physics into two groups of problems. First, here is a diagram for the mechanism of explosion.

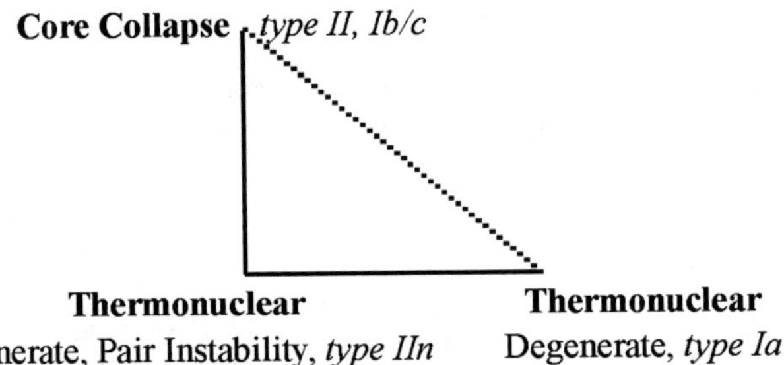

And the next diagram illustrates main ways to produce light during supernova explosions. Here S is entropy, and 'heating' is entropy production.

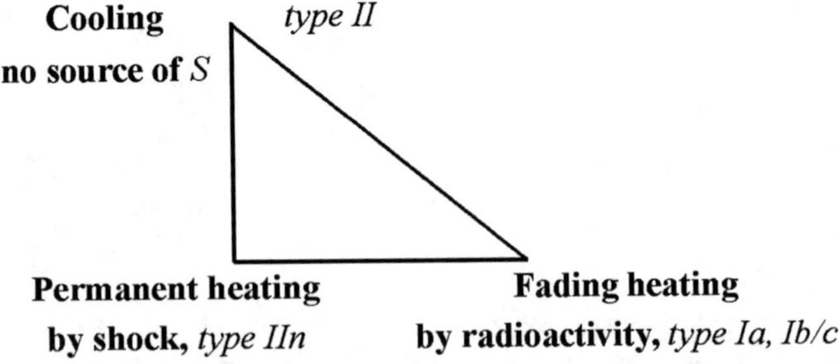

Remarkably, the details of the explosion are not important for explaining the light curves of many supernovae that are born from giant stars that retain their huge hydrogen envelopes. Successful SNII light models were already constructed four decades ago (largely by the work of the Soviet group in Moscow and Riga [5]) due to this insensitivity

TABLE 1. Four kinds of deaths for non-rotating stars

He Core	Main Seq. Mass	Supernova Mechanism
$2 \leq M \leq 40$	$10 \leq M \leq 95$	Fe core collapse to a neutron star or a black hole
$40 < M \leq 60$	$95 < M \leq 130$	Pulsational pair instability followed by Fe core collapse
$60 < M \leq 137$	$130 < M \leq 260$	Pair instability supernova
$137 < M$	$260 < M$	Black hole. Possible GRB ?

to details. The light of SNe II is the manifestation of entropy produced during a short period (hours to days) of the shock propagation in the body of the presupernova star. We cannot exclude the possibility that in some rare cases the SNe II are produced by thermonuclear explosions, not by a collapse, inside hydrogen envelopes. This may be the case for SNe IIn, and especially, SN2006gy. A supernova of type II shines for several months thanks to the heat stored in its body (the shock dies quickly), while in an SNIIn the heat (entropy) is replenished by the shock living several months.

If the hydrogen is lost by a massive star, then we have an SNIb/c. The same (still unknown in details) core-collapse mechanism may lead to their explosions, as in SNe II, however, the light is due now to entropy produced by radioactivity: the decays $^{56}\mathrm{Ni} \to ^{56}\mathrm{Co} \to ^{56}\mathrm{Fe}$ which lead to a slower heating of ejecta. This way of producing light is most important for SNe I of all subtypes. However, some contribution of radioactivity is clearly present in late light curves of type II supernovae as well, and for SN1987A in LMC it was important already before its maximum light, a month after the explosion.

If the radioactive mechanism was responsible for the light of SN2006gy, then the amount of ^{56}Ni must be higher than $10 M_\odot$ [6, 7, 8] . This immediately implies a very large mass for the presupernova star, more than $100 M_\odot$. More important, this implies a huge explosion energy, $(50-80) \times 10^{51}$ erg. One can delineate four kinds of deaths for massive stars, see Table 1. There is some uncertainty about the exact values due to uncertainties in mass-loss, rotation etc. Such massive stars do exist, and they can have powerful explosions in their oxygen cores which experience instability due to the creation of large numbers of electron-positron pairs.

PAIR INSTABILITY SUPERNOVAE

The word ‘instability’ refers here to hydrodynamics, to mechanical equilibrium, not to the process of pair creation which is quite stable and reversible in a thermodynamic sense in stellar interiors. A massive star loses its mechanical stability when pairs are being created because the adiabatic exponent γ goes down at $T \sim 0.1$ MeV , see Fig. 1: the work of contraction is spent in creating new particles and not for raising the momenta of particles that already exist and which provide for the equilibrium pressure.

We can easily estimate the path that leads the star into the domain of the pair-creation instability. From the virial theorem, for a star of mass M and radius R, omitting all coefficients of order unity, $P_c V \sim P_c R^3 \sim G_N M^2/R$. Hence, the pressure P_c in the center is $P_c \simeq G_N M^2/R^4$, while the density $\rho_c \simeq M/R^3$, and they are related as $P_c \simeq G_N M^{2/3} \rho_c^{4/3}$. So, if we have a classical ideal plasma with $P = \mathscr{R} \rho T/\mu$, where $\mathscr{R}$ is the universal gas

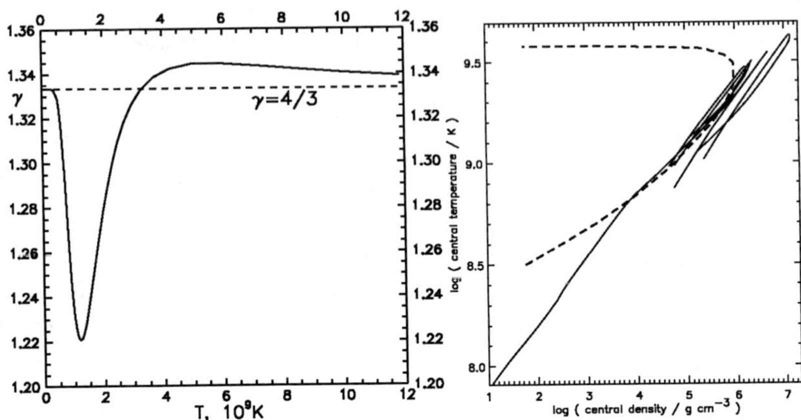

FIGURE 1. Left: adiabatic exponent in the low density asymptotic limit taking into account pair creation [9], see also [10]. Right: evolution track for one of the models [11] with initial $M \sim 103 M_\odot$ (solid line). The approximate boundary of pair-instability is shown by the dashed line

constant, and μ – mean molecular mass, we find

$$T_c \simeq G_N M^{2/3} \rho_c^{1/3} \mu / \mathscr{R}.$$

If we have a relativistic addition of aT^4 to P, the law $T_c \propto \rho_c^{1/3}$ is the same (but the coefficient is a bit different). These relations have been already shown at this conference by Naoki Yoshida and by Marco Limongi. The right panel of Fig. 1 shows how a massive star follows the law $T_c \propto \rho_c^{1/3}$ until entering the domain of pair-instability. We had already computed the light curves of pair-instability supernovae for $M > 130 M_\odot$ (third line in Table 1) some years ago (with S.Woosley and A.Heger), but we do not like them in the case of SN2006gy because we do not see the evidence for a tremendously high explosion energy in that case. If the explosion energy were 2 orders of magnitude higher than for normal SNe, then it should be seen in very broad spectral lines. What we see in SN2006gy is different: it has narrow P Cyg lines (hence, it is type IIn) superimposed on moderately broad emission component ($\sim 5 \times 10^3$ km/s [2]). There is no sign of huge kinetic energy in this event.

Multiple ejections in SNIIn

Supernovae of type IIn are among the most powerful transients in visible light. No radioactive material is needed to explain their light during the first several months: the light is produced by a long living radiative shock in a dense circumstellar envelope. This is the main difference with standard SNe II where the shock breaks out into rarefied interstellar medium and disappears quickly. Spectra and light curves of SNIIn can be explained only when the number density of circumstellar matter at radii of $\sim 10^{15-16}$ cm (where the narrow lines are formed) is unusually high, like $10^{9-10}\,\mathrm{cm}^{-3}$, see Fig. 2. This implies that a large mass (on the order of $M_\odot$ and larger) must be ejected within years, or even months, before the observed SN. The first ejection may have kinetic energy appreciably lower than a standard supernova. The slow motion of its matter explains the narrow lines of a type IIn SN. The supernova itself is an explosion of a normal energy, but inside a dense cloud.

Paper [3] was the first to suggest that an SNIIn had a precursor, a relatively weak explosion ejecting a large slowly moving mass. A dramatically high mass loss needed for the formation of a dense envelope shortly before SNIIn 1994W (Fig. 2) has been derived in [4]. SNIIn 1995G is explained in [12] in a model similar to the one presented in [4]. Double explosions may be observed also for other SN types [13].

According to [14] an $\sim 11 M_\odot$ star might produce strong flashes in the semi-degenerate O-Ne-Mg core a few years prior to SN explosion and the strongest flash could eject most of the hydrogen envelope with velocities ~ 100 km/s. More recent evolutionary computations do not support the conclusion on strong Ne flashes, but this complicated problem deserves further investigation.

Pulsational pair-instability

Another workable mechanism for multiple-explosion SNe has been proposed in our paper [11] to explain SN2006gy. It works for a high initial mass of the presupernova star $\sim 110 M_\odot$ (see Fig. 2). This mechanism is also based on the pair-creation instability, but there is no catastrophic collapse or full explosion of the star.

The second line of Table 1 shows that between 95 and 130 $M_\odot$ a relatively unexplored phenomenon of *pulsational* pair instability supernova [15, 16, 17] occurs. An instability in the mechanical equilibrium is encountered, as in the heavier stars, during the evolution along $T_c \propto \rho_c^{1/3}$ path (Fig. 1). A thermonuclear explosion of oxygen occurs, but the energy released is inadequate to unbind the entire star. It suffices, however, to eject many solar masses of surface material, including the hydrogen envelope, in a series of giant 'pulses' (explosions of various strength).

The binding energy for the hydrogen envelope of $(95-130) M_\odot$ stars is $\sim (0.1-1) \times 10^{49}$ erg while the energy of an explosive pulse is about two orders of magnitude higher. The velocity is in the range $100 \div 5000$ km/s depending on amount of explosive burning and mass ejected. After a pulse, the remaining core contracts searching for a new equilibrium state. It obeys the $T_c \propto \rho_c^{1/3}$ law again, but now the mass is lower, so the track is a bit different (see Fig. 1). The time required for the contraction is sensitive to the strength of the first pulse. If it cools down severely after the pulse expansion, it may be centuries before the star ignites burning again. If it remains hotter than 1.5×10^9 K, it may only take days. After one, two or several explosion pulses the remnant of the massive star continues to live, contrary to other supernovae, and eventually it should collapse.

To explain SN2006gy, we consider the evolution of a star with initial mass $110 M_\odot$. The evolution is calculated using the Kepler code [18, 6] with reduced mass loss. The star encounters pair-instability with a total mass of $75 M_\odot$ (a helium core of $50 M_\odot$) and experiences the first pulse ejecting a cloud with mass $\approx 25 M_\odot$ and $E_{\rm kin} \approx 1.4 \times 10^{50}$ erg.

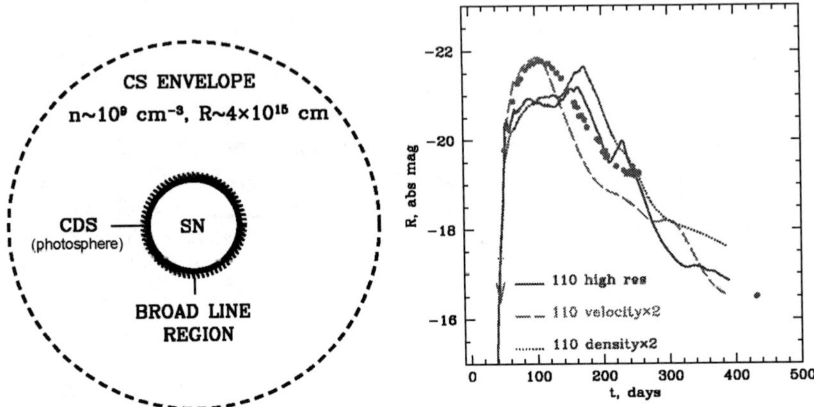

FIGURE 2. Left: SN1994W structure [4] prototypical for SNe IIn. Right: light curve models for SN2006gy. Dots – observations [2], the last observed point is from [21], courtesy M.Tanaka. The solid line is the model discussed in the text (with numerical resolution higher than in [11]), and the dashed one where the velocity of all the ejecta has been multiplied by 2 (hence an artificial increase in the explosion energy to 2.9×10^{51} erg). The dotted line is for the doubled density in ejecta

About 7 years later the remaining core produced the second explosion with $\approx 10 M_{\odot}$ and $E_{\rm kin} \approx 7.2 \times 10^{50}$ erg. Thus, the second ejection was faster and it had sent a shock wave into the massive cloud. The light from this radiating shock we see as SN2006gy (Fig. 2). The shock is still inside the cloud, it has not yet broken out, more than one year after the explosion of SN2006gy.

SOFT X-RAY?

There were arguments against the model of a shock moving into shells from previous explosions for SN2006gy. One can estimate the temperature of the shock using standard formulas for a rarefied medium and find that it must be very high, so SN2006gy must be a powerful source of X-ray emission. *Chandra X-ray observatory* has measured the X-ray flux on 2006 Nov 14 from the direction to SN2006gy. A best fit for X-ray luminosity in (0.5–2 keV) is 1.65×10^{39} erg/s [2] , but it is a few orders of magnitude lower than expected in the naive estimates.

There is no problem with this in our model: the cloud of 25 solar masses is almost transparent to the visible light of the shock because it is almost neutral, but exactly due to this reason it is fully opaque to X-ray light. A large mass lies above the shock, see Fig. 3. The zero of the mass coordinate in Fig. 3 is the inner edge of the ejecta. The radiative shock is located at $M_r \approx 6 M_{\odot}$, or $\log r \approx 15.5$.

If we look into other data on SNe IIn we see that all of them are discovered late in X-rays! See Table 2. This is no wonder in our models: the shock is buried in the material of the first ejection.

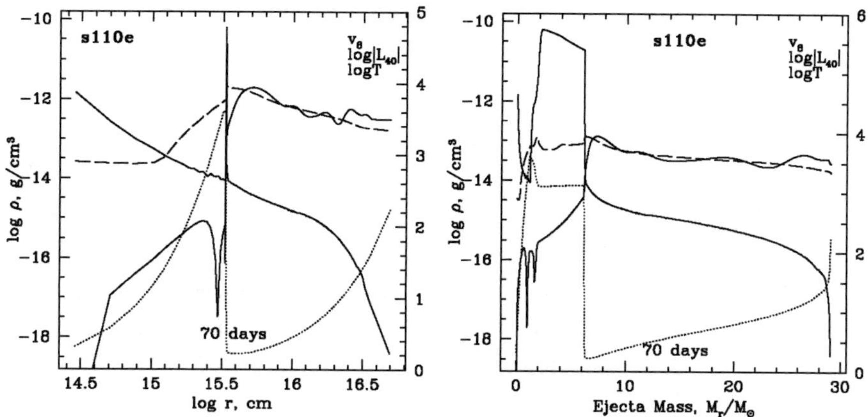

FIGURE 3. Hydrodynamic structure at 70 days as a function of radius (left panel) and of a Lagrangean mass coordinate (right panel). Thick solid line is density (left Y-axis). Thin solid line is luminosity L_{40} (units 10^{40} erg/s). Dotted line is velocity v_8 (units 10^8 cm/s) and dashed line is temperature T (K). The scale for $\log L_{40}$, v_8 and $\log T$ is on the right Y-axis.

TABLE 2. Data on X-ray observations of SNe IIn [20]

SN	galaxy	distance Mpc	discovery day
1986J	NGC 891	9.6	3,300
1988Z	+03-28-022	89	2,370
1994W	NGC 4041	25	1,180
1995N	-2-38-017	24	440
1998S	NGC 3877	17	678
2005ip	NGC 2096	30	490
2005kd	PGC14370	64	450

An alternate explanation for the low X-ray flux is possible. After the work of physicists on nuclear explosions in atmosphere in the 1940s and 1950s, we know that the preheating effect becomes so large in strong supercritical shocks that the *viscous jump* in pressure and density diminishes and completely *disappears.* In *radiation dominated shocks* not only the preheating effect is important; in addition the *momentum transfer* from photons to electrons (and hence to ions, via the electric field) is very large. This also destroys the viscous jump at the shock front. Imshennik and Morozov [19] have found with an accurate accounting of the photon transfer that this happens when $P_r/P_g \simeq 8.5$. This effect implies that postshock temperature may be so low that it does not attain keV range while inside the envelope. All the heat is taken away by 'cold' photons. Thus, the *Chandra* results [2] tell us something about the interaction of the the first pulse ejecta with the ISM, not about the main shock penetrating the pre-ejected shell, which shines in visible light as for SN2006gy.

CONCLUSIONS

1. Radiating shocks are the most probable sources of light in most luminous supernovae like SN2006gy.
2. The medium in which the shining shock propagates is naturally produced in massive star evolution due to violent non-linear pulsations when e^-e^+ pairs become appreciable in the pressure in their interiors.
3. The supercritical radiative shock must be well below the X-ray temperature. Even if the shock becomes hot, the overlaying matter absorbs X-rays for a long time.
4. If SN2006gy is a pulsational-pair-instability SN, a better understanding of mass-loss is needed. Otherwise, one has to find other evolution paths to SNe with double (or multiple) explosions.

ACKNOWLEDGMENTS

My work on SN2006gy was supported by NASA (at UCSC), and by the Russian Foundation for Basic Research and Science Schools (at ITEP). My visit to Tokyo University and to this conference is supported by RESCEU. I am very grateful to S. Woosley and K. Nomoto for their warm hospitality at their Institutions and for fruitful collaboration, to M. Tanaka for sharing the observational data, and to L.Rudnick for valuable comments.

REFERENCES

1. E. Ofek, et al., *ApJ* **659**, L13 (2007).
2. N. Smith, et al., *ApJ* **666**, 1116 (2007).
3. E. Grasberg, D. Nadyozhin, *Soviet Astronomy Letters* **12**, 68 (1986).
4. N. Chugai, S. Blinnikov et al., *MNRAS* **352**, 1213 (2004).
5. E. Grasberg, V. Imshennik, D. Nadyozhin, D. K. *Astrophysics and Space Science* **10**, 28-51 (1971).
6. S. Woosley, A. Heger, and T. Weaver, *Rev.Mod.Phys.* **74**, 1015 - 1071 (2002).
7. K. Nomoto, et al., *American Institute of Physics Conference Series* **937**, 412 (2007).
8. H. Umeda, and K. Nomoto, *ArXiv e-prints* **707**, arXiv:0707.2598 (2007).
9. D. Nadyozhin, *Nauchnye Informatsii* **32**, 3 (1974).
10. S. Blinnikov, N. Dunina-Barkovskaya, and D. Nadyozhin, *ApJS* **106**, 171 (1996).
11. S. Woosley, S. Blinnikov, A. Heger, *Nature* **450**, 390 (2007).
12. N. Chugai and I. Danziger, *Astr. Lett.* **29**, 732 (2003)
13. A. Pastorello, et al., *Nature* **447**, 829 (2007).
14. T. Weaver and S. Woosley, *BAAS* **11**, 724 (1979).
15. Z. Barkat, G. Rakavy, and N.Sack, *Phys.Rev.Lett.* **18**, 379 - 381 (1967).
16. S. Woosley and T. Weaver, "The Physics of Supernovae", in *Radiation Hydrodynamics in Stars and Compact Objects*, IAU Colloq. 89 Proceedings, edited by D. Mihalas and K.-H. Winkler, Springer-Verlag, Lecture Notes in Physics 255, 1986, pp. 91 - 120.
17. A. Heger and S. Woosley, *ApJ* **567**, 532 - 543 (2002).
18. T. Weaver, G. Zimmerman, and S.Woosley, *ApJ* **225**, 1021 - 1029 (1978).
19. V. Imshennik and Yu. Morozov, *Zhurnal Prikladnoj Mekhaniki i Tekhnicheskoj Fiziki* **No. 2**, 8-21 (1964).
20. S. Immler, `http://lheawww.gsfc.nasa.gov/users/immler/` (2007).
21. K. Kawabata, et al. (2008), in preparation.

Multi-Dimensional Simulations of Radiative Transfer in Aspherical Core-Collapse Supernovae

Masaomi Tanaka*, Keiichi Maeda†,**, Paolo A. Mazzali**,‡ and Ken'ichi Nomoto*,†

*Department of Astronomy, Graduate School of Science, University of Tokyo, Tokyo, Japan; mtanaka@astron.s.u-tokyo.ac.jp
†Institute for the Physics and Mathematics of the Universe, University of Tokyo, Kashiwa, Japan
**Max-Planck-Institut für Astrophysik, Garching bei München, Germany
‡Istituto Nazionale di Astrofisica, OATs, Trieste, Italy

Abstract.
We study optical radiation of aspherical supernovae (SNe) and present an approach to verify the asphericity of SNe with optical observations of extragalactic SNe. For this purpose, we have developed a multi-dimensional Monte-Carlo radiative transfer code, SAMURAI (SupernovA MUlti-dimensional RAdIative transfer code). The code can compute the optical light curve and spectra both at early phases ($\lesssim 40$ days after the explosion) and late phases (~ 1 year after the explosion), based on hydrodynamic and nucleosynthetic models. We show that all the optical observations of SN 1998bw (associated with GRB 980425) are consistent with polar-viewed radiation of the aspherical explosion model with kinetic energy 20×10^{51} ergs. Properties of off-axis hypernovae are also discussed briefly.

Keywords: supernovae; gamma-ray bursts; radiative transfer; nucleosynthesis
PACS: 97.60.Bw; 97.10.Ex; 26.30.-k

INTRODUCTION

Although the explosion mechanism of core-collapse supernovae (SNe) is not well understood, there is several observational evidence of non-spherical explosion, obtained by the imaging of *very* nearby SNe, e.g., SN 1987A [1] and Galactic supernova remnants [2]. Even when the imaging is not possible, the detection of polarization from extragalactic SNe [3, 4, 5] suggests that SNe are not spherical. In addition to these studies, spectroscopy of SNe can also give clues of the structure of SN explosion. For example, emission line profiles in the late time spectra ($t \gtrsim 1$ year, where t is the time after the explosion) reveals the asphericity of the explosion [6].

It is well established that a special class of Type Ic SNe [1] are associated with the long gamma-ray bursts (GRBs, see [8] and references therein). This class of SNe is thought to be highly energetic, so called hypernovae (here defined as SNe with ejecta kinetic energy $E_{51} = E_K/10^{51}$ergs > 10; e.g., [9]). The asphericity of hypernovae is of great

[1] SNe that do not show H, He, and strong Si absorption in the early time spectra ($t \lesssim 40$ days) are classified as Type Ic [7].

CP1016, *Origin of Matter and Evolution of Galaxies,*
edited by T. Suda, T. Nozawa, A. Ohnishi, K. Kato, M. Y. Fujimoto, T. Kajino, and S. Kubono

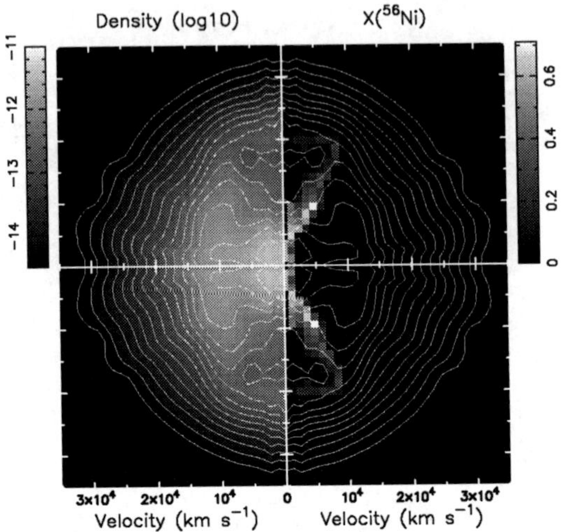

FIGURE 1. Aspherical explosion model A20 with the kinetic energy $E_{51} = E_K/10^{51}$ergs $= 20$. *Left*: Density distribution (log g cm^{-3}) at t=10 days. *Right*: Mass fraction of ^{56}Ni. The velocity can be used as spatial coordinate thanks to the homologous expansion ($r \propto v$).

interest related to the nature of GRBs.

Since GRBs are induced by relativistic jets, hypernovae are also thought to be aspherical. However, the large kinetic energy of hypernovae is estimated by the analysis under the spherical symmetry. No realistic multi-dimensional explosion models have been verified against the observed early phase spectra.

To study the multi-dimensional nature of the SN explosion through the various observational facts, radiative transfer calculations are required to connect observables and hydrodynamics models. In this paper, radiative transfer in SN ejecta is solved with a multi-dimensional Monte-Carlo radiative transfer code, `SAMURAI` (SupernovA MUlti-dimensional RAdIative transfer code), based on hydrodynamic and nucleosynthetic models of hypernovae. The results are compared with observations of SN 1998bw, and implications for off-axis hypernovae are discussed.

EXPLOSION MODELS

We use the results of multi-dimensional hydrodynamic and nucleosynthetic calculations for SN 1998bw [10] as input density and element distributions. Figure 1 shows the density structure and the distribution of ^{56}Ni. Since the original models used a He star as a progenitor, we simply replace the abundance of the He layer with that of the C+O layer. In the hydrodynamic model, energy is deposited aspherically, with more energy in the jet direction (z-axis, defined as $\theta = 0°$). As a result, ^{56}Ni is preferentially synthesized along this direction (right panel of Fig. 1). In this paper, an aspherical model with $E_{51} = 20$ (A20) and a spherical model with $E_{51} = 50$ (F50) are studied. They are constructed based on the models with $E_{51} = 10$ [10, 11].

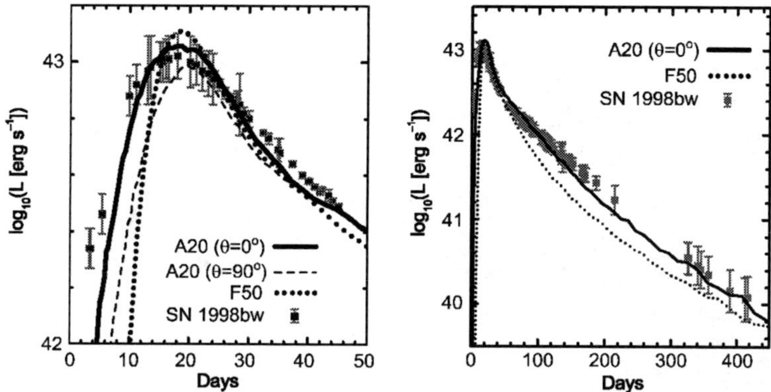

FIGURE 2. The LC of SN 1998bw (points) compared with the results of simulations. *Left*: The LCs at early phases. The LC of the polar-viewed model ($\theta = 0°$) rises earlier than that of the side-viewed model ($\theta = 90°$). *Right*: The LCs including late phases. The LC of the spherical model that explained early phase LC and spectra (F50) fades faster than the observed LC.

THE NUMERICAL CODE

In order to study the detailed properties of the radiation from aspherical SNe, we have unified a SupernovA MUlti-dimensional RAdIative transfer code `SAMURAI`. `SAMURAI` is a combination of 3D codes adopting Monte-Carlo methods to compute the bolometric light curve (LC) [11, 12], and the spectra of SNe from early [13, 14] to late phases [15] [2].

The early phase spectra are calculated as snapshots in the optically thin atmosphere, using the results of the LC simulation as initial conditions. A sharply defined photosphere is assumed as an inner boundary for simplicity. The position of the inner boundary in each direction is determined by averaging the positions of the last scattering photon packets (see Fig. 3 of Maeda et al. [11]). For the computation of ionization and excitation state in the atmosphere, the local physical process same as in the previous 1D code [22, 23]. Line scattering under the Sobolev approximation and electron scattering are taken into account. For line scattering, the effect of photon branching is included as in Lucy [24]. In the simulations, 16 elements are included, i.e., H, He, C, N, O, Na, Mg, Si, S, Ti, Cr, Ca, Ti, Fe, Co and Ni.

LIGHT CURVES

Maeda et al. [11] computed bolometric LCs in 3D space. A common problem in hypernova LCs is that a spherical model reproducing the LC and spectra at early phases ($E_{51} = 50$ for SN 1998bw) declines more rapidly than the observed LC at $t \gtrsim 100$ days [25, 26] (see model F50 in the right panel of Fig. 2). This problem can be solved by

[2] See also [16, 17, 18, 19, 20, 21] for other multi-dimensional codes.

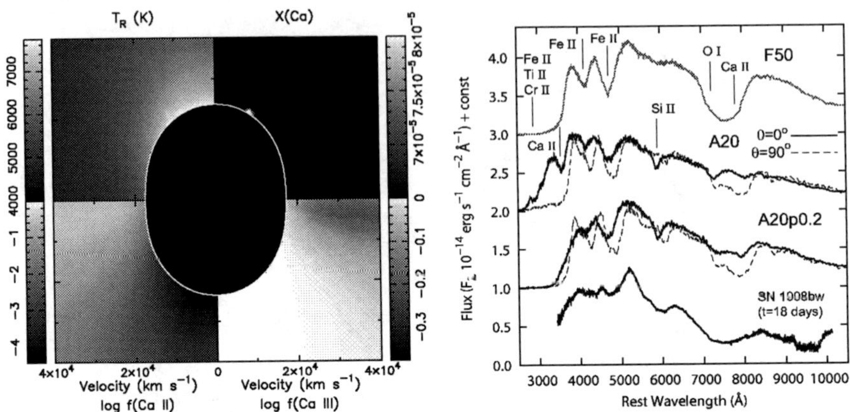

FIGURE 3. *Left*: Temperature structure (*upper left*), Ca mass fraction (*upper right*), ionization fraction of Ca II and Ca III (*lower left* and *lower right*, respectively) in the SN atmosphere at $t = 20$ days. *Right*: The observed spectrum of SN 1998bw at t=18 days compared with the synthetic spectra computed with model F50 (spherical), A20 (aspherical) and A20p0.2 (aspherical + mixing). For aspherical models, the solid and dashed lines show the spectra viewed from polar ($\theta = 0°$) and equatorial ($\theta = 90°$) direction, respectively. The synthetic spectra are scaled to match the observed spectra since the input luminosity is slightly brighter than the observation (Fig. 2, see Tanaka et al. [14] for details). The synthetic spectra are shifted by 3.0, 2.0, 1.0 $\times 10^{-14}$ from top to bottom.

aspherical models with a polar view. In aspherical models, even with a lower kinetic energy ($E_{51} = 10 - 20$, model A20 in Fig. 2) than in the spherical case ($E_{51} = 50$), which allows sufficient trapping of γ-rays at late times, the rapid rise of the LC can be reproduced because of the extended ^{56}Ni distribution [11].

SPECTRA

The right panel of Figure 3 shows the synthetic spectra at $t = 20$ days (around the maximum brightness) for models F50 and A20 compared with the observed spectrum of SN 1998bw. In model A20, all the absorption lines except for Si II $\lambda 6355$ are stronger for larger θ, i.e., for a side view. This is understood by the asphericity of the temperature structure in the SN atmosphere. As shown in the left panel of Figure 3, the temperature near the z-axis is higher than in the equatorial plane by ~ 2000 K (*upper left*), tracing the aspherical distribution of ^{56}Ni. This makes the ionization degree near the z-axis higher (*lower* panels). As a result, all species that have strong lines, i.e., O I, Si II, Ca II, Ti II, Cr II and Fe II, dominate near the equator but not near the z-axis.

The right panel of Figure 4 shows the synthetic spectra at $t = 30$ days for models F50 and A20. The emergent spectra of the aspherical model are not significantly different for different viewing angles (Fig. 4). At this epoch, the temperature structure are still anisotropic, and consequently, the distribution of ionization fractions is also aspherical (see the left panel of Fig. 4). However, since nucleosynthesis occurs entirely near the polar direction in the model, and the photosphere at this epoch is located inside

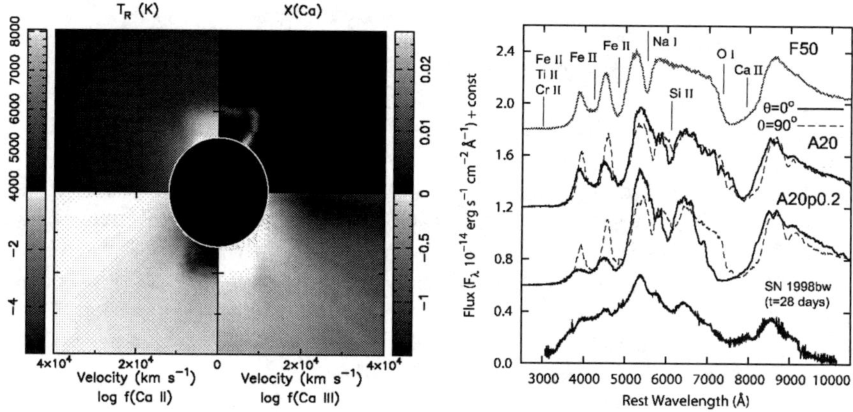

FIGURE 4. Same as Figure 3 but at $t = 30$ days. The synthetic spectra are shifted by 1.8, 1.2, 0.6 $\times 10^{-14}$ from top to bottom.

the region where heavy elements are synthesized in the explosion, the suppression of important ions near the z-axis is compensated by the larger abundance of the heavy elements (see X(Ca) in Fig. 4).

We compare the polar-viewed spectra of model A20 with the observed spectra of SN 1998bw. At $t = 20$ days, the absorptions of O I, Ca II and Fe II/Fe III in the model are weaker than in SN 1998bw. At $t = 30$ days, the Ca II and Fe II lines become strong, although the O I-Ca II absorption at 7000 – 8000 Å is still narrower than in the observed spectrum. In the synthetic polar-viewed spectrum, the peaks around 4000 and 4500 Å are partially suppressed by the high velocity absorption by the extended Fe near the jet, while they are strong in the side-viewed spectrum. The suppression of the peaks is similarly seen in the spectrum of SN 1998bw.

The strengths of the Ca II and Fe II lines at $t = 20$ days can be increased if heavy elements synthesized in the explosion are mixed to outer layers. In SN explosions, Rayleigh - Taylor (R-T) instabilities are expected to occur (see Kifonidis et al. [27] for the case of Type Ic SNe), which could deliver the newly synthesized elements to higher velocities. In Figures 3 and 4, synthetic spectra of model A20p0.2 are also shown. In this model, 20% of synthesized material is assumed to be mixed to the outer layers. The agreement with the observed spectra becomes better especially at $t = 20$ days.

DISCUSSION

We have presented the detailed simulations of optical radiation with realistic jet-like hypernova models. The emergent LC and spectra are different for different viewing angles. The spectral properties are determined by the combination of aspherical abundances and anisotropic ionization states. Although the agreement of the spectra is far from perfect, the spectra of the model with mixing are in qualitative agreement with those of SN 1998bw.

The LC study shows that the kinetic energy of an aspherical model that explains SN 1998bw is $E_{51} = 20$, which is less than that of a well-fitting spherical model ($E_{51} = 50$). The early phase spectra can also be explained by the model with $E_{51} = 20$. However, it should be noted that the higher kinetic energy than the canonical SNe ($E_{51} \sim 1$) is still required.

The simulations enable us to predict the radiation from off-axis hypernovae. The LC viewed off-axis rises more slowly than that of on-axis, and its maximum brightness is fainter (Fig. 2). The spectra viewed off-axis show (1) a slightly lower absorption velocity, (2) stronger peaks around 4000 and 4500 Å (narrower absorption of Fe) and (3) a stronger Na I λ5890 line. However, the spectra still show general appearance of "hypernovae" or "broad-line supernovae". At later phases, off-axis hypernovae would show double-peaked [O I] emission profile [6, 15].

ACKNOWLEDGMENTS

M.T. is supported through the JSPS (Japan Society for the Promotion of Science) Research Fellowship for Young Scientists.

REFERENCES

1. Wang, L., et al. 2002, ApJ, 579, 671
2. Hwang, U., et al. 2004, ApJ, 615, L117
3. Wang, L., et al. 2001, ApJ, 550, 1030
4. Kawabata, K.S., et al. 2002, ApJ, 580, L39
5. Leonard, D.C., et al. 2006, Nature, 440, 505
6. Mazzali, P. A., et al. 2005, Science, 308, 1284
7. Filippenko, A.V., 1997, ARA&A, 35, 309
8. Woosley, S. E. & Bloom, J. S. 2006, ARA&A, 44, 507
9. Nomoto, K., et al. 2006, Nucl. Phys. A, 777, 424 (astro-ph/0605725)
10. Maeda, K., et al. 2002, ApJ, 565, 405
11. Maeda, K., Mazzali, P. A., & Nomoto, K. 2006a, ApJ, 645, 1331
12. Maeda, K. 2006, ApJ, 644, 385
13. Tanaka, M., et al. 2006, ApJ, 645, 470
14. Tanaka, M., et al. 2007, ApJ, 668, L19
15. Maeda, K., et al. 2006b, ApJ, 640, 854
16. Höflich, P, et al. 1996, ApJ, 459, 307
17. Höflich, P, Wheeler, J.C., & Wang, L. 1999, ApJ, 521, 179
18. Thomas, R.C., et al. 2002, ApJ, 567, 1037
19. Kasen, D., et al. 2004, ApJ, 610, 876
20. Kozma, C., et al. 2005, A&A, 437, 983
21. Sim, S.A. 2007, MNRAS, 375, 154
22. Mazzali, P.A. & Lucy, L.B. 1993, A&A, 279, 447
23. Mazzali, P.A. 2000, A&A, 363, 705
24. Lucy, L. B. 1999, A&A, 345, 211
25. Nakamura, T., et al. 2001, ApJ, 550, 991
26. Maeda, K., et al. 2003, ApJ, 593, 931
27. Kifonidis, K., et al. 2000, ApJ, 531, L123

Relativistic EOS of Supernova Matter with Hyperons [1]

A. Ohnishi*, C. Ishizuka*, K. Tsubakihara*, H. Maekawa*, H. Matsumiya*, K. Sumiyoshi† and S. Yamada**

*Department of Physics, Faculty of Science Hokkaido University, Sapporo 060-0810, Japan
†Numazu College of Technology, Numazu 410-8501, Japan
**Science and Engineering, Waseda University, Tokyo 169-8555, Japan

Abstract. We investigate hyperon potentials in nuclear matter through hyperon production reactions, and construct several sets of equation of state (EOS) of nuclear matter including hyperons for numerical simulations of core collapse supernovae.

Keywords: supernova, EOS, hyperon, RMF, DWIA
PACS: 26.50.+x, 26.60.+c, 25.80.Hp, 25.80.Nv

INTRODUCTION

The equation of state (EOS) plays an important role in high density phenomena such as neutron stars and supernova explosions. In the core region of neutron stars, the density is so high that hyperons are expected to admix [1, 2]. Especially, the Σ^- baryon has been expected to appear at lower densities than the lightest hyperon, Λ, since it is negative charged and feels a larger chemical potential than neutral and positively charged baryons. Core-collapse processes leading to supernovae or black holes may involve more extreme conditions at high density and temperature. In order to describe the whole evolution of these processes by numerical simulations, one needs to prepare the set of microphysics under such extreme conditions. One of the most important ingredients is the equation of state (EOS) that contains necessary physical quantities. Until now, the two sets of EOS (Lattimer-Swesty EOS [3] and Shen EOS [4]) have been widely used and applied to numerical simulations of core-collapse supernovae. In these EOSs, hyperons are not included, and their applicability is questionable to the long-time evolution such as the proto-neutron star cooling [5] and the black hole formation [6].

In this work, we investigate hyperon potentials in nuclear matter through hyperon production reactions in the distorted wave impulse approximation (DWIA) with a newly developed local optimal Fermi averaging t-matrix (LOFAt) [7, 8, 9]. Next we include hyperons in the nuclear matter EOS with these potential values [10]. We provide the data table covering a wide range of temperature (T), density (ρ_B), and charge-to-baryon number ratio in hadronic part (Y_C), which enables one to apply to supernova simulations.

---

[1] `http://nucl.sci.hokudai.ac.jp/~chikako/EOS/`

CP1016, *Origin of Matter and Evolution of Galaxies*,
edited by T. Suda, T. Nozawa, A. Ohnishi, K. Kato, M. Y. Fujimoto, T. Kajino, and S. Kubono

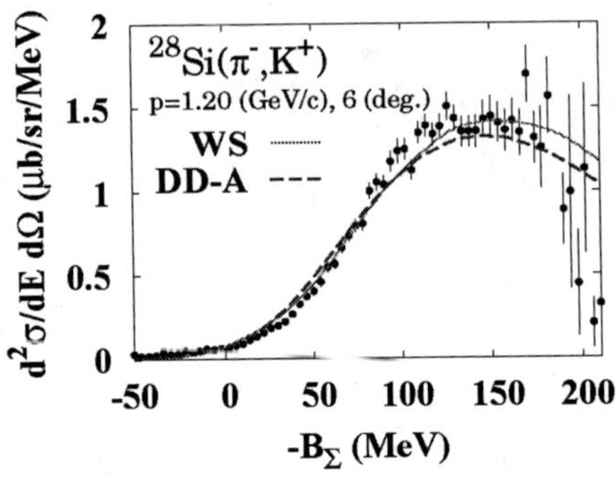

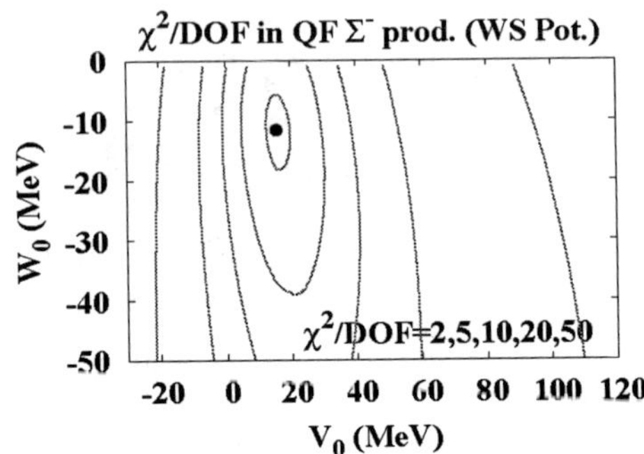

FIGURE 1. Σ^- production spectrum (left) and the χ^2 contour map in the Σ^- production spectrum (right).

HYPERON POTENTIALS IN NUCLEAR MATTER

Hyperon threshold densities and fractions are mainly determined by hyperon potentials in nuclear matter. The Λ-nucleus potential has been known to be around $U_\Lambda = -30$ MeV from single particle energies in the bound state region. It has been naively expected that baryon potential in nuclear matter is proportional to the number of *ud* quarks, suggesting $U_\Sigma \sim U_\Lambda \sim 2/3 U_N \sim -30$ MeV [11], but quark cluster models [12] and a chiral model [13] suggest repulsive potential, $U_\Sigma \sim +30$ MeV. Thus it is very important to determine U_Σ based on real data, not only for astrophysics but also for the understanding of the nature of baryon-baryon interactions.

In extacting Σ and Ξ potentials, presently available data for bound states are so scarce that it is necessary to analyze hyperon production spectra. For Σ, there is only one quasi-bound state ${}^4_\Sigma$He [14], which is too light to extract the potential. For Ξ, no bound state peaks have been specified.

We analyze quasi-free Σ^- production spectra [15] and Ξ^- production spectra [16] in the Green's function method in DWIA [17], in which the hyperon production spectra are represented as,

$$\frac{d^2\sigma}{dE_K d\Omega_K} = \frac{p_K E_K}{(2\pi)^3 v_\pi} R_Y(E_Y)\,, \quad R_Y(E_Y) = -\frac{1}{\pi}\mathrm{Im}\langle i\,|\hat{\mathscr{O}}^\dagger \hat{t}_q^\dagger \frac{1}{E_Y - \hat{H}_Y + i\varepsilon} \hat{t}_q \hat{\mathscr{O}}|i\rangle\,. \quad (1)$$

Here we show the expression in (π, K^+) reactions as an example, and $\hat{\mathscr{O}}$ denotes the operater which converts πN to $K^+ Y$.

In a standard treatment [17], the elementary t-matrix ($\hat{t}_q$) is assumed to be a constant and factorized out from the response function R_Y, then the cross section is found to be represented as the product of a kinematical factor, the Fermi averaging t-matrix ($\bar{t}_q$) and the strength function, $S = -\mathrm{Im}\langle \hat{\mathscr{O}} \hat{G}_Y \hat{\mathscr{O}} \rangle / \pi$. Recently, it has been pointed out that the on-shell kinematics in the Fermi averaging is important to describe hyperon production spectrum shape [18]. When we have potentials for N and Y, it would be also necessary

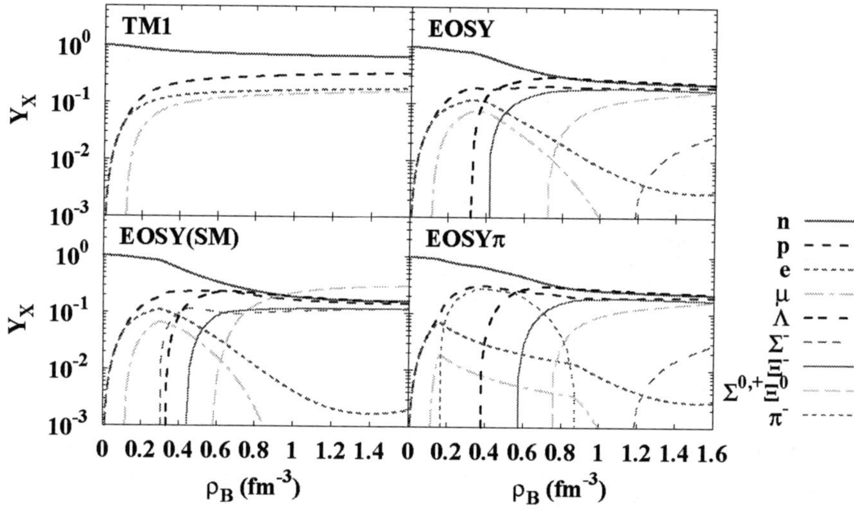

FIGURE 2. Particle number fraction as functions of baryon density in neutron star matter.

to the reaction point dependence [19]. Here we apply the local optimal Fermi averaging t-matrix (LOFAt), $\bar{t}_q(r)$ [7, 8, 9], defined as the Fermi averaging t-matrix under the on-shell condition with hadron single particle local energies containing potential effects.

We show the Σ^- production spectrum in $^{28}\mathrm{Si}(\pi^-,K^+)\Sigma^-X$ in the left panel of Fig. 1, and the χ^2 contour map in the case of Woods-Saxon potential in the right panel of Fig. 1. We find that the Σ^--nucleus potential is well determined to be $U_\Sigma(\rho_0) \simeq (+15\ \mathrm{MeV}, -10\ \mathrm{MeV})$ in the case of the Woods-Saxon potential, which simulates the $t\rho$ approximation. With higher order terms in ρ, $U_Y = \alpha\rho_B + \beta\rho_B^\gamma$, we cannot determine the potential depth only from the quasi-free Σ^- production data, but we find that the spectrum is well described with the potential determined from the Σ^- atomic shift [21]. In the latter case, we need to have more repulsive potential in order to cancel the effects from the attractive potential pocket on the nuclear surface. Thus we may conclude $U_\Sigma > +15$ MeV.

We have also analyzed the Ξ^- production spectra [8], and found that the $U_\Xi \simeq -14$ MeV [16, 20] seems to be reasonable, while we underestimate the absolute value of the production yield for light nuclear target.

EOS WITH HYPERONS AND ASTROPHYSICAL APPLICATIONS

In constructing relativistic EOS with hyperons, we start from the parameter set TM1 [22] and its flavor SU(3) extension [11] for nucleon and hyperon sectors. The RMF parameter set TM1 is determined to describe binding energies and nuclear radii of finite nuclei in

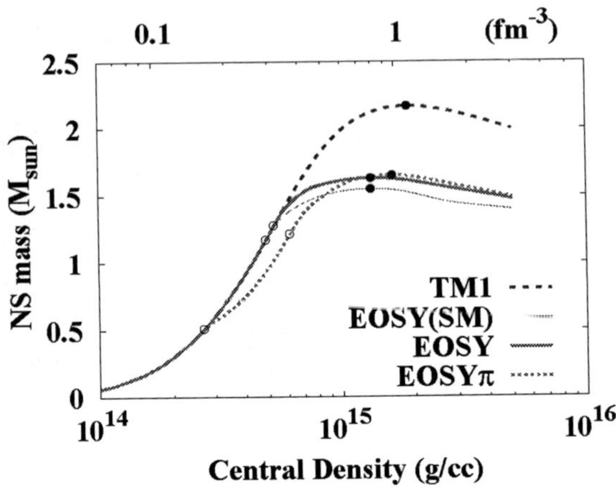

FIGURE 3. Neutron star masses as functions of central density. Filled points show the maximum neutron star masses, and open circles show the threshold densities of hyperons and pions.

a wide mass range. The extension of the RMF to flavor SU(3) has been investigated by many authors. A typical form of the Lagrangian density including hyperons is given as [11],

$$\mathscr{L} = \mathscr{L}_{Free}(B,\sigma,\omega_\mu,\vec{R}_\mu,\zeta,\phi_\mu) - U_\sigma(\sigma) + \frac{1}{4}c_\omega(\omega^\mu\omega_\mu)^2$$
$$-\sum_B \bar{\psi}_B \left(g_{\sigma B}\sigma + g_{\omega B}\not{\omega} + g_{\rho B}\not{\vec{R}}\cdot\vec{t}_B + g_{\zeta B}\zeta - g_{\phi B}\gamma^\mu\phi_\mu \right) \psi_B \,, \tag{2}$$

which contains hidden strangeness ($\bar{s}s$) scalar and vector mesons, ζ and ϕ, in addition to σ, ω and ρ (represented by $\vec{R}^\mu$) mesons.

We adopt the hyperon-vector meson coupling constants based on the SU(6) (flavor-spin) symmetry [11]. Scalar mesons in RMF may partially represent contributions from some other components than $\bar{q}q$, then the hyperon-scalar meson couplings are determined to fit the *recommended* potential strength discussed in the previous section,

$$U_\Sigma^{(N)}(\rho_0) \simeq +30\,\text{MeV}\,, \quad U_\Xi^{(N)}(\rho_0) \simeq -15\,\text{MeV}\,. \tag{3}$$

Now we apply the EOS with hyperons in the present RMF parameter set (EOSY) to neutron star matter [10]. EOSY is softer than the TM1 EOS due to hyperon admixture at $\rho_B \gtrsim 0.3$ fm^{-3}, as shown in Fig. 2. With attractive hyperon potentials (EOSY(SM)), $(U_\Sigma, U_\Xi) = (-30\,\text{MeV}, -28\,\text{MeV})$ [11], Σ^- appears prior to Λ, and π^- may condensate if the πN repulsion is weak (EOS$Y\pi$). The maximum mass is 1.63 $M_\odot$ for EOSY in contrast to 2.17 $M_\odot$ for nucleonic (TM1) EOS. The central density of a typical neutron star having 1.4$M_\odot$ is 0.35 fm^{-3} in nucleonic EOS (TM1), which is a little above the threshold density of Λ in EOSY. In this case, hyperons are limited only in the core region, and the neutron star mass does not get a large reduction as seen in Fig. 3.

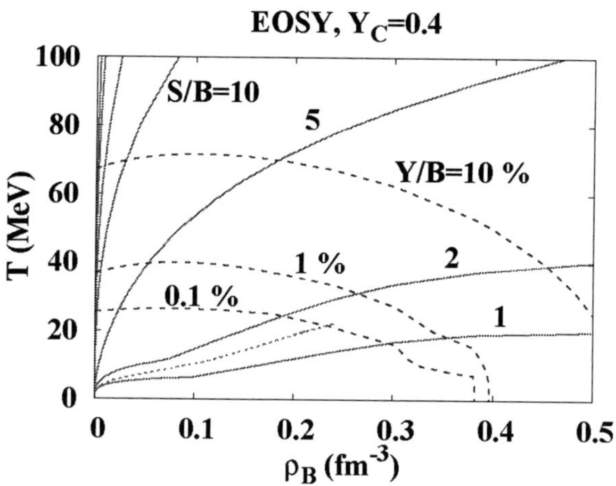

FIGURE 4. Hyperon fraction contours (dashed) and adiabatic paths (solid) in supernova matter at $T = 10$ MeV and $Y_C = 0.4$ in EOSY. The dotted line shows the trajectory of at center during core collapse and bounce.

When we apply the EOS with hyperons to supernova explosions, finite nuclear formation effects are included by adding the differences of the free energy and its derivatives in the Shen EOS and the uniform matter EOS in RMF(TM1) since low density part of EOS is also important. In order to test the EOS table, we calculate the adiabatic collapse of iron core of massive stars with $15M_\odot$ [23] under the spherical symmetry without neutrino-transfer in the same way as Ref. [24]. In Fig. 4, we plot the contour map of the hyperon fraction (sum of strange baryons) in the density-temperature plane. In order to have a significant amount of hyperons, one needs high density or temperature. In supernova core, the entropy per baryon is typically around 1–2 k_B, therefore, one needs high densities 0.3–0.4 fm^{-3} to have 1% mixture of hyperons and 0.45 fm^{-3} for 10%.

We find that in the *model* explosion, caused by the large electron fraction, the maximum density (0.24 fm^{-3}) and temperature (22 MeV) is not high enough for hyperons to appear. However, roles of hyperons must be seen in processes involving higher densties and temperatures or lower electron fractions, such as the thermal evolution of proto-neutron stars [5] and the black hole formation [6].

SUMMARY AND DISCUSSION

In this work, we have investigated hyperon potentials in nuclear matter and extended the Shen EOS by including hyperons with these hyperon potentials.

Hyperon production reactions are analyzed in the distorted wave impulse approximation with a local optimal Fermi averaging t-matrix (LOFAt). We find that the Σ^- and Ξ^- production spectra are well explained with Woods-Saxon potentials. Combined with the Σ^- atom data, we adopt $U_\Sigma = +30$ MeV and $U_\Xi = -15$ MeV as recommended values. These hyperon potentials are included in an $SU_f(3)$ relativistic mean field (RMF) model, which results in the neutron star maximum mass reduction from $2.17M_\odot$ to $1.63M_\odot$ with hyperons. We include finite nuclear formation effects at low densities [4], and the supernova matter EOS table is constructed. Hyperon effects are found to be small in prompt supernova explosions, since the matter does not reach the region of hyperon mixture ($Y_Y > 1\%$ for $T > 40$MeV or $\rho_B > 0.4$ fm^{-3}). Hyperon effects should be seen in processes involving higher densties, higher temperatures or lower electron fractions.

ACKNOWLEDGMENTS

This work is supported in part by the Ministry of Education, Science, Sports and Culture, Grant-in-Aid for Scientific Research under the grant numbers, 15540243, 1707005, and 19540252.

REFERENCES

1. N. K. Glendenning, "Compact Stars: Nuclear Physics, Particle Physics and General Relativity" (Springer-Verlag, Berlin, 2000), and references therein.
2. S. Balberg, A. Gal, Nucl. Phys. A **625** (1997), 435.
3. J. M. Lattimer, F. D. Swesty, Nucl. Phys. A **535** (1991), 331.
4. H. Shen, H. Toki, K. Oyamatsu, K. Sumiyoshi, Prog. Theor. Phys. **100** (1998), 1013.
5. H. Suzuki, "Physics and Astrophysics of Neutrinos", edited by M. Fukugita and A. Suzuki (Springer-Verlag, Berlin, 1994), 763; J. A. Pons *et al.*, Astrophys. J. **513** (1999), 780; Astrophys. J. **553** (2001), 382.
6. K. Sumiyoshi, S. Yamada, H. Suzuki, S. Chiba, Phys. Rev. Lett. **97** (2006), 091101; K. Sumiyoshi, S. Yamada, H. Suzuki, Astrophys. J. **667** (2007), 382.
7. H. Maekawa, K. Tsubakihara, A. Ohnishi A 2007 Euro. Phys. J. A **33** (2007), 269.
8. H. Maekawa, K. Tsubakihara, H. Matsumiya, A. Ohnishi, [arXiv:0704.3929 (nucl-th)].
9. H. Maekawa, PhD Thesis, Hokkaido University, March 2008; H. Maekawa, K. Tsubakihara, H. Matsumiya, A. Ohnishi, in preparation.
10. C. Ishizuka, A. Ohnishi, K. Tsubakihara, K. Sumiyoshi, S. Yamada, [arXiv:0802.2318 (nucl-th)]
11. J. Schaffner, I. Mishustin, Phys. Rev. C **53** (1996), 1416.
12. M. Kohno *et al.*, Nucl. Phys. A **674** (2000) 229.
13. N. Kaiser, Phys. Rev. C **71** (2005) 068201.
14. T. Nagae *et al.*, Phys. Rev. Lett. **80** (1998), 1605; T. Harada, S. Shinmura, Y. Akaishi and H. Tanaka, Nucl. Phys. A **507** (1990), 715; T. Harada, Phys. Rev. Lett. **81** (1998), 5287.
15. H. Noumi *et. al.*, Phys. Rev. Lett. **89** (2002) 072301 [Erratum: Phys. Rev. Lett. **90** (2003) 049902]; P. K. Saha *et. al.*, Phys. Rev. C **70** (2004) 044613.
16. T. Fukuda *et al.*, Phys. Rev. C **58** (1998), 1306; P. Khaustov *et al.*, Phys. Rev. C **61** (2000), 054603.
17. O. Morimatsu and K. Yazaki, Nucl. Phys. A **483** (1988) 493; O. Morimatsu and K. Yazaki, Nucl. Phys. A **435** (1985) 727; T. Harada and Y. Akaishi, Prog. Theor. Phys. **96** (1996) 145; M. T. López-Arias, Nucl. Phys. A **582** (1995) 440; S. Tadokoro, H. Kobayashi and Y. Akaishi, Phys. Rev. C **51** (1995) 2656.
18. T. Harada and Y. Hirabayashi, Nucl. Phys. A **744** (2004), 323; Nucl. Phys. A **759** (2006), 143; Nucl. Phys. A **767** (2006), 206
19. M. Kohno *et al.*, Prog. Theor. Phys. **112** (2004) 895.
20. S. Aoki *et al.*, Phys. Lett. B **355** (1995), 45.
21. C. J. Batty, E. Friedman, A. Gal, Phys. Lett. B **335** (1994), 273.
22. Y. Sugahara, H. Toki, Nucl. Phys. A **579** (1994), 557.
23. S. E. Woosley, T. A. Weaver Astrophys. J. Suppl. **101** (1995), 181.
24. K. Sumiyoshi, H. Suzuki, S. Yamada, H. Toki, Nucl. Phys. A **730** (2004), 227.

Variational Calculations for the Nuclear Equation of State toward Supernova Simulations

H. Kanzawa*, K. Oyamatsu†, K. Sumiyoshi** and M. Takano‡

*Department of Pure and Applied Physics, Science and Engineering, Waseda University, 3-4-1 Okubo Shinjuku-ku, Tokyo 169-8555, Japan
†Department of Media Production and Theories, Aichi Shukutoku University, Nagakute-cho, Aichi 480-1197, Japan
**Numazu College of Technology, Ooka 3600, Numazu, Shizuoka 410-8501, Japan
‡Research Institute for Science and Engineering, Waseda University, 3-4-1 Okubo Shinjuku-ku Tokyo 169-8555, Japan

Abstract. We construct an equation of state (EOS) for uniform nuclear matter at zero and finite temperatures with the variational method starting from the realistic nuclear Hamiltonian composed of the AV18 two-body potential and the UIX three-body interaction. Using the obtained EOS, we perform the Thomas-Fermi calculation of atomic nuclei, and tune adjustable parameters in the EOS for uniform matter to reproduce empirical data of β-stable nuclei, toward a nuclear EOS table for supernova simulations.

Keywords: Nuclear matter, Nuclear EOS, Variational method, Neutron stars, Supernovae
PACS: 21.65.-f, 26.50.+x, 26.60.-c, 97.60.Jd

INTRODUCTION

The nuclear equation of state (EOS) plays important roles in numerical simulations for supernova (SN) explosions and related astrophysical phenomena. At present, the EOS sets available for SN simulations are limited. An example is the one by Lattimer and Swesty[1] and another is by Shen et al.[2]; the latter is extended recently to take into account the hyperon mixing[3] and hadron-quark phase transition[4]. Since these EOSs are obtained in phenomenological frameworks, EOSs based on microscopic many-body approaches starting from the realistic nuclear Hamiltonian are desirable.

This situation motivates us to construct a nuclear EOS for SNe using a variational method. In this paper, we report the present status of this undertaking. First, we construct an EOS for uniform nuclear matter at zero and finite temperatures[5]. Second, we apply the obtained EOS to the Thomas-Fermi calculation of atomic nuclei, and tune parameters in the EOS to reproduce empirical data of β-stable nuclei.

CP1016, *Origin of Matter and Evolution of Galaxies*,
edited by T. Suda, T. Nozawa, A. Ohnishi, K. Kato, M. Y. Fujimoto, T. Kajino, and S. Kubono

UNIFORM NUCLEAR MATTER AT ZERO TEMPERATURE

In this section, we calculate the energy for uniform matter at zero temperature. We decompose the nuclear Hamiltonian H into H_2 and H_3. The two-body Hamiltonian H_2 is given by

$$H_2 = -\frac{\hbar^2}{2m}\sum_{i=1}^{N}\nabla_i^2 + \sum_{i<j}^{N} V_{ij}, \tag{1}$$

where m is the nucleon mass and V_{ij} is the two-body nuclear potential. H_3 is the three-body Hamiltonian caused by the three nucleon interaction (TNI). In this study, we employ the AV18 two-body potential and the UIX three-body potential.

The expectation value of H_2 is evaluated using the Jastrow-type wave function

$$\Psi = \mathrm{Sym}\left[\prod_{i<j} f_{ij}\right]\Phi_{\mathrm{F}}. \tag{2}$$

Here, Φ_{F} is the Fermi-gas wave function at zero temperature, and the correlation function f_{ij} is composed of spin-isospin-dependent central correlation functions $f_{\mathrm{C}ts}(r)$, tensor correlation functions $f_{\mathrm{T}t}(r)$ and spin-orbit correlation functions $f_{\mathrm{LS}t}(r)$ as follows:

$$f_{ij} = \sum_{t=0}^{1}\sum_{s=0}^{1}\left[f_{\mathrm{C}ts}(r_{ij}) + s f_{\mathrm{T}t}(r_{ij}) S_{\mathrm{T}} + s f_{\mathrm{LS}t}(r_{ij})(\boldsymbol{L}\cdot\boldsymbol{s})\right] P_{tsij}. \tag{3}$$

In Eq. (3), S_{T} is the tensor operator, $(\boldsymbol{L}\cdot\boldsymbol{s})$ is the spin-orbit operator and P_{tsij} is the spin-isospin projection operator.

Then, the expectation value of H_2 per nucleon in the two-body cluster approximation, E_2/N, is minimized with respect to variational functions, $f_{\mathrm{C}ts}(r)$, $f_{\mathrm{T}t}(r)$ and $f_{\mathrm{LS}t}(r)$, by solving the Euler-Lagrange equations. Here, we impose two constraints on the variational functions: the extended Mayer's condition and the healing-distance

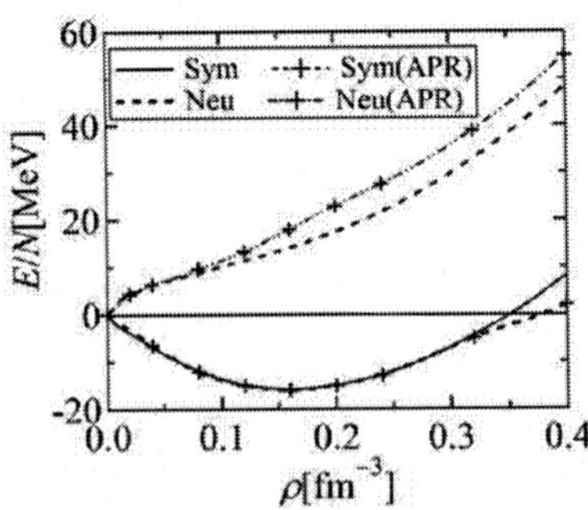

FIGURE 1. Energies per nucleon E/N for symmetric nuclear matter and neutron matter as functions of the density. Corresponding energies obtained by APR are also shown.

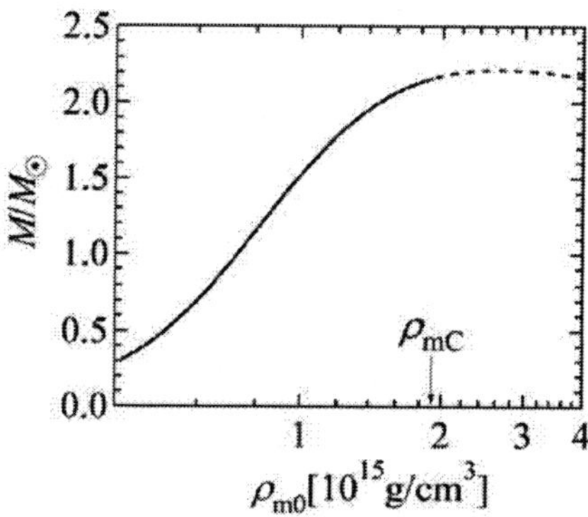

FIGURE 2. Neutron star mass as a function of the central matter density. The dashed line shows the causality-violated solutions.

condition. The healing distance includes an adjustable parameter, the value of which is chosen so that E_2/N is close to the result obtained by Akmal, Pandharipande and Ravenhall (APR)[6] using the Fermi Hypernetted Chain method.

The TNI contribution per nucleon is expressed in the form

$$\frac{E_3}{N} = \alpha \frac{\langle H_3^{\rm R} \rangle}{N} + \beta \frac{\langle H_3^{2\pi} \rangle}{N} + \gamma \rho^2 e^{-\delta\rho}. \tag{4}$$

Here, $H_3^{\rm R}$ and $H_3^{2\pi}$ are the repulsive and two-π-exchange parts of H_3, respectively. In Eq. (4), the brackets represent the expectation values using the Fermi-gas wave function, and ρ is the nucleon number density. The values of the adjustable parameters, α, β, γ and δ, are determined so that the total energy $E/N = E_2/N + E_3/N$ reproduces the empirical saturation conditions. Here, α and β are common to symmetric nuclear matter and neutron matter, while $\gamma = 0$ for neutron matter.

Figure 1 shows E/N for symmetric nuclear matter and neutron matter with the saturation density $\rho_0 = 0.16$ fm^{-3}, saturation energy $E_0/N = -15.8$ MeV, incompressibility $K = 250$ MeV and symmetry energy $E_{\rm sym}/N = 30$ MeV. It is seen that the obtained E/N is in fair agreement with the result by APR.

Figure 2 shows the neutron star mass calculated using the above-obtained E/N as a function of the central mass density $\rho_{\rm m0}$. The maximum mass is $2.22M_\odot$. It is noted that the causality violation occurs at $\rho_{\rm m0}$ larger than the critical density $\rho_{\rm mC} \sim 1.91 \times 10^{15}$ g/cm^3. At the critical density $\rho_{\rm mC}$, the neutron star mass is $2.16M_\odot$.

UNIFORM NUCLEAR MATTER AT FINITE TEMPERATURES

In this section, we calculate the free energy per nucleon F/N for uniform nuclear matter at finite temperatures. Following the method by Schmidt and Pandharipande[7, 8], F/N at temperature T can be expressed as

$$\frac{F}{N}=\frac{E_{\mathrm{T0}}}{N}-T\frac{S_0}{N}. \tag{5}$$

The approximate internal energy E_{T0}/N is the sum of E_{T2}/N and E_3/N. Here, E_{T2}/N is the expectation value of H_2 using the Jastrow-type wave function at finite temperature $\Psi(T)$ in the two-body cluster approximation. $\Psi(T)$ has the form similar to Eq. (2) except that Φ_{F} is the Fermi-gas wave function at finite temperature specified by the averaged occupation probability of the quasi-nucleon states:

$$n(k)=\left\{1+\exp\left[\frac{(\varepsilon(k)-\mu)}{k_{\mathrm{B}}T}\right]\right\}^{-1}, \tag{6}$$

where k is the wave number, and $\varepsilon(k)=\hbar^2k^2/2m^*$ is the quasi-nucleon energy with m^* being the effective mass of the quasi-nucleon. In Eq. (6), μ is determined by the normalization condition of $n(k)$. The TNI contribution E_3/N is assumed to be the same as that at zero temperature. The approximate entropy S_0/N can be expressed as

$$\frac{S_0}{N}=-\frac{k_{\mathrm{B}}}{N}\sum_i\left[\{1-n(k)\}\ln\{1-n(k)\}+n(k)\ln n(k)\right]. \tag{7}$$

Then, F/N is minimized with respect to m^* for each ρ and T.

In Fig. 3, F/N is compared with the result by Friedman and Pandharipande (FP)[9]. At high densities, the present result is higher than that by FP, which implies that the EOS obtained in this study is stiffer than that by FP.

THOMAS-FERMI CALCULATION OF ATOMIC NUCLEI

In order to obtain a complete nuclear EOS for SN simulations, the energy calculation for nonuniform nuclear matter is necessary, because SN matter becomes nonuniform at low densities. We are planning to construct the EOS for nonuniform

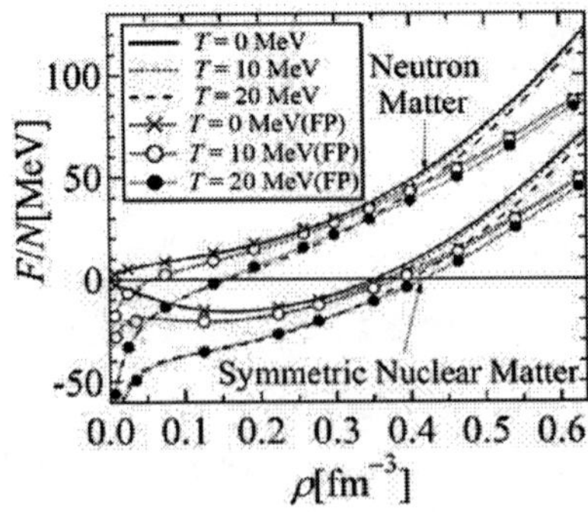

FIGURE 3. Free energies per nucleon at T = 0, 10 and 20 MeV for symmetric nuclear matter and neutron matter as functions of the density. The free energies obtained by FP are also shown.

matter using the Thomas-Fermi (TF) approximation. In this section, before treating nonuniform nuclear matter, we perform the TF calculation of atomic nuclei, and tune the parameters in the EOS to reproduce empirical data.

We employ a simplified TF approximation used by Oyamatsu[10] to express the binding energy $B(N, Z)$ of each nucleus of the proton number Z and the neutron number N as

$$-B(N,Z)=\int d\boldsymbol{r}\varepsilon(\rho_{\rm n}(r),\rho_{\rm p}(r))+F_0\int d\boldsymbol{r}|\nabla\rho(r)|^2+\frac{e^2}{2}\int d\boldsymbol{r}\int d\boldsymbol{r}'\frac{\rho_{\rm p}(r)\rho_{\rm p}(r')}{|\boldsymbol{r}-\boldsymbol{r}'|}. \quad (8)$$

The first term on the right-hand side of Eq. (8) is the gross energy, and the energy density of uniform nuclear matter $\varepsilon(\rho_{\rm n}, \rho_{\rm p})$ for the neutron number density $\rho_{\rm n}$ and the proton number density $\rho_{\rm p}$ is calculated using E/N obtained above. The second term represents the gradient energy with $\rho(r) = \rho_{\rm p}(r) + \rho_{\rm n}(r)$, and F_0 = 70 MeVfm^5. The third term is the Coulomb energy. In this calculation, $\rho_i(r)$ (i = p, n) is parameterized, and then $B(N, Z)$ is maximized with respect to parameters in $\rho_i(r)$ for each nucleus.

The calculated masses, RMS charge radii and proton numbers of several β-stable nuclei are compared with the empirical values given in Ref. [10]. Figure 4(a) shows the difference between the calculated and empirical masses, $\Delta M = M_{\rm cal} - M_{\rm emp}$ (filled circle). Similarly, the difference of the RMS charge radii, $\Delta r_{\rm RMS}$, and the proton numbers of β-stable nuclei, ΔZ_β, are shown in Figs. 4(b) and 5, respectively. As seen in these figures, $\Delta M \sim 8 - 22$ MeV, $\Delta r \sim 0.01 - 0.06$ fm, and $|\Delta Z_\beta| \sim 1.5 - 3.3$.

In order to reduce $|\Delta M|$, $|\Delta r|$, and $|\Delta Z_\beta|$, values of the parameters α, β, γ and δ in Eq. (4) are tuned. Here, the values of β for symmetric nuclear matter and for neutron matter are determined independently. F_0 in Eq. (8) is also treated as an adjustable parameter. The results are shown in Figs. 4 and 5 by open circles. As shown in the figures, $|\Delta M|$ and $|\Delta r|$ decreased significantly: $|\Delta M| < 3$ MeV and $|\Delta r| < 0.01$ fm. The values of $|\Delta Z_\beta|$ also decreased, but not sufficiently, by about 0.5 for heavy nuclei. Correspondingly, $\rho_0 = 0.17$ fm^{-3}, $E_0/N = -16.2$ MeV, $K = 247$ MeV and $E_{\rm sym}/N = 32.4$ MeV are obtained for the EOS of uniform nuclear matter.

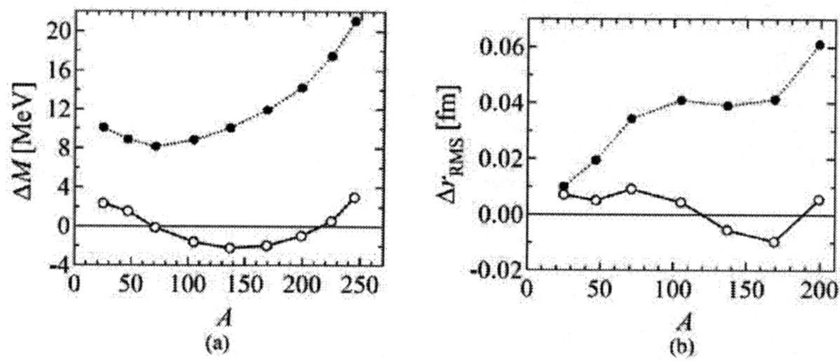

FIGURE 4. Deviation of the masses and RMS charge radii from the empirical values: (a) ΔM; (b) $\Delta r_{\rm RMS}$. Filled circles are for the EOS constructed in the previous section and open circles are for the parameter-tuned EOS.

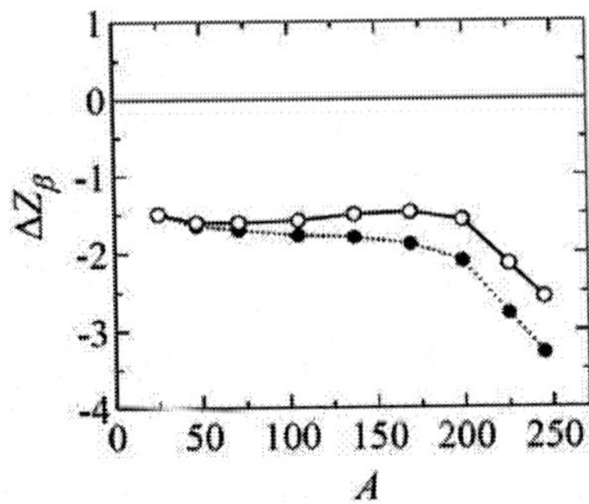

FIGURE 5. Deviation of the proton numbers of β-stable nuclei from the empirical values, ΔZ_β. Filled circles are for the EOS obtained in the previous section and open circles are for the parameter-tuned EOS.

CONCLUDING REMARKS

In this paper, we constructed an EOS for uniform nuclear matter at zero and finite temperatures starting from the realistic Hamiltonian. Then we performed the TF calculation of atomic nuclei, and tuned the parameters in the EOS to reproduce the empirical data of β-stable nuclei. The deviation of the masses and RMS charge radii decreased significantly by tuning the parameters, although the proton numbers of the β-stable nuclei was improved only slightly. We are planning to perform the TF calculation for nonuniform nuclear matter, toward a nuclear EOS table for SN simulations.

ACKNOWLEDGMENTS

This study is supported by a Grant-in-Aid for the 21st century COE program "Holistic Research and Education Center for Physics of Self-organizing Systems" at Waseda University, and Grants-in-Aid from the Scientific Research Fund of the JSPS (Nos. 18540291, 18540295 and 19540252). Some of the numerical calculations were performed with SR8000/MPP and SR11000/J2 at the Information Center of the University of Tokyo.

REFERENCES

1. J. M. Lattimer and F. D. Swesty, *Nucl. Phys. A* **535**, 331-376 (1991).
2. H. Shen, H. Toki. K. Oyamatsu and K. Sumiyoshi, *Prog. Theor. Phys.* **100**, 1013-1031 (1998).
3. C. Ishizuka, A. Ohnishi, K. Tsubakihara, K. Sumiyoshi and S. Yamada, arXiv:0802.2318 [nucl-th].
4. K. Nakazato, K. Sumiyoshi and S. Yamada, *Phys. Rev. D* submitted (2007).
5. H. Kanzawa, K. Oyamatsu, K. Sumiyoshi and M. Takano, *Nucl. Phys. A* **791**, 232-250 (2007).
6. A. Akmal, V. R. Pandharipande and D. G. Ravenhal, *Phys. Rev. C* **58** 1804-1828 (1998).
7. K. E. Schmidt and V. R. Pandharipande, *Phys. Lett. B* **87**, 11-14 (1979).
8. A. Mukherjee and V. R. Pandharipande, *Phys. Rev. C* **75**, 035802 (2007).
9. B. Friedman and V. R. Pandharipande, *Nucl. Phys. A* **361**, 502-520 (1981).
10. K. Oyamatsu, *Nucl. Phys. A* **561**, 431-452 (1993).

11. NEUTRON STAR and HIGH DENSITY MATTER

Equation of state of dense matter for core-collapse supernovae, compact objects and neutrino bursts

Kohsuke Sumiyoshi

Numazu College of Technology (NCT), Ooka 3600, Numazu, Shizuoka 410-8501, Japan

Abstract. This paper reviews the recent progress in the research of equation of state of dense matter for supernova simulations. The equation of state plays an important role to clarify the mechanism of core-collapse supernovae, the formation of compact objects and the properties of supernova neutrinos. We describe the problems of explosion mechanism in current supernova simulations, putting emphasis on the equation of state. We discuss also the topics on the birth of neutron stars (or black holes) and the associated neutrino bursts, which reflect the properties of dense matter.

Keywords: Equation of state, supernova, neutron star, black hole, neutrino
PACS: 13.15.+g, 26.60.Kp, 95.85.Ry, 97.60.Bw, 97.60.Lf

INTRODUCTION

Core-collapse supernovae are the fate of evolution of massive stars in the mass range of 10$\sim$100M$_\odot$ [1, 2]. For stars of $\sim$20M$_\odot$, the gravitational collapse of such massive stars leads to supernova explosions with E_{exp} $\sim$$10^{51}$ erg and leads to the neutron star formations with supernova neutrino bursts [3, 4]. For more massive stars of $\sim$40M$_\odot$, for example, they lead to the black hole formations and may lead to more energetic explosions such as gamma ray bursts [5, 6]. To understand the evolution of galaxies and matters, we need to clarify the mechanism of various explosions, the formation of neutron stars and black holes, the supernova neutrino bursts from various events and the nucleosynthesis of heavy elements from massive stars.

The core-collapse supernovae start from the gravitational collapse of iron core of massive stars [1, 2]. During the initial stage of collapse, the electron capture proceeds due to high electron Fermi energies and produces neutrinos. The neutrinos escape freely at the beginning, but they are trapped at high densities of $\sim$$10^{12}$ g/cm^3 during the collapse. The core bounce occurs at just above the nuclear matter density and the shock wave is launched. The shock wave propagates outward and if it reaches the surface of Fe core, this is the supernova explosion. However, the mechanism of supernova explosion is still elusive despite the extensive numerical studies for decades [3, 4]. The proto-neutron star, which is hot and contains neutrinos, is formed at center and supernova neutrinos are emitted from this object. The nucleosynthesis proceeds in the high temperature environment during this explosion and the central object cools down to the cold neutron star. If the presupernova stars are very massive, they lead to the black hole formation and may become the origin of the energetic phenomena such as hypernovae [5, 6].

CP1016, *Origin of Matter and Evolution of Galaxies*,
edited by T. Suda, T. Nozawa, A. Ohnishi, K. Kato, M. Y. Fujimoto, T. Kajino, and S. Kubono

EQUATION OF STATE FOR SUPERNOVAE

During the episode of core-collapse supernovae explained above, the density becomes high beyond the nuclear matter density, the temperature becomes high over 10 MeV and the matter becomes very neutron rich. Therefore, one has to provide the information of nuclear physics, i.e. the equation of state and the rates of weak reactions regarding neutrinos, at extreme conditions. We focus here on the equation of state (EOS) to discuss the role of dense matter in core-collapse supernovae. The EOS provides the pressure to determine the stellar structure and dynamics, the entropy (temperature) to determine the neutrino energy in the trapped and emission regions, and the composition to calculate the rates of electron captures and neutrino interactions. It is a difficult task to provide the set of physical EOS for supernova simulations since one has to cover the wide range of density, electron fraction and temperature in a consistent framework checked by the experimental data. Even after long-standing efforts, there are limited sets of EOS so far. The Wolff-Hillebrandt EOS is the pioneering work on the construction of EOS table based on the Skyrme Hartree-Fock theory [7]. There are two EOS sets currently available in public. They are the Lattimer-Swesty EOS [8] and the Relativistic EOS (or Shen EOS) [9, 10].

The Lattimer-Swesty EOS (LS-EOS) is the extension of compressible liquid drop model [8]. The energy of uniform matter is given by a function of density, proton fraction and temperature, following the form in the Skyrme Hartree-Fock theory. The parameters in the energy function are fixed by the saturation properties with the values of binding energy, symmetry energy and incompressibilities. They adopt the two-zone model by solving the equilibrium condition for nuclear liquid phase and surrounding gas phase in the Wigner-Seitz cell. The EOS set is provided as a subroutine program and has been frequently used in recent numerical simulations.

Recently, the tables of EOS based on the relativistic many body theories have become available. The relativistic EOS by Shen et al. (Shen-EOS) [9, 10] is constructed by the relativistic mean field (RMF) theory. The RMF theory is based on the relativistic nuclear many body frameworks [11] and has been successfully applied to the studies of nuclear structures and dense matter. We stress that the nuclear interactions to derive Shen-EOS are constrained by the properties of unstable nuclei [12], which are available recently in radioactive nuclear beam facilities in the world. For example, the neutron skin thickness of neutron-rich nuclei, which is sensitive to the nuclear symmetry energy, is well reproduced for isotopes of light nuclei [13, 14]. The local density approximation is utilized to describe the inhomogenous distribution of matter as a mixture of nucleons, alpha-particles and nuclei.

We note that there are recent studies to extend the Shen-EOS table to higher densities and temperatures. The tables of EOS including hyperons have been constructed in the RMF theory extended to flavor SU(3) by Ishizuka et al. [15]. The extension of the Shen-EOS table within the nucleon degree of freedom to the quark degree of freedom has been also done to study the extreme conditions toward the black hole formation by Nakazato et al. [16]. In addition to the RMF framework, the variational approach for nuclear many body problems has been utilized to calculate the EOS of nuclear matter at finite temperature for the construction of EOS table [17]. The reports on these EOS developments can be found in this volume of proceedings and also will be published

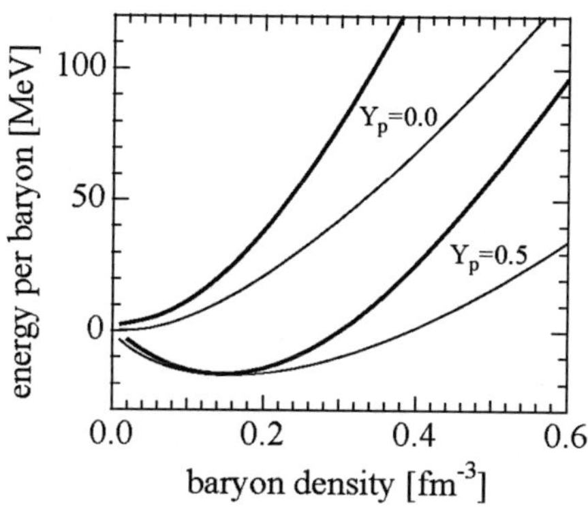

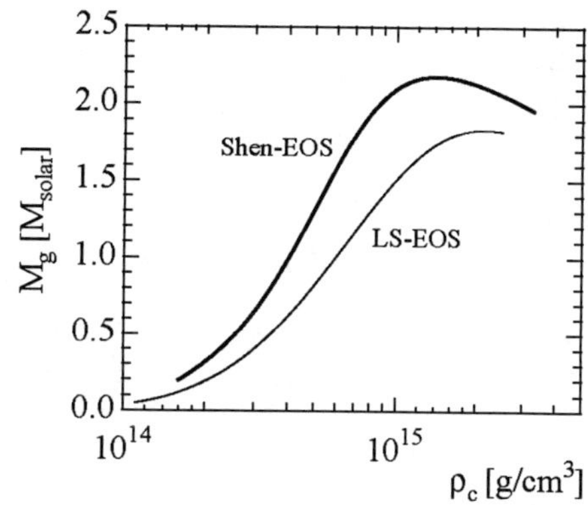

FIGURE 1. Comparisons between Shen-EOS (thick) and LS-EOS (thin) for energy per baryon as a function of baryon density (left) and for neutron star masses as a function of central density (right).

elsewhere.

Having the two sets of EOS (LS-EOS and Shen-EOS) for supernova simulations, we can compare the properties of EOS's and explore the influence of dense matter on the mechanism of core-collapse supernovae [18]. Since the relativistic many body frameworks tend to provide a stiff EOS as compared with the non-relativistic frameworks [19], Shen-EOS is more stiff than LS-EOS (Fig. 1, left). The incompressibility of Shen-EOS is 281 MeV obtained by the calculations within the RMF theory whereas the incompressibility of LS-EOS is a parameter of energy function. We adopt the set of 180 MeV for the incompressibility among three choices in LS-EOS. The EOS set of this value has been popularly used in most of numerical simulations so far. When those two sets of EOS are adopted for cold neutron stars, the maximum masses are $1.8M_\odot$ and $2.2M_\odot$ for LS-EOS and SH-EOS, respectively, due to the difference of stiffness (Fig. 1, right). The difference of stiffness may affect the behavior of core bounce and the evolution of proto-neutron stars.

The density dependence of the symmetry energy tends to be strong in the relativistic many body frameworks [19]. The symmetry energy of Shen-EOS is 36.9 MeV and is larger than the value (29.3 MeV) of LS-EOS. We remark that those values in Shen-EOS are not inputs to determine the interactions, but the outcome after the fitting of nuclear interactions to experimental data of nuclei including neutron-rich ones. This difference can lead to different compositions in supernova cores and may change the rates of electron captures and neutrino reactions in the supernova mechanism.

SUPERNOVA SIMULATIONS WITH THE EOS TABLES

Recently, the numerical simulations by solving exactly the neutrino-radiation hydrodynamics have been done under the spherical symmetry (1D) [20, 21, 22, 23]. The critical studies on microphysics have become possible to investigate its influence on supernova dynamics in 1D simulations without the uncertainty came from the approximate neutrino transfer. The two EOS sets (LS-EOS and Shen-EOS) have been adopted for recent simulations to clarify the influence of dense matter [23]. For both cases, the numerical simulations do not show any sign of successful explosion (Fig. 2, left) although there are important differences found in the profiles at core bounce stage and at later phase. We remark here that there have been systematic studies on the incompressibility and the symmetry energy using the analytic form for EOS [24, 25, 26]. A smaller incompressibility is preferable to have a large initial shock energy, however, no explosion is obtained within the reasonable range constrained by nuclear experiments and neutron star masses [24, 26]. A larger symmetry energy is preferable to have a smaller reduction of lepton fraction through electron captures on free protons [25, 26] while the symmetry energy is constrained by nuclear masses and neutron skins. Shen-EOS has the larger symmetry energy constrained by the neutron skins and provides a larger bounce core than that for LS-EOS, however, no explosion is obtained in 1D simulations [23].

Since the explosion is not obtained in 1D simulations at the moment and the observational facts suggest, the asymmetry is thought to be necessary to have the successful explosion. There have been extensive studies on multi-dimensional (multi-D) simulations (See [3, 4] for references), however, we note that the neutrino transfer have been approximately solved due to the lack of computing resources. In the two recent articles based on the sophisticated 2D simulations by two groups, they reported global asymmetries due to the hydrodynamical instabilities called SASI and explosions at $\sim$500 ms after the core bounce [27, 28]. They claimed different mechanisms, i.e. the acoustic mechanism and the neutrino heating mechanism. The situation is still controversial since they use different progenitors, microphysics and numerical schemes. Therefore, follow-up studies to clarify the mechanism are necessary and one needs to perform multi-D neutrino transfer simulations together with careful analyses on microphysics.

TOWARD THE SUCCESSFUL EXPLOSION

Nuclear physics is important even in the multi-D simulations with the hydrodynamical instabilities. This is because the nuclear physics sets the initial launch of shock wave. The initial shock energy can be estimated by the gravitational binding energy of bounce core and is proportional to the lepton fraction and central density at center. It is preferable to have a bounce core with a large mass and a small radius. The shock wave must propagate through the Fe core up to the surface against the energy loss due to the dissociation of Fe nuclei. This energy loss is significant and amounts to the order of 10^{51} erg, which is comparable to the explosion energy, for $\sim$0.1M$_\odot$ difference in mass. The difference of this size has been found in the 1D simulations when we adopt the two sets of EOS due to the difference of symmetry energy [23]. Although this effect is not decisive for the explosion in 1D calculations, it may be helpful in the marginal situation in multi-D simulations.

Even after the launch of shock wave at bounce and the following stall of the shock wave with the hydrodynamical instabilities, the nuclear physics plays important roles to set the condition for the revival of shock wave. One of the most important processes is the neutrino heating. Neutrinos are emitted from the surface of nascent proto-neutron

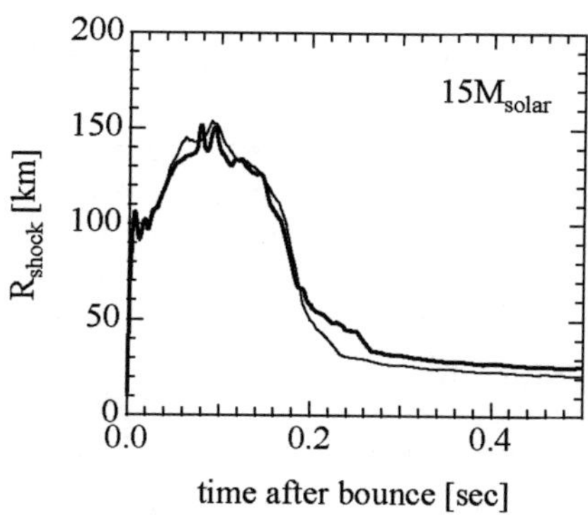

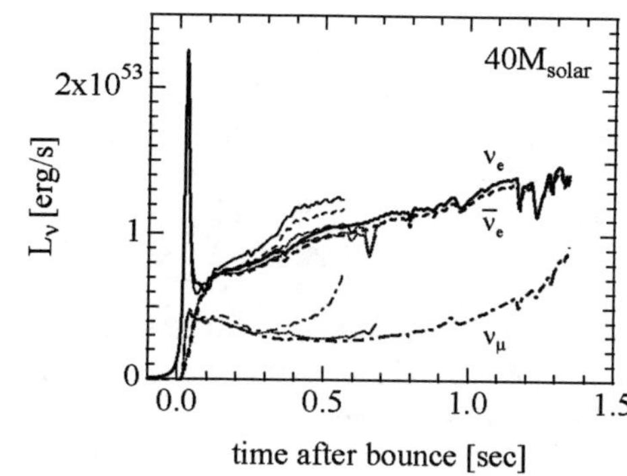

FIGURE 2. The position of shock wave (left) and the neutrino luminosities for three species (right) as a function of time after bounce in the collapse of the $15M_\odot$ and $40M_\odot$ stars with Shen-EOS (thick) and LS-EOS (thin). The case of hyperon EOS is shown in the right panel by dotted lines.

star and a portion of them interacts with the material just behind the stalled shock wave. They heat the material behind the shock wave by dropping energy and may help to push the shock wave outward again. The heating rate is sensitive to the details in nuclear physics, being proportional to the amount of targets (mostly nucleons and possibly nuclei), neutrino energy and luminosity. The properties of supernova neutrinos depend on the properties of proto-neutron stars. Lastly, the EOS determines the profile of proto-neutron stars such as temperature. Therefore, the EOS is crucial to set the condition for the revival of shock wave.

Regarding the neutrino heating, the process via ^{4}He in supernova cores attracts attention recently. Inelastic neutrino-^{4}He scattering can deposit the energy to the matter in the heating region and might help the shock revival [29]. In fact, it has been demonstrated that the shock revival might occur if neutrino-^{4}He heating rates are 3–10 times larger than the standard value with a sufficient perturbation [30]. According to their simulations, ^{4}He appears abundantly just behind the shock wave in addition to nucleons and the contribution from ^{4}He becomes significant in the heating region for the case of the enhanced rates for neutrino-^{4}He. One should keep in mind that the neutrino heating depends crucially on the neutrino-reaction rates and the fraction of ^{4}He. The neutrino-^{4}He reactions have been studied by W. Haxton [29, 31] and are recently examined in further detail [32, 33].

Recent studies report that light clusters other than ^{4}He may appear in supernova cores and may contribute to the neutrino heating as well [34, 35, 36]. A=3 nuclei can appear in dense matter and the energy transfer rates via neutrino reactions may dominate over those of ^{4}He [35]. In the recent microscopic calculations to treat the formation of light clusters, deuterons, tritons and ^{3}He as well as ^{4}He appear between the surface of proto-neutron star and the shock position inside a supernova core [36]. Those light clusters may contribute to extra heating and cooling processes in addition to the process via nucleons and might affect the shock dynamics.

Another important application of EOS to core-collapse supernovae is the prediction of supernova neutrinos. After the core bounce, the nascent proto-neutron star cools down by emitting neutrinos from its surface for $\sim$20 s as detected in the event of SN1987A. There have been extensive studies to examine the properties of supernova neutrinos such as time profiles of average energies and luminosities regarding EOS [2, 37, 38, 39]. In the comparison of the two EOS sets, the average energies and luminosities of neutrinos turned out to be low in the case of Shen-EOS as compared to the case of LS-EOS [39]. This is because Shen-EOS is stiff and, hencc, the central density and temperature of proto-neutron star are low. We remark that the EOS difference appears long after the evolution for $\sim$10 s.

Recent studies reveal that neutrino bursts from black hole formation have clear characteristics to probe the EOS [40, 41]. In the case of stars (ex. $40M_{\odot}$) more massive than those for ordinary supernovae, the mass of Fe core is too large to have explosion intrinsically and those massive stars lead to the black hole formation. Starting from the Fe-core collapse, the core bounce occurs and the shock wave is launched without successful propagation. Since the accretion of material is intense, the proto-neutron star at center becomes massive toward the maximum mass supported by the EOS. The dynamical re-collapse occurs beyond the mass limit and the black hole is formed. This kind of black hole formations may lead to hypernovae or faint supernovae, depending on ejected energies and rotation rates [42].

In the scenario of black hole formations via accreting proto-neutron stars, it is interesting to note that the associated neutrino burst has a sudden termination at the timing of black hole formation. Since the maximum mass of proto-neutron stars is determined by the EOS, the timing of the black hole formation depends crucially on the EOS. In the numerical studies of a $40M_{\odot}$ star by neutrino-radiation hydrodynamics [40, 41], the duration of neutrino emission is found to be short ($\sim$1 s) with the dependence on the stiffness of EOS. The average energies and luminosities of neutrinos increase quickly toward the black hole formation (Fig. 2, right). These characteristics of neutrino bursts can be used to search for the signal of black hole formation and to probe the properties of dense matter.

During the evolution of the accreting proto-neutron star, the density and temperature at center become very high because of the contraction according to the increasing mass. Therefore, new degrees of freedom such as hyperons and quarks may definitely appear during the collapse to the black hole. The appearance of new phases may shorten the duration of neutrino burst further because the maximum mass becomes smaller due to the softening of EOS. When we adopt the hyperonic EOS table [15], the re-collapse occurs much earlier than the case of Shen-EOS because of hyperon mixture [43]. The neutrino signal in the case of the hyperonic EOS table is very similar to the case of Shen-EOS, but the duration is clearly different (Fig. 2, right). The black hole formation adopting the quark matter EOS table, other EOS parameters, different models of progenitors are currently under investigation and numerical results will be reported elsewhere.

SUMMARY

The current research on supernovae by the sophisticated numerical simulations of neutrino-radiation hydrodynamics require the precise knowledge of nuclear physics at extreme conditions. The equation of state (EOS) is one of important keys among them to clarify the mechanism of core-collapse supernovae.

The recent progress of EOS table for supernova simulations enables us to investigate the influence of EOS on supernovae in a quantitative manner. Through the comparison between the Shen-EOS and the Lattimer-Swesty EOS, it has been clarified that the compositional change as well as the stiffness of EOS may affect the dynamics of core bounce and shock propagation. In the challenging studies of multi-dimensional simulations of supernovae with hydrodynamical instabilities, the EOS plays basic roles to set the supernova conditions which determine the fate of shock wave. The shock revival may be assisted by the neutrino-related reactions in dense matter with all possible compositions. New compositions, which have not been considered so far in supernova simulations, might change reaction rates, contribute to extra neutrino heating and help the shock propagation.

The role of EOS is essential to follow the thermal evolution of nascent proto-neutron stars and the profiles of associated neutrino emissions. The characteristics of neutrino signals from compact objects reflect the properties of dense matter, therefore, they can be a probe of EOS. The black hole formation from massive stars has the unique signature in neutrino bursts. The different duration of neutrino signal toward the black hole formation can be used to extract the information of EOS at high densities and temperatures. The emergence of hyperons and quarks at high densities and temperatures may affect significantly the collapse of proto-neutron stars to the black hole.

ACKNOWLEDGMENTS

This contribution is based on the research projects with S. Yamada, K. Nakazato and H. Suzuki for the numerical simulations and with H. Shen, M. Takano, K. Oyamatsu, A. Ohnishi, C. Ishizuka and H. Toki for the tables of equation of state. The author is grateful to the collaborators for their continuous works and encouragement. The author thanks also N. Tominaga, H. Umeda, S. Fujimoto and T. Kajino for discussions on stellar models and their applications. The numerical simulations were performed at NAOJ, JAEA and YITP. This work is partially supported by the Grants-in-Aid for the Scientific Research (18540291, 18540295, 19540252) of the MEXT of Japan.

REFERENCES

1. H. A. Bethe, *Rev. Mod. Phys.* **62**, 801 (1990).
2. H. Suzuki, "Supernova Neutrinos," in *Physics and Astrophysics of Neutrinos*, edited by M. Fukugita, and A. Suzuki, Springer-Verlag, Tokyo, 1994, p. 763.
3. K. Kotake, K. Sato, and K. Takahashi, *Rep. Prog. Phys.* **69**, 971 (2006).
4. H.-T. Janka, K. Langanke, A. Marek, G. Martínez-Pinedo, and B. Müller, *Phys. Rep.* **442**, 38 (2007).
5. A. Heger, C. Fryer, S. Woosley, N. Langer, and D. Hartmann, *Astrophys. J.* **591**, 288 (2003).

6. K. Maeda, and K. Nomoto, *Astrophys. J.* **598**, 1163 (2003).
7. W. Hillebrandt, K. Nomoto, and R. G. Wolff, *Astron. Astrophys.* **133**, 175 (1984).
8. J. M. Lattimer, and F. D. Swesty, *Nucl. Phys.* **A535**, 331 (1991).
9. H. Shen, H. Toki, K. Oyamatsu, and K. Sumiyoshi, *Nucl. Phys.* **A637**, 435 (1998).
10. H. Shen, H. Toki, K. Oyamatsu, and K. Sumiyoshi, *Prog. Thoer. Phys.* **100**, 1013 (1998).
11. R. Brockmann, and R. Machleidt, *Phys. Rev.* **C42**, 1965 (1990).
12. Y. Sugahara, and H. Toki, *Nucl. Phys.* **A579**, 557 (1994).
13. T. Suzuki, et al., *Phys. Rev. Lett.* **75**, 3241 (1995).
14. Y. Sugahara, K. Sumiyoshi, H. Toki, A. Ozawa, and I. Tanihata, *Prog. Thoer. Phys.* **96**, 1165 (1996).
15. C. Ishizuka, A. Ohnishi, K. Tsubakihara, K. Sumiyoshi, and S. Yamada, in preparation (2008).
16. K. Nakazato, K. Sumiyoshi, and S. Yamada, submitted to Astrophys. J. (2008).
17. H. Kanzawa, K. Oyamatsu, K. Sumiyoshi, and M. Takano, *Nucl. Phys.* **A791**, 232 (2007).
18. K. Sumiyoshi, H. Suzuki, S. Yamada, and H. Toki, *Nucl. Phys.* **A730**, 227 (2004).
19. K. Sumiyoshi, K. Oyamatsu, and H. Toki, *Nucl. Phys.* **A595**, 327 (1995).
20. M. Rampp, and H.-T. Janka, *Astrophys. J.* **539**, L33 (2000).
21. A. Mezzacappa, M. Liebendoerfer, O. E. B. Messer, W. R. Hix, F.-K. Thielemann, and S. W. Bruenn, *Phys. Rev. Lett.* **86**, 1935 (2001).
22. T. A. Thompson, A. Burrows, and P. Pinto, *Astrophys. J.* **539**, 865 (2003).
23. K. Sumiyoshi, S. Yamada, H. Suzuki, H. Shen, S. Chiba, and H. Toki, *Astrophys. J.* **629**, 922 (2005).
24. E. Baron, J. Cooperstein, and S. Kahana, *Phys. Rev. Lett.* **55**, 126 (1985).
25. S. W. Bruenn, *Astrophys. J.* **340**, 955 (1989).
26. F. D. Swesty, J. M. Lattimer, and E. S. Myra, *Astrophys. J.* **425**, 195 (1994).
27. R. Buras, H.-T. Janka, M. Rampp, and K. Kifonidis, *Astron. Astrophys.* **457**, 281 (2006).
28. A. Burrows, E. Livne, L. Dessart, C. D. Ott, and J. Murphy, *Astrophys. J.* **640**, 878 (2006).
29. W. C. Haxton, *Phys. Rev. Lett.* **60**, 1999 (1988).
30. N. Ohnishi, K. Kotake, and S. Yamada, *Astrophys. J.* **667**, 375 (2006).
31. S. E. Woosley, D. H. Hartmann, R. D. Hoffman, and W. C. Haxton, *Astrophys. J.* **356**, 272 (1990).
32. T. Suzuki, C. Satoshi, T. Yoshida, T. Kajino, and T. Otsuka, *Phys. Rev.* **C74**, 034307 (2006).
33. D. Gazit, and N. Barnea, *Phys. Rev. Lett.* **98**, 192501 (2007).
34. C. J. Horowitz, and A. Schwenk, *Nucl. Phys.* **A 776**, 55 (2006).
35. E. O'Connor, D. Gazit, C. J. Horowitz, A. Schwenk, and N. Barnea, *Phys. Rev.* **C75**, 055803 (2007).
36. K. Sumiyoshi, and G. Röpke, submitted to Phys. Rev. Lett. (2007).
37. A. Burrows, *Astrophys. J.* **334**, 891 (1988).
38. J. A. Pons, S. Reddy, M. Prakash, J. M. Lattimer, and J. A. Miralles, *Astrophys. J.* **513**, 780 (1999).
39. H. Suzuki, H. Kogure, F. Tomioka, K. Sumiyoshi, S. Yamada, and H. Shen, *Nucl. Phys.* **718**, 703 (2003).
40. K. Sumiyoshi, S. Yamada, H. Suzuki, and S. Chiba, *Phys. Rev. Lett.* **97**, 091101 (2006).
41. K. Sumiyoshi, S. Yamada, and H. Suzuki, *Astrophys. J.* **667**, 382 (2007).
42. K. Nomoto, N. Tominaga, H. Umeda, K. Maeda, T. Ohkubo, J. Deng, and P. A. Mazzali, "Hypernovae, Black-Hole-Forming Supernovae, and First Stars," in *The Fate of the Most Massive Stars*, edited by R. Humphreys, and K. Stanek, 2005, vol. 332 of *Astronomical Society of the Pacific Conference Series*, p. 374.
43. K. Sumiyoshi, S. Yamada, and H. Suzuki, in Proceedings of the International Symposium on Neutrino Astrophysics: Frontiers of Neutrino Astrophysics (2008), in press.

The accretion phase of core collapse supernovae

T. Fischer [1,†], P. Gögelein*, M. Liebendörfer†, A. Mezzacappa** and K.-F. Thielemann†

*Institut für Theo. Phys., Universität Tübingen, D-72076 Tübingen, Germany
†Departement Physik, Universität Basel, CH-4056 Basel, Switzerland
**Physics Division, Oak Ridge National Laboratory, Oak Ridge, TN 37831

Abstract. We present data from recent core collapse simulations, using spherically symmetric general relativistic radiation hydrodynamics featuring a three-flavor neutrino Boltzmann transport solver and an equation of state (eos) for hot and dense nuclear matter. A nuclear reaction network has recently been implemented, which allows us to investigate explosion models into the neutrino wind phase several seconds after core bounce. In the absence of an earlier explosion, the continuous accretion of outer layers of the progenitor star lead eventually to the collapse of the protoneutron star (PNS) at the very center and the formation of a black hole.

Keywords: supernovae, explosions, black hole formation

INTRODUCTION

An initially stable progenitor star in the mass range of $8-75\,M_{\odot}$, becomes gravitationally unstable at the end of its stellar life. Photodisintegration of heavy nuclei as well as electron captures on free protons and nuclei, reduce the pressure and the core of the progenitor starts to collapse. Density and temperature increase during collapse, nuclei start to dissociate above the neutron drip line ($\rho \sim 10^{12} g/cm^3$) and the electron-neutrino luminosity increases. The collapse continues until nuclear density ($\rho \sim 2\times 10^{14} g/cm^3$) where matter is not compressible any further so that the collapse halts. A sound wave forms, which turns quickly into a shock front as soon as it hits the outer still infalling layers and turns into an accretion shock. Below neutrino trapping density, many electron-neutrinos are produced and released in the so called neutrino burst within $10-20\,ms$ after bounce. This neutrino energy loss plus the dissociation of heavy infalling nuclei turn the shock into a standing accretion shock (SAS) at $50\,ms$ after bounce. The SAS expands to a few $100\,km$ initially and propagates inward due to the continued infall of heavy nuclei and neutrino cooling. On the other hand, at the very center forms a protoneutron star (PNS) with an initial radius of $\sim 50\,km$. The present proceeding reports on the two possible fates of stalled shocks, either a successful shock revival and the consequential explosion or a failed shock revival and a consequential black hole formation via PNS collapse during the accretion phase.

The results presented here are based on the spherically symmetric general relativistic radiation hydrodynamics code Agile-Boltztran, an implicit three-flavor neutrino and anti-neutrino Boltzmann transport solver [1], [2], [3], [4], [5], [6]. It solves the general

[1] Project funded by the Swiss National Foundation, grant no. 106627/1

CP1016, *Origin of Matter and Evolution of Galaxies*,
edited by T. Suda, T. Nozawa, A. Ohnishi, K. Kato, M. Y. Fujimoto, T. Kajino, and S. Kubono

relativistic transport equation coupled to the microphysical interactions [7] and [8].

The proceeding is organised as follows. In section 2, we discuss long term (several $10\,s$ after bounce) explosion models while section 3 is devoted to failed explosions and the PNS accretion phase that eventually leads to the formation of a black hole.

EXPLOSIONS

The idea of reviving a once stalled shock again via neutrino reactions behind and ahead of the shock by absorption and scattering, could be the driving explosion mechanism of a massive star leading to so-called neutrino driven explosions [9], [10], [11], [12], [13]. However, the absence of any explosion using spherically symmetric core collapse simulations of progenitors more massive then the $ONeMg$-core of $8\,M_{\odot}$, makes it necessary to enhance the neutrino heating artificially to achieve explosion models in spherical symmetric simulations with accurate neutrino transport for nucleosynthesis analysis, see for example [14], [15]. The region of low density and high entropy between the SAS and the PNS surface is thereby subject to artificially enhanced heating and cooling rates such that detailed balance remains unaffected. After the explosion had been launched, the neutrino luminosities decay exponentially. As the shock expands to the outer edge of the Si-layer, the further shock propagation is self-dynamically supported by the internal energy of nuclear O-burning, and no artificially enhanced neutrino heating is necessary anymore. In order to treat the energy benefit from the nuclear burning processes carefully, we couple the low temperature regime of our explosive radiation-hydrodynamics model to a nuclear reaction network [16], that follows the dynamical evolution of 18 mostly α-nuclei to calculate the nuclear energy release. The eos for the high temperature ($T > 0.5\,MeV$) regime is thereby handled via an eos for hot and dense nuclear matter [17]. Next to the energy benefit from nuclear reactions in general, the nuclear reaction network becomes necessary to handle the correct abundances as the matter temperature behind the explosion shock decreases during the shock expansion and nuclei (mostly ^4He) start to freeze out. These He-nuclei are found to extend from the PNS surface up to the explosion shock. This region is subject to neutrino heating, since the neutrino luminosities are still high enough after $\sim 2s$ after bounce. There, a secondary shock forms, the so called neutrino wind termination shock [18], with local matter velocities even larger then those of the explosion shock (see Figure 1a). In that region, the entropy rises from $S = 50\,k_B$ up to $S \geq 200\,k_B$ (see Figure 1b) as the termination shock propagates outward and the density decreases rather quickly while the ν-temperatures remain similar. However, although high entropies are reached and the matter behind the explosion shock is found to be slightly proton rich with a constant $ye \sim 0.53$, the low density makes the benefit for nucleosynthesis analysis questionable with respect to the r-process [19]. In Figure 2 we plot the time evolution of the radial shells, where the formation of the termination shock can be observed after $\sim 3s$ post-bounce. While the termination shock is supported by neutrino heated accelerated mass from behind, after about $15s$ post-bounce the neutrino wind slows down and the shock starts to decline. If it will totally disappear, needs further investigation. The results presented here are in qualitative agreement with similar work recently published on the subject of long term post bounce evolution [18].

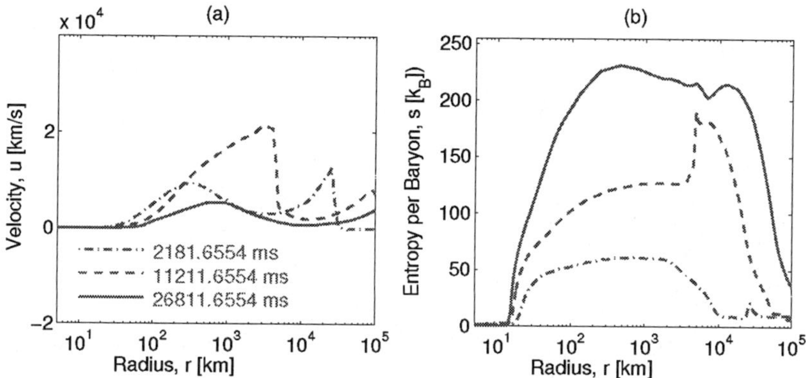

FIGURE 1. Radial velocity and entropy profiles are plotted for different snap shots during the neutrino wind phase of a $10M_{\odot}$ progenitor [20] core collapse simulation. The data shown here are preliminary due to a low resolution of the neutrino propagation angle.

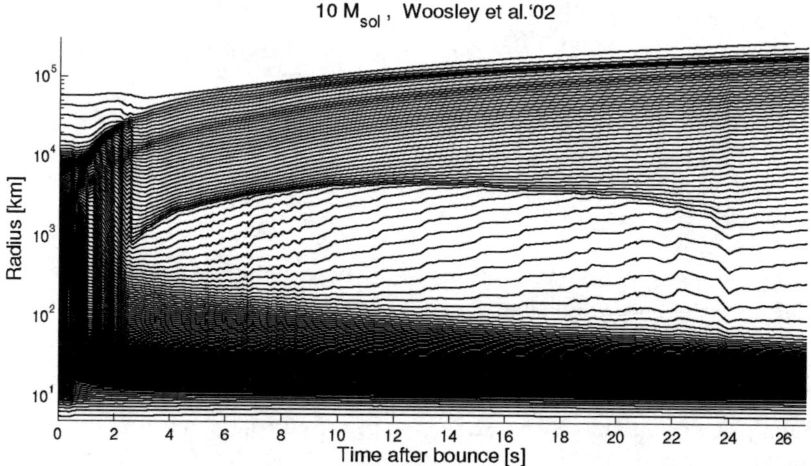

FIGURE 2. Mass shell plot as a function of time, illustrating the contracting PNS at the center, the outgoing explosion shock and the low density neutrino wind region within between.

FAILED EXPLOSIONS AND PNS ACCRETION

While, as discussed above, spherically symmetric core collapse models, including sophisticated input physics and Boltzmann neutrino transport, fail to explain the neutrino driven explosion mechanism, such models are perfect to investigate accretion phenomena in detail. Furthermore, if the mass of the accreting PNS exceeds the critical limit (depending on the eos), neutron pressure and nuclear forces fail to keep the PNS stable against gravity and the PNS collapses to a black hole. During the accretion phase,

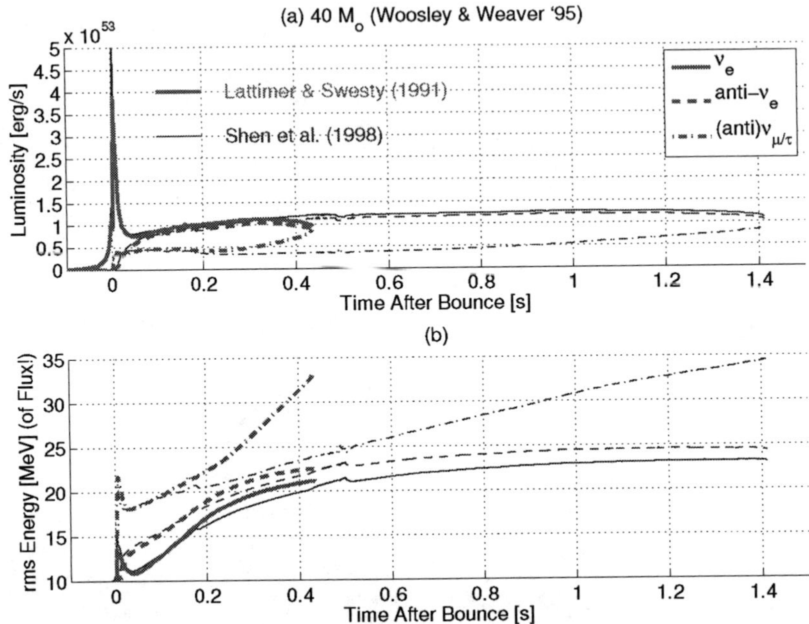

FIGURE 3. Neutrino luminosities and rms-energies for a $40M_\odot$ progenitor, illustrating the increasing (μ/τ)(anti)neutrino luminosity during the accretion phase while comparing two different eos.

the (anti)(μ,τ)-neutrino luminosity increase has been emphasis and illustrated at the example of a $40M_\odot$ progenitor model [6]. We extended this study and investigated the connection between the (anti)(μ,τ)-neutrino luminosity increase and the ongoing compactification of the PNS during the accretion phase, which is related to the maximum increase of temperature and electron fraction at the (μ,τ)-neutrino spheres. Figure 3a shows the neutrino luminosities of a $40M_\odot$ progenitor [21] and Figure 3b shows the mean neutrino energies, taken at $500\,km$ distance. There, the (μ,τ)-neutrino luminosity increase can be recognised after the contraction of the PNS has reached a certain level of compactness, after $\sim 250\,ms$ and $\sim 1\,s$ after bounce (using the eos described in [22] and [17] respectively). The latter one represents a much stiffer eos, such that the PNS accretion time is extended to $\sim 1\,s$ before becoming gravitationally unstable and collapsing to a black hole. Our results are in agreement with data presented in a detailed study on the subject of an eos comparison during the accretion phase of a $40M_\odot$ progenitor model [23].

SUMMARY AND OUTLOOK

The spherically symmetric core collapse explosion model presented here $(10M_\odot)$, using Boltzmann neutrino transport, an eos for hot and dense nuclear matter as well as an nuclear reaction network, covers a range of post bounce evolution more then $\sim 25\,s$. Such investigations of post bounce long term evolution of core collapse simulations of massive progenitors, might be of certain interest for explosive nucleosynthesis analysis. However, the illustrated features need further investigation before becoming certain, for example the formation of the neutrino wind termination shock, the evolution of the slightly proton rich region between PNS and explosion shock and the evolution of the termination shock with time as matter becomes more and more transparent. In disagreement with observations, the low explosion energies achieved in such models, may also rise the question of some missing aspect(s) of core collapse supernovae modelling in general. Next to explosions with artificially enhanced neutrino heating rates, we compare the accretion phase of two failed core collapse supernovae of a massive progenitor ($40\,M_{\odot}$) using different eos, before collapsing to a black hole. Special attention has been devoted to the increasing μ/τ(anti)-neutrino luminosities during PNS contraction as well as to the extended PNS accretion time scale for a stiff eos.

REFERENCES

1. A. Mezzacappa, and S. W. Bruenn, **405**, 637–668 (1993).
2. A. Mezzacappa, and S. W. Bruenn, **405**, 669–684 (1993).
3. A. Mezzacappa, and S. W. Bruenn, **410**, 740–760 (1993).
4. A. Mezzacappa, and O. E. B. Messer, *Journal of Computational and Applied Mathematics* **109**, 281–319 (1999).
5. M. Liebendörfer, S. Rosswog, and F.-K. Thielemann, **141**, 229–246 (2002).
6. M. Liebendörfer, O. E. B. Messer, A. Mezzacappa, S. W. Bruenn, C. Y. Cardall, and F.-K. Thielemann, **150**, 263–316 (2004).
7. P. J. Schinder, and S. L. Shapiro, **50**, 23–37 (1982).
8. S. W. Bruenn, **58**, 771–841 (1985).
9. H. A. Bethe, and J. R. Wilson, **295**, 14–23 (1985).
10. S. W. Bruenn, and W. C. Haxton, **376**, 678–700 (1991).
11. A. Mezzacappa, J. M. Blondin, O. E. B. Messer, and S. W. Bruenn, "Core Collapse Supernovae: Modeling Requirements and Surprises," in *Origin of Matter and Evolution of Galaxies*, edited by S. Kubono, W. Aoki, T. Kajino, T. Motobayashi, and K. Nomoto, 2006, vol. 847 of *American Institute of Physics Conference Series*, pp. 179–189.
12. H.-T. Janka, **368**, 527–560 (2001).
13. H.-T. Janka, R. Buras, F. S. Kitaura Joyanes, A. Marek, M. Rampp, and L. Scheck, *Nuclear Physics A* **758**, 19–26 (2005).
14. C. Fröhlich, P. Hauser, M. Liebendörfer, G. Martínez-Pinedo, F.-K. Thielemann, E. Bravo, N. T. Zinner, W. R. Hix, K. Langanke, A. Mezzacappa, and K. Nomoto, **637**, 415–426 (2006).
15. C. Fröhlich, W. R. Hix, G. Martínez-Pinedo, M. Liebendörfer, F.-K. Thielemann, E. Bravo, K. Langanke, and N. T. Zinner, *New Astronomy Review* **50**, 496–499 (2006).
16. F.-K. Thielemann, F. Brachwitz, P. Höflich, G. Martinez-Pinedo, and K. Nomoto, *New Astronomy Review* **48**, 605–610 (2004).
17. H. Shen, H. Toki, K. Oyamatsu, and K. Sumiyoshi, *Progress of Theoretical Physics* **100**, 1013–1031 (1998).
18. A. Arcones, H.-T. Janka, and L. Scheck, **467**, 1227–1248 (2007).
19. T. Kuroda, S. Wanajo, and K. Nomoto, **672**, 1068–1078 (2008).
20. S. E. Woosley, A. Heger, and T. A. Weaver, *Reviews of Modern Physics* **74**, 1015–1071 (2002).
21. S. E. Woosley, and T. A. Weaver, **101**, 181–+ (1995).
22. J. M. Lattimer, and F. Douglas Swesty, *Nuclear Physics A* **535**, 331–376 (1991).
23. K. Sumiyoshi, S. Yamada, and H. Suzuki, **667**, 382–394 (2007).

Equation of State for Hadron-Quark Mixed Phase and Stellar Collapse

Ken'ichiro Nakazato*, Kohsuke Sumiyoshi† and Shoichi Yamada*,**

*Department of Physics, Waseda University, 3-4-1 Okubo, Shinjuku, Tokyo 169-8555, Japan
†Numazu College of Technology, Ooka 3600, Numazu, Shizuoka 410-8501, Japan
**Advanced Research Institute for Science & Engineering, Waseda University, 3-4-1 Okubo, Shinjuku, Tokyo 169-8555, Japan.

Abstract. We construct an equation of state (EOS) including the hadron-quark phase transition by the Gibbs conditions for finite temperature. We adopt the equation of state based on the relativistic mean field theory for the hadronic phase taking into account pions and the MIT bag model of the deconfined 3-flavor strange quark matter for the quark phase.

The gravitational collapse of a massive star is computed using our equation of state and we find that the interval time from the bounce to the black hole formation, namely the duration time of the neutrino emission, becomes shorter for the model with pions and quarks. This fact implies that we may be able to probe observationally the EOS of hot dense matter in future.

Keywords: black hole physics — hydrodynamics — neutrinos — equation of state — dense matter
PACS: 21.65.-f, 12.39.Ba, 97.60.-s, 95.30.-k

INTRODUCTION

It has been theoretically suggested that hadronic matter undergoes a deconfinement transition to quark matter at high temperature and/or high density. It is also inferred that the transition occurs inside compact stars. Stars with a quark matter central region and a hadronic matter mantle are called hybrid stars. This transition may play an important role in the gravitational collapse of a star, such as a core collapse supernova. In addition, there is a possibility of forming a black hole by the transition because the compact stars have maximum masses. In this case, the EOS with finite temperature and including neutrinos is needed whereas the EOS can be calculated under zero temperature and neutrino-less β equilibrium for hybrid stars. So far, the stellar collapse including the hadron-quark phase transition with a finite temperature EOS and neutrinos has not been studied. In this study, we construct an EOS with the hadron-quark mixed phase for finite temperature and perform the numerical simulation of the stellar collapse using this EOS. Further details of this study are seen in our lately submitted manuscript [1].

FORMULATIONS

In this section, we construct the EOS for hadron, quark and their mixture as a function of baryon mass density ρ_B, electron fraction Y_e and temperature T. We assume that the pure hadronic matter exists under the transition density and the pure quark matter exists for a higher density than that of the end point of the mixed phase.

CP1016, *Origin of Matter and Evolution of Galaxies*,
edited by T. Suda, T. Nozawa, A. Ohnishi, K. Kato, M. Y. Fujimoto, T. Kajino, and S. Kubono

EOS for Hadronic Matter and Quark Matter

We adopt the Shen EOS [2, 3] for the pure hadronic matter. This EOS is based on the relativistic mean field theory and an inhomogeneity of the matter is also taken into account using Thomas-Fermi approximation for the regime with $T \lesssim 15$ MeV and subnuclear densities. Moreover, we add the effects of charged pions, $\pi^{\pm}$, and neutral pions, π^0, to the Shen EOS. Although it is nontrivial to deal pions interacting with nucleons, we treat pions in the minimum model so as to examine their qualitative effects. Our model may overestimate the population of pions [4] because the pions are handled as an ideal boson gas with their rest mass in vacuum, $m_{\pi^{\pm}}$. Their chemical potentials are set as $\mu_{\pi^0} = 0$ and $\mu_{\pi^-} = -\mu_{\pi^+} = \mu_n - \mu_p$, where μ_p and μ_n are those of proton and neutron, respectively. In addition, the charged pions are condensed in our model when their chemical potential is equal to $m_{\pi^{\pm}}$ and the condensed pions contribute not to the pressure but to the energy density by their rest masses.

As for the pure quark matter, we adopt the MIT bag model [5] as the pure quark matter. This is a phenomenological model which describes the nature of the confinement and the asymptotic freedom of quarks. In this model, free quarks are confined in the "bag" and this "bag" has a positive potential energy per unit volume [5]. Thus the thermodynamical potential is expressed as $\Omega_Q = \Omega_0 + BV$, where V and Ω_0 are the volume of the "bag" and the thermodynamical potential of free quarks as ideal fermions, respectively. The bag constant, B, is a parameter characterizing this model, and we investigate its dependence later. From thermodynamical relations, we can calculate the number density, pressure, energy density and so on. As for the current quark mass, we adopt $m_u c^2 = 2.5$ MeV, $m_d c^2 = 5$ MeV and $m_s c^2 = 100$ MeV in this study.

EOS for Mixed Phase

In order to construct the EOS for hadron-quark mixed phase, we employ the method in [6]. Here, we assume that the equilibrium is achieved not only by the strong interactions but also by the weak interactions and neutrinos are completely trapped owing to high density. Thus the following reactions are in chemical equilibrium and the relations of the chemical potentials are given as:

$$\begin{aligned}
&p + e^- \leftrightarrow n + \nu_e, && \\
&n \leftrightarrow u + 2d, && \mu_p + \mu_e = \mu_n + \mu_\nu, && (1)\\
&p \leftrightarrow 2u + d, \quad \Longrightarrow && \mu_n = \mu_u + 2\mu_d, && (2)\\
&d \leftrightarrow u + e^- + \bar{\nu}_e, && \mu_p = 2\mu_u + \mu_d, && (3)\\
&s \leftrightarrow u + e^- + \bar{\nu}_e, && \mu_d = \mu_s. && (4)\\
&u + d \leftrightarrow u + s, &&
\end{aligned}$$

Since two phases are also in mechanical equilibrium, we require the condition,

$$P_H = P_Q, \tag{5}$$

where P_H and P_Q are the pressures of hadronic and quark phases, respectively. The baryon mass density, ρ_B, and the electron fraction, Y_e, of the mixed phase are related to the baryon mass densities and the electric charge fractions (Y_C) of the hadronic phase (H) and quark phase (Q) as,

$$\rho_B = (1-\chi)\rho_{B,H} + \chi\rho_{B,Q}, \quad (6)$$

$$Y_e\rho_B = (1-\chi)Y_{C,H}\rho_{B,H} + \chi Y_{C,Q}\rho_{B,Q}, \quad (7)$$

where χ is a volume fraction of the quark phase.

RESULTS

In this section, we show the features of the EOS described in the preceding section and perform the numerical simulation of the stellar collapse using this EOS. In the following, we choose a mixed EOS with the hadronic matter including pion and the quark matter of the bag constant $B = 250$ MeV fm^{-3} ($B^{1/4} = 209$ MeV), as a reference model.

Features of EOS

First of all, we examine the composition of our EOS for neutron ster (NS) matter to compare with other EOS's in the previous studies using similar schemes. In FIGURE 1, we show the particle fractions, $Y_i \equiv \frac{n_i}{n_B}$, where n_B and n_i represent the baryon number density and number density of the particle i, respectively, for the reference model of NS matter. In this figure, we find that the current results are consistent with the previous studies, especially for the pion in [4] and for the hadron-quark mixed phase in [6]. It is noted that the EOS, which is consistent with the previous studies at zero temperature, is

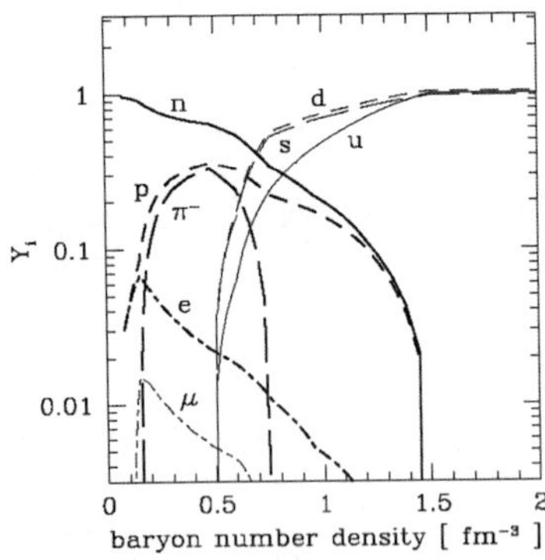

FIGURE 1. Particle fractions for the reference model of NS matter. The particle fraction, Y_i, is defined as $\frac{n_i}{n_B}$, where n_i represents the number density of the particle i. It is noted that muons whose chemical potential is same as that of electrons are contained in the NS matter model.

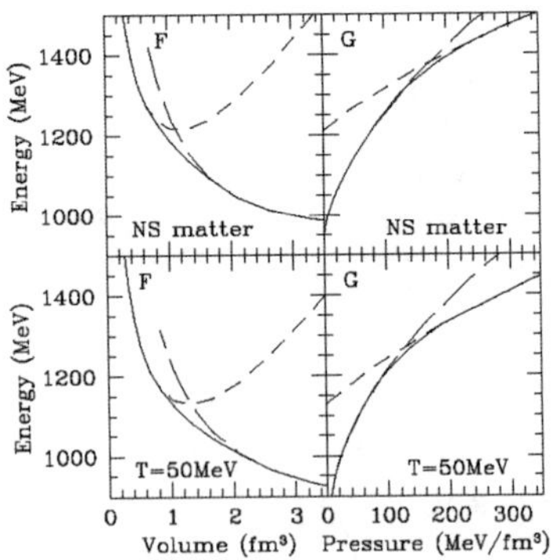

FIGURE 2. EOS's for the reference models of mixed matter (solid lines), pure hadronic matter (long-dashed lines) and pure quark matter (short-dashed lines). The upper left and upper right panels show the Helmholtz free energies per baryon (F) as a function of a specific volume and the Gibbs free energies per baryon (G) as a function of the pressure, respectively, for the NS matter. The lower two panels are the same as upper two panels but for SN matter with $T = 50$ MeV and $Y_l = 0.1$.

extended to finite temperature in our study.

In FIGURE 2, we compare the EOS for the reference model with those for pure hadronic and quark matters. This figure illustrates both for NS matter and SN matter, which has finite temperature and can be used for the simulations for stellar collapses such as supernovae (SN). For SN matter, we show the case of $T = 50$ MeV and the electron type lepton fraction $Y_l = 0.1$, where Y_l is defined as the sum of the electron fraction, Y_e, and the electron-type neutrino fraction, Y_{ν_e}. When the neutrinos are fully trapped, Y_l is conserved for each fluid element in the stellar core. In the upper left and lower left panels of FIGURE 2, we show the Helmholtz free energies per baryon as a function of the specific volume. If the EOS is thermodynamically stable, this function is convex downward. This feature is fulfilled for our models. In the upper right and lower right panels of FIGURE 2, on the other hand, we show the Gibbs free energies per baryon as a function of the pressure. We can recognize that the energy of the mixed phase is always lower than that of the pure hadronic and quark phases.

In FIGURE 3, we show the temperature dependences of the transition density and the critical baryon chemical potential for our EOS of the SN matter with $Y_l = 0.1$. These phase diagrams are given not only for the reference model but also for a model without pions while the bag constant of both models is 250 MeV fm^{-3}. The baryon chemical potential in these figures, μ_B, is the same as the neutron chemical potential in the preceding sections, μ_n. We can see that the reference model has larger transition density and baryon chemical potential than those of the model without pions. This is because pions make the EOS softer; in other words, pressure gets lower for a fixed baryon number density. Thus the condition of equilibrium (5) is satisfied for larger density. Incidentally, though the existence of an end point of the transition line is suggested in the high temperature regime (150 MeV $\lesssim T \lesssim$ 200 MeV), our model cannot reproduce this critical point in principle and plots for $T \gtrsim$ 150 MeV are not given in FIGURE 3. However, the critical baryon chemical potential drops dramatically with the temperature in the regime $T \gtrsim$ 100 MeV, which is consistent with properties suggested recently.

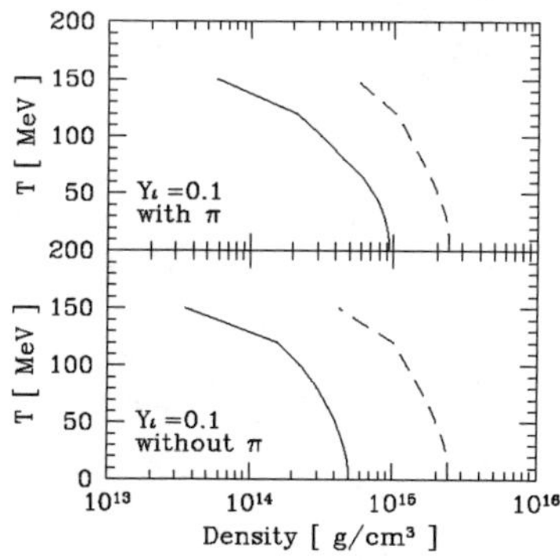

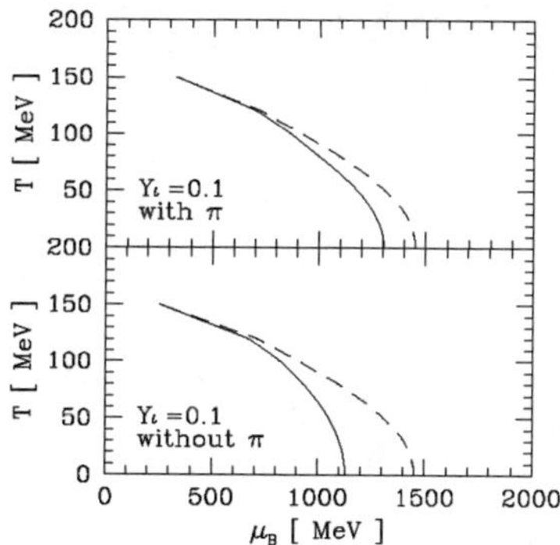

FIGURE 3. Phase diagrams of our EOS of the SN matter with $Y_l = 0.1$ for $T < 150$ MeV. Solid lines represent boundaries of hadronic matter and mixed matter and dashed lines do those of mixed matter and quark matter. For the upper left and upper right panels, an EOS with pions is used while an EOS without pions is used for the lower left and lower right panels. $B = 250$ MeV fm^{-3} is chosen for the bag constant for all panels.

We examine the maximum mass of hybrid stars constructed by our EOS for NS matter. In the following, we refer to the EOS's without pions and quarks (the original Shen EOS [2, 3]), without pions but with quarks, with pions and without quarks and the model with pions and quarks (the reference model) as OO, OQ, PO and PQ, respectively. The bag constant is $B = 250$ MeV fm^{-3} for the models with quarks. In FIGURE 4, we show the mass-radius trajectories for our EOS's. We can see that the pion population and the hadron-quark transition lower the maximum mass because they soften the EOS. For instance, the maximum mass of model OO is 2.2 solar masses ($M_\odot$) while those of models PO, OQ and PQ are $2.0M_\odot$, $1.8M_\odot$ and $1.8M_\odot$, respectively. In the right panel, we show the dependence on B for the models with pions. We can see that the maximum mass becomes lower when the bag constant becomes small. This is because the phase transition density is lower for smaller bag constants. It is also noted that the reference model ($B = 250$ MeV fm^{-3}) is consistent with the recent measurements of compact star masses.

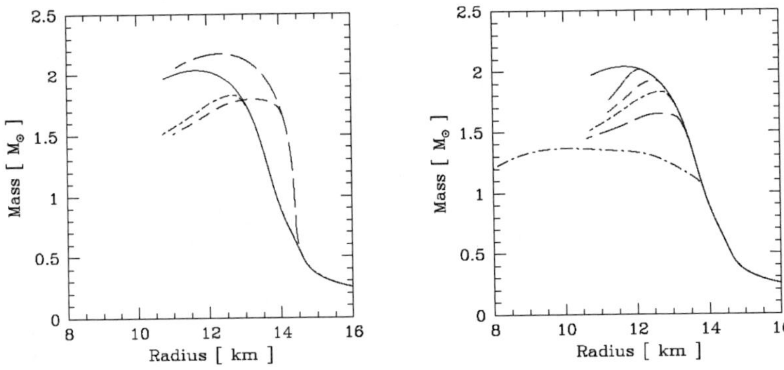

FIGURE 4. Mass-radius trajectories for our EOS's of NS matter. In the left panel, long-dashed, short-dashed, solid and dot-dashed lines represent the models OO (the original Shen EOS [2, 3]), OQ, PO and PQ (the reference model), respectively, where the definition of the models is given in the text. In the right panel, we show the results of the models with pions, and the solid line is same as that in the left panel. Other lines correspond, from bottom to top, to the models with $B = 150, 200, 250, 300, 400$ MeV fm^{-3}.

Stellar Collapse with Hadron-Quark Phase Transition

We perform the numerical simulation of the stellar collapse using the EOS's examined in the preceding section. In this study, we set the bag constant $B = 250$ MeV fm^{-3} for all the models with quarks. A result of an evolutionary calculation for a Population III star with $M = 100M_\odot$ [7] is chosen as the initial model of our simulation. In [8], the numerical simulation of its collapse has been already done under the Shen EOS and we follow the scheme of the reference. We solve the general relativistic hydrodynamics and neutrino transfer equations simultaneously under spherical symmetry. Incidentally, we have assumed for the EOS that the electron-type neutrinos are in equilibrium with other particles in the hadron-quark mixed phase and the pure quark phase. Hence after the phase transition occurs, we do not compute the neutrino distribution functions and assume that they are Fermi-Dirac functions for all species conserving the electron-type lepton fraction, Y_l, for each fluid element. Moreover, we neglect the entropy variation from the neutrino transport. We can justify this gap in the neutrino treatments because the density is high enough for neutrinos to be trapped anyway at the phase transition. It is also noted that we can compute the collapse up to the apparent horizon formation, which is a sufficient condition for the formation of a black hole.

In FIGURE 5, we show the time profiles of the central baryon mass density. These models have a bounce owing to thermal nucleons at subnuclear density ($\sim 1.4 \times 10^{14}$ g cm^{-3}) and then recollapse to a black hole. The contribution of pions makes a difference at $\sim 2 \times 10^{14}$ g cm^{-3} and that of quarks does so for larger density. We can also see that the effect of quarks begins to work suddenly at the transition density whereas that of pions does gradually. This is because the thermal pions appear before the pion condensation. Since the EOS becomes softer owing to the contribution of pions and quarks, the interval time becomes shorter. Model OO (the original Shen EOS) takes 20% longer to recollapse than model PQ (the reference model in this study) does. We note that the duration of neutrino emission is almost same as the interval time from the bounce to the apparent horizon formation [9]. Using this difference of the interval time, we may be able to probe observationally the EOS of hot dense matter in future.

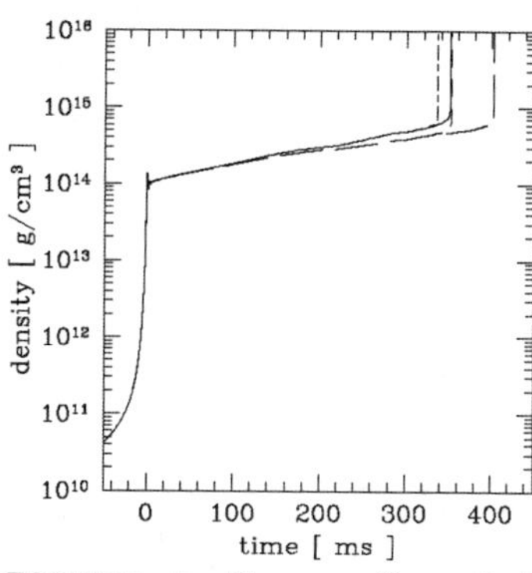

FIGURE 5. Time profiles of the central baryon mass density. The notation of the lines is same as in the left panel of FIGURE 4 and the time is measured from the bounce.

ACKNOWLEDGMENTS

We are grateful to Hideyuki Umeda for providing a progenitor model. We would like to thank Akira Ohnishi and Chikako Ishizuka for fruitful discussions. Numerical computations were performed on the supercomputers in NAOJ, JAERI, YITP and KEK. This work was partially supported by JSPS Research Fellowship, Grants-in-Aid for the Scientific Research from the Ministry of Education, Science and Culture of Japan and the 21st-Century COE Program at Waseda University.

REFERENCES

1. K. Nakazato, K. Sumiyoshi, and S. Yamada, *submitted* (2008).
2. H. Shen, H. Toki, K. Oyamatsu, and K. Sumiyoshi, *Nucl. Phys. A* **637**, 435 (1998a).
3. H. Shen, H. Toki, K. Oyamatsu, and K. Sumiyoshi, *Prog. Theor. Phys.* **100**, 1013 (1998b).
4. N. K. Glendenning, *Astrophys. J.* **293**, 470 (1985).
5. A. Chodos, R. L. Jaffe, K. Johnson, C. B. Thorn, and V. F. Weisskopf, *Phys. Rev. D* **9**, 3471 (1974).
6. N. K. Glendenning, *Phys. Rev. D* **46**, 1274 (1992).
7. K. Nomoto, N. Tominaga, H. Umeda, K. Maeda, T. Ohkubo, J. Deng, and P. A. Mazzali, *ASP Conf. Ser.* **332**, 374 (2005).
8. K. Nakazato, K. Sumiyoshi, and S. Yamada, *Astrophys. J.* **666**, 1140 (2007).
9. K. Sumiyoshi, S. Yamada, and H. Suzuki, *Astrophys. J.* **667**, 382 (2007).

Hypernuclei and dense matter in RMF model with chiral SU(3) potential

K.Tsubakihara, H.Matsumiya, H.Maekawa and A.Ohnishi

Department of Physics, Faculty of Science, Hokkaido University, Sapporo 060-0810, Japan

Abstract. We present a RMF model including chiral SU(3) potential which is suggested form strong coupling limit of lattice QCD and we discuss nuclear star in this chiral SU(3) RMF model and show an effect to nuclear star maximum mass by introducing this potential. The calculation results in soft EOS which imcomressibility is as similar as empirical value. However, when we mind an appearance of hyperon, estimated nuclear star maximum mass underestimates the observable data. We suppose we have to investigate re-stiffen mechanism in high baryon density phase.

Keywords: Nuclear matter, Hypernuclei
PACS: 21.65.-f, 21.80.+a

INTRODUCTION

In constructing the dense matter equation of state (EOS) [1, 2], it is desired to respect both chiral symmetry and hypernuclear physics [3]. In dense matter, strangeness is expected to play a decisive role and the partial restoration of chiral symmetry would modify the hadron properties.

For chiral symmetry side, We have recently developed a chiral SU(2) symmetric RMF model [?] with logarithmic sigma potential in the form of $-\log\sigma$, which is derived in the strong coupling limit (SCL) of the lattice QCD [5]. In this model, the energy density in vacuum at zero temperature is evaluated as,

$$U_\sigma = -a\log(\det MM^\dagger) + \frac{b}{2}\mathrm{tr}(MM^\dagger) - c\,\sigma \sim -a_\sigma \log\sigma + \frac{b_\sigma}{2}\sigma^2 - c_\sigma\,\sigma$$

where M denotes the SU(2) meson matrix, $M = (\sigma + i\pi\cdot\tau)/\sqrt{2}$. In this SCL model, we can describe not only symmetric nuclear matter but also bulk properties of finite nuclei.

On the other hand, from a viewpoint of hypernuclear physics , we develop a extended chiral SU(3) RMF model which include both of chiral symmetry and hypernuclear physics informations [6]. We determine the hyperon-meson coupling constants in this chiral SU$_f$(3) RMF model by fitting existing data. We can reproduce the separation energies of single Λ hypernuclei (S_Λ) and the $\Lambda\Lambda$ bond energy ($\Delta B_{\Lambda\Lambda}$) in $^{6}_{\Lambda\Lambda}$He by choosing the coupling constants appropriately in a reasonable parameter range. The EOS of symmetric matter is found to be softened by the scalar meson with hidden strangeness, $\zeta = \bar{s}s$, which couples with σ through the determinant interaction.

In addition, it seems essential to consider isovector-type interaction because the balance between ρ_n and ρ_p is extremely upset in neutron star medium. In RMF model, it is covered in coupling between isovector-vector meson, regarded as ρ, and baryon. However, another isovector channel due to isovector-scalar meson, a_0, is neglected so

CP1016, *Origin of Matter and Evolution of Galaxies*,
edited by T. Suda, T. Nozawa, A. Ohnishi, K. Kato, M. Y. Fujimoto, T. Kajino, and S. Kubono

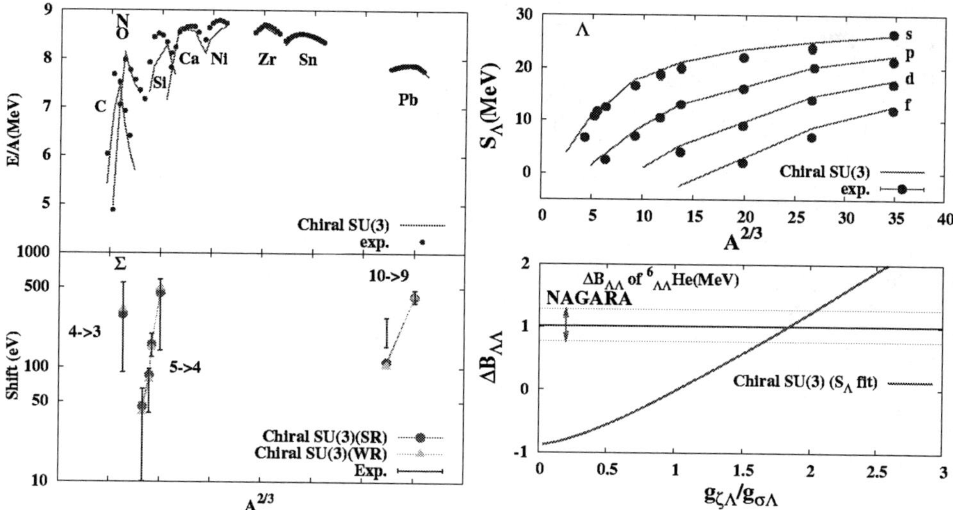

FIGURE 1. Binding energy of normal nuclei, s_Λ of Λ hypernuclei, $\Delta B_{\Lambda\Lambda}$ of ${}^{6}_{\Lambda\Lambda}\mathrm{He}$ and atomic shift of Σ^-. All parameters are determined by fixing these experimental data.

far since its effect tends to be canceled when we fix coupling constant between meson and baryon. We suppose we distinguish a difference whether a_0 meson is included or not.

In this work, we discuss nuclear star matter in this chiral SU(3) RMF model including a_0 meson coupling and investigate the effect to nuclear star maximum mass by introducing them.

MODEL DESCRIPTION

In this paper, we adopts a RMF Lagrangian which treat a_0 meson explicitly and reads,

$$\begin{aligned}\mathscr{L} =& \mathscr{L}_{\mathrm{Free}}(\psi_i, \bar{\psi}_i, \sigma, \zeta, \omega, \rho, \phi) + \mathscr{L}_{\mathrm{EM}} + V_\chi + \frac{D_\omega}{4}\omega^4 \\ &+ \sum_i \bar{\psi}_i \left[g_{\sigma i}\sigma + g_{\zeta i}\zeta - \gamma_\mu \left(g_{\omega i}\omega^\mu + g_{\rho i}\rho^\mu + g_{\phi i}\phi^\mu \right) \right] \psi_i \end{aligned} \tag{1}$$

where we introduce Chiral SU(3) potential extended form chiral SU(2) one by adding $\mathrm{V}_{mboxA}(1)$ anomaly term and written as,

$$V_\chi = -\frac{a}{2}\log\det\left(M^\dagger M\right) + b\mathrm{tr}\left(M^\dagger M\right) + C_\sigma\sigma + C_\zeta\zeta + d\left(\det M + \det M^\dagger\right)$$
$$= -a\left[\left\{\log\left(1+\frac{\sigma}{f_\pi}\right) + \frac{\sigma}{f_\pi} - \frac{\sigma^2}{2f_\pi}\right\} + \frac{1}{2}\left\{\log\left(1+\frac{\zeta}{f_\zeta}\right) + \frac{\zeta}{f_\zeta} - \frac{\zeta^2}{2f_\zeta}\right\}\right] . \quad (2)$$

In determining B–M_v coupling constants, we assume that vector current conservation is in good agreement, thus we briefly use $\mathrm{SU}_f(3)$ symmetry relation for B–M_v coupling constants,

$$\mathscr{L}_{\mathrm{BM}} = \sqrt{2}\{g_s\,\mathrm{tr}\,(M)\,\mathrm{tr}\,(\bar{B}B) + g_1\,\mathrm{tr}\,(\bar{B}MB) + g_2\,\mathrm{tr}\,(\bar{B}BM)\} . \quad (3)$$

and then hyperon–vector meson couplings are given as

$$g_{\omega\Lambda} = \frac{5}{6}g_{\omega N} - \frac{1}{2}g_{\rho N},\ g_{\phi\Lambda} = \frac{\sqrt{2}}{3}\left(g_{\omega N} + 3g_{\rho N}\right) , \quad (4)$$

$$g_{\omega\Sigma} = g_{\rho\Sigma} = \frac{g_{\phi\Xi}}{\sqrt{2}} = \frac{1}{2}(g_{\omega N} + g_{\rho N}) , \quad (5)$$

$$g_{\omega\Xi} = g_{\rho\Xi} = \frac{g_{\phi\Sigma}}{\sqrt{2}} = \frac{1}{2}(g_{\omega N} - g_{\rho N}) . \quad (6)$$

Here, parameters in this model are $g_{\omega N}$, $g_{\rho N}$, $g_{a_0 N}$, $g_{\sigma Y}$, $g_{\zeta Y}$, D_ω and m_σ and they are assigned by reproducing symmetric nuclear matter, normal nuclei, and hypernuclear data such as s_Λ and $\Delta B_{\Lambda\Lambda}$.

After fitting these data, neutron star matter EOS is computed with β-equilibrium and charge neutrality With this EOS, we calculate hadronic Star mass by solving TOV equation and check whether the results support observable data or not.

RESULTS

In fig.1, with the chiral SU(3) potential, this RMF model can explain experimental data of B/A and charge rms radii (normal nuclei), S_Λ and $\Delta B_{\Lambda\Lambda}$ (Λ hypernuclei) and atomic shift of Σ^- (Σ hyperatom) in reasonable parameter range. Then, EOS of symmetric nuclear matter gets softer than TM model [7] or SU(2) version [**?**] as shown in fig.2. Here, we briefly show $g_{aN} = 0$ case and we can reproduce these data in same degree.

After fixing the parameters in previous procedure, we calculate neutron star matter EOS with same parameter. We reveal the results in fig.3 and from this results, we find that smaller E/V even in high density reflects smaller incompressibility of this model. We also realize that a_0 meson's effect be very small even in high ρ_B.

In addition, we introduce a degree of freedom of Λ hyperon explicitly and calculate NS matter EOS with parameter sets determined from Λ hypernuclear data. When we compute NS mass with this EOS, calculated NS mass presented in fig. 4 underestimates observed data and results in $1.3\mathrm{M}_\odot$. We can note that a_0 meson also has no effect for NS mass same as EOS results.

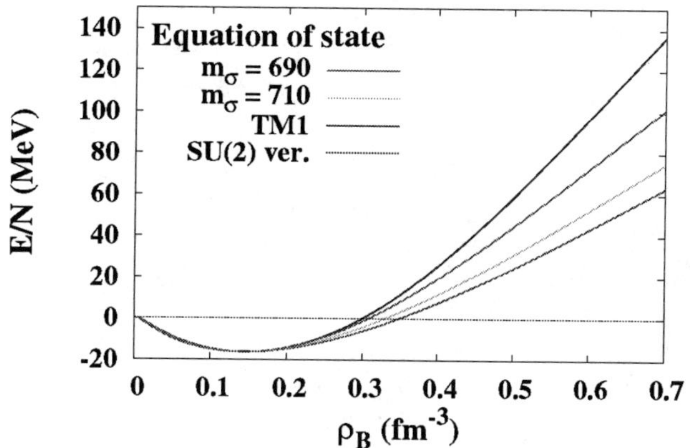

FIGURE 2. Symmetric nuclear matter EOS of TM1, chiral SU(2) ver. and chiral SU(3) one.

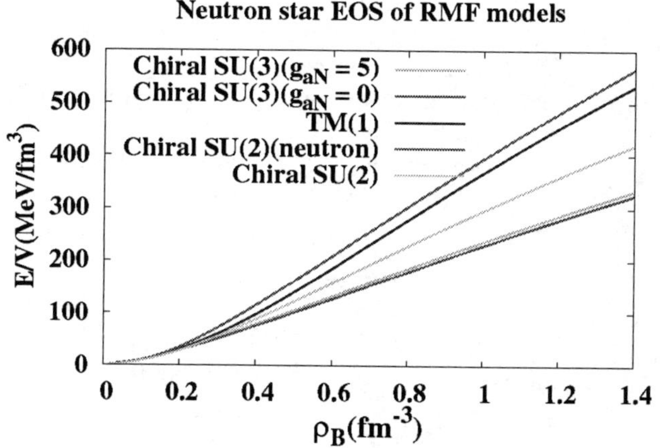

FIGURE 3. Calculated NS matter EOS with TM1, chiral SU(2) ver. and chiral SU(3) one. We also briefly show calculated results including a_0 meson explicitly.

SUMMARY AND DISCUSSION

In this work, we investigate how large effect hyperon and a_0 meson have in RMF model with chiral SU(3) potential derived from Strong Coupling Limit lattice QCD.

For a_0 meson side, on both EOS in high ρ_B and NS maximum mass, We find there is no signal in the results whether a_0 meson is considered or not when we fix model parameters so as to reproduce experimental data, In addition, when we calculate NS mass with parameter sets determined from Λ hypernuclear data, it results in $1.3M_\odot$.

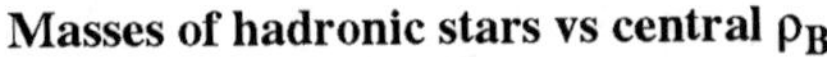

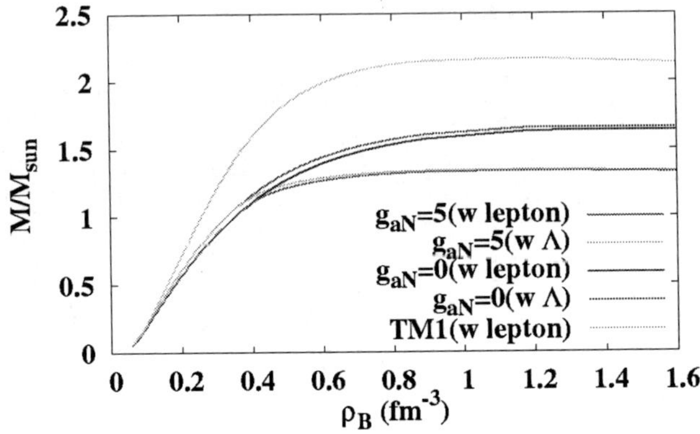

FIGURE 4. Calculated NS mass with TM1, chiral SU(2) ver. and chiral SU(3) one.

Sym. nuclear matter around ρ_0 is in good agreement with Friedman–Pandaripaunde EOS but this EOS is too soft in high ρ_B to support NS mass. Thus we suppose we have to some introduce re-stiffen mechanisms in high ρ_B. For example, $\sigma\omega$ coupling [8, 9] leads stiff EOS in high ρ_B and the experiment [10] suggests ω mass reduction in nuclear medium caused by partial chiral symmetry restoration. From this point of view, we may solve this problem by taking this coupling partially.

ACKNOWLEDGMENTS

This work is supported in part by the Ministry of Education, Science, Sports and Culture, Grant-in-Aid for Scientific Research under the grant numbers, 15540243 and 1707005.

REFERENCES

1. Lattimer and Swesty, Nucl. Phys. A **535** (1991) 331.
2. H. Shen, H. Toki, K. Oyamatsu and K. Sumiyoshi, Prog. Theor. Phys. **100** 1013.
3. Yamamoto, Nishizaki and Takatsuka, Nucl. Phys. A **691** (2001) 432.
4. K. Tsubakihara and A. Ohnishi, Prog. Theor. Phys. **117** (2007) 903.
5. N. Kawamoto and J. Smit, Nucl. Phys. B **192** (1981) 100.
6. K. Tsubakihara, H. Maekawa and A. Ohnishi, Eur. Phys. J. A **33** (2007) 269.
7. Y. Sugahara and H. Toki, Nucl. Phys. A **579** (1994) 557.
8. Y. Ogawa, H. Toki, S. Tamenaga, H. Shen, A. Hosaka, S. Sugimoto and K. Ikeda, Prog. Theor. Phys. **111** (2004) 75.
9. J. Boguta, Phys. Lett. B 120 (1983) 34.
10. Naruki et al., Phys. Rev. Lett. **96** (2006) 092301.

Approximate Energy Expression for Nuclear Matter with Tensor Structure Functions

Kazunori Tanaka[a,b] and Masatoshi Takano[a]

[a]*Research Institute for Science and Engineering, Waseda University, 3-4-1, Okubo, Shinjuku-ku, Tokyo 169-8555, Japan*
[b]*Department of Pure and Applied Physics, Graduate School of Science and Engineering, Waseda University, 3-4-1, Okubo, Shinjuku-ku, Tokyo, 169-8555, Japan*

Abstract. We refine the approximate energy expression for neutron matter by taking into account three-body cluster terms with tensor correlations. The energy expression is an explicit functional of two-body distribution functions and used conveniently in the variational method. The energy expression proposed previously does not include kinetic energy terms caused by noncentral correlations sufficiently, and the obtained energy is too low. In this study, we refine the energy expression by introducing the main part of the missing three-body cluster terms that are related to the tensor correlations. The refined energy expression automatically guarantees necessary conditions on tensor structure functions. The obtained energy per neutron with the v6'potential is considerably higher than that of the old energy expression.

Keywords: Nuclear matter; Neutron matter; Variational method; Nuclear EOS
PACS: 21.65.-f; 21.65.Cd; 21.65.Mn; 24.10.Cn

INTRODUCTION

The nuclear equation of state (EOS) is one of the most important inputs in neutron star studies. In order to obtain a reliable nuclear EOS, many-body calculations starting from the realistic nuclear Hamiltonian are necessary. We have been studying a variational method with approximate energy expressions[1]. In this method, the energy per particle is expressed explicitly as a functional of variational functions. Then, the Euler-Lagrange equations are derived directly from the energy expression, and fully minimized energies are obtained by solving them. In the case of liquid ^{3}He in which ^{3}He atoms interact through two-body central forces, the obtained energy is reasonable. However, in the case of nuclear matter in which two-body noncentral forces are taken into account, the obtained energies for neutron matter and symmetric nuclear matter are too low, and noncentral distribution functions have unrealistic long tails[2]. This energy lowering is considered to be mainly due to inappropriateness of the kinetic-energy expression caused by the noncentral correlations. Therefore, as the first step of the refinement, we improve the energy expression with respect to the tensor correlation. In constructing the refined energy expression, we rely on necessary conditions on the tensor structure function.

CP1016, *Origin of Matter and Evolution of Galaxies*,
edited by T. Suda, T. Nozawa, A. Ohnishi, K. Kato, M. Y. Fujimoto, T. Kajino, and S. Kubono

APPROXIMATE ENERGY EXPRESSION FOR NEUTRON MATTER

In this paper, we consider spin-unpolarized neutron matter at zero-temperature, and start from the following Hamiltonian:

$$H = -\sum_{i=1}^{N} \frac{\hbar^2}{2m} \nabla_i^2 + \sum_{i<j}^{N} V_{ij}, \tag{1}$$

where m is the neutron mass, and, in the case of the V14-type potential, the two-body potential V_{ij} is expressed as

$$V_{ij} = \sum_{s=0}^{1} \left\{ V_{Cs}(r_{ij}) + s\left[V_T(r_{ij}) S_{Tij} + V_{SO}(r_{ij}) (\boldsymbol{s} \cdot \boldsymbol{L}_{ij}) \right] + V_{qLs}(r_{ij}) \boldsymbol{L}_{ij}^2 + s V_{qSO}(r_{ij}) (\boldsymbol{s} \cdot \boldsymbol{L}_{ij})^2 \right\} P_{sij}, \tag{2}$$

where S_{Tij} is the tensor operator and P_{sij} is the spin projection operator.

The approximate energy expression for neutron matter is constructed using the spin-dependent radial distribution functions $F_s(r)$ (s = 0, 1), tensor distribution function $F_T(r)$ and spin-orbit distribution function $F_{SO}(r)$ defined as follows:

$$F_s(r_{12}) = \Omega^2 \sum_{\text{spin}} \int \Psi^*(x_1, \cdots, x_N) P_{s12} \Psi(x_1, \cdots, x_N) d\boldsymbol{r}_3 \cdots d\boldsymbol{r}_N, \tag{3}$$

$$F_T(r_{12}) = \Omega^2 \sum_{\text{spin}} \int \Psi^*(x_1, \cdots, x_N) S_{T12} \Psi(x_1, \cdots, x_N) d\boldsymbol{r}_3 \cdots d\boldsymbol{r}_N, \tag{4}$$

$$F_{SO}(r_{12}) = \Omega^2 \sum_{\text{spin}} \int \Psi^*(x_1, \cdots, x_N) (\boldsymbol{s} \cdot \boldsymbol{L}_{12}) \Psi(x_1, \cdots, x_N) d\boldsymbol{r}_3 \cdots d\boldsymbol{r}_N, \tag{5}$$

where Ω is the volume of the system and Ψ is the wave function. These distribution functions are regarded as independent variational functions.

The structure functions $S_{Cn}(k)$ (n = 1, 2) are also necessary for the energy expression. They are defined as

$$S_{C1}(k) \equiv \frac{1}{N} \left\langle \left| \sum_{i=1}^{N} \exp(i\boldsymbol{k} \cdot \boldsymbol{r}_i) \right|^2 \right\rangle = 1 + S_1(k) + S_0(k) \geq 0, \tag{6}$$

$$S_{C2}(k) \equiv \frac{1}{3N} \left\langle \left| \sum_{i=1}^{N} \boldsymbol{\sigma}_i \exp(i\boldsymbol{k} \cdot \boldsymbol{r}_i) \right|^2 \right\rangle = 1 + \frac{1}{3} S_1(k) - S_0(k) \geq 0. \tag{7}$$

Here, $S_s(k)$ is defined as the Fourier transform of $F_s(r)$:

$$S_s(k) = \rho \int [F_s(r) - F_s(\infty)] \exp(i\boldsymbol{k} \cdot \boldsymbol{r}) d\boldsymbol{r}, \tag{8}$$

with ρ being the number density.

Using these functions, the approximate energy expression is proposed in Ref. [2] as follows:

$$\frac{E}{N}=\frac{3}{5}E_{\mathrm{F}}+2\pi\rho\int_0^\infty\left\{\left[\sum_{s=0}^{1}F_s(r)V_{\mathrm{C}s}(r)\right]+F_{\mathrm{T}}(r)V_{\mathrm{T}}(r)+F_{\mathrm{SO}}(r)V_{\mathrm{SO}}(r)\right.$$
$$\left.+\left[\sum_{s=0}^{1}F_{\mathrm{qL}s}(r)V_{\mathrm{qL}s}(r)\right]+F_{\mathrm{qSO}}(r)V_{\mathrm{qSO}}(r)\right\}r^2dr$$
$$+\frac{\pi\hbar^2\rho}{2m}\int_0^\infty\sum_{s=0}^{1}\left\{F_{\mathrm{C}s}(r)\left[\frac{1}{F_{\mathrm{C}s}(r)}\frac{dF_{\mathrm{C}s}(r)}{dr}-\frac{1}{F_{\mathrm{F}s}(r)}\frac{dF_{\mathrm{F}s}(r)}{dr}\right]^2\right\}r^2dr$$
$$+\frac{2\pi\hbar^2\rho}{m}\int_0^\infty\left\{8\left\{\left[\frac{dg_{\mathrm{T}}(r)}{dr}\right]^2+\frac{6}{r^2}\left[g_{\mathrm{T}}(r)\right]^2\right\}F_{\mathrm{F1}}(r)+\frac{2}{3}\left[\frac{dg_{\mathrm{SO}}(r)}{dr}\right]^2F_{\mathrm{qF1}}(r)\right\}r^2dr$$
$$-\frac{\hbar^2}{16\pi^2m\rho}\int_0^\infty\sum_{n=1}^{2}(2n-1)\frac{(S_{\mathrm{C}n}(k)-1)(S_{\mathrm{C}n}(k)-S_{\mathrm{CF}}(k))^2}{S_{\mathrm{C}n}(k)/S_{\mathrm{CF}}(k)}k^4dk. \qquad (9)$$

The first term on the right-hand side of Eq. (9) is the one-body kinetic energy term with E_{F} being the Fermi energy. The second term represents the potential energy, where $F_{\mathrm{qL}s}(r)$ and $F_{\mathrm{qSO}}(r)$ are expressed using auxiliary functions, $F_{\mathrm{C}s}(r)$, $g_{\mathrm{T}}(r)$ and $g_{\mathrm{SO}}(r)$. These auxiliary functions are related to $F_s(r)$, $F_{\mathrm{T}}(r)$ and $F_{\mathrm{SO}}(r)$; their definitions as well as explicit expressions of $F_{\mathrm{qL}s}(r)$ and $F_{\mathrm{qSO}}(r)$ are given in Ref. [2]. The remaining terms in Eq. (9) represent the kinetic energy caused by the correlation between neutrons. Here, $F_{\mathrm{F}s}(r)$ and $S_{\mathrm{CF}}(k)$ are $F_s(r)$ and $S_{\mathrm{C}n}(k)$ in the case of the Fermi gas, respectively. $F_{\mathrm{qF}s}(r)$ is a distribution function related to the quadratic orbital angular momentum, whose explicit functional form is shown in Ref. [2]. In the energy expression, the central, tensor and spin-orbit potential energies are expressed exactly. It is also noted that the energy expression automatically guarantees necessary conditions on $S_{\mathrm{C}n}(k)$, i.e., the inequalities in (6) and (7).

REFINEMENT OF THE ENERGY EXPRESSION

As mentioned above, the calculated energy of neutron matter obtained with the energy expression Eq. (9) is too low, and the noncental distribution functions, $F_{\mathrm{T}}(r)$ and $F_{\mathrm{SO}}(r)$, have unrealistic long tails. In order to refine the energy expression, we consider the kinetic energy caused by the noncentral correlations. In the following, we only consider the central and tensor forces and corresponding correlations, as the first step of the refinement.

First, we assume the Jastrow-type wave function:

$$\Psi(x_1,\cdots,x_N)=\mathrm{Sym}\left[\prod_{i<j}f_{ij}\right]\Phi(x_1,\cdots,x_N). \qquad (10)$$

Here, Sym[ ] is a symmetrizer and Φ is the Fermi-gas wave function. The correlation function f_{ij} is expressed as

$$f_{ij}=\sum_{s=0}^{1}\left[f_{\mathrm{C}s}(r_{ij})+sf_{\mathrm{T}}(r_{ij})S_{\mathrm{T}ij}\right]P_{sij}, \tag{11}$$

where $f_{\mathrm{C}s}(r)$ and $f_{\mathrm{T}}(r)$ are spin-dependent central and tensor correlation functions, respectively.

Next, we cluster-expand both the expectation value of the Hamiltonian per neutron $\langle H\rangle/N$ and the approximate energy expression E/N given in Eq. (9). Then, we find that the one-body and two-body cluster terms in $\langle H\rangle/N$ are included in E/N exactly. Namely, $\Delta E/N = \langle H\rangle/N - E/N$ consists of the three-body and higher-order cluster terms. We evaluate the main part of the three-body cluster terms in $\Delta E/N$, and express it using the structure functions as

$$\frac{\Delta E_3}{N}=-\frac{\hbar^2}{32m\pi^2\rho}\int_0^\infty\left\{(S_{\mathrm{C}2}(k)-1)[S_{\mathrm{T}}(k)]^2-\frac{1}{18}[S_{\mathrm{T}}(k)]^3\right\}k^4dk. \tag{12}$$

Here, $S_{\mathrm{T}}(k)$ is the tensor structure function defined as

$$S_{\mathrm{T}}(k)=4\pi\rho\int_0^\infty F_{\mathrm{T}}(r)j_2(kr)r^2dr. \tag{13}$$

It is noted that $\Delta E_3/N$ goes to negative infinity through the variational procedure, i.e., it is a harmful term. In order to convert this harmful term into harmless one, we introduce the following necessary conditions on the tensor structure function:

$$S_{\mathrm{CT}1}(k)=\frac{1}{k^2N}\left\langle\left|\sum_{i=1}^{N}(\boldsymbol{\sigma}_i\cdot\boldsymbol{k})\exp(i\boldsymbol{k}\cdot\boldsymbol{r}_i)\right|^2\right\rangle=S_{\mathrm{C}2}(k)-\frac{1}{3}S_{\mathrm{T}}(k)\geq 0, \tag{14}$$

$$S_{\mathrm{CT}2}(k)=\frac{1}{2k^2N}\left\langle\left|\sum_{i=1}^{N}(\boldsymbol{\sigma}_i\times\boldsymbol{k})\exp(i\boldsymbol{k}\cdot\boldsymbol{r}_i)\right|^2\right\rangle=S_{\mathrm{C}2}(k)+\frac{1}{6}S_{\mathrm{T}}(k)\geq 0. \tag{15}$$

Then, we propose the refined approximate energy expression for neutron matter as follows:

$$\frac{E_{\mathrm{new}}}{N}=\frac{3}{5}E_{\mathrm{F}}+2\pi\rho\int_0^\infty\left\{\left[\sum_{s=0}^{1}F_s(r)V_{\mathrm{C}s}(r)\right]+F_{\mathrm{T}}(r)V_{\mathrm{T}}(r)\right\}r^2dr$$
$$+\frac{\pi\hbar^2\rho}{2m}\int_0^\infty\sum_{s=0}^{1}\left\{F_{\mathrm{C}s}(r)\left[\frac{1}{F_{\mathrm{C}s}(r)}\frac{dF_{\mathrm{C}s}(r)}{dr}-\frac{1}{F_{\mathrm{F}s}(r)}\frac{dF_{\mathrm{F}s}(r)}{dr}\right]^2\right\}r^2dr$$

$$
+\frac{2\pi\hbar^2\rho}{m}\int_0^\infty 8\left\{\left[\frac{dg_{\rm T}(r)}{dr}\right]^2+\frac{6}{r^2}\left[g_{\rm T}(r)\right]^2\right\}F_{\rm FI}(r)r^2dr
$$
$$
-\frac{\hbar^2}{16\pi^2 m\rho}\int_0^\infty \frac{(S_{\rm CI}(k)-1)(S_{\rm CI}(k)-S_{\rm CF}(k))^2}{S_{\rm CI}(k)/S_{\rm CF}(k)}k^4dk
$$
$$
-\frac{\hbar^2}{16\pi^2 m\rho}\int_0^\infty \sum_{n=1}^{2} n\frac{(S_{{\rm CT}n}(k)-1)(S_{{\rm CT}n}(k)-S_{\rm CF}(k))^2}{S_{{\rm CT}n}(k)/S_{\rm CF}(k)}k^4dk. \qquad (16)
$$

The last term on the right-hand side of Eq. (16) includes $\Delta E_3/N$ properly and the denominators in the integrands are introduced so that the last term becomes harmless. Furthermore, the energy expression $E_{\rm new}/N$ automatically guarantees the inequalities in (14) and (15).

NUMERICAL CALCULATIONS AND DISCUSSION

Using the refined energy expression Eq. (16), the Euler-Lagrange equations for $F_{{\rm C}s}(r)$ and $g_{\rm T}(r)$ are derived, and solved numerically with the v6' potential[3] as the two-body nuclear potential. As shown in Fig. 1, the obtained energy is higher than the result of the old energy expression. In this figure, the result of the auxiliary field diffusion Monte Carlo (AFDMC) calculation[4] is also plotted; our result is somewhat lower than that by AFDMC.

The tensor distribution function $F_{\rm T}(r)$ is shown in Fig. 2. The unrealistic long tail of $F_{\rm T}(r)$ disappears in the case of the refined energy expression. It is also noted that $S_{\rm CT2}(k)$ obtained with the old energy expression violates the inequality in (15), which implies that the necessary conditions on $S_{{\rm CT}n}(k)$ play important roles in this variational calculation.

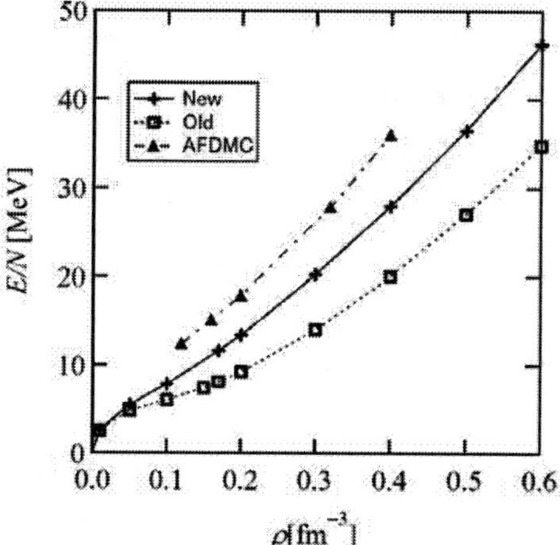

FIGURE 1. Energies per neutron for neutron mater with the v6'potential. The solid curve is for the refined energy expression and the dotted curve with squares is for the old energy expression. The curve with triangles is the result by AFDMC.

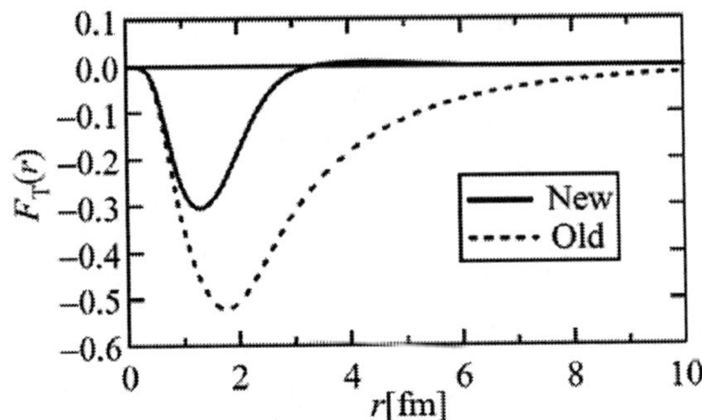

FIGURE 2. The tensor distribution function $F_T(r)$.

Refinement of the energy expression for symmetric nuclear matter with respect to the tensor correlations is now in progress. Toward the realistic nuclear EOS, further studies such as refinement of the energy expression with respect to the spin-orbit correlation and inclusion of the three-body nuclear force are necessary.

ACKNOWLEDGEMENTS

This study is supported by a Grand-in-Aid for the 21st century COE program "Holistic Research and Education Center for Physics of Self-organizing Systems" at Waseda University, and Waseda University Grants for Special Research Projects 2003A-628 and 2007A-902.

REFERENCES

1. M. Takano and M. Yamada, *Prog. Theor. Phys.* **91**, 1149 (1994).
2. M. Takano and M. Yamada, *Prog. Theor. Phys.* **100**, 745 (1998).
3. B. S. Pudliner, V. R. Pandharipande, J. Carlson, S. C. Pieper and R. B. Wiringa, *Phys. Rev.* C **56**, 1720 (1997).
4. A. Sarsa, S. Fantoni and K. Schmidt, *Phys. Rev. C* **68**, 24308 (2003).

12. SPECIAL SESSION: RI BEAMS FOR NUCLEAR ASTROPHYSICS

ORNL Radioactive Beams for Stellar Explosion Studies

Michael S. Smith

Physics Division, Oak Ridge National Laboratory, Oak Ridge, Tennessee, 37831-6354, USA

Abstract. Thermonuclear reactions on unstable nuclei generate the energy that power nova explosions and X-ray bursts. In these explosions and others such as supernovae, these reactions serve to synthesize nuclei that (via their decay) can serve as tracers of the explosion mechanism. A powerful approach to improve our understanding of these explosions is to utilize beams of radioactive nuclei for direct and indirect measurements of these reactions. We are pursuing this approach at the Holifield Radioactive Ion Beam Facility (HRIBF) at Oak Ridge National Laboratory (ORNL) to study reactions in the rp-process (with beams of ^{17,18}F) and the r-process (with beams of ^{82}Ge, ^{84}Se ^{130,132}Sn, ^{134}Te). These measurements are combined with synergistic data evaluations and element synthesis calculations. Highlights of recent results are presented.

Keywords: nuclear astrophysics, nucleosynthesis, r-process, rp-process, radioactive beam, stellar explosions, nova, supernova, X-ray burst, thermonuclear reaction rates, nuclear data, visualization, simulations, waiting point, reaction network
PACS: 26.30.-k, 26.30.Ca, 26.50.+x, 95.30.-k, 97.10.Cv, 97.30.Qt, 97.80.Gm, 97.80.Jp

STELLAR EXPLOSIONS AND UNSTABLE NUCLEI

Unstable nuclei can be synthesized and – before they decay – undergo subsequent reactions in the extreme temperature and density conditions found in stellar explosions. Nucleosynthesis and thermonuclear energy generation in explosive environments are therefore quite different than in non-exploding stars. The paucity of experimental information on the structure and reactions of unstable nuclei, however, forces explosion simulations to rely primarily on theoretical reaction rate estimates which can, in some cases, be incorrect by orders of magnitude.

The availability of beams of radioactive nuclei has, fortunately, opened a new frontier in studies of novae, supernovae, X-ray bursts, and other stellar explosions [1]. A powerful approach has been developed which involves direct measurements of the astrophysical reaction of interest with beams of unstable nuclei *combined with* indirect determinations via transfer, scattering, and other measurements. We are pursuing this approach at the Holifield Radioactive Ion Beam Facility (HRIBF) [2, 3] at Oak Ridge National Laboratory (ORNL) to study reactions in the rp-process (with beams of ^{17,18}F) and the r-process (with beams of ^{82}Ge, ^{84}Se ^{130,132}Sn, ^{134}Te). Such measurements are contributing to a new empirical foundation for studies of these cataclysmic events.

CP1016, *Origin of Matter and Evolution of Galaxies*,
edited by T. Suda, T. Nozawa, A. Ohnishi, K. Kato, M. Y. Fujimoto, T. Kajino, and S. Kubono

ORNL HOLIFIELD RADIOACTIVE ION BEAM FACILITY

The Isotope Separator On-Line (ISOL) technique [4, 1] is used at HRIBF to produce beams of unstable proton- and neutron-rich nuclei. Beams of over 175 species can be accelerated with intensities greater than 10^3 particles per second (pps), 60 of them with intensities greater than 10^6 pps [3]. There are 32 proton-rich beams (e.g., ^{17,18}F) produced by light-ion bombardment of high-temperature oxide targets (e.g., HfO) fabricated in a fibrous structure. The rest are neutron-rich unstable nuclides produced via proton-induced fission reactions occurring when during the proton bombardment of a UC target with a foam-like structure. The ability to reaccelerate these n-rich beams to energies above the Coulomb barrier (e.g., to 4.5 MeV/u) is unique in the world. The purity of some beams are significantly enhanced by molecular transport in the 25 MV electrostatic post-accelerator [5], while other beams are fully stripped at the post-accelerator exit to 100% purity. Devices such as the Daresbury Recoil Separator [6], the Recoil Mass Separator [7], clover Ge arrays, the Silicon Detector Array (SIDAR) [8] are used to detect the products of reactions with these radioactive beams.

MEASUREMENTS WITH NEUTRON-RICH BEAMS FOR THE r-PROCESS

The rapid neutron capture process or r-process involves a series of neutron captures on neutron-rich unstable nuclei interspersed with β-decays. Believed to be responsible for the formation of approximately half of the elements heavier than iron, it is the least well understood nucleosynthesis process [9]. The r-process is believed to occur in core collapse supernovae – specifically, in the high-entropy bubble above a newly-born proto-neutron star [9, 10]. There is now striking observational evidence of abundances of heavy elements on the surface of individual metal-poor stars in the halo of our galaxy that match those predicted by some r-process simulations [11].

Nuclear structure information (masses, lifetimes, level structure, decay properties) for thousands of nuclei out to the neutron drip line is needed as input for simulations of the r-process. Additionally, nuclear reaction information, especially near the N = 50 and 82 closed neutron shells where the abundances peak, is important in the r-process, especially in the later cooling phases [12]. Since neutron capture reactions on neutron-rich unstable nuclei cannot be measured directly, transfer reactions such as (d,p) can be used to "simulate" neutron capture and to provide level information necessary to determine the rates of these important reactions. At HRIBF, we have carried out the first such neutron-transfer reaction measurements on nuclei in the r-process path at N=50 (^{82}Ge and ^{84}Se [13, 14]) and N=82 (^{130}Sn(d,p)^{131}Sn, ^{132}Sn(d,p)^{133}Sn [15], and ^{134}Te(d,p)^{135}Te). Analysis of the N=82 experiments are still in progress. For ^{130}Sn(d,p)^{131}Sn (Fig. 1), a 81 μg/cm^2 $(CD_2)_n$ target was bombarded with a 630 MeV, 2×10^5 pps beam of radioactive ^{130}Sn nuclei. Protons from the (d,p) reaction were detected in the SIDAR annular array of silicon detectors [8] supplemented with elements of the new Oak Ridge - Rutgers University Barrel Array (ORRUBA) [16], and the ^{131}Sn heavy recoils were detected in coincidence in a carbon-foil microchannel plate detector. Our experimental energy resolution was approximately 250 keV, primarily limited by the

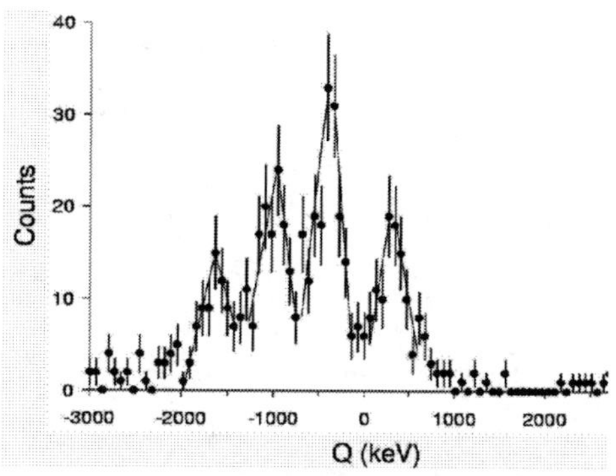

FIGURE 1. (a) Preliminary Q-value spectrum for the ^{130}Sn(d,p)^{131}Sn reaction at 4.8 MeV/u.

thickness of the target needed to obtain a reasonable yield. Four strong single-particle levels in ^{131}Sn were measured for the first time, with statistical (systematic) uncertainties in excitation energy of approximately 25 (30) keV. Angular distributions of protons in coincidence with ^{131}Sn recoils were measured and will be used to determine the spins, parities, and spectroscopic strengths of these levels by comparison to DWBA calculations. Preliminary results [17] suggest that a $(1/2)^-$ and $(3/2)^-$ level are both below the neutron capture threshold in ^{131}Sn, agreeing with the FRDM mass model predictions but disagreeing with the RMFT model [18]. This increases the ^{130}Sn(n,γ)^{131}Sn direct capture cross section by a factor of 500 above that calculated using the RMFT masses [18].

MEASUREMENTS WITH PROTON-RICH BEAMS FOR THE rp-PROCESS

Proton capture reactions on proton-rich unstable nuclei generate the thermonuclear energy that powers nova explosions and X-ray bursts. The relevant reaction sequences are the Hot CNO cycle, the rp-process, and the αp-process [1]. The structure properties and reactions of proton-rich nuclei are needed to determine the nuclear burning occurring in these outbursts. Calculations of the synthesis of the long-lived radioactive nucleus ^{18}F are particularly important, as the characteristic radiation signature of the decay of this nucleus in the outer envelope of the explosion can help diagnose the explosion mechanism [19]. A series of measurements of reactions that produce and destroy the ^{18}F nova observable have been made at HRIBF, including: direct measurements of ^{18}F(p,α)^{15}O at the 330-keV [20] and 665-keV [21] resonances; measurements of ^{18}F(p,p) with both thin [22] and thick [23] targets; measurement of the ^{17}F(p,p) reaction [8]; and measurements of ^{18}F(d,p)^{19}F [24] and ^{18}F(d,n)^{19}Ne [25]. Recently, the interference between three $3/2^+$ resonances in the ^{18}F(p,α)^{15}O reaction was measured for the first time [26], reducing the uncertainty of the higher temperature rate by 37%.

A novel experimental technique has been developed to measure the energy and strength of resonances in (p,α) reactions in inverse kinematics. A differentially pumped

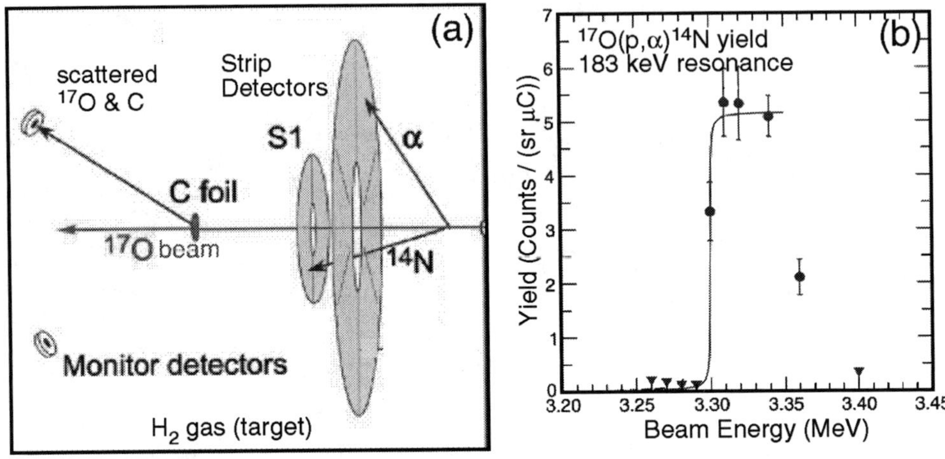

FIGURE 2. (a) Experimental setup for $^{17}O(p,\alpha)^{14}N$; (b) reaction yield over 183 keV resonance.

windowless gas target is used with arrays of Si detectors in the target gas to measure both the light and heavy ions from the reaction in coincidence (Fig. 2a.) This approach was used with a stable ^{17}O beam (Fig. 2) to measure the properties of the very important 183-keV resonance in the $^{17}O(p,\alpha)^{14}N$ reaction (Fig. 2b) [27]. This approach will be utilized in the future to measure properties of resonances in the $^{18}F(p,\alpha)^{15}O$ reaction.

SYNERGISTIC DATA EVALUATIONS & ELEMENT SYNTHESIS CALCULATIONS

To fully explore the astrophysical implications of laboratory measurements, it is essential to combine the latest results with those measured previously and with theoretical calculations, a "data evaluation", to obtain a best value. This result can then be processed (e.g., into reaction rates) and inserted into an astrophysics simulation code. We have followed this approach in a recent evaluation [28] of the properties of twenty-eight ^{18}F + p levels in ^{19}Ne. This work incorporated results from the series of HRIBF measurements with ^{18}F beams described above along with previous knowledge of ^{19}Ne and analog levels in ^{19}F. The new level information was converted into reaction rates and incorporated into a post-processing nucleosynthesis code [29] utilizing the **Computational Infrastructure for Nuclear Astrophysics** [30], a free online system at `nucastrodata.org`. The new ^{18}F + p reaction rates (lower by a factor of 3 from previous estimates) result in a factor of 1.6 increase in the ^{18}F production in novae. The impact of new $^{17}O(p,\alpha)^{14}N$ and $^{17}O(p,\gamma)^{18}F$ reaction rates based on our $^{17}O(p,\alpha)^{14}N$ measurement described above was also investigated with this online software suite. The ^{18}F production was found to decrease by up to a factor of 10 in a model of a nova explosion on a 1.15 solar mass white dwarf, with substantially less impact on models of explosions on more massive white dwarf stars [27]. Recent developments in this online system include the addition of numerous new novae and X-ray burst simulations, new tools to help process evaluations into rate libraries [31], and a tool to enable custom searches for rp-process waiting points [32].

SUMMARY

At the Holifield Radioactive Ion Beam Facility at ORNL, beams of reaccelerated, neutron-rich unstable nuclei [^{82}Ge, ^{84}Se ^{130,132}Sn, ^{134}Te] that are unique in the world are being used to make indirect measurements of the structure of and reactions on nuclei near the closed N=50 and N=82 closed neutron shells. Such measurements are providing an empirical foundation for studies of the r-process in supernovae. Beams of radioactive ^{17,18}F are also being used to improve our understanding of novae and X-ray bursts. These measurements are being combined with data evaluations and element synthesis calculations using the online system at `nucastrodata.org`, the **Computational Infrastructure for Nuclear Astrophysics**.

ACKNOWLEDGMENTS

This paper describes the work of the RIBENS Collaboration (Radioactive Ion Beams for Explosive Nucleosynthesis Studies), specifically work led by: R.L. Kozub [^{130}Sn(d,p)^{131}Sn]; K.L. Jones [^{132}Sn(d,p)^{133}Sn]; S.D. Pain [^{134}Te(d,p)^{135}Te]; J.C. Blackmon and B.H. Moazen [^{17}O(p,α)^{14}N]; S.D. Pain and J.C. Blackmon (ORRUBA); K. Chae and D.W. Bardayan [^{18}F(p,α)^{15}O]; E.J. Lingerfelt and K. Buckner (Computational Infrastructure for Nuclear Astrophysics); W.R. Hix (element synthesis calculations); T. Sunayama (waiting point finder); and C.D. Nesaraja [^{18}F(p,α)^{15}O evaluations]. ORNL is managed by UT-Battelle, LLC, for the U.S. Department of Energy under contract DE-AC05-00OR22725.

REFERENCES

1. M.S. Smith, K.E. Rehm, *Ann. Rev. Nucl. Part. Sci.* **51** (2001) 91.
2. B.A. Tatum, J.R. Beene, in Proc. Particle Accelerator Conference (PAC'05), Knoxville, TN, 16-20 May (2005), Knoxville, TN; Proc. PAC 05, pg. 2556, ed. E. Horak, IEEE Catalog 05CH37623C, Piscataway, NJ, 2005, p. 3641.
3. http://www.phy.ornl.gov/hribf/
4. H. Ravn, *Phys. Rep.* **54**, 201 (1979).
5. D. Stracener *et al.*, Proceedings XXVI Symposium on Nuclear Physics, Taxco, Mexico, Jan. 6-9, 2003, Rev. Mex. Fis. (Suppl.) **49** 92-96 (2003).
6. R. Fitzgerald *et al.*, *Nucl. Phys. A* **748**, 351 (2005).
7. C.J. Gross *et al.*, *Nucl. Instrum. Methods Phys. Res. A* **450**, 12-29 (2000).
8. D.W. Bardayan *et al.*, *Phys. Rev. C* **62**, 0155804 (2000).
9. Y.-Z. Qian, G.J. Wasserburg, *Phys. Rept.* **442**, 237 (2007).
10. S.E. Woosley *et al.*, *Astrophys. J.* **433**, 229 (1994).
11. C. Sneden et al., *Ap. J.* **591**, 936 (2003).
12. R. Surman, J. Engel, *Phys. Rev. C*, **64**, 035801 (2003).
13. J.S. Thomas *et al.*, *Phys. Rev. C* **71**, 021302(R) (2005).

14. J.S. Thomas *et al.*, *Phys. Rev. C* **76**, 044302 (2007).
15. K.L. Jones et al., *Acta Physica Polonica B* **38**, 1205 (2007).
16. S.D. Pain et al., Nucl. Instrum. Meth. B **261**, 1122 (2007).
17. R.L. Kozub et al., in preparation (2008).
18. T. Rauscher et al., *Phys. Rev. C* **57**, 2031 (1998).
19. M. Hernanz, J. Jose, New Astronomy Rev. **48**, 35 (2003).
20. D.W. Bardayan *et al.*, Phys. Rev. Lett. **89**, 262501 (2002).
21. D.W. Bardayan *et al.*, Phys. Rev. C **63**, 065802 (2001).
22. D.W. Bardayan *et al.*, Phys. Rev. C **62**, 042802(R) (2000).
23. D.W. Bardayan *et al.*, Phys. Rev. C **70**, 015804 (2004).
24. R.L. Kozub *et al.*, Phys. Rev. C **71**, 032801(R) (2005).
25. C.R. Brune et al., Bull. Am. Phys. Soc., BAPS.2006.APR.L8.4 (2006).
26. K.Y. Chae et al., Phys. Rev. C **74**, 012801 (2006).
27. B.H. Moazen et al., Phys. Rev. C **75**, 065801 (2007).
28. C.D. Nesaraja et al., *Phys. Rev. C* **75**, 055809 (2007).
29. W.R. Hix, F.-K. Thielemann, *J. Comp. Appl. Math*, **109**, 321 (1999).
30. M.S. Smith et al., in Proc. Int. Symp. Nuclear Astrophysics - Nuclei in the Cosmos IX, Proceedings of Science, June 2006, http://pos.sissa.it//archive/conferences/028/180/NIC-IX 180.pdf.
31. M.S. Smith et al., *"Thermonuclear Reaction Rate Libraries and Software Tools for Nuclear Astrophysics Research"*, this proceedings.
32. T. Sunayama et al., *"Waiting Points in Nova and X-ray burst Nucleosynthesis"*, this proceedings.

Nuclear Structure of ^{8}B Studied by Proton Resonance Scatterings on ^{7}Be

H. Yamaguchi*, Y. Wakabayashi*, S. Hayakawa*, G. Amadio*, S. Kubono*, H. Fujikawa*, A. Saito†, J.J. He**, T. Teranishi‡, Y.K. Kwon§, S. Nishimura¶, Y. Togano‖, N. Iwasa††, K. Inafuku††, M. Niikura*, D.N. Binh* and L.H. Khiem‡‡

*Center for Nuclear Study (CNS), University of Tokyo, RIKEN campus, 2-1 Hirosawa, Wako, Saitama 351-0198, Japan
† Department of Physics, Graduate School of Science, University of Tokyo 7-3-1 Hongo, Bunkyo-ku, Tokyo 113-0033, Japan
**School of Physics, The University of Edinburgh Mayfield Road, Edinburgh EH9 3JZ, UK
‡Department of Physics, Kyushu University, 6-10-1 Hakozaki, Fukuoka 812-8581, Japan
§Department of Physics, Chung-Ang University, Seoul 156-756, South Korea
¶The Institute of Physical and Chemical Research (RIKEN), 2-1 Hirosawa, Wako, Saitama 351-0198, Japan
‖Department of Physics, Rikkyo University, Tokyo 171-8501, Japan
††Department of Physics, Tohoku University, Aoba, Sendai, Miyagi 980-8578, Japan
‡‡ Institute of Physics and Electronics, Vietnam Academy of Science and Technology 18 Hoang Quoc Viet St., Nghia do, Hanoi, Vietnam

Abstract. A new measurement of the proton resonance scattering on ^{7}Be was performed up to the excitation energy of 6.8 MeV using the low-energy RI beam facility CRIB (CNS Radioactive Ion Beam separator) at the Center for Nuclear Study (CNS) of the University of Tokyo. The excitation function of ^{8}B above 3.5 MeV was successfully measured for the first time, providing important information about the reaction rate of ^{7}Be$(p,\gamma)^8$B, which is the key reaction in the solar ^{8}B neutrino production. For more intensive experimental studies with RI beams, the development of a cryogenic gas target system is ongoing at CNS. In this paper a preliminary result of the ^{7}Be experiment and the present status of the development of the target system are presented.

Keywords: proton resonance scattering, astrophysical S-factor, solar neutrino problem
PACS: 25.40.Ny, 27.20.+n, 21.10.Hw, 26.65.+t

INTRODUCTION

The astrophysical S-factor $S_{17}(E)$ of the ^{7}Be$(p,\gamma)^8$B reaction is one of the most important parameters in the standard solar model, because its value at the energy of the solar center is directly related to the flux of the ^{8}B neutrino, which is the dominant component of the solar neutrinos detected by many neutrino observatories on the earth. It is claimed that S_{17} should be determined with a precision better than about 5% in the energy region below 300 keV in order to test the solar model by comparing the theoretical prediction for the ^{8}B neutrino flux with the observations [1]. For that reason great efforts were spent by many experimental groups [2, 3, 4, 5, 6, 7]. The precision of the existing data, however, is still limited because of the very small cross section of the ^{7}Be$(p,\gamma)^8$B reaction in such a low energy region.

CP1016, *Origin of Matter and Evolution of Galaxies*,
edited by T. Suda, T. Nozawa, A. Ohnishi, K. Kato, M. Y. Fujimoto, T. Kajino, and S. Kubono

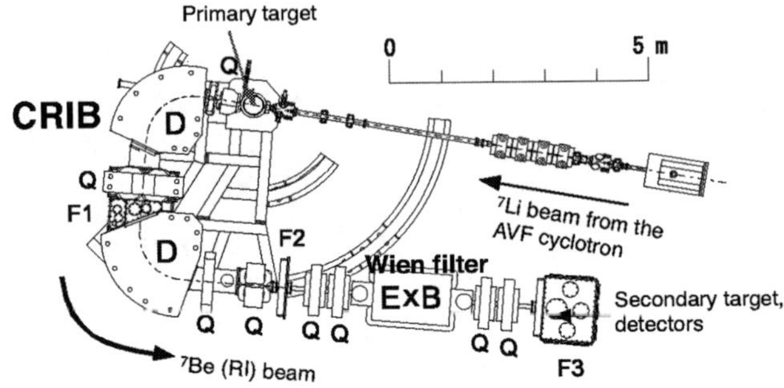

FIGURE 1. CRIB overview. D, Q, and F represent a dipole magnet, a quadrupole magnet, and a focal plane, respectively.

To evaluate S_{17} at low energies, one needs information about the nuclear structure of ^{8}B which, however, has been poorly known until recently. Only the lowest two excited states at 0.77 MeV and 2.32 MeV were clearly observed in the previous experiments. It has been reported that a wide resonance is observed at around 3 MeV, and it was explained as a low-lying 2s state [8]. The reason why a 2s state appears at such a low energy is also an interesting subject [9, 10, 11]. This kind of wide states may affect the ^{7}Be(p,γ)^{8}B reaction rate in the energy region far below 1 MeV. In another measurement [12], the wide state was not directly observed; nevertheless they concluded that their spectrum was consistent with the existence of the state, if it is located at 3.5 MeV with a width of 4 MeV or more. Thus we intended to measure the resonances of ^{8}B to evidently observe the 3.5 MeV resonance reported in the previous measurements, and also to explore the unknown region E >3.5 MeV.

METHOD

The measurement was performed at CRIB [13, 14]. CRIB can produce RI beams with the in-flight method, using primary heavy-ion beams from the AVF cyclotron of RIKEN (K=70). Figure 1 shows the whole structure of CRIB. The primary beam used in this measurement was ^{7}Li^{3+} of 8.76 MeV/u, with the beam current of about 100 pnA. The RI-beam production target was pure hydrogen gas, which was enclosed in an 8-cm-long cell, at 760 Torr and room temperature ($\sim$ 300 K). The ^{7}Be beam energy used in this measurement was 53.8 MeV, corresponding to the maximum center-of-mass energy of 6.7 MeV. We obtained 100% pure ^{7}Be^{4+} beam of 3×10^{5} particles per second, at the target.

We used an experimental method for the proton elastic resonance scattering established at CRIB [15]. A main feature of this method is the thick target [16], which enables us to measure the cross section of various excitation energies at the same time.

The targets and detectors for the scattering experiment were placed in a vacuum chamber located at the downstream of the Wien filter. Figure 2 shows a schematic view of the experimental setup in the chamber. Two PPACs (Parallel-Plate Avalanche

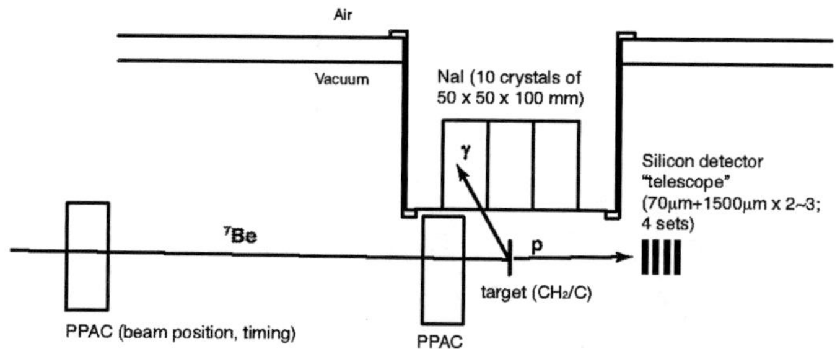

FIGURE 2. Arrangement of the detectors and targets in the experimental chamber.

Counters [17]) measured the timing and position of the incoming ^{7}Be beam. The timing information was used for making event triggers, and also for the particle identification with the time-of-flight (TOF) method. The position of the beam and the incident angle at the target were determined by extrapolating the positions measured by the PPACs. The targets were 39-mg/cm^2-thick polyethylene (CH_2), and 54-mg/cm^2-thick carbon, both of which were sufficiently thick to stop all of the ^{7}Be beam. Carbon foils were used for evaluating backgrounds coming from carbon atoms in the polyethylene target. Multi-layered silicon detector sets, referred to as ΔE-E telescopes, were used for measuring the energy and angular distribution of the recoil protons. They were placed about 23 cm distant from the target, and they covered scattering angle of up to 45 degrees in the laboratory frame. Each ΔE-E telescope consisted of a ΔE counter and two or three E counters, all of which had an area of 50 mm $\times$ 50 mm. Each of the ΔE counters was about 70 μm thick and divided into 16 strips for both sides. The E counters, which were 1.5 mm thick, were placed behind the ΔE counters. The recoil proton energy was 23 MeV at maximum, and the silicon detectors were sufficiently thick to stop all the recoil protons in them. With these ΔE-E telescopes, we identified recoil protons from other particles. NaI detectors were used for measuring 429 keV gamma rays from inelastic scatterings, $p(^7\text{Be}, {}^7\text{Be}^*)p$, to the first excited state at 429 keV (J^π=1/2$^-$). Each NaI crystal has a geometry of 50 mm $\times$ 50 mm $\times$ 100 mm. In this measurement we used ten crystals covering $\sim$20% of the total solid angle.

Compared to the previous measurements with similar methods [8, 12], the present measurement has the following advantages. First, we used a high-energy ^{7}Be beam which enabled us to measure the excitation function of ^{8}B up to the excitation energy of 6.8 MeV. Second, our measurement covered almost full range of the scattering angle between 0 and 45 degrees in laboratory frame, and thus we obtained complete information about the angular distribution. Third, the inelastic scattering cross section was measured with NaI detectors in order to evaluate the excitation function precisely.

RESULT

Excitation functions were obtained by calculating the cross section from the numbers of the proton events. An excitation function for the scattering angles of 0–5 degree is

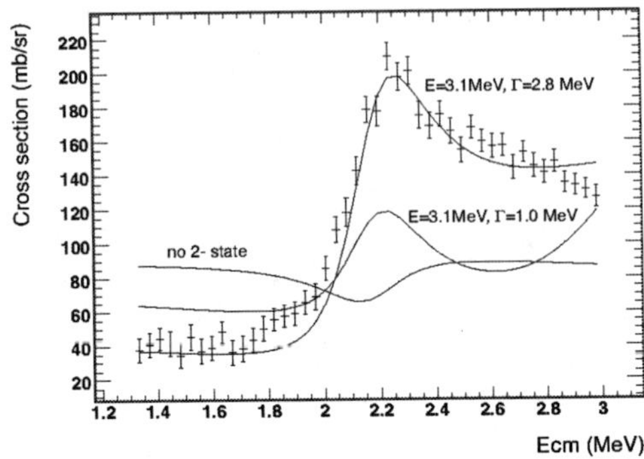

FIGURE 3. Preliminary analysis result on the proton excitation function of ^{7}Be below 3 MeV, measured at 0–5 degrees. R-matrix fit results with the known parameters for the 3^+ state at 2.3 MeV and three different parameter sets (excitation energy E and width Γ) for the 2^- state are illustrated. The excitation energy of ^{8}B, E_{ex} equals to E_{cm}+0.1375 MeV.

shown in Figure 3. The cross section roughly agrees with the past measurements [8, 12] in the energy region around the known resonance at 2.3 MeV. The 3.5 MeV resonance, which was reported as a wide resonance, was not clearly seen, but its energy and width can be determined by evaluating the interference with the known resonance at 2.3 MeV. A preliminary analysis result shown in the figure indicates that the peak observed at 2.3 MeV is possibly enhanced by a wide state that locates at the energy of 3 MeV or higher.

DEVELOPMENT OF THE CRYOGENIC GAS TARGET SYSTEM

A thick and thermostable gas target is necessary for the production of intense RI beams, which is applicable to the measurements of low cross section reactions such as ^{7}Be(p,γ)^{8}B. To determine the S_{17} at 1 MeV with 5% or better precision, a ^{7}Be beam of more than 10^7 particles per second will be needed. We developed a cryogenic gas target system to meet the requirements. The performance of the target system is discussed in this section.

The main features of our cryogenic gas target system are as follows.

1. Liquid-nitrogen cooling
   The gas is enclosed in a cell, which has a shape of 80 mm-long cylinder with a diameter of 20 mm. The cell have 2 μm-thick Havar foils as beam windows. The target gas and window foils are cooled by liquid nitrogen. When the target is cooled to the liquid nitrogen temperature (77 K), it is nearly four times thicker compared to the room temperature target at the same pressure.
2. Forced gas circulation
   The target gas can be forced to flow by circulating it with a pump in an enclosed system that includes the target cell. It has been reported that the effective target thickness is significantly reduced when the heating power of the incident beam

exceeds a value of about 10 mW/mm [18]. The density-reduction effect is expected to be suppressed by circulating the gas with a flow rate of the order of 10 slm (standard liters per minute). In the present system, it is possible to circulate the hydrogen gas with the flow rate of up to 55 slm.

3. Oxygen concentration monitoring
   Since a large quantity of the hydrogen gas is stored in the present system, an explosion by the air leakage must be avoided. Considering the explosive limits of hydrogen, the concentration of oxygen in the hydrogen gas must be lower than 5%. We have installed an oxygen analyzer for monitoring the concentration during the run.

Figure 4 shows the whole structure of the target system. The target gas cell is shown at the bottom of the figure, located in a vacuum chamber. There is a gas circulation line

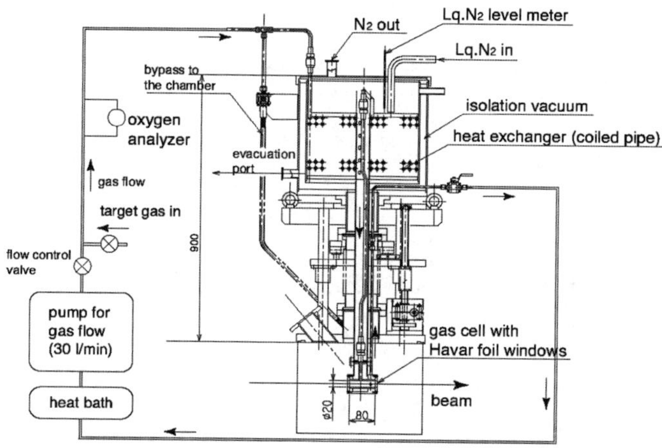

FIGURE 4. Design of the cryogenic gas target system.

through the target cell and the top Dewar, most of which consists of 3/8-inch-diameter pipes. A diaphragm pump (Iwaki Co., Ltd., APN-P450NST) was installed in the line to maintain a gas circulation rate of 30 slm. A heat exchanger (coiled-pipe type) in the top dewar cools the injected gas down to near 77 K, as the gas passes through it.

We have performed some basic tests of the cryogenic target system producing a ^{7}Be beam at CRIB, and we obtained the beam intensity of 2×10^8 particles per second using the target of 2.3 mg/cm^2-thick hydrogen gas at 85–90 K.

SUMMARY

We have studied the proton resonance scattering of ^{7}Be by using a pure ^{7}Be beam produced at CRIB. The excitation function of ^{8}B was measured up to the excitation energy of 6.8 MeV, using the thick-target method. The excited state of ^{8}B at 3.5 MeV was reported as an unexpectedly wide state, and thus it is important to determine its parameters. According to the current preliminary results, our measurement is expected to provide good information on this excited state.

A cryogenic gas target system was developed for the RI beam production at CRIB. The basic features were tested using heavy-ion beams, and we successfully obtained the ^{7}Be beam of 2×10^8 particles per second.

ACKNOWLEDGMENTS

We are grateful to RIKEN accelerator staff for their help. This work was supported by the Grant-in-Aid for Young Scientists (B) (Grant No. 17740135) of JSPS.

REFERENCES

1. E.G. Adelberger *et al.*, *Rev. of Mod. Phys.* **70**, 1265–1291 (1998).
2. B. Filippone, A. Elwyn, C. Davids, and D. Koetke, *Phys. Rev. C* **28**, 2222 (1983).
3. F. Hammache, G. Bogaert, P. Aguer, C. Angulo, S. Barhoumi, L. Brillard, J. F. Chemin, G. Claverie, A. Coc, M. Hussonnois *et al.*, *Phys. Rev. Lett.* **80**, 928 (1998).
4. F. Strieder, L. Gialanella, G. Gyürky, F. Schümann, R. Bonetti, C. Broggini, L. Campajola, P. Corvisiero, H. Costantini, A.D'Onofrio *et al.*, *Nucl. Phys. A* **696**, 219 (2001).
5. F. Schümann, F. Hammache, S. Typel, F. Uhlig, K. Sümmerer, I. Böttcher, D. Cortina, A. Förster, M. Gai, H. Geissel *et al.*, *Phys. Rev. Lett.* **90**, 232501 (2003).
6. A. Junghans, E. Mohrmann, K. A. Snover, T. D. Steiger, E. Adelberger, J. Casandjian, H. Swanson, L. Buchmann, S. Park, A. Zyuzin *et al.*, *Phys. Rev. C* **68**, 065803 (2003).
7. R. Cyburt, B. Davids, and B. Jennings, *Phys. Rev. C* **70**, 045801 (2004).
8. V. Gol'dberg, G. Rogachev, M. Golovkov, V. Dukhanov, I. Serikov, and V. Timofeev, *JETP Lett.* **67**, 1013–1017 (1998).
9. A. van Hees and P. Glaudemans, *Z. Phys. A* **314**, 323 (1983).
10. A. van Hees and P. Glaudemans, *Z. Phys. A* **315**, 223 (1984).
11. F. Barker and A. Mukhamedzhanov, *Nucl. Phys. A* **673**, 526–532 (2000).
12. G. Rogachev, J. Kolata, F. Becchetti, P. DeYoung, M. Hencheck, K. Helland, J. Hinnefeld, B. Hughey, P. Jolivette, L. Kiessel, . M. L. H.-Y. Lee, T. O'Donnell, G. Peaslee, D. Peterson, D. Roberts, P. Santi, and S. Shaheen, *Phys. Rev. C* **64**, 061601(R) (2001).
13. S. Kubono, Y. Yanagisawa, T. Teranishi, S. Kato, T. Kishida, S. Michimasa, Y. Ohshiro, S. Shimoura, K. Ue, S. Watanabe, and N. Yamazaki, *Eur. Phys. J.* **A13**, 217 (2002).
14. Y. Yanagisawa, S. Kubono, T. Teranishi, K. Ue, S. Michimasa, M. Notani, J. He, Y. Ohshiro, S. Shimoura, S. Watanabe, N. Yamazaki, H. Iwasaki, S. Kato, T. Kishida, T. Morikawa, and Y. Mizoi, *Nucl. Instrum. Meth. Phys. Res., Sect. A* **539**, 74–83 (2005).
15. T. Teranishi, S. Kubono, S. Shimoura, M. Notani, Y. Yanagisawa, S. Michimasa, K. Ue, H. Iwasaki, M. Kurokawa, Y. Satou, T. Morikawa, A. Saito, H. Baba, J. Lee, C. Lee, Z. Fulop, and S. Kato, *Phys. Lett. B* **556**, 27–32 (2003).
16. K. Artemov, O. Belyanin, A. Vetoshkin, R. Wolskj, M. Golovkov, V. Gol'dberg, M. Madeja, V. Pankratov, I. Serikov, V. Timofeev, V. Shadrin, and J. Szmider, *Sov. J. Nucl. Phys* **52**, 408 (1990).
17. H. Kumagai, A. Ozawa, N. Fukuda, K. Sümmerer, and I. Tanihata, *Nucl. Instrum. Meth. Phys. Res., Sect. A* **470**, 562 (2001).
18. J. Göerres, K. Kettner, H. Kräwinkel, and C. Rolfs, *Nucl. Instrum. Meth. Phys. Res* **177**, 295 (1980).

Direct measurement of ^{8}Li + d reactions of astrophysical interest

T. Hashimoto*, H. Ishiyama†, Y. X. Watanabe†, Y. Hirayama†, N. Imai†, H. Miyatake†,*, S. C. Jeong†, M.-H. Tanaka†, N. Yoshikawa†, T. Nomura†, S. Mitsuoka*, K. Nishio*, T. K. Sato*, S. Ichikawa*, H. Ikezoe*, A. Osa**, M. Matsuda**, S. K. Das‡, Y. Mizoi‡, T. Fukuda‡, A. Sato§ and T. Shimoda§

*Advanced Science Research Center, JAEA, Tokai, Ibaraki, 319-1195, Japan
†IPNS, KEK, Tsukuba, Ibaraki, 305-0801, Japan
**Tokai Research and Development Center, JAEA, Tokai, Ibaraki, 319-1195, Japan
‡Lab. of Phys. , Osaka Electro-Communication University, Neyagawa, Osaka 572-8530, Japan
§Dept. of Phys., Osaka University, Toyonaka, Osaka 560-0043, Japan

Abstract. The excitation functions of the ^{8}Li(d, t)^{7}Li and the ^{8}Li(d, p)^{9}Li reactions were directly measured using ^{8}Li beams with E_{cm} = 0.3, 0.4, 0.5, 0.7, 0.8, 1.0, 1.1, and 1.2 MeV irradiating CD_2 targets. These beam energies covered the Gamow peaks for 1 - 3 $\times$ 10^9 K. Large cross sections were observed at around E_{cm} = 0.8 MeV for the ^{8}Li(d, t)^{7}Li and E_{cm} = 1.0 MeV for the ^{8}Li(d, p)^{9}Li, implying the existence of a resonant state of ^{10}Be at the excitation energy of around 22.4 MeV. Enhanced with the present work, the astrophysical reaction rates of the ^{8}Li(d, t)^{7}Li and the ^{8}Li(d, p)^{9}Li reactions are higher by one and two orders of magnitude than the previously evaluated ones, respectively.

Keywords: Nuclear astrophysics; Nucleosynthesis; Radioactive nuclear beams
PACS: 25.10.+s; 25.60.-f; 26.30.+k

INTRODUCTION

In recent decades, nuclear reactions involving the unstable nucleus ^{8}Li are thought to be key reactions to synthesize heavier elements under the high temperatures T_9 = 1 - 3 (T_9 = 10^9 K) and neutron-rich environments in the universe [1, 2]. Continuous efforts have been made to experimentally determine the reaction rates of the nuclear reactions induced by ^{8}Li [3, 4, 5, 6, 7, 8, 9, 10, 11, 12, 13]. Among such reactions, the ^{8}Li(d, t) and (d, α) reactions destruct ^{8}Li nuclei, and the ^{8}Li(d, p) reaction creates heavy elements through ^{8}Li nuclei. Although direct cross-section measurements of these reactions have been performed [12, 13], the measured energy regions were higher than the Gamow peaks for T_9 = 1 - 3.

The excitation function for the light-mass system in the energy region of astrophysical interest is often dominated by isolated resonant states of the compound system, since the level density of the compound nucleus is not so high. If there are one or more unknown resonances in the energy region of interest, the extrapolation of the measured cross sections to in lower energies is not feasible.

The low-energy cross sections of the ^{8}Li(d, t), (d, α), and (d, p) reactions can be enhanced by the highly excited states of ^{10}Be which have recently been observed [14, 15]. In this paper, we present the measured reaction rates of the ^{8}Li(d, t)^{7}Li reaction in

CP1016, *Origin of Matter and Evolution of Galaxies*,
edited by T. Suda, T. Nozawa, A. Ohnishi, K. Kato, M. Y. Fujimoto, T. Kajino, and S. Kubono

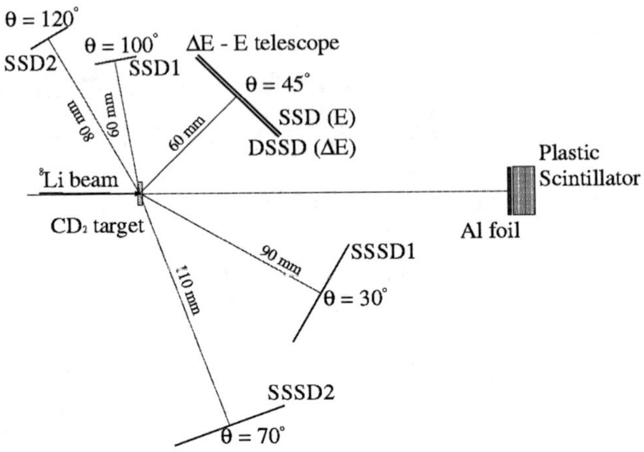

FIGURE 1. Top view of the detector setup. The recoiled protons and tritons were measured by the ΔE - E telescope and SSD1 and SSD2. The scattered ^{7}Li and ^{9}Li were measured by the SSSD1 and SSSD2 in coincidence with the SSD1 and SSD2. The beam dose was measured by the plastic scintillator. The ΔE - E telescope consisted of the DSSD (ΔE) and SSD (E).

the temperature region of T_9 = 1 - 3. In addition, we present the preliminary result of ^{8}Li(d, p)^{9}Li reaction in the same temperature.

EXPERIMENT

The ^{8}Li beams of 7 different energies, E = 0.18, 0.24, 0.30, 0.40, 0.46, 0.60, and 0.75 MeV/nucleon, were provided by Tokai Radioactive Ion Accelerator Complex (TRIAC) [16]. These energies correspond to E_{cm} = 0.3, 0.4, 0.5, 0.7, 0.8, 1.0, and 1.1 MeV, respectively. In addition, the ^{8}Li beam of E = 0.85 MeV/nucleon (E_{cm} = 1.2 MeV) was provided by a recoil mass separator (RMS) [17] used as an in-flight secondary-beam separator [18] at Japan Atomic Energy Agency (JAEA).

In the case of experiment at TRIAC, the typical intensity, energy resolution, and diameter of the ^{8}Li beam were 1×10^{5} particles per second (pps), 0.85% in 1σ and 3.4 mm in 1σ, respectively. For the beam provided by the RMS, those parameters were 6×10^{3} pps, 4.3% in 1σ, and 2.6 mm in 1σ, respectively.

Deuterated polyethylene (CD_2) was employed as the deuteron target. The thicknesses were 130 μg/cm^2 and 250 μg/cm^2 for the experiments at TRIAC and at the RMS, respectively.

The detector system was designed in order to measure the total energy and scattered angle of emitted particles. This system was composed of a ΔE - E silicon telescope, two solid state silicon detectors (SSD1 and SSD2), two single-sided silicon strip detectors (SSSD1 and SSSD2) and a plastic scintillator, as shown in Fig. 1.

The ΔE - E telescope was placed to cover backward angles in the center-of-mass (cm) system and to perform identification of light particles. The active area of the ΔE - E telescope was 50×50 mm^2. The telescope consisted of a double-sided silicon strip

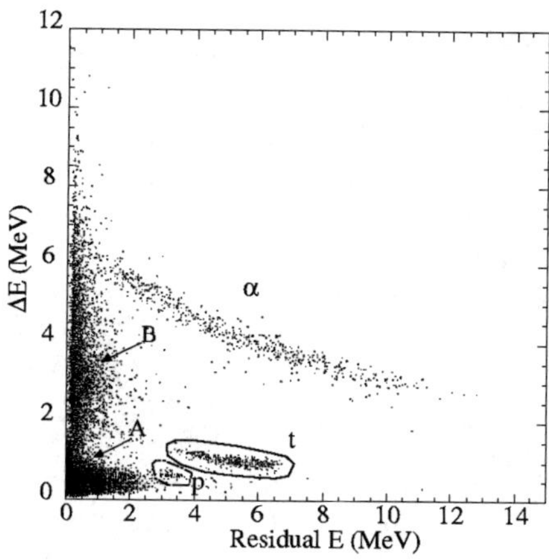

FIGURE 2. Typical ΔE - E plots at E_{cm} = 1.0 MeV. Detected α-particles, tritons and protons are indicated as 'α', 't' and 'p', respectively. The area surrounded by solid lines and dash-dotted lines denote the gate accepting for tritons and protons, respectively. The group 'A' and 'B' are background events (See text).

detector (DSSD) of 43 μm thickness and a 375 μm thick SSD. The ΔE detector had 16 horizontal and 16 vertical strips on the front and the rear side, respectively.

At the forward angles, the event selection was performed by coincidence measurement of a triton and a ^{7}Li particle. The two SSDs were placed upstream of the target to measure the triton energy. The SSSDs were placed on the opposite side of the target to the SSDs to measure the energy and scattered angle of the ^{7}Li particle. The size of each SSD and each SSSD were 18 × 18 mm^2 and 50 × 50 mm^2, respectively.

The beam intensity was monitored by a plastic scintillator during the measurement. The plastic scintillator was covered by a 50 μm thick aluminum plate and placed downstream of the target. Since the ^{8}Li particles and α particles fed from ^{8}Li were stopped in the aluminum foil, the scintillator detected only β rays fed from ^{8}Li. The total dose of the ^{8}Li beam was determined from the yield of β rays.

ANALYSIS

Figure 2 shows a typical ΔE - E plot obtained by the ΔE - E telescope. Protons, tritons, and α particles were separated clearly. Deuterons were not observed in this figure, since they were stopped in the ΔE detector; the kinetic energy of the elastically scattered deuteron is less than 2.6 MeV and the deuteron did not pass through the ΔE detector. In this figure, two groups of background were also observed, which were labeled as 'A' and 'B'. The group 'A' is β-rays emitted from ^{8}Li. The group 'B' is 2α particles in coincidence with β-rays fed from ^{8}Li, which was elastically scattered by ^{12}C in the CD_2 target and was stopped in the ΔE detector.

The tritons surrounded by the solid lines in Fig. 2 were selected. As for t - ^{7}Li coincidence events detected by the SSDs and SSSDs, we required two kinematical conditions of the ^{8}Li(d, t)^{7}Li reaction. Firstly, we selected SSDs and SSSDs events, in which the measured triton and ^{7}Li energies were consistent with the kinematics of the reaction. For example, in the case at E_{cm} = 1.0 MeV, the triton energies measured by SSDs should be 0.9 to 2.0 MeV,. The corresponding ^{7}Li energy measured by SSSDs should be 5.0 to 7.5 MeV. Secondly, the momentum conservation must hold in the lab system; parallel and transverse components of the particle (triton, ^{7,8}Li) momenta with respect to the beam direction should be conserved. Therefore the momentum deviation from the conservation law was used for selecting the reaction events. In the case of E_{cm} = 1.0 MeV, the transverse momenta and the parallel momentum deviations were restricted within $\pm$40 MeV/c and $\pm$25 MeV/c, respectively, by the taking into account the detector energy and angular resolutions.

For the analysis of of ^{8}Li(d, p)^{9}Li reaction, the protons surrounded by the dash-dotted lines in Fig. 2 were selected. In the low-energy region less than E_{cm} = 0.7 MeV, the analysis was not able to be performed, since the proton events overlapped with the 'A' background events. The analysis of ^{8}Li(d, α)^{6}He reaction is in progress.

RESULT

The excitation function of the ^{8}Li(d, t)^{7}Li reaction is shown in the left panel of Fig. 3 together with the previous results [13]. The present experiment provides new data in the low-energy region below E_{cm} = 1.5 MeV and covers the Gamow peaks corresponding to T_9 = 1 - 3. The present cross section at E_{cm} = 1.2 MeV is almost the same value as the previous result at E_{cm} = 1.5 MeV [13]. An enhancement in the excitation function of the ^{8}Li(d, t)^{7}Li reaction at around E_{cm} = 0.8 MeV was observed. The angular distributions of triton measured at various E_{cm} are shown in Fig. 4. Since all the angular distributions

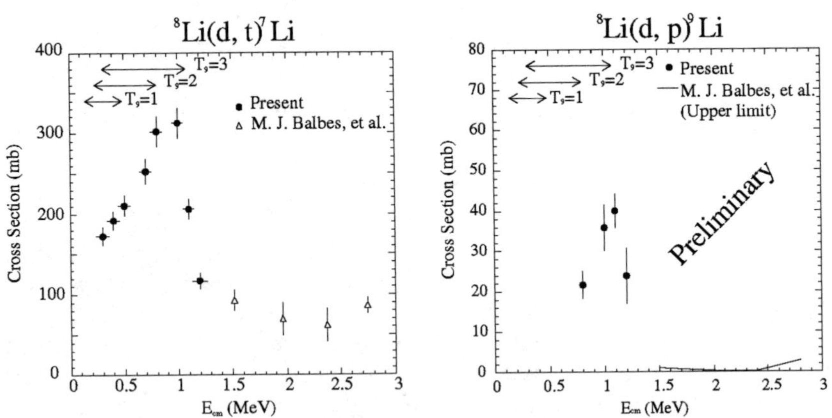

FIGURE 3. The excitation functions of the ^{8}Li(d, t)^{7}Li (right) and the ^{8}Li(d, p)^{9}Li reactions (left). The closed circles show the present results and the open triangles and solid line show the previous results [13]. The inset figure of right side figure shows the ratio of the partial cross section to the first excited state of ^{7}Li to the total cross section. Energy regions of astrophysical interest are indicated by the arrows.

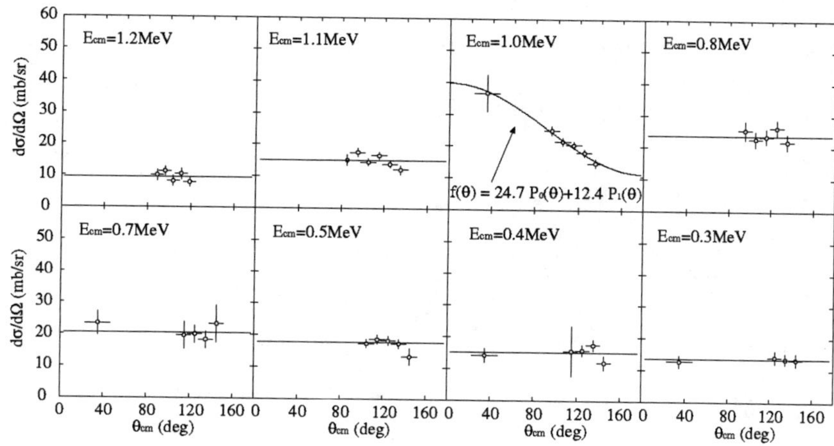

FIGURE 4. Angular distributions of the $^8Li(d, t)^7Li$ reaction at eight reaction energies. The horizontal axis indicates the scattering angle in cm system, and the vertical axis is the differential cross section in units of mb/sr. The solid curve at E_{cm} = 1.0 MeV is the result of fitting with Legendre polynomials, while the solid lines at other energies indicate weighted mean values.

except that at E_{cm} = 1.0 MeV were isotropic, the compound reaction process seems to dominate the enhancement at around E_{cm} = 0.8 MeV. This energy corresponds to the excitation energy of 22.4 MeV in ^{10}Be, where an excited state and triton decaying from that state were observed [14, 15].

The preliminary excitation function of the $^8Li(d, p)^9Li$ reaction is shown in the right panel of Fig. 3. The cross sections were obtained by using the data of ΔE-E telescope only and assuming the isotropic angular distributions. We could clearly see a resonance-like structure at around E_{cm} = 1.0 MeV, which may correspond to the resonance state located at around 22.4 MeV in ^{10}Be [14, 15].

Those resonance-like structures in the excitation functions have strong influence on the reaction rates of the $^8Li(d, t)^7Li$ and the $^8Li(d, p)^9Li$ reactions in the astrophysical environments. The resultant reaction rates are compared with a previous evaluation [13] in Fig. 5. The present $^8Li(d, t)^7Li$ reaction rates were about one order of magnitude higher than the previous rates [13], and the $^8Li(d, p)^9Li$ reaction rates were about two orders of magnitude higher than the the previous rates [13]. These enhancements are due to the observed resonance-like structures. The effect of the present results in the network calculation of explosive nucleosynthesis, for example, in the inhomogeneous Big Bang and in supernova explosions, will be studied under the collaboration with theorists.

SUMMARY

We have performed the direct measurement of the $^8Li + d$ reactions at TRIAC. The excitation functions of $^8Li(d, t)^7Li$ and $^8Li(d, p)^9Li$ reactions were measured in the energy regions of E_{cm} = 0.3 - 1.2 MeV and E_{cm} = 0.7 - 1.1 MeV, respectively. Two resonance-like structures were observed at E_{cm} = 0.8 MeV in the $^8Li(d, t)^7Li$ excitation function

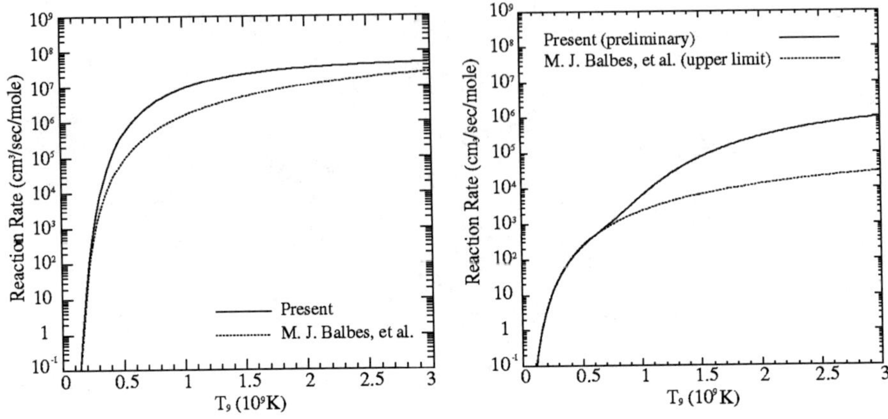

FIGURE 5. Reaction rates of the ^{8}Li(d, t)^{7}Li and the ^{8}Li(d, p)^{9}Li reactions as a function of temperature. The horizontal axis indicates the temperature T_9 (10^9 K), and the vertical axis is the reaction rate in units of cm^3/s/mole. The solid curves indicates the present result and the dashed curves gives the result of Balbes et al. [13].

and at E_{cm} = 1.0 MeV in the ^{8}Li(d, p)^{9}Li excitation function. Both of the resonance-like structures seem to correspond to the same excited state located at 22.4 MeV in ^{10}Be and make strong influences on the astrophysical reaction rates.

REFERENCES

1. Kajino, T. et al., *Astrophys. J.* **359**, 267 (1990).
2. Terasawa, M. et al., *Astrophys. J.* **562**, 470 (2001).
3. Paradellis, T. et al., *Z. Phys.* **A 337**, 211 (1990).
4. Boyd, R. N. et al., *Phys. Rev. Lett.* **68**, 1283 (1992).
5. Gu, X. et al., *Phys. Lett.* **B 243**, 31 (1995).
6. Mizoi, Y. et al., *Phys. Rev.* **C 62** 065801 (2000).
7. Cherubini, S. et al., *Eur. Phys. J.* **A 20** 355 (2004).
8. Ishiyama, H. et al., *Phys. Lett.* **B 640** 82 (2006).
9. Kobayashi, H. et al., *Phys. Rev.* **C 67** 015806 (2003).
10. Li, Z. H. et al., *Phys. Rev.* **C 71** 052801(R) (2005).
11. Becchetti, F. D. et al., *Nucl. Phys.* **A 584** 507 (1992).
12. Şahin, L. et al., *Phys. Rev.* **C 65** 038801 (2002).
13. Balbes, M. J. et al., *Nucl. Phys.* **A 584** 315 (1995).
14. Nelson R. O. and Michaudon A. *Nucl. Sci. Eng.* **140** 195 (2002).
15. Fletcher, N. R. et al., *Phys. Rev.* **C 68** 024316 (2003).
16. Miyatake, H. et al., *Nucl. Instrum. Methods.* **B 204** 746 (2003).
17. Ikezoe, H. et al., *Nucl. Imstrum. Methods.* **A 376** 420 (1996).
18. Ishiyama, H. et al., *Nucl. Instrum. Methods* **A 560** 366 (2006).

13. WEAK INTERACTION and NEUTRINO PHYSICS

Neutrino-Nucleus Reactions and Nucleosynthesis

Toshio Suzuki*, Satoshi Chiba†, Takashi Yoshida**, Michio Honma‡,
Koji Higashiyama§, Hideyuki Umeda¶, Ken'ichi Nomoto¶,||,
Toshitaka Kajino††,** and Takaharu Otsuka‡‡

*Department of Physics, College of Humanities and Sciences, Nihon University, Sakurajosui 3-25-40, Setagaya-ku, Tokyo 156-8550, Japan,
Center for Nuclear Study, University of Tokyo, Hirosawa, Wako-shi, Saitama, 351-0198, Japan
†Advanced Science Research Center, Japan Atomic Energy Agency, 2-4 Shirakata-Shirane, Tokai, Naka-gun, Ibaraki 319-1195, Japan
**National Astronomical Observatory of Japan, Mitaka, Tokyo 181-8588, Japan
‡Center for Mathematical Sciences, University of Aizu, Aizu-Wakamatsu, Fukushima, Japan
§Department of Physics, Chiba Institute of Technology, Narashino, Chiba 275-0023, Japan
¶Department of Astronomy, School of Science, University of Tokyo, Tokyo 113-0033, Japan
||Institute for the Physics and Mathematics of the Universe, University of Tokyo, Chiba 277-8582, Japan
††Department of Astronomy, Graduate School of Science, University of Tokyo, Hongo, Bunkyo-ku, Tokyo 113-0033, Japan
‡‡Department of Physics and Center for Nuclear Study, University of Tokyo, Hongo, Bunkyo-ku, Tokyo 113-0033, Japan
RIKEN, Hirosawa, Wako-shi, Saitama 351-0198, Japan

Abstract. Neutrino-induced reactions on ^{12}C, ^{4}He as well as Fe and Ni isotopes are studied based on new shell model Hamiltonians for *p*-shell and *fp*-shell. Gamow-Teller and spin-dipole transitions are investigated, and applied to neutrino-nucleus reactions induced by both DAR and supernova neutrinos. The reaction cross sections are found to be enhanced compared with conventional Hamiltonians as well as previous calculations. The production yields of ^{7}Li and ^{11}B during supernova explosions are found to be enhanced, and the effects of neutrino oscillations and implications of the enhancement on the constraint on temperature for $\nu_{\mu,\tau}$ and $\bar{\nu}_{\mu,\tau}$ are discussed. Production of other light elements such as ^{10}Be and ^{10}B by neutrino processes is also discussed. Neutral current reactions on Ni and Fe isotopes induced by supernova neutrinos are investigated. Effects of neutrino-induced reactions on the production yields of heavy elements such as Mn are discussed.

Keywords: shell model, neutrino-nucleus reactions, nucleosynthesis
PACS: 21.60.Cs, 25.30.Pt, 24.60.Dr, 26.30.Jk

NEUTRINO-INDUCED REACTIONS ON ^{12}C AND ^{4}HE

Recently, new shell model Hamiltonians for *p*-shell nuclei were obtained[1, 2] based on the Cohen-Kurath[3] and Millener-Kurath[4] interactions. A new Hamiltonian, SFO[2], properly takes into account important roles of spin-isospin interactions, especially, the tensor interaction. Systematic improvements in the descriptions of the magnetic moments of the *p*-shell nuclei as well as Gamow-Teller (GT) transitions, for example, in ^{12}C and ^{14}N have been obtained[2]. The tensor components of the interaction have proper character consistent with those of the $\pi+\rho$ exchange potentials, that is, attractive between $j_> = \ell+\frac{1}{2}$ and $j_< = \ell-\frac{1}{2}$ orbits while repulsive between $j_>$ and $j_>$ ($j_<$ and $j_<$) orbits. This property is quite important for the shell evolution, namely, the change of

CP1016, *Origin of Matter and Evolution of Galaxies*,
edited by T. Suda, T. Nozawa, A. Ohnishi, K. Kato, M. Y. Fujimoto, T. Kajino, and S. Kubono

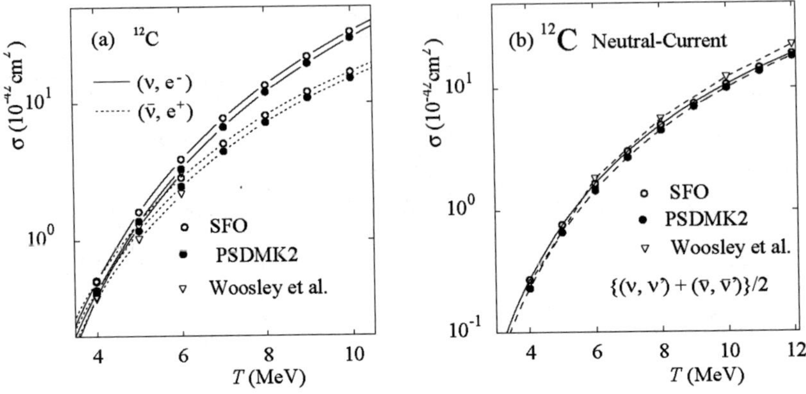

FIGURE 1. Calculated (a) charge-exchange and (b) neutral current reaction cross sections for ^{12}C induced by supernova neutrinos with temperature T obtained by shell model calculations with the use of the SFO and PSDMK2 Hamiltonians. Previous results in ref. [11] are also shown for comparison.

the shell structure toward drip-lines[5, 6]. These improvements in the Hamiltonian are essential for the study of neutrino-nucleus reactions, which are induced mainly by GT and spin-dipole transitions.

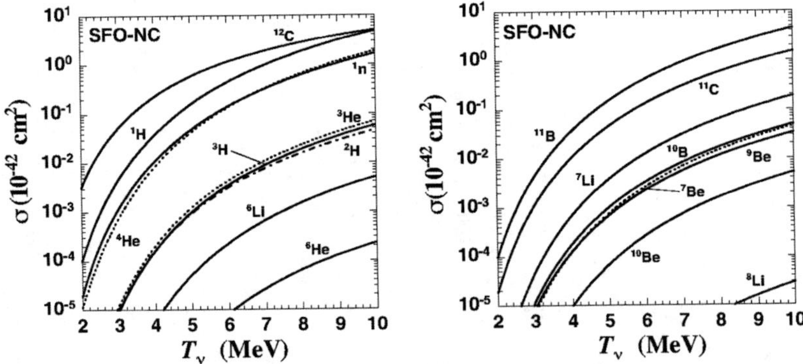

FIGURE 2. Calculated neutral current reaction cross sections for ^{12}C induced by supernova neutrinos with temperature T_ν obtained by shell model calculations with the use of the SFO Hamiltonian. Knocked-out particles and residual nuclei are denoted in the left and right figures, respectively.

We study neutrino-induced reactions on ^{12}C based on the new improved shell model Hamiltonian, SFO[2]. Experimental cross section data are available for ^{12}C. The exclusive charge-exchange and neutral current reaction cross sections for ^{12}C induced by the

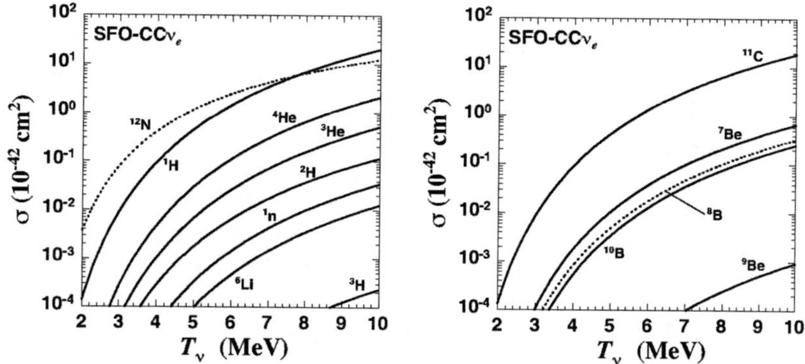

FIGURE 3. Calculated reaction cross sections for ^{12}C $(\nu, e^- \text{ X})$ B induced by supernova neutrinos with temperature T_ν obtained by shell model calculations with the use of the SFO Hamiltonian. Knocked-out particles X and residual nuclei B are denoted in the left and right figures, respectively.

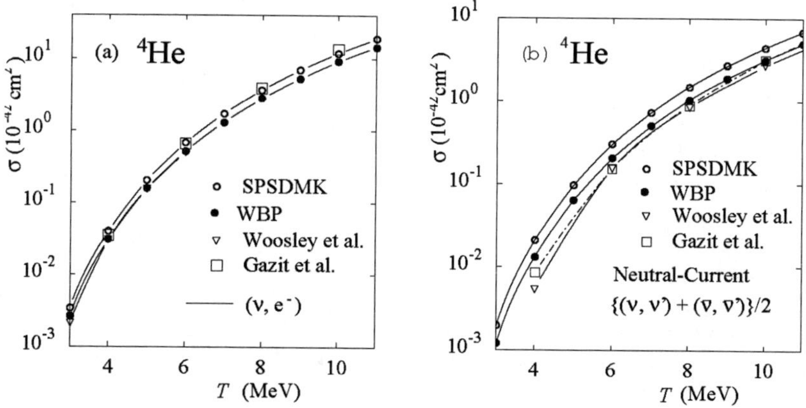

FIGURE 4. Calculated charge-exchange (a) and neutral current (b) reaction cross sections for ^{4}He induced by supernova neutrinos with temperature T_ν obtained by shell model calculations with the use of the WBP and SPSDMK Hamiltonians. Previous results in ref. [11] and recent ones in ref. [14] are also shown for comparison.

GT transition are well reproduced by the SFO Hamiltonian for the DAR neutrinos[6]. Calculated charge-exchange cross section, $\sigma = 9.06\times10^{-42}$ cm^2, is consistent with the observed one, $\sigma = 8.9\pm0.3\pm0.9\times10^{-42}$ cm^2 [7]. For the neutral current reactions, calculated cross sections, $\sigma = 9.76$ and 2.68×10^{-42} cm^2, are also consistent with the observed values, $\sigma = 10.4\pm1.0\pm0.9$ and $3.2\pm0.5\pm0.4\times10^{-42}$ cm^2, for (ν_e, ν_e') +

($\bar{\nu}_\mu$, $\bar{\nu}_\mu$') and (ν_μ, ν_μ') reactions, respectively[8]. Here, the axial-vector coupling constant is a little bit reduced , g_A^{eff}/g_A^{free} = 0.95, to reproduce the exclusive GT transition strength[9]. The inclusive charge-exchange reaction is also well reproduced with a quenching for the axial-vector coupling constant, g_A^{eff}/g_A^{free} = 0.70 (0.90) for the 2^- (0^-, 1^- etc.) states, which is consistent with the electron scattering data [10]. A rather large quenching for the 2^- states is needed to reproduce both the M2 form factor in ^{12}C and the inclusive neutrino-induced reaction cross section. Calculated inclusive reaction cross section, $\sigma = 13.5\times10^{-42}$ cm^2, is consistent with the observed value, $\sigma = 13.2\pm0.5\pm1.3\times10^{-42}$ cm^2[7].

Reactions induced by supernova neutrinos have been investigated, and the cross sections are found to be enhanced compared with those by the conventional shell model Hamiltonians (e. g. PSDMK2[4]) as well as previous calculations[11] as shown in Fig. 1. Reaction cross sections for the γ decay, proton, neutron, deuteron, ^{3}He, t and α knock-out processes including multi-particle knock-out processes are shown in Fig. 2 and 3 for supernova neutrinos with temperature T_ν. The branching ratios for the processes have been calculated by the Hauser-Feshbach theory.

Now, we study neutrino-induced reaction on ^{4}He as they play important roles in the nucleosynthesis of light elements during supernova explosions. Calculated cross sections for neutrino-induced reactions on ^{4}He are shown in Fig. 4 for the WBP[12] and the SPSDMK[4, 13] Hamiltonians. Dominant contributions come from spin-dipole transitions. Calculated results are found to be enhanced compared with the previous calculations[11]. They are also enhanced compared with the recent ab-initio calculations with AV8' interaction[14] for the neutral current reactions. In case of the charge-exchange reactions, cross sections by ref. [14] are larger than those by the shell model calculations except for the SPSDMK Hamiltonian at $T \leq 6$ MeV. The cross sections in ref. [14] have steeper temperature dependence than those by the shell model cases. This is due to the more spreading of the spin-dipole strength in ref. [14] than the shell model cases. When we take into account the effects of the spreading, the temperature dependence of the cross sections gets closer to those in ref. [14]. In case of ^{4}He, proton and neutron knock-out processes are the dominant ones for both the charge-exchange and the neutral current reactions.

SYNTHESIS OF LIGHT ELEMENTS

Nucleosynthesis path of light elements ^{7}Li and ^{11}B during supernova explosions is shown in Fig. 5. The reactions ^{4}He (ν, ν'p) ^{3}H is important for the production of ^{7}Li as well as ^{11}B through ^{7}Li (α, γ) ^{11}B in the He/C layer. The reaction ^{12}C (ν, ν'p) ^{11}B is important for the production of ^{11}B in the O/Ne and O/C layers. The enhancement of the ν-^{4}He and ν-^{12}C reaction cross sections is found to lead to the increase of the production yields of ^{7}Li and ^{11}B during supernova explosions [6] about by 30% and 20% (80% and 50%), respectively, in case of WBP+SFO (SPSDMK+PSDMK2) compared with those for the previous case[16].

The enhancement of the production yields of ^{11}B and ^{7}Li leads to a slight decrease of possible $\nu_{\mu,\tau}$ and $\bar{\nu}_{\mu,\tau}$ temperature in supernovae. From the estimates of the production

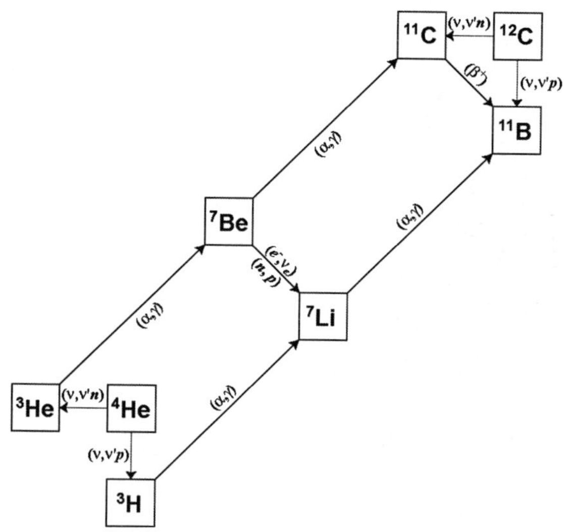

FIGURE 5. Nucleosynthesis path of light elements ^{7}Li and ^{11}B during supernova explosions.

yield of ^{11}B in supernovae in galactic chemical evolution models and total neutrino energy produced by supernova explosions, constraints on the neutrino temperatue, 4.5 MeV $\leq T_{\nu_{\mu,\tau}} \leq$ 6.4 MeV and 4.4 MeV $\leq T_{\nu_{\mu,\tau}} \leq$ 6.1 MeV are obtained for WBP+SFO and SPSDMK+PSDMK2 cases , respectively. These values are a bit smaller than a constraint, 4.8 MeV $\leq T_{\nu_{\mu,\tau}} \leq$ 6.6 MeV, previously obtained[15] with the use of the cross sections[11].

When there are neutrino oscillations, charge-exchange reactions on ^{4}He and ^{12}C become important and ^{7}Li and ^{11}B are produced through the production of ^{7}Be and ^{11}C. The yield ratio $N(^7\text{Li})/N(^{11}\text{B})$ becomes sensitive to the mixing angle θ_{13} for the normal neutrino mass hierarchy, which can be used to determine the lower limit of θ_{13} as well as if the neutrino mass hierarchy is normal or inverted[16].

Besides ^{7}Li and ^{11}B, light elements such as ^{10}Be, ^{10}B, ^{9}Be and ^{6}Li are also produced by neutrino processes as wee see from Figs. 2 and 3. ^{10}Be is mainly produced through ^{12}C (ν, $\nu' x$) ^{10}Be in both the O-rich and He/C layers. Charge-exchange reactions ^{12}C ($\bar{\nu}_e, e^+ x$) ^{10}Be are also important if the temperature of $\bar{\nu}_e$ is large. ^{10}B is produced mainly by ^{12}C ($\nu, \nu' x$) ^{10}B in the O-rich layer. Most of ^{9}Be is produced through ^{12}C ($\nu, \nu' x$) ^{9}Be [17].

NEUTRINO-INDUCED REACTIONS ON FE AND NI ISOTOPES

Charge-Exchange Reaction on ^{56}Fe

Neutrino-induced reaction on Fe and Ni isotopes have been studied with the use of a new shell-model Hamiltonian for the fp-shell, GXPF1J [18]. The Hamiltonian

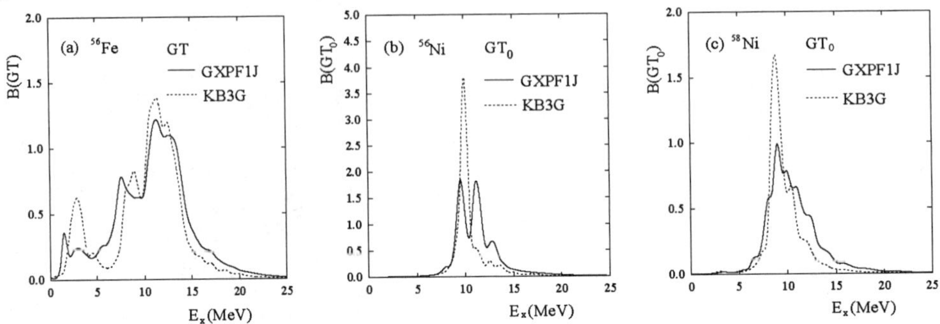

FIGURE 6. (a) Calculated B(GT) strengths for ^{56}Fe obtained by the GXPF1J and KB3G Hamiltonisns. Calculated non-charge-exchange $B(GT_0)$ strengths for (b) ^{56}Ni and (c) ^{58}Ni.

reproduces rather well M1 strength of ^{48}Ca, ^{50}Ti, ^{52}Cr and ^{54}Fe as well as GT strength in ^{58}Ni [19, 18]. The Gamow-Teller strengths in ^{54,56}Fe and $^{56-64}$Ni are found to be more fragmented than those obtained by the KB3G Hamiltonian[20]. The strength for ^{56}Fe is shown in Fig. 6(a). The summed B(GT) value for ^{56}Fe with a universal quenching factor of 0.74 is 9.47 and 8.86 for the GXPF1J and KB3G Hamiltonian, respectively, which are close to the observed value, B(GT) = 9.9±2.4[21]. Reaction cross section for ^{56}Fe (ν, e^-) ^{56}Co induced by the DAR neutrinos is measured. Calculated cross section for the GT transitions, $\sigma = 1.60\ (1.56) \times 10^{-40}$cm^2 for the GXPF1J (KB3G) case, gives about half of the observed value, $\sigma = 2.56 \pm 1.08 \pm 0.43 \times 10^{-40}$ cm^2[22].

Neutral Current Reactions on Ni and Fe Isotopes

Cross sections induced by the supernova neutrinos are also studied for the Gamow-Teller transitions. In case of neutral-current reactions, the fragmentation of GT_0 (T_f = T, T+1) strengths for the GXPF1J Hamiltonian gets more pronounced compared with the KB3G one than the charge-exchange cases. Strengths for ^{56}Ni and ^{58}Ni are shown in Figs. 6(b) and 6(c), respectively. The neutral-current reaction cross sections are found to be enhanced for the GXPF1J compared with the KB3G, especially for high ν temperatures as shown in Fig. 7.

Nucleosynthesis of Mn in Population III Supernovae

Now, we study particle knock-out processes in neutral current reactions on ^{56}Ni, and discuss production of ^{55}Mn through a path, ^{56}Ni $(\nu, \nu' p)$ ^{55}Co (e^-, ν_e) ^{55}Fe (e^-, ν_e) ^{55}Mn in supernova explosions. Calculated reaction cross sections are shown in Fig. 8

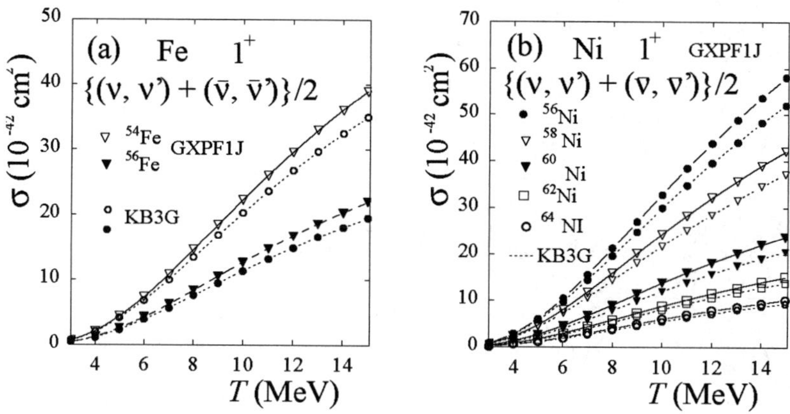

FIGURE 7. Neutral current reaction cross sections for (a) Fe and (b) Ni isotopes induced by supernova neutrinos with temperature T. Calculated results obtained by the GXPF1J and KB3G Hamiltonians are shown.

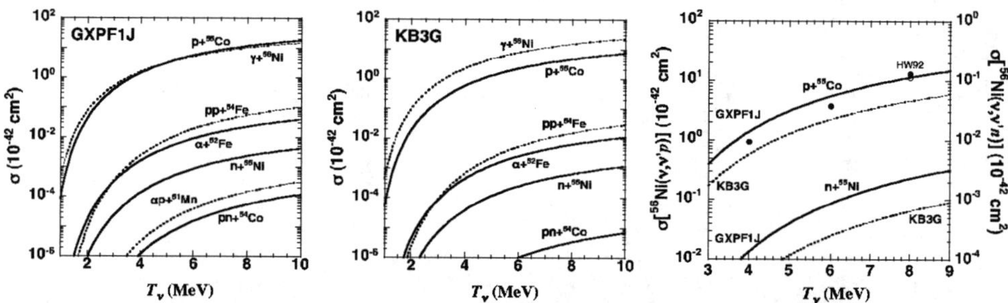

FIGURE 8. Neutral current reaction cross sections for ^{56}Ni induced by supernova neutrinos with temperature T_ν obtained by the GXPF1J (left) and KB3G (middle) Hamiltonisnas. Knocked-out particles and residual nuclei are denoted. Nucleon knock-out cross sections are compared with those of ref.[23] which are denoted as HW92 (right).

for both GXPF1J and KB3G Hamiltonians and compared with previous nucleon knock-out cross sections[23]. Proton knock-out cross sections are found to be much enhanced for the GXPF1J case. This is due to the fact that the strength is more fragmented and a large fraction of the strength is found in the proton emission channel ($E_x \geq 10$ MeV) in case of the GXPF1J Hamiltonian (see Fig. 6(b)). Note that the transition to the ground state of ^{55}Co (7/2$^-$) at E_x = 7.2 MeV is hindered due to its f-wave nature.

Production yields of elements during supernova explosions for a population III star with mass 15 $M_\odot$[24] are investigated. Total ν energy of $E_\nu = 3\times10^{53}$ ergs and ν temperatures of $(T_{\nu_e}, T_{\bar{\nu}_e}, T_{\nu_{\mu,\tau}}) = (4, 4, 6)$ MeV are assumed. Cross sections of ref. [23] are used except for neutral current reactions on ^{56}Ni. Enhancement of the production yield of ^{55}Mn is obtained when the neutrino processes are included as shown in Table 1. The yield for the GXPF1J Hamiltonian is enhanced compared with those for other Hamiltonians.

TABLE 1. Production yields of Mn as well as the logarithmic value of the yield ratio over Fe relative to the solar abundances [Mn/Fe] in a supernova explosion model of a population III star with 15 $M_\odot$

Model	GXPF1J	GXPF1J × 2	KB3G	KB3G × 2	HW02	No ν
$M(^{55}\mathrm{Co})/(10^{-4}M_\odot)$	4.16	5.00	3.56	4.00	3.80	2.29
$M(\mathrm{Mn})/(10^{-4}M_\odot)$	4.31	5.16	3.72	4.16	3.96	2.30
[Mn/Fe]	-0.25	-0.17	-0.32	-0.27	-0.29	-0.53

REFERENCES

1. T. Otsuka, R. Fujimoto, Y. Utsuno, B. A. Brown, M. Honma and T. Mizusaki, *Phys. Rev. Lett* **87**, 082502 (2001).
2. T. Suzuki, R. Fujimoto and T. Otsuka, *Phys. Rev.* C **67**, 044302 (2003).
3. S. Cohen and D. Kurath, *Nucl. Phys.* **73**, 1 (1965).
4. D.J. Millener and D. Kurath, *Nucl. Phys.* **A255**, 315 (1975).
5. T. Otsuka, T. Suzuki, R. Fujimoto, R. Grawe and Y Akaishi, *Phys. Rev. Lett* **95**, 232502 (2005).
6. T. Suzuki, S. Chiba, T. Yoshida, T. Kajino and T. Otsuka, *Phys. Rev.* C **74**, 034307 (2006).
7. L. B. Auerbach *et al.* (LSND Collaborations), *Phys. Rev.* C **64**, 065501 (2001).
8. B. E. Bodmann *et al.* (KARMEN Collaborations), *Phys. Lett.* **B332**, 251 (1994);
   B. A. Armbruster *et al.*, (KARMEN Collaboration), *Phys. Lett.* **B423**, 15 (1998).
9. D. H. Wilkinson and B. E. F. Macefield, *Nucl. Phys.* **A232**, 58 (1974).
10. T. E. Drake, E. L. Tomusiak and H. S. Caplan, *Nucl. Phys.* **A118**, 138 (1968);
    A. Yamaguchi, T. Terasawa, K. Nakahara and Y. Torizuka, *Phys. Rec.* C **3**, 1750 (1971);
    C. Gaarde *et al.*, *Nucl. Phys.* **A422**, 189 (1984).
11. S. E. Woosley, D. H. Hartmann, R. D. Foffman and W. C. Haxton, *Astrophys. J.* **356**, 272 (1990).
12. S. E. Warburton and B. A. Brown, *Phys. Rev.* C **46**, 923 (1992).
13. OXBASH, B. A. Brown, A. Etchegoyen and W. D. M. Rae, *MSU Cyclotron Laboratory Report* No. 524 (1986).
14. D. Gazit and N. Barnea, *Phys. Rev.* C **70**, 048801 (2004); *Phys. Rev. Lett* **98**, 192501 (2007).
15. T. Yoshida, T. Kajino and D. H. Hartmann, *Phys. Rev. Lett* **94**, 231101 (2005).
16. T. Yoshida *et al.*, *Phys. Rev. Lett* **96**, 091101 (2006); *Astrophys. J.* **649**, 319 (2006).
17. T. Yoshida, T. Suzuki, S. Chiba, T. Kajino, H. Yokomakura, K. Kimura, A. Takamura and D. H. Hartmann, to be submitted to Astrophys. J.
18. M. Honma *et al.*, *Phys. Rev.* C **65**, 061301 (2002); ibid. C **69**, 034335 (2004); *Journal of Physics: Conference Series* **20**, 7 (2005).
19. Y. Fujita *et al.*, *Eur. Phys. J* A **13**, 411 (2002).
20. A. Poves, J. Sánchez-Solano, E. Caurier and F. Nowacki, *Nucl. Phys.* **A694**, 157 (2001).
21. J. Rapaport *et al.*, *Phys. Rev.* C **24**, 335 (1981).
22. E. Kolbe, K. Langanke and G. Martinez-Pinedo, *Phys. Rev.* C **60**, 052801 (1999).
23. R. D. Hoffman and S. E. Woosley, Neutrino interaction cross sections and branching ratios, 1992, http://www-phys.llnl.gov/Research/RRSN/nu-csbr/neu-rate.html
24. T. Yoshida, H. Umeda and K. Nomoto, *Astrophys. J* **672**, 1043 (2008).

Neutrino Masses and Mixing

K.Ishikawa and Y.Tobita

Department of Physics, Hokkaido University, Sapporo 060-0810, Japan

Abstract. We report (1) the current status of neutrino parameters and (2) our recent work on implications of particle's coherence, which are weakly related each others. In the first part, current status of the neutrino parameters obtained from oscillation experiments and their prospects are briefly reviewed. From various oscillation experiments, existence of three mass scales have been confirmed. One value of the difference of mass squared is around $10^{-3}eV^2$ and another is around $10^{-5}eV^2$. Although mixing angles are partly found, one important angle, θ_{13} is left unknown.

In the second part, implications of coherence length of particles in the scattering of ultra-high energy cosmic rays (UHCR) with cosmic background radiations (CBR) is discussed. Although coherence length is regarded usually irrelevant to observations, its role is important in several situations of recent experiments which include that of the ultra-high energy charged particles. Here we discuss the scattering of UHCR with CBR.

Keywords: Neutrino parameters,coherence,UHCR,CBR
PACS: 14.60.Pq,13.85.Tp

NEUTRINO MASSES AND MIXING

Three families of quarks and leptons

In the standard model of high energy physics,fundamental interactions of matters are described by non-Ableian gauge theories. Forces are mediated by the gauge fields and interactions of matters with gauge fields are determined by the gauge principle and are understood well. There are three families of matters and the masses and mixing angles are the free parameters which can not be determined uniquely by the gauge principle. The masses and mixing angles are determined actually by the coupling constants with Higgs fields and vacuum expectation values of the Higgs fields. A principle for their coupling is unknown now. They are determined by the experiments.

Experimentally the masses of quarks and charged leptons were known well, but neutrino's masses were unknown for a long time. Recently it was found that the masses of neutrinos are extremely small and are very different from others.

In the present article, we summarize the current status of the neutrino masses and mixing angles, first, and we discuss implications of coherence length of particles, second.

Neutrino oscillations

Neutrino masses and mixing angles are obtained from oscillation experiments. A flavor eigenstate $|\nu_\alpha\rangle$ is written by a linear combinations of mass eigenstates $|\nu_i\rangle$ and

CP1016, *Origin of Matter and Evolution of Galaxies,*
edited by T. Suda, T. Nozawa, A. Ohnishi, K. Kato, M. Y. Fujimoto, T. Kajino, and S. Kubono

a 3×3 unitary matrix $U_{\alpha i}$ as,

$$|\nu_\alpha\rangle = \sum_i U^*_{\alpha i}|\nu_i\rangle. \tag{1}$$

Evolution of the mass eigenstate is given by the proper time τ and its mass m_i as, $|\nu_i(\tau)\rangle = e^{-im_i\tau}|\nu_i(0)\rangle$. The one particle energy is written in the super-relativistic region as $E(p) = p + \frac{m^2}{2p}$ and the above phase factor is expressed by using the time and distance, (t, L), during which the neutrino propagate, as

$$e^{-im\tau} = e^{-i(E(p)t-pL)} = e^{-i\frac{m^2}{2p}L}, t = L, \tag{2}$$

in the unit $c = 1$. So the evolution of flavor eigenstates are given by

$$|\nu_\alpha(L)\rangle = \sum_i U^*_{\alpha i} e^{-i(\frac{m_i^2}{2E})L} U_{i\beta}|\nu_\beta\rangle \tag{3}$$

and the flavor changing transition probability at the distance L is

$$P(\nu_\alpha \to \nu_\beta) = \delta_{\alpha\beta} - 4\sum_{i>j} R(U^*_{\alpha i}U_{\beta i}U_{\alpha j}U^*_{\beta j})\sin^2(\Delta m^2_{ij}(\frac{L}{4E})) + 2\sum Im(U^*_{\alpha i}U_{\beta i}U_{\alpha j}U^*_{\beta j})\sin(2\Delta m^2_{ij}(\frac{L}{4E})) \tag{4}$$

The phase in the above probability is proportional to the ratio between the distance L and the energy E and the value becomes

$$\Delta m^2_{ij}(\frac{L}{4E}) = 1.27\Delta m^2_{ij}(eV^2)\frac{L(km)}{E(GeV)} \tag{5}$$

using the physical units given in the parenthesis. It should be noted that the scale where the above phase becomes order 1 is macroscopic even though the physics is of microscopic. Eq.(4) is the transition probability formula of the neutrino in the vacuum. In the matter, coherent interaction of neutrino with matter gives an additional effect and the oscillation formula is modified. This is MSW effect and is important for the solar neutrino case.

The MNSP matrix U is parameterized in the particle data group (PDG) [1]as

$$U = \begin{pmatrix} c_{13}c_{12} & s_{12}c_{13} & s_{13} \\ -s_{12}c_{23} - s_{23}s_{13}c_{12} & c_{23}c_{12} - s_{23}s_{13}s_{12} & s_{23}c_{13} \\ s_{23}s_{12} - s_{13}c_{23}c_{12} & -s_{23}c_{12} - s_{23}c_{12} - s_{13}s_{12}c_{23} & c_{23}c_{13} \end{pmatrix}$$

where $c_{ij} = \cos\theta_{ij}, s_{ij} = \sin\theta_{ij}$, and $\theta_{12} = \theta_{sol}, \theta_{23} = \theta_{atm}$

To derive the formula Eq.(4), we used the neutrino's one particle wave functions. This is verified because, (1)neutrino interact with matter very weakly, (2)the positions of neutrino production and detection are locally well defined. The second point is connected with the topic which we will discuss in the second part.

Solar neutrino experiments

group	target	unit	experiment	expectation	experiment/SSM
Homstake [3]	Cl^{37}	SNU	2.56±0.23	8.5	0.30±0.03
SAGE [4]	Ga^{71}	SNU	66.9±5.2	131	0.51±0.04
Gallex [5]	Ga^{71}	SNU	69.3±5.5	131	0.53±0.04
SK [6]	water	$106cm^2/s$	2.35±0.08	5.79	0.41±0.02
SNO(pure) [7]	D_2O	$106cm^2/s$	1.76±0.11	5.79	0.30±0.02
SNO(salt) [8]	D_2O	$106cm^2/s$	5.21±0.47	5.79	0.88±0.08

TABLE 1. Summary of solar neutrino experiments

Solar neutrino experiments are summarized in the Table 1. Five experiments from Homstake to SNO(pure) count the electron neutrino events, whereas the SNO(salt) experiment counts the whole neutral current events. Deficit of solar neutrino have been detected in the above five experiments and the ratio between the experimental value and standard solar model(SSM) are less than about 0.5. The last value of the SNO(salt) experiment shows no deficit and the whole flux is consistent with the standard solar model. Thus experiments show that the electron neutrino flux has a deficit but the total neutrino flux has no deficit. The fact that the electron neutrino flux has deficit and the total neutrino flux has no deficit show that the electron neutrino is converted to other neutrinos during propagation from the sun and the reason of deficit is the flavor oscillation.

Atmospheric, solar, accelerator, and reactor experiments

From asymmetry of the atmospheric electron neutrino and muon neutrino events, neutrino oscillations were found by super-K in 1998 [9]. Data have accumulated. Super-K is used also for a long base line experiment, K2K. K2K [10] has observed 108 neutrino events, whereas $150^{11.6}_{-10.0}$ have been expected for no oscillations. A longer base line experiment, MINOS [11], also found a deficit of neutrino flux.

These experiments gave parameters

$$1.4\times 10^{-3}eV^2 \le |m_3^2 - m_1^2| \le 3.3\times 10^{-3}eV^2 \tag{6}$$

$$0.34 \le \sin^2(\theta_{atom}) \le 0.66. \tag{7}$$

Solar neutrino experiments of the previous section and a long distance reactor experiment, KamLAND [12], gave,

$$m_2^2 - m_1^2 = (8.0\pm 0.3)\times 10^{-5}eV^2 \tag{8}$$

$$\sin^2(\theta_{sol}) = 0.30\pm 0.03. \tag{9}$$

Short distance reactor experiment, CHOOZ [13], gave

$$\sin^2(2\theta_{rct}) \le 0.20 \quad \text{for} \quad |m_3^2 - m_1^2| = 2.0\times 10^{-3}eV^2 \tag{10}$$

$$\sin^2(2\theta_{rct}) \le 0.16 \quad \text{for} \quad |m_3^2 - m_1^2| = 2.5\times 10^{-3}eV^2 \tag{11}$$

$$\sin^2(2\theta_{rct}) \le 0.14 \quad \text{for} \quad |m_3^2 - m_1^2| = 3.0\times 10^{-3}eV^2. \tag{12}$$

Global fits

From all data, current 3σ intervals of Δm^2_{ij} and $\sin^2\theta_{ij}$ [2] are given as,

$$\begin{aligned}&\Delta m^2_{21} = 7.1 - 8.3[10^{-5}eV^2],\ \sin^2\theta_{12} = 0.26 - 0.40\\&\Delta m^2_{31} = 2.0 - 2.8[10^{-3}eV^2],\ \sin^2\theta_{23} = 0.34 - 0.67\\&\sin^2\theta_{13} \leq 0.050.\end{aligned} \tag{13}$$

Summary of masses and mixing

Masses and mixing angles

From oscillation experiments, it is confirmed that the neutrinos have masses. Among three masses, two masses are close each other with the difference $\Delta m^2 \approx 8.0 \times 10^{-5}eV^2$ and the third one has a larger difference with $\Delta m^2 \approx 2.5 \times 10^{-3}eV^2$. Concerning the order of three masses, there are two possible patterns. In the first pattern, which is called the standard mass hierarchy, the degenerate two masses are lighter than the other. In the second pattern, which is called the inverted mass hierarchy, the degenerate two masses are heavier than the other. It is the future experiments that decide which hierarchy is realized.

Open problems

Despite of the above successes, there left many open problems concerning neutrino parameters. Some of them are the following:

1.Determination of absolute values of masses.
2.Determination of other component of mixing matrix U.
3.Is neutrino Dirac or Majorana?
4.Is neutrino used for astronomical observation?
5.Other implications and theory.

Many experiments are in progress or under considerations. So, more progresses will be made in the future.

COSMIC BACKGROUND RADIATIONS OF FINITE COHERENCE LENGTH

Implications of coherence length of the particles are studied in the second part. When precision measurements of the neutrino parameters will be made, this problem would become important. Here, instead of the neutrino scattering, we study scattering of ultra-high energy cosmic rays (UHCR) with cosmic background radiation (CBR).

GZK bound

Flux of UHCR beyond 10^{20} eV is expected to be suppressed by its pion production collision with CBR. Observation of UHCR became possible recently by several experiments [14, 15, 16, 17, 18]. A possible signal of UHCR around this energy region has been found [17], although experiments are controversial.

Greisen, Zatsepin and Kuz'min (GZK) [19, 20] predicted that the flux of UHCR is suppressed from the inelastic π production of UHCR with the CBR. The inelastic cross section becomes sizable around the energy $10^{20}eV$ where the Δ resonance contributes. The mean free path beyond this energy becomes the order of 20 Mpc, which is much less than the size of the universe. Hence the UHCR can not propagate long distance in the universe, then. So the flux of cosmic ray should be suppressed beyond $10^{20}eV$, known as GZK bound. For the estimation of the total cross section, the cross section between gamma and nucleon collisions in laboratory frame is used.

Most CBR may have been produced before the decoupling time in the early universe and may have finite coherence length. We assume that CBR has a finite coherence length and study its implications in this article. Because an effect of the finite coherence length is negligible in the ordinary scattering of high energy physics, it has not been taken into account in the previous works on GZK bound. We show, however, that the effect is not negligible in certain situations [21].

Wave packet

Wave packets which have finite spatial extensions are needed actually for asymptotic conditions of scattering processes to be satisfied. In most situations of high energy physics, however, the wave packet effect is negligible. We discuss a counter example, where the wave packet plays important roles, in the second half of the article.

Scattering amplitudes for particles expressed by the wave packets of a finite spatial extension in relativistic field theory have been formulated in Ref.[22]. Wave packet has a finite extension in the coordinate space, in the momentum, and in the energy. Amplitudes give the characteristic behavior of the wave packet scattering. We apply this generalized S-matrix defined by the wave packets for the analysis of UHCR.

Gaussian wave packets, i., e., coherent states, are used for expressing the particles with finite coherence length. Spherically symmetric states in three spatial dimensions are defined as $\langle\vec{x}|\vec{P}_0,\vec{X}_0\rangle = N_3\exp(i\frac{\vec{P}_0}{\hbar}(\vec{x}-\vec{X}_0) - \frac{1}{2\sigma}(\vec{x}-\vec{X}_0)^2), N_3^2 = (\pi\sigma)^{-\frac{3}{2}}, \frac{\vec{P}_0}{\hbar} = \vec{k}_0$ The same states are expressed in the momentum representation as $\langle\vec{p}|\vec{P}_0,\vec{X}_0\rangle = N_3\sigma^{3/2}\exp(-i\vec{p}\cdot\vec{X}_0 - \frac{\sigma}{2}(\vec{P}_0-\vec{p})^2)$ and satisfy the normalization condition in the momentum representation. For the asymmetric wave packets, the asymmetric σ is used.

The set of functions for an arbitrary value of σ satisfy the completeness condition, despite the fact that states of different values of the momenta and coordinates are not orthogonal.

The time evolution of the free wave is determined by the creation and annihilation operators, $a(\vec{p})^\dagger$ and $a(\vec{p})$ and the free Hamiltonian,

$$H_0 = \int d^3pE((\vec{p}))a^\dagger(\vec{p})a(\vec{p}),\ E(\vec{p}) = \sqrt{p^2+m^2} \tag{14}$$

$$[a(\vec{p},t),a^\dagger(\vec{p}',t')]\delta(t-t') = \delta(\vec{p}-\vec{p}')\delta(t-t'). \tag{15}$$

The operator for the wave packets, $A(\vec{P}_0,\vec{X}_0,T_0,t)$,and its conjugate that annihilate and create wave packet states with certain boundary condition in t are defined by linear combinations of the operators $a(\vec{p},t)$ and its conjugate.

The wave packet spreads with time and the spreading velocity in the transverse direction , v_T, and the longitudinal direction, v_L,are given by

$$v_T = \sqrt{\frac{2}{\sigma}}\frac{1}{E(\vec{P}_0)}, v_L = \sqrt{\frac{2}{\sigma}}\frac{m^2}{(E(\vec{P}_0))^3}. \tag{16}$$

The v_T depends on the energy and the v_L depends on the energy and the mass. Massive wave packet spreads in both directions. After a macroscopic time, if any wave packet of the massive particle spreads to huge size, these waves may be treated as plane waves approximately.

For the massless wave, however, wave packet spreads only in the transverse direction, and wave packets do not spread and its size is kept fixed in the longitudinal direction even after the macroscopic period. This is because the velocity is constant for the massless case regardless of the momentum. Thus, wave packet nature of the massless particle in the longitudinal direction is kept for the long period. We study implications of the wave packet for the typical massless particle, photon, in the following.

Resonance in the wave packet scattering

The square of the center of mass energy, S, is defined by

$$S = (M_p^2 + 2E_pP_\gamma - 2|\vec{P}_p||\vec{P}_\gamma|\cos\theta) \simeq (M_p^2 + 2E_p \cdot P_\gamma(1-\cos\theta)), \tag{17}$$

where $(P_\gamma,\vec{P}_\gamma),(E_p,\vec{P}_p),\theta$ are four momenta of the photon and the proton and the collision angle. The mass and width of the Δ, M_Δ and Γ, are $M_\Delta = 1232[\text{MeV}],\Gamma = 120[\text{MeV}]$.

Breit-Wigner partial wave amplitude is

$$f_l(\theta) = \frac{\sqrt{2l+1}}{p}\frac{\Gamma/2}{\sqrt{S}-M_\Delta+i\Gamma/2}. \tag{18}$$

We assume that the coherent photon wave function of the photon-proton system of the momentum $\vec{p}_0(|p_0| = h\nu)$ is expressed by,

$$\psi = \left(\frac{1}{\sqrt{2\pi\sigma^2}}\right)^3 \exp -\frac{(\vec{p}-\vec{p}_0)^2}{2\sigma^2} \tag{19}$$

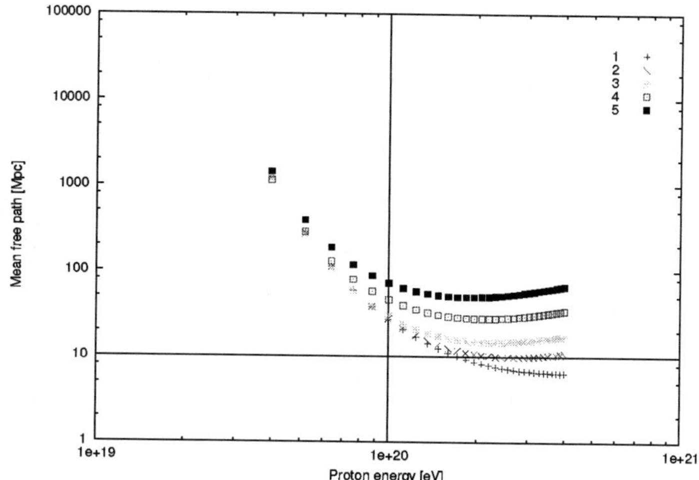

FIGURE 1. Proton energy dependence of cross section is given for plane wave (denoted by 1), σ = 2.725[K] (2), σ = 5.450[K] (3), σ = 13.625[K] (4) and σ = 27.250[K] (5).

where σ shows the width. Actually a charged particle is combined with coherent soft photons and collinear hard photons in order to cure the infra-red divergence or the mass singularity caused by massless particle, photon [23, 24]. So in a system of ultra-high energy charged particle, the energy and the momentum of the charged particle spreads. The wave function of Eq.(19), $\psi(\vec{p})$, should be understood as to include the coherent spreading of the center of mass energy due to the infrared divergence.

Using the amplitude for the wave packet, the average cross section is given by,

$$\sigma_{CMB} = \frac{\left| \int d\nu U(\nu) \ \int d^3p \ \psi(\vec{p}) \frac{4\pi}{p} \frac{\left(\frac{\Gamma}{2}\right)}{\sqrt{S} - M_\Delta - i\frac{\Gamma}{2}} \right|^2}{\int d\nu U(\nu)}. \tag{20}$$

where $U(\nu)$ is given by$U(\nu, T = 2.725[K]) = \frac{8\pi h\nu^3}{c^3} \frac{1}{e^{\frac{h\nu}{k_B T}} - 1}$ where parameters h, c, k_B, T, are Plank constant, speed of light , Boltzman constant, and temperature. Using the cross section, we calculate the average mean free path of UHCR.

The result is given in Fig. 1. The mean free path varies depending on the coherence length. So Lorentz invariance is effectively violated by the particle's coherence [25, 26, 27].

Summary

We found that the mean free path of UHCR in the 10^{20} eV region varies depending on the coherence length of CBR. It is clear from the Figure that the mean free path of

UHCR becomes longer if the CBR has a finite coherence length. The effect becomes important in the pion production threshold energy region. The coherence length will supply useful information.

ACKNOWLEDGMENTS

This work was partially supported by the special Grant-in-Aid for Promotion of Education and Science in Hokkaido University provided by the Ministry of Education, Science, Sports and Culture, a Grant-in-Aid for Scientific Research on Priority Area (Dynamics of Superstrings and Field Theories, Grant No. 13135201), and a Grant-in-Aid for Scientific Research on Priority Area (Progress in Elementary Particle Physics of the 21st Century through Discoveries of Higgs Boson and Supersymmetry, Grant No. 16081201) provided by the Ministry of Education, Science, Sports and Culture, Japan.

REFERENCES

1. Particle Data Group, W. -H. Yao et al, J. Phy. **G33**, (2006) and 2007 partial update. Abbreviations of experiments and whole summary can be found in this article.
2. M. Maltoni, et al, "Status of global fits to neutrino oscillations"; hep-ph/0405172v6
3. R. Davis, Prog. Part. Nucl. Phys. **32**, 13, (1994)
4. SAGE Coll., J. N. Abdurashitov et al, Phys. Rev. **C60**, 055801, (1999)
5. GALLEX Coll., W. Hampel et al, Phys. Lett. **B447**, 127, (1999)
6. Super-Kamiokande Coll., S. Fukuda et al, Phys. Lett. **B539**, 179, (2002)
7. SNO Coll., Q. R. Ahmad et al, Phys. Rev. Lett. **89**, 011301, (2002)
8. SNO(s) Coll., Q. R. Ahmad et al, Phys. Rev. Lett. **89**, 041801, (2003)
9. Super-K Coll., S. Fukuda et al, Phys. Rev. Lett. **81**, 1562, (1998)
10. K2K Coll., M. H. Ahn et al., Phys. Rev. Lett. **90**, 041801, (2003)
11. MINOS Coll., J. Nelson, Talk at Neutrino 2006, 13-19 June 2006, Santa Fe, New Mexico, USA (2006)
12. KamLAND Coll., K. Eguchi et al, Phys. Rev. Lett. **90**, 021802, (2003)
13. CHOOZ Coll., M. Apollonio et al., Phys. Lett. **B466**, 415, (1999)
14. Lawrence, M., A., R. J. O. Reid, and A. A. Watson, J. Phys. G **17**, 733, (1991)
15. Afanasiev, B., et al., 1993, in Proceedings of the Tokyo Workshop on Technique fot the study of the Extremely High Energy Cosmic Ray Research, University of Tokyo, Tokyo, Japan, p35.
16. Bird. D., et al., Astrophys. J. **424**, 491, (1994)
17. Takeda, M., et al., Phys. Rev. Lett. **81**, 1163(1998); Takeda, M., et al., in Proceedings of the 26th International Cosmic Ray Conference, Salt Lake City, Vol. 3, p.252, (1999)
18. The Pierre Auger Collaboration, Science, **318**, no. 5852, pp. 938 - 943, (2007)
19. Greisen, K., Phys. Rev. Lett. **16**, 748, (1966)
20. Zatsepin, Z. T., and V. A. Kuz'min, Zh. Eksp. Teor. Fiz. Pis'ma Red **4**, 144, (1966)
21. K. Ishikawa and Y. Tobita, arXiv:0801.3124[hep-ph], (2008)
22. K. Ishikawa and T. Shimomura, Prog. Theor. Phys. **114**, 1201-1234, (2005).
23. D. R. Yennie, S. C. Frautschi, and H. Suura, Ann. Phys. (N. Y) **13**, 379, (1961); V. Chung, Phys. Rev. **140**, B1110, (1965); L. Faddeev and P. Kulish, Theor. Math. Phys. **4**, 745, (1971)
24. T. Kinoshita, J. Math. Phys. **3**, 650, (1962); T. D. Lee and Nauenberg, Phys. Rev. **133**, B1549, (1964)
25. Humitaka Sato and Takao Tati, Prog. Theor. Phys. **47**, 1788-1790, (1972).
26. G. Amelino-Camilia et al., Nature, **393**, 763, (1998).
27. S. Coleman and Sheldon L. Glashow, Phys. Rev. **D59**, 116008, (1999); D. Colladay and V. A. Kostelecky, Phys. Rev. **D58**, 116002, (1998)

Gamow-Teller Transitions in Proton-Rich pf-shell Nuclei -relevance to supernovae explosions-

Y. Fujita and RCNP, Osaka, High Resolution (^{3}He,t) Collaboration*, B. Rubio†, W. Gelletly** and β-decay collaboration‡

*Department of Physics, Osaka University, Toyonaka, Osaka 560-0043, Japan
†IFIC, CSIC-University of Valencia, E-46071 Valencia, Spain
**Department of Physics, University of Surrey, Guildford GU2 7XH, Surrey, UK
‡ Valencia-Surrey-Osaka-Leuven-GSI-Santiago-Lund-Legnaro-Istanbul

Abstract. Gamow-Teller (GT) transitions starting from unstable pf-shell nuclei play important roles in neutrino-induced reactions that happen under the extremely high temperature conditions in core-collapse (type II) supernovae. In the β decay, it is difficult to obtain GT strengths B(GT) to higher excited states, but accurate half-lives can be measured. On the other hand, high-resolution (^{3}He,t) charge-exchange reactions at 0° and at 420 MeV yield cross-sections, that are proportional to B(GT) values, for individual transitions up to high excitation. Assuming isospin symmetry, we performed a unique analysis to determine absolute B(GT) values for the $T_z = \pm 1 \rightarrow 0$ analogous GT transitions. Further β-decay studies for unstable pf-shell nuclei to obtain accurate half-lives and feeding ratios are in progress.

Keywords: isospin symmetry, charge-exchange reaction, β decay
PACS: 25.55.Kr, 23.40.-s, 27.40.+z

INTRODUCTION

In the core-collapse stage of type II supernovae, weak-interaction processes of pf-shell nuclei play important roles. Therefore, studies of electron capture and β decay caused by charged currents and neutrino-nucleus scattering involving neutral currents are of great astrophysical interest [1]. The charged-current processes are dominated by Fermi and Gamow-Teller (GT) transitions, but our knowledge of the important GT transitions is very poor. Direct information on the GT transition strength B(GT) can be derived from β-decay measurements. Pioneering β-decay studies were performed on several far-from-stability pf-shell nuclei (e.g. ^{46}Cr [2], ^{50}Fe [3], and ^{54}Ni [4]). However, B(GT) values with large uncertainties have been derived for at most a few low-lying states. Note that the study of the feeding to a higher excited state in β decay is difficult, because the phase-space factor (f-factor) decreases rapidly with the excitation energy.

Charge-exchange (CE) (^{3}He,t) reactions allow access to GT transitions at higher excitation energies. In particular, it was shown that measurements at around 0° and at an intermediate beam energy of 140 MeV/nucleon provide a good probe of GT transitions. This is due to the fact that (a) GT excitations are dominant, and (b) there is a close proportionality between the GT cross-sections at 0° and the B(GT) values [5, 6, 7]

$$\sigma^{\rm GT}(0°) = \hat{\sigma}^{\rm GT}(0°) B({\rm GT}), \qquad (1)$$

CP1016, *Origin of Matter and Evolution of Galaxies*,
edited by T. Suda, T. Nozawa, A. Ohnishi, K. Kato, M. Y. Fujimoto, T. Kajino, and S. Kubono

where $\hat{\sigma}^{GT}(0°)$ is the GT unit cross-section at 0° for a specific mass A system. Therefore, the study of B(GT) values can reliably be extended up to high excitation energy, if at least one absolute B(GT) value is available from β decay.

Studies of GT strengths in pf-shell nuclei using (p,n) and (n,p) reactions at intermediate energies started in the 1980s. Due to their limited energy resolutions of $\approx$ 300 keV and $\approx$ 1 MeV, respectively, it was not easy to calibrate the unit cross-section $\hat{\sigma}^{GT}(0°)$ by using B(GT) values from β decay. In addition, B(GT) values are poorly known for pf-shell nuclei.

The development of precise beam matching techniques [8] realized an energy resolution of $\approx$ 30 keV in intermediate energy (^{3}He,t) reactions at 0° [9]. With this one order-of-magnitude better resolution, we can now study GT and Fermi states that were unresolved in the pioneering (p,n) reactions. In addition, new techniques developed for β-decay studies promote the re-investigation of very unstable pf-shell nuclei. By exploiting these excellent measurements, we present here a unique analysis to determine absolute B(GT) values by combining the precise strength distribution from the (^{3}He,t) reaction with the decay Q-value and lifetime from the mirror β decay.

ISOSPIN SYMMETRY AND STANDARD B(GT)

Under the assumption that isospin T is a good quantum number, an analogous structure is expected for nuclei with the same mass A but with different T_z (isobars). In the "$T = 1$ triplet", GT and also Fermi transitions from the $J^\pi = 0^+$ ground states (g.s.) of the $T_z = \pm 1$ even-even nuclei to 1^+ states (GT states) and the 0^+ state in the $T_z = 0$ odd-odd nucleus are analogous transitions (see Fig. 1). In the pf-shell region, $T_z = +1 \to 0$ transitions can be studied via (^{3}He,t) reactions on stable $T_z = +1$ target nuclei, and the analogous $T_z = -1 \to 0$ transitions can be studied via β decays. Assuming that the analogous GT transitions have the same B(GT) values, the absolute B(GT) values from β decays can in principle be used for the normalization. Then the study of B(GT) distributions can be extended by the (^{3}He,t) reactions to higher excitation energies overcoming the limits imposed by the Q-values in β decays. However, due to large uncertainties in the β-decay B(GT) values, this idea was not practical until now. Thus, a new idea was requested to obtain the absolute B(GT) values.

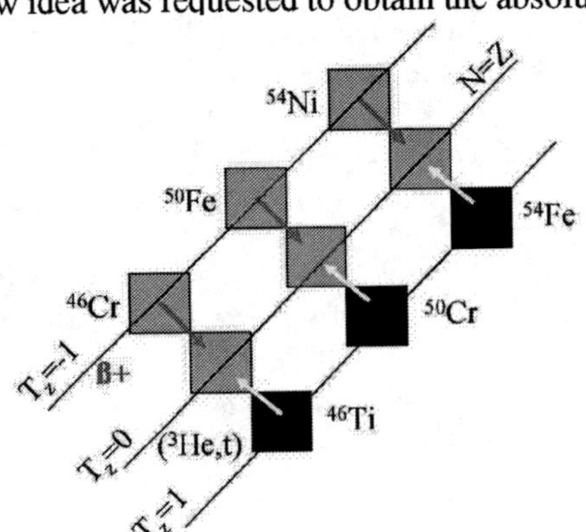

FIGURE 1. Isospin symmetry relationship in mass $A = 46, 50$ and 54 systems. The (^{3}He,t) reaction studies the $T_z = +1 \to 0$ GT transitions starting from stable nuclei ^{46}Ti, ^{50}Cr, and ^{54}Fe (see the indication by arrows). On the other hand, the β decay allows the study of the analogous $T_z = -1 \to 0$ GT transitions from ^{46}Cr, ^{50}Fe, and ^{54}Ni.

β DECAY AND THE (^{3}He,t) REACTION

Partial half-lives t_F and t_i of the β decay to the Fermi and ith GT state multiplied by the relevant f-factors f_F and f_i are related to the reduced Fermi and GT transition strengths B(F) and B_i(GT), respectively

$$f_F t_F = K/B(F)(1-\delta_c) \quad \text{and} \quad f_i t_i = K/\lambda^2 B_i(GT), \tag{2}$$

where $K = 6147.8 \pm 1.6$, $\lambda = g_A/g_V = -1.270 \pm 0.003$, and δ_c is the Coulomb correction factor [10]. The Fermi strength is concentrated in the transition to the isobaric analog state of the g.s. of the mother nucleus (IAS), and has the value $|N-Z|$. Uncertainties in B(GT) values originate from uncertainties in the decay Q-value, the total half-life $T_{1/2}$, and the branching ratios (feeding ratios) determining t_i. The accurate determination of the feeding ratios to higher excited states is more difficult due to the smaller f-factors. On the other hand, in the studies of analogous GT transitions using (^{3}He,t) reactions, relative transition strengths to these higher excited states can be obtained accurately from the $\sigma^{GT}(0°)$ values. It should be noted that the β-decay feeding ratios can be deduced using these values and f-factors that are calculated from the decay Q-value. Absolute B(GT) values can then be deduced in the "merged analysis" by further combining the total half-life $T_{1/2}$ of the β decay.

MERGED ANALYSIS

The inverse of $T_{1/2}$ is the sum of the inverse of the partial half-lives t_F of the Fermi transition to the IAS and the t_is of the GT transitions to the ith GT states

$$(1/T_{1/2}) = (1/t_F) + \sum_{i=GT} (1/t_i). \tag{3}$$

Applying Eq. (2), t_F and also the t_is can be eliminated,

$$1/T_{1/2} = 1/K \left[B(F)(1-\delta_c) f_F + \sum_{i=GT} \lambda^2 B_i(GT) f_i \right]. \tag{4}$$

In order to relate the strengths of GT and Fermi transitions in a CE reaction, we introduce the ratio R^2 of unit GT and Fermi cross-sections at 0°

$$R^2 = \hat{\sigma}^{GT}/\hat{\sigma}^{F} = [\sigma_i^{GT}/B_i(GT)]/[\sigma^F/B(F)(1-\delta_c)]. \tag{5}$$

Assuming isospin symmetry, this ratio R^2 is expected to be the same for the $T_z = \pm 1 \to 0$ transitions. Eliminating B_i(GT) by using R^2, we get

$$1/T_{1/2} = [B(F)(1-\delta_c)/K\sigma^F] \left[\sigma^F f_F + (\lambda^2/R^2) \sum_{i=GT} \sigma_i^{GT} f_i \right], \tag{6}$$

where $B(F) = 2$ and $\delta_c \approx 0.005$ for the β decay of $T_z = -1$ pf-shell nuclei [10].

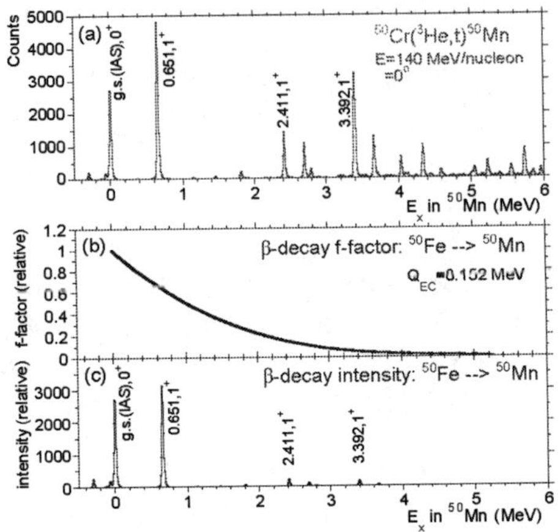

FIGURE 2. (a) The ^{50}Cr(^{3}He,t)^{50}Mn spectrum for events with scattering angles $\Theta \leq 0.5°$. Major $L = 0$ states are indicated by their excitation energies in MeV. (b) The f-factor for the ^{50}Fe β decay, normalized to unity at $E_x = 0$ MeV. (c) The estimated ^{50}Fe β-decay energy spectrum that is obtained by multiplying the f-factor with the ^{50}Cr(^{3}He,t) spectrum. Note that the IAS is stronger by a factor of R^2/λ^2 in the real β-decay measurement (see text).

Eq. (6) shows that the inverse of $T_{1/2}$ is proportional to the sum of the intensities of the observed Fermi and GT states weighted by f-factors, where a further correction factor λ^2/R^2 is needed for the GT intensities to compensate for the differences in the coupling constants in the Fermi and GT β decays and unit Fermi and GT cross-sections in the (^{3}He,t) reaction. Now we concentrate on the $A = 50$ system, where a rather accurate $T_{1/2}$ is known for the decay of ^{50}Fe. By assuming good isospin symmetry, the energy spectrum of GT transitions in the ^{50}Fe β decay can be estimated by multiplying the ^{50}Cr(^{3}He,t) spectrum [Fig. 2(a)] with the f-factor [Fig. 2(b)] as shown in Fig. 2(c). This predicted β-decay spectrum shows that more than an order-of-magnitude better sensitivity was needed to detect the transitions to the second and higher excited GT states in the measurement of the ^{50}Fe β decay. The estimated feedings to these excited GT states, however, amount in total to about 20% of the feeding to the first 0.651 MeV GT state, although each of them is small.

By solving Eq. (6), we get a value of $R^2 = 7.5 \pm 2.0$. The error mainly comes from the uncertainty in the $T_{1/2}$ value in the β decay measurement and also from the Q_β-value used for the f-factor calculation. The absolute B(GT) values were calculated using Eq. (5). Owing to the newly estimated feedings to higher excited GT states, the B(GT) value of the first GT state decreased by about 20% from the earlier β-decay value of 0.60(16) to 0.50(13). Here again the uncertainty originates mainly from the $T_{1/2}$ value and the Q value. The details of the merged analysis are given in Ref. [11].

β-DECAY MEASUREMENTS OF PROTON RICH NUCLEI

Accurate β-decay $T_{1/2}$ values play important roles in the merged analysis. In addition, accurate measurement of branching ratios give us further information on the isospin symmetry of transitions. These studies are in progress at Louvain-la-Neuve and at GSI.

Firstly, we extended our study to the $A = 54$ system, in which $T_z = \pm 1 \rightarrow 0$ mirror transitions are measured in the ^{54}Fe(^{3}He,t)^{54}Co reaction and the ^{54}Ni β decay, respectively. The $T_{1/2}$ value of the β decay was studied at the CRC, Louvain-la-Neuve. The proton-rich nucleus ^{54}Ni was produced by the ^{54}Fe(^{3}He,$3n$) fusion evaporation reaction at the beam energy of 45 MeV. Nickel was selectively ionized in a laser ion source using a two-step ionization scheme, and $A = 54$ nuclei were selected by the Leuven isotope separator facility (LISOL) [12, 13, 14]. The mass-separated ions were implanted on a tape system surrounded by three plastic β detectors and two MINIBALL Ge detectors. The $T_{1/2}$ value of ^{54}Ni was determined by observing the decrease in the intensity of the delayed 937 keV γ rays from the 937 keV, first $J^{\pi} = 1^{+}$ state. The fitted decay curve for this 937 keV γ ray gave a half-life of 114 ± 5 ms, a longer value than in Ref. [4] (106 ± 12 ms). A longer half-life and also the finding of higher excited states in the ^{54}Fe(^{3}He,t)^{54}Co reaction made the GT transition strength to the $E_x = 937$ keV state smaller. In the merged analysis, we tentatively obtain $B(\mathrm{GT}) = 0.48(5)$; this value is smaller by about 30% compared to the previous β-decay value of 0.68(16) [4]. The ^{54}Fe(p,n)^{54}Co reaction at $E_p = 135$ MeV and 0° was also used to study the GT transition to this state [15]. Compared to their B(GT) value of 0.74(5), which was derived using their own systematics, our new value is smaller by 35%.

In July, 2007, the β decays of ^{42}Ti, ^{46}Cr, ^{50}Fe, and ^{54}Ni were studied as part of the RISING stopped beam campaign [16] at the FRagment Separator (FRS), GSI, Darmstadt. The experimental goals were to measure the half-lives as well as B(GT) values for these decays as a function of energy as precisely as possible for comparison with the strengths measured in the charge exchange reactions. Beams of ^{42}Ti, ^{46}Cr, ^{50}Fe, and ^{54}Ni were produced by the fragmentation process from a primary 680 MeV/nucleon ^{58}Ni beam of 0.1nA on a Be target. Each beam was well separated by the FRS facility and ions were implanted into an active stopper system consisting of three layers of double-sided silicon strip detectors (DSSDs) with an area 50×50 mm and 16×16 strips. They were surrounded by the RISING array composed of EUROBALL cluster Ge detectors in close geometry. The overall γ-ray detection efficiency was about 15% at 1.33 MeV. Due to the high production rate for ^{54}Ni and the good detection efficiency of the RISING setup, high-energy delayed γ rays were seen [Fig. 3(b)] at the energies corresponding to those of GT states observed in the ^{54}Fe(^{3}He,t) measurement [Fig. 3(a)]. A good symmetry is suggested between the mirror transitions, which supports the basis of the merged analysis. In addition, it seems that these GT states decay mainly by direct γ transitions to the 0^{+} g.s.

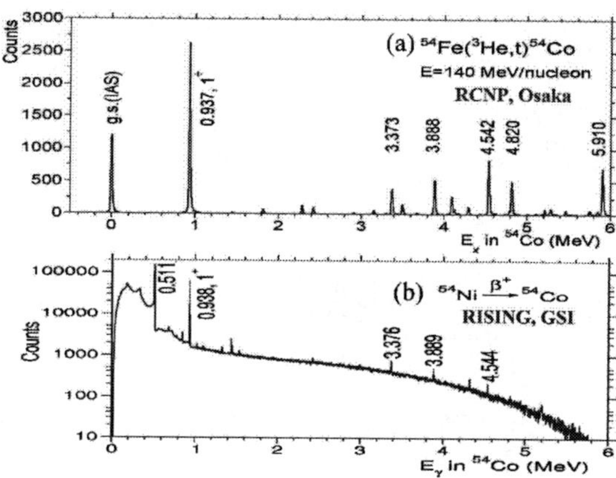

FIGURE 3. (a) The ^{54}Fe(^{3}He,t)^{54}Co spectrum for events with scattering angles $\Theta \leq 0.5°$. Major excited $L = 0$ states, most probably the GT states, are indicated by energies in MeV. They are excited by the $T_z = +1 \rightarrow 0$ GT transitions. (b) On-line γ-ray spectrum at the RISING, GSI measured in coincidence with the β particles from the ^{54}Ni decay. The existence of γ-ray peaks and CE-reaction peaks at corresponding energies suggests a good mirror symmetry of $T_z = -1 \rightarrow 0$ and $T_z = +1 \rightarrow 0$ GT transitions.

PROSPECTS

We performed (^{3}He,t) measurements on $T_z = +1$ pf-shell nuclei to study $T_z = +1 \rightarrow 0$ GT transitions. With an energy-resolution of about 30 keV, discrete GT states were identified. The unknown "energy spectra" of the $T_z = -1 \rightarrow 0$ β decays of exotic nuclei were estimated by multiplying the f-factor calculated from the Q-value of the decay. By further combining the half-life values obtained in the β-decay measurements, absolute values of GT transition strengths B(GT) were derived.

This "merged analysis" for the determination of absolute GT strengths can be extended also to $T = 2$ and even higher T systems, thus allowing us to obtain the GT strength distributions not only in stable nuclei, but also in proton-rich exotic nuclei to which experimental access is not easy. Note that the uncertainty in the GT strength obtained is mainly due to the errors in the $T_{1/2}$ and the decay Q value. We are now making an effort to measure accurate $T_{1/2}$ values for proton-rich nuclei at FRS and ISOL facilities. In addition, accurate Q-value measurements are in progress at trap facilities (see e.g. [17]). A better knowledge of these quantities will make the merged analysis more fruitful as a means of determining the GT strengths in proton-rich, exotic pf-shell nuclei, which are needed to deduce the astrophysical transition rates under extreme conditions.

The high-resolution (^{3}He,t) experiments were performed at RCNP, Osaka University. The β-decay measurements were performed at CRC, Louvain la Neuve and GSI, Darmstadt. The authors are grateful to the accelerator groups of these facilities. This work was supported in part by Monbukagakusho, Japan under Grant No. 18540270, the Spanish MEC under Grant No. EPA2005-03993, the Brix-IAP Research Program (P06/23) and FWO-Vlaanderen, Belgium. Y.F. and B.R. acknowledge support of the Japan-Spain collaboration programme by JSPS and CSIC. The β-decay experiments were partly supported through EURONS No. 506065.

REFERENCES

1. K. Langanke *et al.*, *Phys. Rev. Lett.* **93**, 202501 (2004).
2. T.K. Onishi *et al.* *Phys. Rev. C* **72**, 024308 (2005).
3. V.T. Koslowsky *et al.* *Nucl. Phys.* **A624**, 293 (1997).
4. I. Reusen *et al.* *Phys. Rev. C* **59**, 2416 (1999).
5. T.N. Taddeucci *et al.*, *Nucl. Phys.* **A469**, 125 (1987).
6. Y. Fujita *et al.*, *Phys. Rev. C* **67**, 064312 (2003).
7. Y. Fujita *et al.*, *Phys. Rev. C* **75**, 057305 (2007).
8. Y. Fujita *et al.*, *Nucl. Instrum. Meth. Phys. Res. B* **126**, 274 (1997);
9. H. Fujita *et al.*, *Nucl. Instrum. Meth. Phys. Res. A* **484**, 17 (2002).
10. J.C. Hardy and I.S. Towner, *Phys. Rev. C* **71**, 055501 (2005), *ibid. Nucl. Phys. News* **16**, 11 (2006).
11. Y. Fujita *et al.*, *Phys. Rev. Lett.* **95**, 212501 (2005).
12. Yu. Kudryavtsev *et al.*, *Nucl. Instrum. Meth. Phys. Res. B* **179**, 412 (2001),
13. M. Facina *et al.*, *Nucl. Instrum. Meth. Phys. Res. B* **226**, 401 (2004), and references therein.
14. Yu. Kudryavtsev *et al.*, *Nucl. Instrum. Meth. Phys. Res. B* **114**, 350 (1996).
15. B.D. Anderson *et al.*, *Phys. Rev. C* **41**, 1474 (1990).
16. "Stopped Beam" RISING experimental campaign at GSI, spokespersons: P.H. Regan, J. Gerl, and H.J. Wollersheim.
17. A. Kankainen *et al.*, *Eur. Phys. J. A* **29**, 271 (2006).

Neutrino Mass Bounds from $0\nu\beta\beta$ Decays and Large Scale Structures

Y.-Y. Keum*,†, K. Ichiki** and T. Kajino*,‡

*Theory Division, National Astronomical Observatory, Mitaka, Tokyo 181-8588, Japan
†Department of Physics, National Taiwan University, Taipei, Taiwan 10672, R.O.C.
**Research Center for the Early Universe, University of Tokyo, 7-3-1 Hongo, Bunkyo-ku, Tokyo 113-0033, Japan
‡Department of Astronomy, Graduate School of Science, University of Tokyo, Bunkyo-ku, Tokyo 113-0033, Japn

Abstract. we investigate the way how the total mass sum of neutrinos can be constrained from the neutrinoless double beta decay and cosmological probes with cosmic microwave background (WMAP 3-year results), large scale structures including 2dFGRS and SDSS data sets. First we discuss, in brief, on the current status of neutrino mass bounds from neutrino beta decays and cosmic constrain within the flat ΛCMD model. In addition, we explore the interacting neutrino dark-energy model, where the evolution of neutrino masses is determined by quintessence scalar filed, which is responsable for cosmic acceleration today. Assuming the flatness of the universe, the constraint we can derive from the current observation is $\sum m_\nu < 0.87$eV at the 95 % confidence level, which is consistent with $\sum m_\nu < 0.68$eV in the flat ΛCDM model.

Keywords: Neutrino Masses, Interacting Neutrino Dark-Energy, Neutrinoless Double Beta Decays
PACS: 98.80.Jk, 98.80.Cq, 98.80.-k

There are three well known ways to get the direct information on the absolute mass of neutrinos by using: Tritium β-decay experiment, neutrinoless double beta decay experiment, and astrophysical observations.

(A) Neutrinoless Double Beta Decays:

The standard method for the measurement of the absolute value of the neutrino mass is based on the detailed investigation of the high-energy part of the β-spectrum of the decay of tritium:

$$^3H \longrightarrow {}^3He + e^- + \bar{\nu}_e \tag{1}$$

This decay has a small energy release ($E_0 \simeq 18.6keV$) and a convenient life time ($T_{1/2} =$ 12.3 years). Since the flavour eigenstates are different from mass eigenstates in neutrino sector, in general, electron neutrino can be expressed as $\nu_{eL} = \sum_i U_{ei}\,\nu_{iL}$, where ν_i is the field of neutrino with mass m_i, and U is the unitary mixing matrix. Neglecting the recoil of the final nucleus, the resulting spectrum can be analyzed in term of a single mean-

[1] Email: yykeum@phys.ntu.edu.tw

CP1016, *Origin of Matter and Evolution of Galaxies*, edited by T. Suda, T. Nozawa, A. Ohnishi, K. Kato, M. Y. Fujimoto, T. Kajino, and S. Kubono squared electron neutrino mass $\langle m_\beta \rangle^2 = \sum_j m_j^2 |U_{ej}|^2 = m_1^2|U_{e1}|^2 + m_2^2|U_{e2}|^2 + m_3^2|U_{e3}|^2$ If the neutrino mass spectrum is practically degenerate: $m_1 \simeq m_2 \simeq m_3$, the neutrino mass can be measured in these experiments. Present-day tritium experiments Mainz[1] and Troitsk[2] gave the following results:

$$m_1^2 = (-1.2 \pm 2.2 \pm 2.1)\, eV^2 \quad \text{(Mainz)}, \tag{2}$$

$$= (-2.3 \pm 2.5 \pm 2.0)\, eV^2 \quad \text{(Troitsk)}. \tag{3}$$

This value corresponds to the upper bound $m_1 < 2.2eV$ $(95\% C.L.)$ Another useful method is by using the neutrinoless double beta decay. The search for neutrinoless double β-decay $(A,Z) \longrightarrow (A,Z_2) + e^- + e^-$ for some even-even nuclei is the most sensitive and direct way of investigating the nature of neutrinos with definite masses. In this process, total lepton number is violated ($\Delta L = 2$) and is allowed only if massive neutrinos are Majorana particles. The rate of $0\nu\beta\beta$ is approximately

$$\frac{1}{T_{1/2}^{0\nu}} = G_{0\nu}(Q_{\beta\beta}, Z)\, |M_{0\nu}|^2\, \langle m_{\beta\beta} \rangle^2, \tag{4}$$

where $G^{0\nu}$ is the phase space factor for the emission of the two electrons, $M_{0\nu}$ is nuclear matrix elements, and $< m_{\beta\beta} >$ is the effective majorana mass of the electron neutrino: $\langle m_{\beta\beta} \rangle \equiv |\sum_i U_{ei}^2 m_i|$. The most stringent lower bounds for the time of life of $0\nu\beta\beta$-decay were obtained in the Heidelberg-Moscow[3] and IGEX[4] ^{76}Ge experiments:

$$T_{1/2}^{0\nu} \geq 1.9 \cdot 10^{25} years \quad (90\% C.L.) \quad \text{Heidelberg} - \text{Moscow}, \tag{5}$$

$$T_{1/2}^{0\nu} \geq 1.57 \cdot 10^{25} years \quad (90\% C.L.) \quad \text{IGEX}. \tag{6}$$

Taking into account different calculation of the nuclear matrix elements, from these results the following upper bounds were obtained for the effective Majorana mass: $|m_{\beta\beta}| < (0.35 - 1.24)\, eV$. Many new experiments (including CAMEO, CUORE, COBRA, EXO, GENIUS, MAJORANA, MOON and XMASS experiments) on the search for the neutrinoless double β-decay are in preparation at present. In these experiments the sensitivities $|m_{\beta\beta}| \simeq (0.1 - 0.015)\, eV$ are expected to be achieved. It is very difficult to confirm the normal hierarchy pattern of neutrino mass when $m_1 < 1.7 \cdot 10^{-3}\, eV$, however for the inverted case, it can be detected if $m_3 < 8.9 \cdot 10^{-3}\, eV$ and $m_{ee} > 0.012\, eV$.

(B) Cosmological Constrains within the Standard Cosmology:

Within the standard cosmological model, the relic abundance of neutrinos at present epoch was come out straightforwardly from the fact that they follow the Fermi-Dirac distribution after freeze out, and their temperature is related to the CMB radiation temperature T_{CMB} today by $T_\nu = (4/11)^{1/3} T_{CMB}$ with $T_{CMB} = 2.726$ K, providing $n_\nu = 6\zeta(3)/11\pi^2 T_{CMB}^3$, where $\zeta(3) \simeq 1.202$, which gives $n_\nu \simeq 112 cm^{-3}$ for each family of neutrinos at present. By now the massive neutrinos become non-relativistic, and their contribution to the mass density (Ω_ν) of the universe can be expressed as $\Omega_\nu h^2 =$

$\Sigma/93.14eV$. where Σ stands for the sum of the neutrino masses. In this relation, we include the effect of three neutrino oscillation [5]. We should notice that when obtaining the limit of neutrino masses one usually assumes:

- the standard spatially flat ΛCDM model with adiabatic primordial perturbations,
- they have no non-standard interactions,
- neutrinos decoupled from the thermal background at the temperatures of order 1 MeV.

These simple conditions can be modified from several effects: due to a sizable neutrino-antineutrino asymmetry, due to additional light scalar field coupled with neutrinos [8], and due to the light sterile neutrino [9]. However, analysis of WMAP and 2dFGRS data gave independent evidence for small lepton asymmetries [10, 11], and such a scenario with a light scalar field is strongly disfavored by the current CMB power spectrum data [12]. We will not therefore take into account such non-standard couplings of neutrinos in the following. In addition, current cosmological observations are sensitive to neutrino masses $0.1\,\text{eV} < \Sigma < 2.0\,\text{eV}$. In this mass scale, the mass-square differences are small enough and all three active neutrinos are nearly degenerate in mass. Therefore we take the assumption of degenerate mass hierarchy. Even if we consider different mass hierarchy pattern, it will be very difficult to distinguish such hierarchy patterns from cosmological data alone [13].

After neutrinos decoupled from the thermal background, they stream freely and their density perturbations are damped on scale smaller than their free streaming scale. Consequently the perturbations of cold dark matter (CDM) and baryons grow more slowly because of the missing gravitational contribution from neutrinos. The free streaming scale of relativistic neutrinos grows with the hubble horizon. When the neutrinos become non-relativistic, their freestreaming scale shrinks, and they fall back into the potential wells. The neutrino density perturbation with scales larger than the freestreaming scale resumes to trace those of the other species. Thus the free streaming effect suppresses the power spectrum on scales smaller than the horizon when the neutrinos become non-relativistic. The co-moving wavenumber corresponding to this scale is given by $k_{nr} = 0.026\,(m_\nu/1\,eV)^{1/2}\,\Omega_m^{1/2}\,h\text{Mpc}^{-1}$, for degenerated neutrinos, with almost same mass m_ν. The growth of fourier modes with $k > k_{nr}$ will be suppressed because of neutrino free-streaming. The power spectrum of matter fluctuations can be written as $P_m(k,z) = P_*(k)\,T^2(k,z)$, where $P_*(k)$ is the primordial spectrum of matter fluctuations, to be a simple power law $P_*(k) = A\,k^n$, where A is the amplitude and n is the spectral index. Here the transfer function $T(k,z)$ represents the evolution of perturbation relative to the largest scale. If some fraction of the matter density (e.g., neutrinos or dark energy) is unable to cluster, the speed of growth of perturbation becomes slower. Because the contribution to the fraction of matter density from neutrinos is propotional to their masses (Eq. ()), the larger mass leads to the smaller growth of perturbation. The suppression of the power spectrum on small scales is roughly proportional to f_ν [14]: $\Delta P_m(k)/P_m(k) \simeq -8f_\nu$. where $f_\nu = \Omega_\nu/\Omega_M$ is the fractional contribution of neutrinos to the total matter density. This result can be understood qualitatively from the fact that only a fraction $(1-f_\nu)$ of the matter can cluster when massive neutrinos are present [15]. Analyses of CMB data are not sensitive to neutrino masses if neutrinos behave

TABLE 1. Recent cosmological neutrino mass bounds (95% C.L.)

Cosmological Data Set	Σ bound (2σ)	References
CMB (WMAP-3 year alone)	< 2.0 eV	Fukugita et al.[16]
LSS[2dFGRS]	< 1.8 eV	Elgaroy et al.[17]
CMB + LSS[2dFGRS]	< 1.2 eV	Sanchez et al.[18]
"	< 1.0 eV	Hannestad[19]
CMB + LSS + SN1a	< 0.75 eV	Barger et al.[20]
"	< 0.68 eV	Spergel et al.[21]
CMB + LSS + SN1a + BAO	< 0.62 eV	Goobar et al.[22]
"	< 0.58 eV	
CMB + LSS + SN1a + Ly-α	< 0.21 eV	Seljak et al.[23]
CMB + LSS + SN1a + BAO + Ly-α	< 0.17 eV	Seljak et al.[23]

as massless particles at the epoch of last scattering. According to the analytic consideration in [24], since the redshift when neutrino becomes non-relativistic is given by $1+z_{nr} = 6.24 \cdot 10^4 \Omega_\nu h^2$ and $z_{rec} = 1088$, neutrinos become non-relativistic before the last scattering when $\Omega_{nu} h^2 > 0.017$ (i.e. $\Sigma > 1.6eV$). Therefore the dependence of the position of the first peak and the height of the first peak on $\Omega_\nu h^2$ has a turning point at $\Omega_\nu h^2 \simeq 0.017$. This value also affects CMB anisotropy via the modification of the integrated Sachs-Wolfe effect due to the massive neutrinos. However an important role of CMB data is to constrain other parameters that are degenerate with Σ. Also, since there is a range of scales common to the CMB and LSS experiments, CMB data provides an important constraint on the bias parameters. We summarize some of the recent cosmological neutrino mass bounds within the flat-ΛCDM model in table 1.

(C) Neutrino Mass Bounds in Interacting Neutrino Dark-Energy Model

With our previous works [25, 26, 27], we investigate the cosmological implication of an idea of the dark-energy interacting with neutrinos [28, 29]. For simplicity, we consider the case that dark-energy and neutrinos are coupled such that the mass of the neutrinos is a function of the scalar field which drives the late time accelerated expansion of the universe.

In our scenario, Equations for quintessence scalar field are given by

$$\ddot{\phi} \; + \; 2\mathscr{H}\dot{\phi} + a^2 \frac{dV_{\rm eff}(\phi)}{d\phi} = 0\,, \qquad V_{\rm eff}(\phi) = V(\phi) + V_{\rm I}(\phi)\,, \tag{7}$$

$$V_{\rm I}(\phi) \; = \; a^{-4} \int \frac{d^3q}{(2\pi)^3} \sqrt{q^2 + a^2 m_\nu^2(\phi)} f(q)\,, \qquad m_\nu(\phi) = \bar{m}_i e^{\beta \frac{\phi}{M_{\rm pl}}}\,, \tag{8}$$

where $V(\phi)$ is the potential of quintessence scalar field, $V_{\rm I}(\phi)$ is additional potential due to the coupling to neutrino particles [29, 30], and $m_\nu(\phi)$ is the mass of neutrino coupled to the scalar field, where we assume the exponential coupling with a coupling parameter β. $\mathscr{H}$ is $\dot{a}/a$, where the dot represents the derivative with respect to the conformal time τ.

Energy densities of mass varying neutrino (MaVaNs) and quintessence scalar field are described as

$$\rho_\nu = a^{-4}\int \frac{d^3q}{(2\pi)^3}\sqrt{q^2+a^2m_\nu^2}f_0(q)\,, \tag{9}$$

$$3P_\nu = a^{-4}\int \frac{d^3q}{(2\pi)^3}\frac{q^2}{\sqrt{q^2+a^2m_\nu^2}}f_0(q)\,, \tag{10}$$

$$\rho_\phi = \frac{1}{2a^2}\dot\phi^2+V(\phi)\,, \qquad P_\phi = \frac{1}{2a^2}\dot\phi^2-V(\phi)\,. \tag{11}$$

From equations (9) and (10), the equation of motion for the background energy density of neutrinos is given by

$$\dot\rho_\nu+3\mathcal{H}(\rho_\nu+P_\nu) = \frac{\partial \ln m_\nu}{\partial\phi}\dot\phi(\rho_\nu-3P_\nu)\,. \tag{12}$$

Here we consider three different types of the quintessence potential: (1) inverse power law potentials (Model I), (2) SUGRA type potential models (Model II), (3) exponential type potentials (Model III), which are given, respectively:

$$V(\phi) = M^4\left(\frac{M_{pl}}{\phi}\right)^\alpha \;;\; M^4\left(\frac{M_{pl}}{\phi}\right)^\alpha e^{3\phi^2/2M_{\rm pl}^2} \;;\; M^4 e^{-\alpha(\frac{\phi}{M_{pl}})}\,. \tag{13}$$

The coupling between cosmological neutrinos and dark energy quintessence could modify the CMB and matter power spectra significantly. It is therefore possible and also important to put constraints on coupling parameters from current observations. For this purpose, we use the WMAP3 [31, 32] and 2dFGRS [33] data sets.

The flux power spectrum of the Lyman-α forest can be used to measure the matter power spectrum at small scales around $z < 3$ [34, 35]. It has been shown, however, that the resultant constraint on neutrino mass can vary significantly from $\sum m_\nu < 0.2$eV to 0.4eV depending on the specific Lyman-α analysis used [36]. The complication arises because the result suffers from the systematic uncertainty regarding to the model for the intergalactic physical effects, i.e., damping wings, ionizing radiation fluctuations, galactic winds, and so on [37]. Therefore, we conservatively omit the Lyman-α forest data from our current analysis.

Because there are many other cosmological parameters than the MaVaNu parameters, we follow the Markov Chain Monte Carlo(MCMC) global fit approach [38] to explore the likelihood space and marginalize over the nuisance parameters to obtain the constraint on parameters we are interested in. Our parameter space consists of

$$\vec{P} \equiv (\Omega_b h^2, \Omega_c h^2, H, \tau, A_s, n_s, m_i, \alpha, \beta)\,, \tag{14}$$

where $\omega_b h^2$ and $\Omega_c h^2$ are the baryon and CDM densities in units of critical density, H is the hubble parameter, τ is the optical depth of Compton scattering to the last scattering surface, A_s and n_s are the amplitude and spectral index of primordial density fluctuations, and (m_i, α, β) are the parameters of MaVaNs. We find no observational signature which

TABLE 2. Global analysis data within 2σ deviation for different types of the quintessence potential.

Quantites	Model I	Model II	Model III	WMAP-3 data (ΛCDM)
α	< 4.38	$0.10 - 11.82$	< 1.41	—
β	< 1.12	< 1.36	< 1.53	—
$\Omega_B h^2[10^2]$	2.09–2.36	2.09–2.35	2.08–2.34	2.23 ± 0.07
$\Omega_{CDM} h^2[10^2]$	9.87 – 12.30	9.85–12.40	9.84–12.33	12.8 ± 0.8
H_0	58.39 – 72.10	58.55–71.70	58.99–71.58	72 ± 8
Z_{re}	6.13 – 14.94	4.00–14.78	6.64–14.78	—
n_s	0.92 – 0.99	0.92–0.98	0.92–0.98	0.958 ± 0.016
$A_s[10^{10}]$	18.25 – 23.41	18.20–23.32	18.33-23.27	—
$\Omega_Q[10^2]$	57.43 – 75.60	57.59–75.02	58.45–75.05	71.6 ± 5.5
$Age/Gyrs$	13.59 – 14.40	13.59–14.35	13.61–14.36	13.73 ± 0.16
$\Omega_{MVN} h^2[10^2]$	< 0.95	< 0.91	< 0.84	$< 1.97(95\%C.L.)$
τ	0.031–0.143	0.028–0.139	0.032–0.140	0.089 ± 0.030

favors the coupling between MaVaNs and quintessence scalar field, and obtain the upper limit on the coupling parameter as shown in table 2.

$$\beta < 0.46, 0.47, 0.58\,(1\sigma);\ [1.12,\ 1.36,\ 1.53\,(2\sigma)], \tag{15}$$

and the present mass of neutrinos is also limited to

$$\Omega_\nu h^2_{\rm today} < 0.0044,\ 0.0048,\ 0.0048\ \ (1\sigma);\ [0.0095,\ 0.0090,\ 0.0084\ \ (2\sigma)], \tag{16}$$

for models I, II and III, respectively. When we apply the relation between the total sum of the neutrino masses M_ν and their contributions to the energy density of the universe: $\Omega_\nu h^2 = M_\nu/(93.14eV)$, we obtain the constraint on the total neutrino mass: $M_\nu < 0.45\ eV(68.5\%C.L.)$ $[0.87\ eV(95\%C.L.)]$ in the interacting neutrino dark-energy model.

Beyond the scope of our current analysis, there are other possibilities in cosmological probes of neutrino masses: (1) the evolution of cluster abundance with redshit may provide further constraints on neutrino masses, (2) the Lyman-α forest provides constaints on the matter power spectrum on scale of $k \sim 1hMpc^{-1}$, where the effect of massive neutrinos is most viable, (3) Deep and wide weak lensing survey will make it possible, in the future, to perform weak lensing tomography of the matter density field. The combination of weak lensing tomography and high-precision CMB-polarization experiments may reach sensitivities down to the lower bound of 0.06 eV on the sum of the neutrino masses [39, 40]. In this case, normal hierarchy pattern will be detectable.

ACKNOWLEDGMENTS

Y.Y.K's work is partially supported by Grants-in-Aid for NSC in Taiwan, and Center for High Energy Physics/KNU in Korea. K.I's work is supported by Grant-in-Aid for JSPS Fellows. T.K's work is supported by the Grant-in-Aid for Scientific Research (17540275) of the Ministry of Education, Culture, Sports, Science and Technology of Japan, the JSPS Core-to-Core Program, International Research Network for Exotic Femto System (EFES), and the Mitsubishi Foundation.

REFERENCES

1. J.Bonne et al., Nuch. Phys. B (Proc. Suppl.), 91 (2001) 273.
2. Ch. Weinheimer, Nucl. Phys. B (Proc. Suppl.), 118 (2003) 279.
3. L. Baudis et al. [Heidelberg-Moscow Collaboration], Phys. Rev. Lett. **83**, 41 (1999).
4. C. E. Aalseth et al. [IGEX Collaboration], Phys. Rev. **D 65**, 092007 (2002); Phys. Rev. **D 70**, 078302 (2004).
5. G. Mangano et al., Nucl. Phys. **B729** (2005) 221.
6. S. M. Bilenky, A. Faessler, and F. Simkovic, Phys. Rev. **D 70**:033003 (2004).
7. V. A. Rodin, A. Faessler, F. Simkovic, and P. Vogel, Phys. Rev. **C 68**:044302 (2003).
8. J. F. Beacom, N. F. Bell and S. Dodelson, Phys. Rev. Lett. **93** 121302 (2004).
9. S. Dodelson, A. Melchiorri and A. Slasar, Phys. Rev. Lett. **97** 04031 (2006).
10. S. Hannestad, JCAP **05** 004 (2003).
11. E. Pierpaoli, Mon. Not. Roy. Astron. Soc. **342** L63 (2003).
12. S. Hannestad, JCAP **0502** 011 (2005); arXiv astro-ph/0411475.
13. A Slosar, Phys. Rev. **D73** 123501 (2006).
14. W. Hu, D. J. Eisenstein and M. Tegmark, Phys. Rev. Lett. **80**, 5255 (1998) [arXiv:astro-ph/9712057].
15. J. R. Bond, G. Efstathiou and J. Silk, Phys. Rev. Lett. **45** 1980 (1980).
16. M. Fukugita, K. Ichikawa, M. Kawasaki and O, Lahav, Phys. Rev. **D 74**, 027302 (2006).
17. O. Elgaroy et al., Phys. Rev. Lett**89** 061310 (2002); O. Elgaroy and O. Lahav, JCAP 0304 (2003) 004.
18. A. G. Sanchez et al., Mon. Not. Roy. Astron. Soc. 366 (2006) 189.
19. S. Hannestad, JCAP 0305 (2003) 004.
20. V. Barger, D. Marfatia, and A. Tregre, Phys. Lett. **B 595** 55 (2004).
21. D. N. Spergel et al. [WMAP Collaboration], astro-ph/0603449.
22. A. Goodbar, S. Hannestad, E. Mortsell, and H. Tu, J. Cosmol. Astropart. Phys. 06 (2006) 019.
23. U. Seljak, A. Slosar, and P. McDonald, J. Cosmol Astropart. Phys. 10 (2006) 014.
24. K. Ichikawa, M. Fukugita and M. Kawasaki, Phys. Rev. **D71** 043001 (2005).
25. K. Ichiki and Y.-Y. Keum, arXiv:astro-ph/0705.2134 (to be published in JCAP).
26. Y.-Y. Keum, Mod. Phys. Lett. **A22**:2131-2142 (2007).
27. K. Ichiki and Y.-Y. Keum, "*Neutrino Masses from Cosmological Probes in interacting Neutrino Dark-Energy Models*", arXiv:hep-ph/0803.2274.
28. D. B. Kaplan, A. E. Nelson and N. Weiner, Phys. Rev. Lett. **93**:091801, (2004); R. D. Peccei, Phys. Rev. **D71**:023527 (2005).
29. R. Fardon, A. E. Nelson and N. Weiner, JCAP 0410:005, 2004; [arXiv:astro-ph/0309800].
30. X. J. Bi, P. h. Gu, X. l. Wang and X. M. Zhang, Phys. Rev. **D69**:113007 (2004); [arXiv:hep-ph/0311022].
31. G. Hinshaw *et al.*(WMAP collaboration), arXiv:astro-ph/0603451.
32. L. Page *et al.*(WMAP collaboration), arXiv:astro-ph/0603450.
33. S. Cole *et al.* [The 2dFGRS Collaboration], Mon. Not. Roy. Astron. Soc. **362**, 505 (2005) [arXiv:astro-ph/0501174].
34. P. McDonald, J. Miralda-Escude, M. Rauch, W. L. W. Sargent, T. A. Barlow, R. Cen and J. P. Ostriker, Astrophys. J. **543**, 1 (2000) [arXiv:astro-ph/9911196].
35. R. A. C. Croft *et al.*, Astrophys. J. **581**, 20 (2002) [arXiv:astro-ph/0012324].
36. A. Goobar, S. Hannestad, E. Mortsell and H. Tu, JCAP **0606**, 019 (2006) [arXiv:astro-ph/0602155].
37. P. McDonald, U. Seljak, R. Cen, P. Bode and J. P. Ostriker, Mon. Not. Roy. Astron. Soc. **360**, 1471 (2005) [arXiv:astro-ph/0407378].
38. A. Lewis and S. Bridle, Phys. Rev. **D66**, 103511 (2002).
39. Y. S. Song and L. Knox, Phys. Rev. bf D 70 (2004) 063510.
40. J. Lesgourgues and S. Pastor, Phys. Rept. **429**, 307 (2006).

14. NUCLEAR ABUNDANCES IN SUPERNOVAE

Gas and Dust Layers from Cas A's Explosive Nucleosynthesis

Lawrence Rudnick

Department of Astronomy, University of Minnesota, Minneapolis, MN 55455, USA

Abstract. Our group has developed a new picture of the structure of Cas A's explosion using 5-40 micron images and spectra from the Spitzer Space Telescope. In this picture, two roughly spherical shocks (forward and reverse) were initially set up by the outer layers of the exploding star. Deeper layers were ejected in a highly flattened structure with large protrusions in the plane of the flattening; some of these are visible as jets. As these aspherical deeper layers encounter the reverse shock at different locations, they become visible across the electromagnetic spectrum, with different nucleosynthesis layers visible in different directions. In the infrared, we see the gas lines of Ar, Ne, O, Si, S, and Fe at different locations, along with higher ionization states of the same elements visible in the optical and X-ray parts of the spectrum. These different nucleosynthesis layers appear to have formed characteristic types of dust, the deep layers producing dust rich in silicates, while dust from the upper layers is dominated by Al_2O_3 and carbon grains. In addition, we see circumstellar dust heated by its encounter with the forward shock. We estimate the total dust mass currently visible that was formed in the explosion to be $\sim$0.02-0.05 $M_\odot$. Rough extrapolations of these measurements to SNe in high redshift galaxies may be able to account for the lower limit of their observed dust masses. There is a large amount of gas, and presumably dust, that is currently not visible at any wavelength, including both the cooled post-reverse-shock ejecta and the material which has not yet encountered the reverse shock, where some select infrared emission is apparent.

Keywords: Infrared emission, supernova remnants,explosive nucleosynthesis, dust
PACS: 95.85.H, 98.58.C. 95.30.W, 26.30.E, 97.60.B, 98.38M

INTRODUCTION

Cassiopeia A is one of the best-studied supernova remnants, and has long been a rich source of information due to its youth, proximity and brightness across the electromagnetic spectrum. It is a member of the "oxygen-rich" class, with a distribution of heavy elements indicative of a core-collapse supernova. With a likely explosion date of 1681$\pm$19 [13], it is young enough that the motions of the shocks and expanding ejecta can be directly observed over timescales of several years. Detailed studies of the spatial and temporal variations in composition, temperature, and ionization state together with the dynamics promise to uncover the history of Cas A's explosion and nucleosynthesis.

In this short review, based on the work of our Spitzer Cas A team [1], I will first describe a dynamical picture of the explosion, synthesized from optical, radio, X-ray and infrared observations. Then I will summarize our new findings from infrared imaging

[1] T. DeLaney [MIT], J. Ennis [Minnesota], H. Gomez [Wales], T. Kozasa [Hokkaido], W. Reach and J. Rho [Spitzer Space Center], L. Rudnick [Minnesota], J.D. Smith [Arizona, Toledo], and A. Tappe [Spitzer, Harvard-CfA].

CP1016, *Origin of Matter and Evolution of Galaxies*,
edited by T. Suda, T. Nozawa, A. Ohnishi, K. Kato, M. Y. Fujimoto, T. Kajino, and S. Kubono

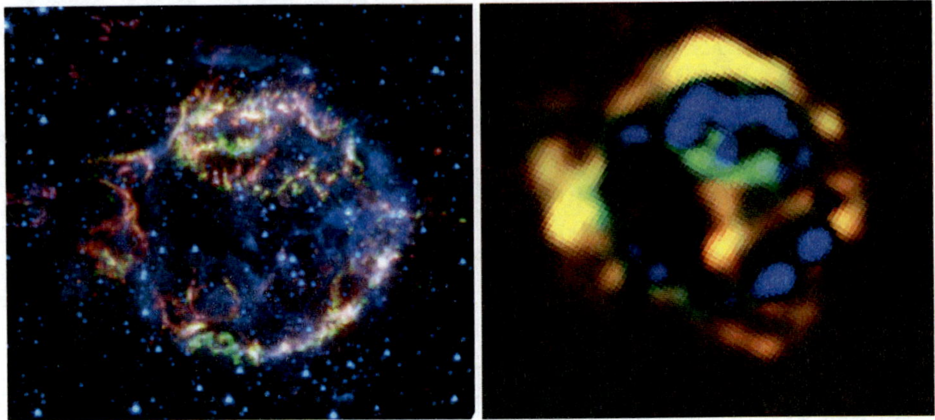

FIGURE 1. Spitzer infrared images of Cas A showing the varied compositions of gas and dust at different locations in the remnant. Colors described in text. Left: Combination of the four Spitzer IRAC camera channels, including both gaseous line and continuum emission. Reproduced from Fig. 2 of [10]. Right: Combination of three continuum dominated regions of the spectrum from the Spitzer IRS, showing differences in dust composition.

and spectroscopy with the Spitzer telescope which show that the nucleosynthesis layers have been largely preserved in each direction. In addition, we now have direct evidence for the ejecta that have not yet reached the reverse shock; Cas A has much more in store for us over the next few hundred years.

DYNAMICS OF THE EXPLOSION

The overall structure of Cas A is a bright irregular ring of emission (see Figure 1) seen at all wavelengths, composed of stellar ejecta surrounded by a faint plateau of emission with a thin outer boundary visible in X-rays, identified with the outer shock [15]. The bright ring is then identified with the reverse shock, at a current radius of ~2 pc, using a distance of 3.4 kpc to the remnant [30]. The reverse shock moves inward with respect to the ejecta but outward with respect to our external frame at the present epoch.

Given its youth and proximity, Cas A has allowed detailed studies of its proper motion in the optical, radio and X-ray regimes. The optical picture is the most straightforward; its "fast-moving knots" of ejecta showing a homologous expansion with a well-defined origin in space and time [13]. The undecelerated expansion rate is ~0.3%/yr, which translates to ~5000 km/s at the position of the bright ring (reverse shock). At X-ray and radio wavelengths, the situation becomes more complicated, with a turbulent component to the motions [5] and mean expansion rates of 0.2%/yr and 0.1%/yr, respectively. This seems paradoxical, since the bright rings are coincident in the optical, X-ray and radio images, but appear to be moving at different speeds.

The resolution of this paradox comes with the recognition that the bright ring marks the position of the large scale reverse shock, but what we actually measure are the

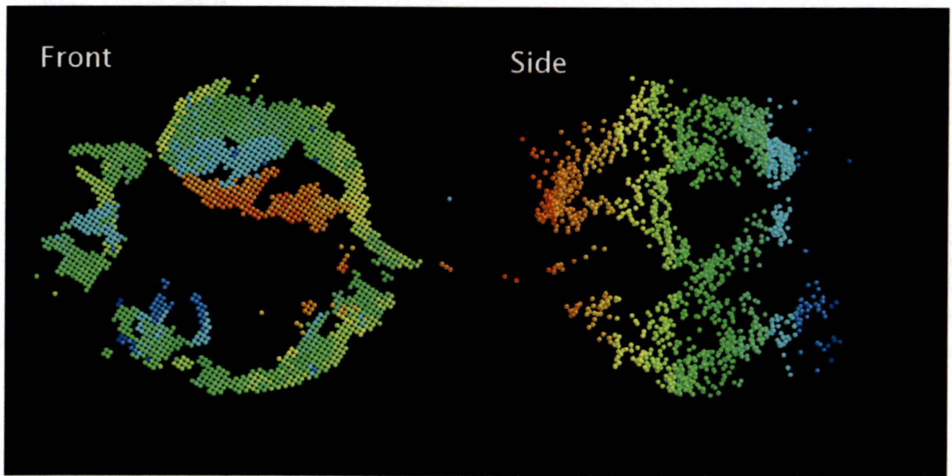

FIGURE 2. Images of Cas A in the [Ar II] line around 7μm, color-coded and three-dimensional reconstruction using Doppler velocities. Velocities range from -5000 km/s (blue) to 7000 km/s (red). In the side view, the observer is to the right. Analysis and diagrams courtesy of T. DeLaney (MIT).

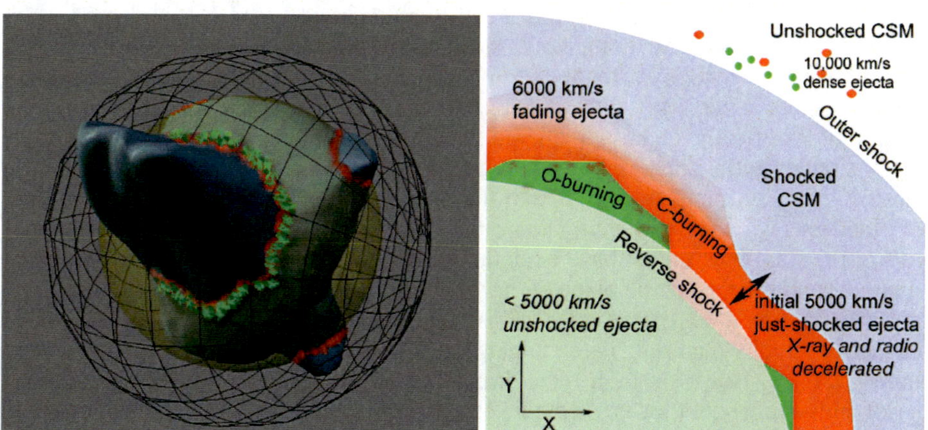

FIGURE 3. Left: Cartoon of the Cas A explosion. The wire frame is the outer shock, which shows no ejecta emission. The transparent yellow sphere is the reverse shock. The blue-gray structure represents deep layer ejecta. Colored rings mark the emission from ejecta at the positions where they encounter the reverse shock. Right: Cross section illustrating these same structures labelled according to their observed emission (Reproduced from Fig. 10 of [10])

motions of freely-expanding clumps of material as they encounter the reverse shock. At this time and place they are compressed, heated – perhaps decelerated – and become visible at different wavelengths. After a short period of time ($\sim$30 years) their emission fades, leaving behind a bright ring of emission whose inner boundary is the reverse shock. Indeed, newly shocked ejecta have been detected with multiple epoch HST images [27]. The different speeds through the reverse shock then reflect different degrees of deceleration of the material radiating at different wavelengths. Although the details of these interactions have not been worked out, the idea of coupling deceleration to various emission processes has been recognized for many years [1].

The most important implication of this picture is that any picture of Cas A represents only a very selective snapshot. The ejecta from the explosion emerged with a wide range of velocities, and currently are found all the way from their origin to $\sim$4 pc. However, given the position of the reverse shock today, almost all of the material that is visible is that fraction that emerged at $\sim$5000 km/s. This highly selective view needs to be incorporated into all calculations of the progenitor and ejected masses and the nucleosynthetic yields.

The description above is essentially a one-dimensional (radial) characterization of the explosion dynamics and its relation to the observations. However, the addition of Doppler velocities from our Spitzer observations also forces us to re-draw our picture of the explosion structure. Figure 2 shows the results of our 3-D reconstruction of the positions of the argon ejecta [6]. The reverse shock is clearly not illuminated uniformly; instead it is seen as a series of ring-like structures on a spherical surface. Most of the emission is found in a broad band close to the plane of the sky; the front and back surfaces (as seen by us) have very little emission. Ring-like structures on a spherical surface have also been seen in the optical emitting material [30, 24].

We have developed a cartoon, shown in Figure 3, of how such structures could be produced. First, we recognize that the heavy elements from the inner layer of the star are efficient emitters in the X-ray, optical and infrared. In our cartoon, the outer layers of the star explode spherically, creating the large scale forward and reverse shocks that appear roughly circular in the plane of the sky. Aside from synchrotron radiation, there is little optical, X-ray or infrared emission from these structures except where they encounter heavy elements which are efficient radiators. These heavy elements arise from the inner layers of the star, which must have exploded very asymmetrically, with high velocities in a broad band roughly in the plane of the sky, and smaller velocities towards and away from us. This broad band is itself very asymmetric, with large protruding fingers of even higher velocity material. These fingers are partially visible as the well-known "jet" features [19, 16]. However, their most dramatic effect is to create the observed bright rings of emission where they cross the reverse shock. By contrast, little emission is seen towards the center of the remnant (in the plane of the sky) since little of the inner-layer material has yet reached the reverse shock in those directions.

Supernova explosion models are beginning to deal with the large-scale asymmetries such as can be set up by rotation and instabilities [18, 14]. Whether the geometry we propose for Cas A's explosion can arise naturally, or would require very special conditions, remains to be seen.

NUCLEOSYNTHESIS LAYERS

Remarkably, given the very strong asymmetries in the explosion, it appears that along each radial line from the initial explosion, the nucleosynthetic layering has remained

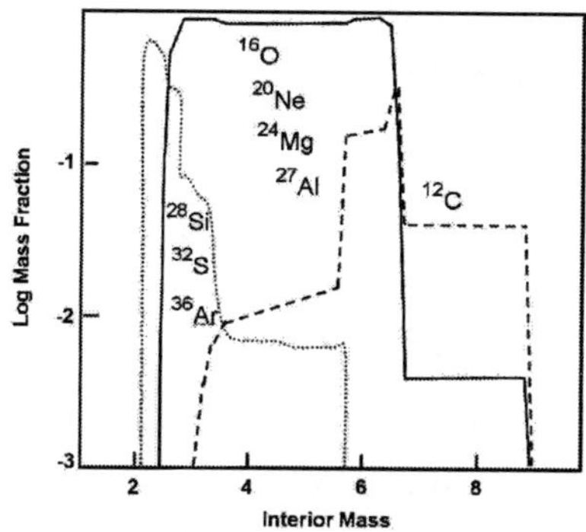

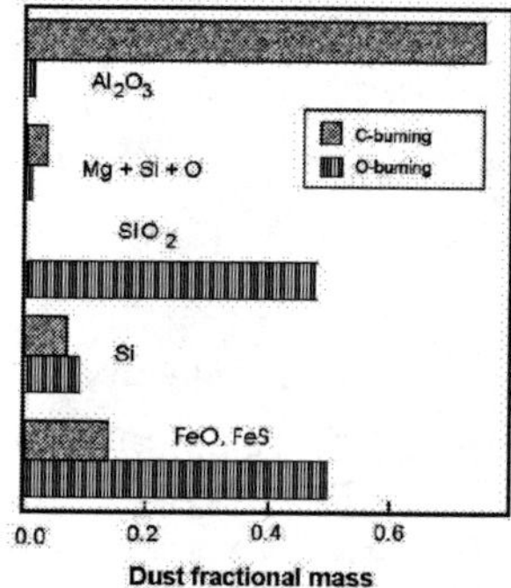

FIGURE 4. Left: Composition of nuclear burning layers, adapted from [34] for a 25 $M_\odot$ star. Three representative distributions are shown, for the He- (dashed), C- (solid) and O- (dots) burning products. Right: Chart showing two representative, but not unique models of the composition of two primary classes of ejecta dust, reflecting different burning layers [31].

largely intact, despite some mixing in small-scale knots. Looking back to Figure 1 (left), we see a wide variety of colors representing different ratios of emission in the various IRAC bands. The predominant red/orange hues are from regions where IRAC channel 4 has a large contribution from [Ar II] at 7μm, as confirmed by our IRS spectra [10]. Spectra of the yellow/green regions show a much higher ratio of [Ne II]/[Ar II].

Large variations in the brightness of different elements are also seen in the optical [12] and X-ray [16]. However, such variations could have been due to a variety of causes in addition to local composition, such as temperature, density and ionization state. Comparison of the variations in the different wavebands revealed the underlying pattern. The crescent-shaped green feature slightly east of south in Figure 1 has (relatively) strong [Ne II] emission in the infrared, increasing [O III] emission in the optical, and a distinct gap in the silicon-dominated X-ray emission. A similar crescent-shaped feature towards the north also shows O VIII emission in the X-ray. We therefore suggest that these are locations where the carbon-burning layers (the solid line in Figure 4, left) are now encountering the reverse shock, producing a multi-temperature, multi-ionization state plasma dominated in the gaseous state by oxygen and neon.

Looking now at the strong [Ar II] (red) regions in Figure 1 (left), we find that these match very closely to the places where helium- and hydrogen-like ionization states of Si and S are seen in the X-ray and, e.g., [S III] is seen in the optical [12]. Thus, we again have multiple temperatures and ionization states in the same region, but now dominated by the products of oxygen burning instead of the carbon-burning products.

A more detailed discussion of the segregation of these various nucleosynthetic layers is presented in [10]. The overall picture that emerges is that the wide spatial variations in composition seen in infrared, optical and X-ray studies are *not* due to local differences in ionization age or temperature, but instead reflect very specific asymmetries in the geometry of the underlying explosion. These asymmetries are ones in which the nucleosynthetic layers have remained mostly intact along each radial direction, but the velocity profiles along different radial directions vary over a range of approximately five.

NEWLY FORMED DUST

The extent to which supernovae are net producers of dust is a long-standing question, one key to the overall recycling of material in the Galaxy. Recently, very large quantities of dust have been found in high-redshift galaxies and quasars [17, 3]. For supernovae to be the source of dust in z=6 quasars [9, 25], ($\sim$1 $M_\odot$/SN) is required. Theoretical investigations [21, 33, 28] suggest that $\sim$0.1 - 1.0 $M_\odot$ can be formed in the ejecta of Type II SNe. However, observations of nearby supernovae and remnants find much lower values, e.g., SN 1987A ($\sim 10^{-3} M_\odot$) [11] and the Crab nebula ($< 10^{-2} M_\odot$) [32].

The presence of significant quantities of dust in Cas A's ejecta was demonstrated through ISO observations [23, 2, 7]. Our new Spitzer IRS data have now shown that dust is almost always visible wherever ejecta are seen at other wavelengths (see Figure 1, right), and with different compositions at different locations [31]. Heated circumstellar dust is also seen from the forward shock region.

The presence of the dust wherever higher temperature plasmas are present is an important finding. It means that to first order, the dust can survive its passage through shocks and resist sputtering for at least several decades in the hostile post-reverse shock environment. How it does this, and on what spatial scale the dust and 10^7K plasma are segregated, are matters for future work. There is likely a significant amount of cooler ejecta dust between the forward and reverse shocks, from material that crossed the reverse shock earlier. However, like the X-ray and optical emission, it is not currently visible. The similarity in the width of the X-ray and dust bright rings suggests that the dust is being continually heated by the surrounding X-ray gas.

The colors in Figure 1 (right) represent different shaped dust continuum spectra. There is a clear segregation between different regions, indicating different compositions and physical conditions in the dust, as discussed in detail by [31] and as seen in Figure 4. Around the outside, the dust emission (yellow in Figure 1, right) shows a broad peak around 26μm, and is likely due to small circumstellar carbon grains heated by the forward shock. The bulk of the bright ring (blue) shows continuum emission with a strong peak around 21μm, indicating the presence of silicates such as SiO_2 and magnesium proto-silicates. Green regions lack the strong silicate signature, and are likely dominated by Al_2O_3 and C grains.

These varying compositions in the dust reflect the same nucleosynthetic layering seen in the gaseous emission. Strong silicate dust emission, for example, is seen where silicon and sulfur gas are seen in the optical and X-ray and argon is seen in the infrared – all reflecting the composition of the oxygen-burning layers. Similarly, Al_2O_3 predominates where neon and oxygen gas are seen – the products of carbon burning. The segregation of these different dust compositions provides unambiguous evidence for production in the aftermath of the supernova explosion, and before any significant mixing can take place. Since the destruction of grains depends on their composition and size [29], future detailed comparisons between the fine-scale distributions of different types of dust and X-ray gas may allow us to better understand the physics of these processes.

We made estimates of the total mass in currently visible ejecta dust by combining the results from the different observed compositions. A mass of freshly formed dust between 0.02 and 0.05 $M_\odot$ is required to explain the emission up to 70μm. Most of this mass is in colder dust, 40 – 150 K, and the derived mass depends strongly on its assumed composition, with higher masses when Fe or FeO dust are used to explain the long-wavelength emission. We cannot usefully constrain the presence of even colder dust associated with Cas A [8, 22]. At best, the amount of dust we observe can account for <20% of that expected [28, 33] from a 15 $M_\odot$ progenitor. However, even this amount of dust is interesting from a cosmological standpoint, since injection of such quantities of dust for 2 billion years up to a redshift of 4 could provide the low-end estimates of $\sim 10^8$ $M_\odot$ observed for submm galaxies and quasars [17, 4].

INSIDE THE REVERSE SHOCK

One surprising result of our Spitzer observations is the presence of substantial emission from the central regions of the remnant, in material that has not yet reached the reverse shock. We see this in the lines of [Si II] (35μm), [S III] (33 μm), and [Fe II] or [O IV] (26 μm) where UV radiation from the reverse shock might provide the needed ionization energy. In Figure 5, we see the dramatic transition of Si to silicate dust at the reverse shock. At that point, the [S II] disappears, and heated silicate dust and strong X-ray emission appears, e.g., from Si XIII. Over the next 300 years, the reverse shock will reach the center of the remnant, with the remnant eventually shrinking in apparent size and producing bright X-ray and dust emission from the inner oxygen and silicon burning layers of the star. The front and back surfaces, now dark, will greatly increase in brightness. Unfortunately, few of us will be around to verify these expectations.

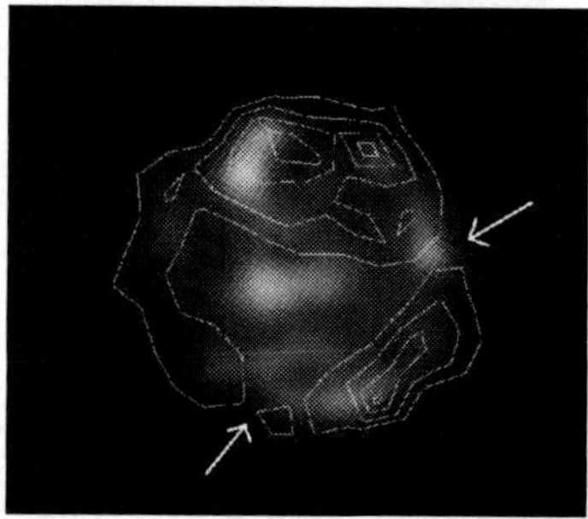

FIGURE 5. Two images of the silicon in Cas A. The contours show the location of the dust emission with strong emission from silicates. The grey scale shows the emission from gaseous [Si II]. Arrows indicate imminent reverse shock encounters with the oxygen-burning layers, where X-ray and optical Si emission and silicate dust will become visible in the future.

CONCLUDING REMARKS

Cas A is beginning to provide a much richer picture of the dynamics of supernova explosions and how these are tied to the observations at various wavelengths. This work has emphasized that the emission we see is virtually all immediately past the reverse shock – a very restricted snapshot. When these observations are integrated fully into models including all the currently invisible material, we will have a very powerful tool for examining such key issues as nucleosynthetic and dust yields.

ACKNOWLEDGMENTS

Thanks to many students and collaborators over the years who have each contributed critical pieces of this emerging puzzle. The analysis of Spitzer data was supported by NASA through an award issued by JPL/Caltech. Thanks also to the conference organizers for their gracious invitation and support.

REFERENCES

1. M. C. Anderson, T. W. Jones, L. Rudnick, I. Tregillis & H. Kang, *Astroph. Journal* **421**, L31–L34 (1994).
2. R. Arendt, E. Dwek & S. Moseley, *Astroph. Journal 521*, 234 (1999).
3. F. Bertoldi et al., *Astron. & Astroph.* **409**, L47 (2003).
4. R. Chini & E. Kruegel *Astron. & Astroph.* **288**, L33–L36 (1994).
5. T. DeLaney, L. Rudnick, R. Fesen, T. W. Jones, R. Petre & J. A. Morse. *Astroph. Journal* **613**, 343–348 (2004).
6. T. DeLaney, J. D. Smith, L. Rudnick, J. Ennis, W. Reach, J. Rho, T. Kozasa & H. Gomez, *Bull. of the AAS* **37**, 1435, and in preparation.
7. T. Douvion, P. Lagage & E. Pantin, *Astron. & Astroph.* **369**, 589–593 (2001).
8. L. Dunne et al., *Nature* **424**, 285–287 (2003).
9. E. Dwek, F. Galliano & A. P. Jones, *Astroph. Journal* **662**, 927 (2007).
10. J. Ennis, L. Rudnick, W. Reach, J.D. Smith, J. Rho, T. DeLaney, H. Gomez, and T Kozasa, *Astroph. Journal* **652**, 376–386 (2006).
11. B. Ercolano, M. Barlow & B. Sugerman *Mon. Notices Royal Ast. Soc.* **375**, 753 (2007).
12. R. Fesen et al., *Astron. Journal* **122**, 2644–2661 (2001).
13. R. Fesen et al., *Astroph. Journal* **645**, 283–292 (2007).
14. C. L. Fryer & P. A. Young, *Astroph. Journal* **659**, 1438–1448 (2007).
15. E. Gotthelf, B. Koralesky, L. Rudnick, T. W. Jones, U. Hwang & R. Petre, *Astroph. Journal* **552**, L39–L43 (2001).
16. U. Hwang et al., *Astroph. Journal* **615**, L117–L120 (2004).
17. K. Isaak et al., *Mon. Notices Royal Astron. Soc.* **329** 149–162 (2002).
18. H.-Th. Janka, K. Langanke, A. Marek, G. Martinez-Pinedo & B. Mueller *Physics Rep.* **442**, 38–74 (2007).
19. K. Kamper & S. van den Bergh *Astroph. Journal Suppl. Ser.* **32**, 351–366 (1976).
20. T. Kozasa, H. Hasegawa & K. Nomoto *Astroph. Journal* **344** 325–331 (1989).
21. T. Kozasa, H. Hasegawa & K. Nomoto *Astron. & Astroph.* **249** 474 (1991).
22. O. Krause et al., *Nature* **432**, 596–598 (2004).
23. P. Lagage et al., *Astron. & Astroph.* **315**, L273 (1996).
24. S. S. Lawrence et al., *Astron. Journal* **109**, 2635 (1995).
25. R. Maiolino et al., *Memorie d. Soc. Astro. Italiana* **77**, 643 (2006)
26. H. Morgan & M Edmunds, *Mon. Notices Royal Astron. Soc.* **343**, 427–442 (2003).
27. J. Morse et al., *Astroph. Journal* **614**, 727–736 (2004).
28. T. Nozawa, T. Kozasa, H. Umeda, K. Maeda, K. Nomoto *Astroph. Journal* **598**, 795 (2003).
29. T. Nozawa, T. Kozasa, A. Habe, E. Dwek, H. Umeda, N. Tominaga, K. Maeda & K. Nomoto *Astroph. Journal* **666**, 955–966 (2007).
30. J. Reed, J. Hester, A. Fabian, & P. Winkler, *Astroph. Journal* **440**, 706 (1995).
31. J. Rho, T. Kozasa, W. T. Reach , J. D. Smith, L. Rudnick, T. DeLaney, J. A. Ennis, H. Gomez, & A. Tappe, *Astroph. Journal*, 271 (2008).
32. T. Temim et al., *Astron. Journal* **132**, 161 (2006).
33. P. Todini & A. Ferrara *Mon. Notices Royal Astron. Soc* **325**, 726 (2001).
34. S. Woosley & T. Weaver, *Astroph. Journal Suppl. Ser.* **101** 181 (1995).

Suzaku Results of SN 1006: Chemical Abundances of the "youngest" Galactic Type Ia Supernova Remnant

Katsuji Koyama

Department of physics Kyoto University, Kitashirakawa, Sakyo-ku, Kyoto 606-8502

Abstract. SN 1006 is one of the supernova remnants (SNR) recorded in the Japanese diary "Meigetsuki". From the historical records including Meigetsuki, we conclude that SN 1006 was the brightest type Ia supernova remnant. We report on the observations of SN 1006 with the X-ray Imaging Spectrometers (XIS) on board the 5-th Japanese X-ray satellite Suzaku. We found that the ionization age of SN 1006 is the youngest among any Galactic SNRs, hence is the best SNR to study early phase of type Ia. In the X-ray spectrum, we found the K-shell emission lines from heavy elements, in particular that from iron, for the first time. The X-ray emitting plasma is highly over-abundant in heavy elements, hence are likely due to ejecta. The abundance pattern agrees well to the theoretical prediction of type Ia supernova.

Keywords: ISM: individual(SN 1006) --supernova remnants --X-Rays: spectra
PACS: 90

HISTORICAL RECORED OF SUPER NOVA IN JAPAN

I will start my talk from the Japanese historical records of supernova events. In 8 November 1230, Teika Fujiwata wrote some historical guest stars in the diary Meigetsuki [1], the ancient samples of guest stars. Guest stars are transient star-like objects such as comets, novae and supernovae. Out of the 8 recorded events, we can identify three events are supernovae. The remnants are now called as 3C58, Crab Nebula and SN1006. The description of the Crab nebula, the remnant of SN 1054 is:

後冷泉院天喜二年 四月中旬以降丑時、客星出觜參度、見東方、孛天關星、大如歳星

This means that " in 1054, a guest star appeared in between the Orion and Taurus constellations. It was like Jupiter." 3C58, the remnant of SN 1182 is:

高倉院治承五年六月廿五日、庚午、亥時、客星見北方、近王良星守傳舍星

CP1016, *Origin of Matter and Evolution of Galaxies*,
edited by T. Suda, T. Nozawa, A. Ohnishi, K. Kato, M. Y. Fujimoto, T. Kajino, and S. Kubono

This means that " in 1181, a guest star appeared at the northern sky near Cassiopeia β." I will focus on the remnant SN 1006. Bellow is the Japanese description on the SN 1006 event.

一條院 寛弘三年 四月二日 葵酉 夜以降 騎官中 有大客星 如熒惑 光明動燿 連夜正見南方 或云 騎陣将軍星本体 増変光

The English translation is "in May 1st 1006, a great guest star appeared in Kikan, the oriental name of constellations Lupus and Centaurus. The great guest star was bright like Mars, and was visible in the southern sky in every night."

As I will discuss later, SN1006 is the brightest historical SN. So a question is why like Mars? In the midnight of Kyoto city on the day of the on-set of SN1006 (1006/05/01), Mars and Antares were at the southern sky. Then a bright SN appeared near Mars, it suffered from reddening due to the low elevation angle. Our old Japanese observer, therefore said that SN1006 was similar to Mars.

HISTORICAL RECORD OF SN 1006 IN THE WORLD

Many historical records of SN 1006 are found in other countries of Asia and Arabia [2]. In short, these are; (1) in Kyoto Japan, the day of on-set of the SN explosion, it was like Mars, (2) in Baghdad, it was similar to Venus on May 3-rd, (3) in Egypt, it was similar to a quarter of the moon on May 5-th, (5) in China, it was like a half moon on May 6-th and (6) there was a record of possible SN 1006 after a few months as bright as Mars. Since the maximum apparent magnitude m of Mars, Venus and the full moon are $m = -3$, $m = -4.7$ and $m = -12.6$, respectively, we can convert m of SN 1006 in each epoch as about $m = -3$ on May 1, $m = -4.7$ on May 3, $m = -8$ on May 5, $m = -10$ on May 6 and $m = -3$ after a few month. Thus the rise time of the light curve was a few days and reached at the peak luminosity of $m = -10$, then decayed with a time scale of a few month. This profile of the light curve is similar to that of Type Ia SN. At present, we can determine the distance of SN1006 using a field star behind the SNR as 2 kpc [3]. Then the peak luminosity of $m = -10$ is converted to the absolute magnitude $M = -20$. This luminosity is exactly that of a type Ia SN.

THE SUZAKU SATELLITE AND OBSERVATIONS OF SN 1006

The last year, 2006 was one millennium after the SN1006 event. We therefore took memorial X-ray picture by Suzaku [4], the 5th Japanese X-ray astronomical satellite launched on July 10-th 2005. The nick name Suzaku is a "red angel bird" in the Oriental Mythology, living in the southern sky from the emperor's Palace. Figure 1 is a wall painting of Suzaku, the "red angel bird", in the old famous tomb Kitora Kofun at the Asuka (ASCA) village, where the ancient culture of Japan had been originated.

FIGURE 1. The wall painting of Suzaku, "the red angel bird" in the Kitora Kofun, the old famous Japanese style tomb.

Suzaku has the Imaging X-ray mirrors XRT [5] and the X-ray CCD cameras (XIS) on the focal planes [6]. In 2006, the southern sky bird, Suzaku saw the southern sky and took a millennium-memory picture of SN 1006 in X-rays. We got a very nice X-ray image and spectrum from SN 1006 as are given in figures 2a and 2b. Figure 2b is the thermal spectrum from hot plasma inside SN 1006. We discovered heavy elements such as Ar, Ca and Fe for the first time.

In normal case, we can determine the chemical compositions using the conventional plasma code. The plasma code has been made based on the laboratory plasma physics. However this spectrum is very unusual, and hence can not be applied by the conventional plasma code. What is unusual?, what is the difference between this plasma and those in the other SNRs ?.

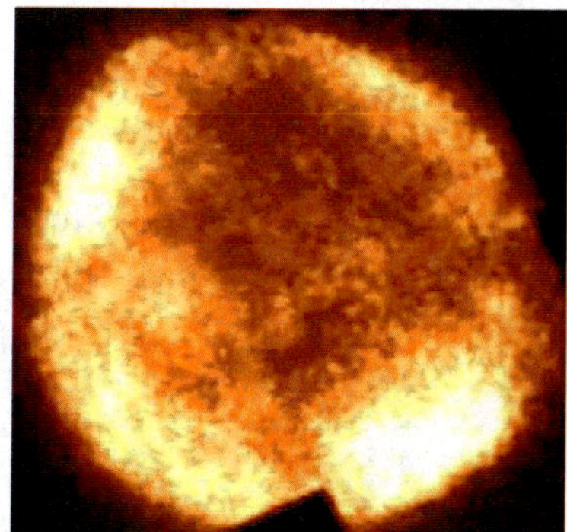

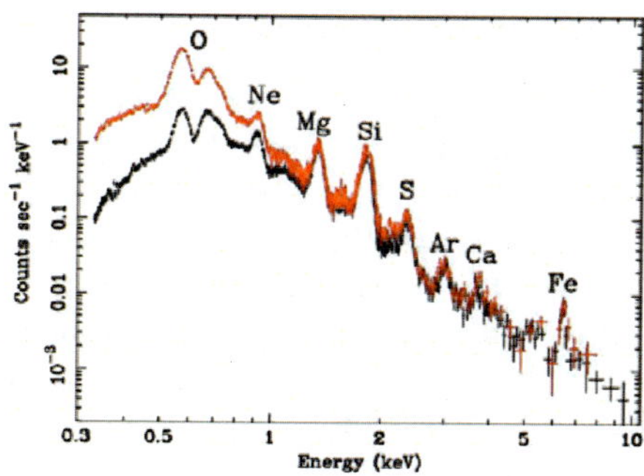

FIGURE 2. The X-ray image of SN 1006 in the oxygen band (0.5--0.6 keV) (2a) and the X-ray spectrum inside SN 1006 (2b). These figures are taken from Yamaguchi et al. 2008 [7].

As an example, I will show the unusual aspect using this oxygen band. Figure 3 is a close-up view of the X-ray spectrum near the oxygen K-shell transition lines. In the high temperature plasma, all the outer shell electrons of the oxygen atom are gone, hence oxygen atom is either H-like with only on electron in the inner K-shell or He-like with two electrons in the K-shell. Therefore the high temperature plasma emits K-shell X-rays from He-like and H-like oxygen at energies of 570 eV and 650 eV. If we see this spectrum from the other SNRs, most of the X-ray astronomers believe that these two-lines are K-shell transitions (Kα) from He-like and H-like oxygen. This, in fact, is true for any other SNRs. However by careful energy scale calibration, we found that lower energy line is surely Kα of He-like oxygen. However the higher energy line is not Kα of H-like oxygen but Kβ of He-like oxygen. We further found Kγ and Kδ lines from He-like oxygen for the first time. No conventional plasma code includes such lines. In order to emit such high level transition lines, we need high electron temperature like 1.5 keV. In this high temperature plasma, oxygen is almost fully ionized, hence no He-like K shell line is emitted. Thus we need very low ionization temperature of about 0.15 keV with higher energy electrons like in a 1.5-keV temperature plasma.

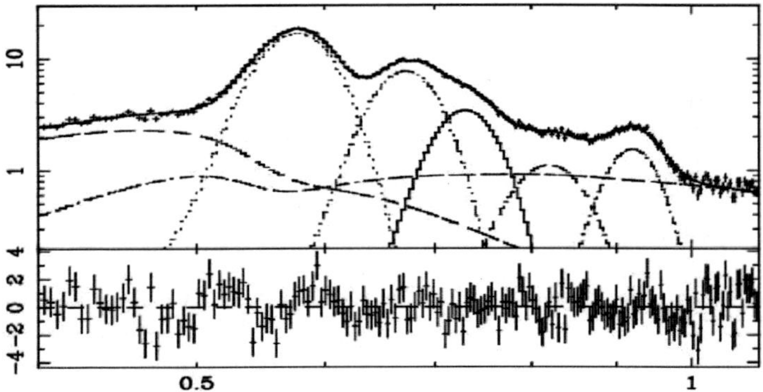

FIGURE 3. A closed-up view of the oxygen band of the SN1006 spectrum. The lowest 4 lines are Kα, Kβ, Kγ and Kδ of He-like oxygen. The highest energy line is Kα of He-like neon. This figure is taken from Yamaguchi et al. 2008 [7].

Figure 4 is a schematic picture of the thermal evolution of the SNR plasma. In the initial state, just after the shock heating, the proton temperature must be 1000 times higher than electron temperature, because the mass is 1000 times larger. Then the thermal energy of proton is transferred to electron and the electron temperature becomes high. The high energy electron ionized the atoms. These processes go with 2-body collisions, and hence can be parameterized by the product ***nt***, where ***n*** is the plasma density and ***t*** is the time after the shock heating. Since the plasma density of SN1006 is fur lower than any other SNRs, the thermal ages of SN1006 comes here. Thus SN1006 is the "youngest" SNR, younger than Cas A, Tycho and Kepler's SNRs.

Taking into account of these unusual conditions, we can fit the spectrum. The best-fit spectrum is shown in figure 5. In the fitting, we need 3 plasma components. One is a plasma with solar abundance, hence would be inter stellar origin. The other two are over-abundant plasma, hence would be due to the SNR ejecta (ejecta 1 and ejecta 2). Ejecta 2 has a very low ionization parameter, hence has been heated recently by the reverse shock. In other words, this plasma comes from inner region of the SNR. While ejecta 1 is from a outer region of the SNR, and hence has been heated early by the reverse shock.

The best-fit abundances relative to oxygen in units of solar abundance ratio are shown in figures 6a and 6b, for the ejecta 1 and 2 respectively. The solid histograms are the theoretical type Ia W7 model by Nomoto et al. (1984) [8]. We see that the

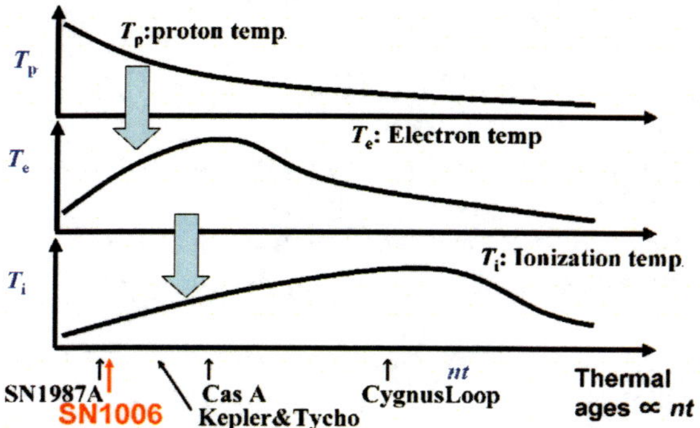

FIGURE 4. A schematic picture of the thermal evolution of the SNR plasma.

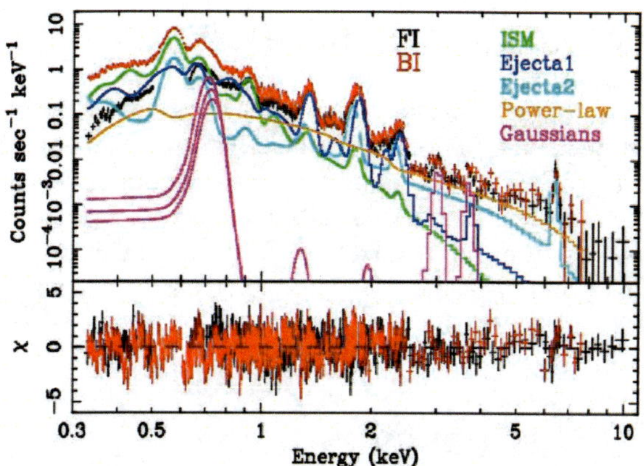

FIGURE 5. The best-fit 3-plasma model. ISM is a plasma with the solar abundance. Ejecta 1 and 2 are over-abundant plasmas. Ejecta 2 has a very low ionization parameter, which has been heated recently, hence comes from an inner region of the SNR. While ejecta 1 has a moderate ionization parameter, hence comes from an outer region of the SNR. This figure is taken from Yamaguchi et al. 2008 [7].

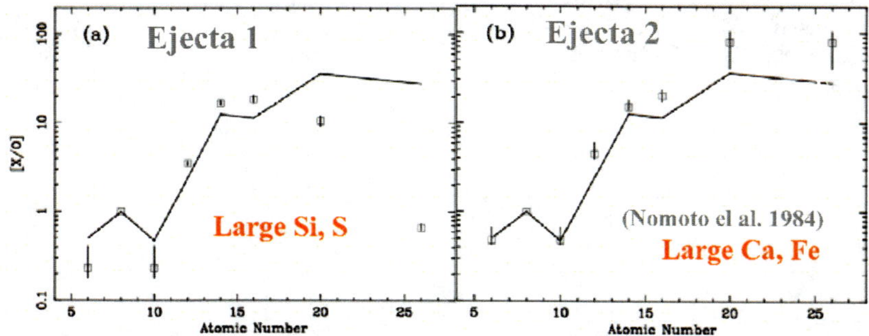

FIGURE 6. The best-fit abundances relative to oxygen for the ejecta 1 (a) and 2 (b). These figures are taken from Yamaguchi et al. 2008 [7].

abundances of ejacta 2 extremely resemble to those of the W7 model, which strongly supports that SN 1006 is a remnant of type Ia SN. Since ejecta 1 is from a outer part of the SNR, hence lack heavier elements like Ca and Fe. Other than these two elements, the abundance pattern is similar to the model W7.

CONCLUSION

The conclusion of the Suzaku observations on SN 1006 are as follows;

(1) We discovered Ar, Ca and Fe Lines in the spectrum of SN 1006 for the first time.
(2) We found a plasma in extremely non-equilibrium between electron temperature and ionization temperature.
(3) The ejecta consists of two plasma components: One has almost identical abundance profile to that of the theoretical prediction of type Ia SNR, and the other has lower abundance in Fe. The former would be due to an inner part of the SNR and the latter is due to an outer part.

ACKNOWLEDGMENTS

The author thanks all of the Suzaku team members for their nice operations, supports and useful information of Suzaku. The XIS members performed careful calibration, and hence provide me with good performances of XIS. Particular thanks are due to H. Yamaguchi for his elaborated data analysis on SN 1006. This work is supported by Grant-in-Aids from the Ministry of Education, Culture, Sports, Science and Technology (MEXT) of Japan, the 21st Century COE "Center for Diversity and Universality in Physics" and Scientific Research A .

REFERENCES

1. T. Fujiwara, 1230, Meigetsuki Vol 52
2. F. R. Stephenson D. H. Clark, D. F. Crawford 1977 MNRAS 180, 567
3. C. C. Wu, D. M. Crenshaw, R. A. Fesen, A. J. S. Hamilton, C. L. Sarazin 1993, ApJ, 416, 247
4. K. Mitsuda, K., et al. 2007, PASJ, 59, 1
5. P. J. Serlemitsos et al. 2007, PASJ, 59, 9
6. K. Koyama et al.2007, PASJ, 59, 23
7. H. Yamaguchi et al. 2008, PSAJ, in press
8. K. Nomoto, F. K. Thielemann, K. Yokoi 1984, ApJ, 286, 644

Supernovae contributions to metals in intra-cluster medium observed with *Suzaku*

Kosuke Sato*, Kazuyo Tokoi†, Kyoko Matsushita*, Yoshitaka Ishisaki†, Noriko Y. Yamasaki**, Manabu Ishida** and Takaya Ohashi†

*Department of Physics, Tokyo University of Science,
1-3 Kagurazaka, Shinjuku-ku, Tokyo 162-8601, Japan
†Department of Physics, Tokyo Metropolitan University,
1-1 Minami-Osawa, Hachioji, Tokyo 192-0397, Japan
**Institute of Space and Astronautical Science (ISAS),
Japan Aerospace Exploration Agency,
3-1-1 Yoshinodai, Sagamihara, Kanagawa 229-8510, Japan

Abstract. We studied the properties of the intra-cluster medium (ICM) in two clusters of galaxies (AWM 7 and Abell 1060) and two groups (HCG 62 and NGC 507) observed with the 5th Japanese X-ray astronomical satellite, *Suzaku*. We measured for the first time precise cumulative ICM metal masses and distributions for various elements, such as O, Ne, Mg, Si, S, Fe, within 0.1 and $\sim 0.3\, r_{180}$. Because of the good XIS sensitivity to emission lines, especially below 2 keV, we directly measured O and Mg line intensities and obtained abundances. Comparing our results with supernova nucleosynthesis models, the number ratio of Type II (SNe II) to Type Ia (SNe Ia) is estimated to be ~ 3.5, assuming the metal mass in the ICM is represented by the sum of products synthesized in SNe Ia and SNe II. Normalized by the K-band luminosities of present galaxies, and including the metals in stars, the integrated number of past SNe II explosions is estimated to be close to or somewhat higher than the star formation rate determined from Hubble Deep Field observations.

Keywords: X-ray, intra-cluster medium
PACS: 98.65.-r, 98.65.Hb

INTRODUCTION

The metal abundances of intra-cluster medium (ICM) have a lot of information to understand the chemical history and evolution of clusters. Because clusters are the largest gravitationally bound systems in the universe, they are expected to confine all the metals provided by member galaxies since the cluster formed. The amount of each metal in the ICM is almost equal to the integrated sum of that synthesized by supernovae type Ia (SNe Ia) and type II (SNe II). The expected elemental products of each type have been intensively studied by several authors [3, 8]. X-ray measurements of the mass of each element in the ICM enable us to examine the past occurrence numbers and the ratio of SNe II to SNe Ia. These numbers also reflect the past stellar initial mass function (IMF) and star formation rate (SFR) in clusters [9].

Based not only on the low and stable background but also on the good sensitivity to emission lines below ~ 1 keV of the *Suzaku* XIS [4], a reliable determination of O and Mg abundances to cluster outer regions has become feasible [7, 10, 11, 12].

We use $H_0 = 70$ km s^{-1} Mpc^{-1} = $0.7\, h_{100}$, $\Omega_\Lambda = 1 - \Omega_M = 0.73$, and the virial radius is assumed to be $r_{180} = 1.95\, h_{100}^{-1} \sqrt{k\langle T\rangle/10\ \mathrm{keV}}$ Mpc [6] in this paper. Errors are 90%

CP1016, *Origin of Matter and Evolution of Galaxies*,
edited by T. Suda, T. Nozawa, A. Ohnishi, K. Kato, M. Y. Fujimoto, T. Kajino, and S. Kubono

confidence region for a single interesting parameter.

METAL MASS DETERMINATION

We selected two clusters, AWM 7 and Abell 1060 (hereafter A 1060), and two groups, HCG 62 and NGC 507, that were observed with *Suzaku*. Detailed information and the analysis method for each object are described in Sato et al. [9, 10, 11] and Tokoi et al. [12]. We excluded Ni and Ne abundances in our analysis in this paper since these two elements were not reliably determined due to the strong and complex Fe-L line emissions. For the first time, we determined the radial abundance profiles of O, Mg, Si, S and Fe, out to $\sim 0.3\ r_{180}$ for each object. Note that the O abundance in the ICM is strongly affected by the foreground Galactic emission, because we cannot resolve the cluster and local lines. Combining the abundance profile obtained with *Suzaku* and the X-ray-luminous gas mass profile with *XMM-Newton*, we calculated cumulative metal mass within $0.1\ r_{180}$ and the whole observed region. Errors of the metal mass plotted in Figure 1 are taken from the statistical errors of each elemental abundance in the spectral fits at $\sim 0.1\ r_{180}$.

MASS PRODUCTS BY SNE IA AND II

In order to examine the SNe Ia and SNe II contribution to the ICM metals, the elemental mass pattern of O, Mg, Si, S and Fe was examined within $0.1\ r_{180}$ and the whole observed region. The mass patterns were fit by a combination of average SNe Ia and SNe II yields per supernova, as shown in Figure 1. The fit parameters were chosen to be the integrated number of SNe Ia ($N_{\rm Ia}$) and the number ratio of SNe II to SNe Ia ($N_{\rm II}/N_{\rm Ia}$), because $N_{\rm Ia}$ could be well constrained due to relatively small errors in the Fe abundance. The SNe Ia and II yields were taken from Iwamoto et al. [3] and Nomoto et al. [8], respectively. We assumed a Salpeter IMF for stellar masses from 10 to 50 $M_\odot$ with the progenitor metallicity of $Z = 0.02$ for SNe II, and W7, WDD1 or WDD2 models for SNe Ia. Figure 1 and 2 summarizes the fit results.

All the objects exhibit similar features. The abundance patterns were better represented by the W7 SNe Ia yield model rather than WDD1. The number ratio of SNe II to SNe Ia with W7 is ~ 3.5 as shown in Figure 2, while the ratio with WDD1 is ~ 2.5. The WDD2 model gave the results quite similar to those of W7. Almost 3/4 of the Fe and $\sim 1/4$ of the Si is synthesized by SNe Ia, in the W7 model, as demonstrated in bottom panels of Figure 1.

We note that most of the fits were not formally acceptable based on the χ^2 values. The 90% and 99% confidence values for 3 degrees of freedom are $\chi^2 = 6.25$ and 11.34, respectively. The models adapted here are probably too simplified, such as assumptions of uniform IMF and SNe yields at all times, spherical symmetry in the ICM, and rather crude treatment of the Galactic emission, etc.

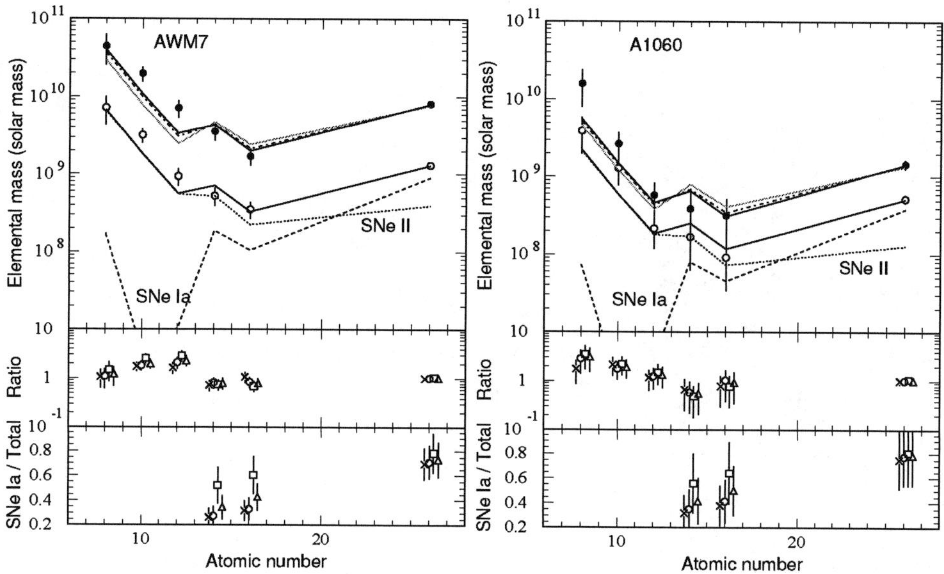

FIGURE 1. Fit results of each elemental mass for AWM 7 and A 1060. Top panels show the mass within the whole observed region (filled circle) and within 0.1 r_{180} (opened circle) fit by $[N_{\rm Ia}\,\{(\text{SNe Ia yield}) + (N_{\rm II}/N_{\rm Ia})\,(\text{SNe II yield})\}]$. Solid, light gray solid, and dotted-dashed lines show the fitting results for W7, WDD1, and WDD2, respectively. Dashed and dotted lines corresponds to the contributions of SNe Ia (W7) and SNe II within 0.1 r_{180}, respectively. Ne (atomic number = 10) is excluded in the fit. Mid and bottom panels indicate ratios of data points to the best-fit, and fractions of the SNe Ia contribution to total mass in the best-fit model for each element, respectively. Circle, square, triangle, and cross correspond to the results for W7, WDD1, WDD2, and W7 ($r < 0.1\ r_{180}$), respectively.

COMPARISON WITH OTHER STUDIES

de Plaa et al. [2] also found that the number ratio of core-collapse SNe (SNe II+Ib+Ic) to SNe Ia (using W7) is ~ 3.5, based on *XMM-Newton* observations of 22 clusters within 0.2 r_{500} for elements between Si and Ni. They also examined the fit including the average O and Ne abundances for two clusters observed with the XMM Reflection Grating Spectrometers (RGS), however RGS observations are limited to the central cluster region where the abundance is strongly affected by the cD galaxy. The number ratio of SNe II to SNe Ia they have obtained is consistent with our results.

Tsujimoto et al. [13] determined that the number ratio of SNe II to SNe Ia for our Galaxy is ~ 6.7 based on stellar metallicity, while it is 3.3–5 for the Large and Small Magellanic Clouds. These results are slightly higher than those for our clusters and groups. It is plausible that spiral galaxies in the Local Group maintain star forming activity until recently. We also note that our results are derived for metals only in the ICM, therefore metals in member galaxies (including stars) which accounts for $\sim 50\%$

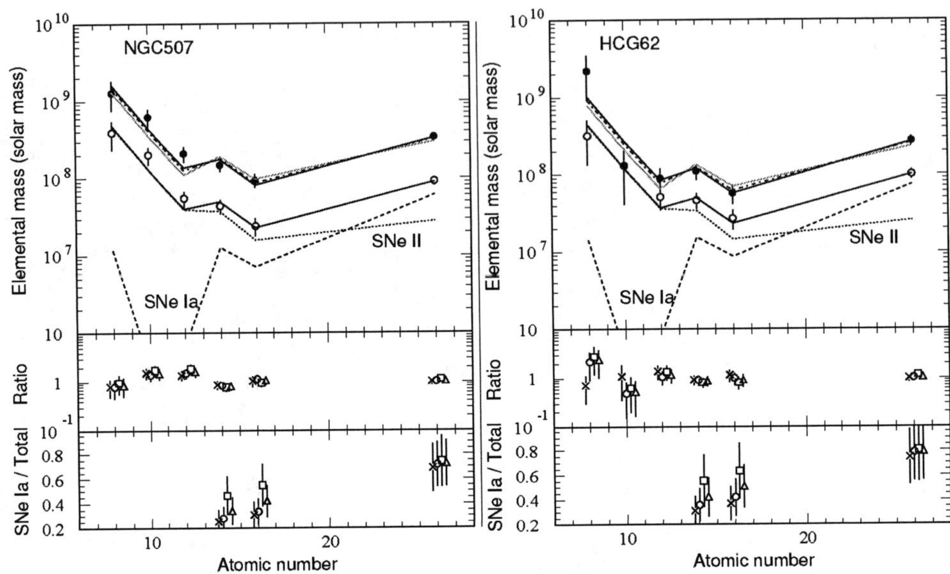

FIGURE 1. Continued. Fit results of each elemental mass for NGC 507 and HCG 62.

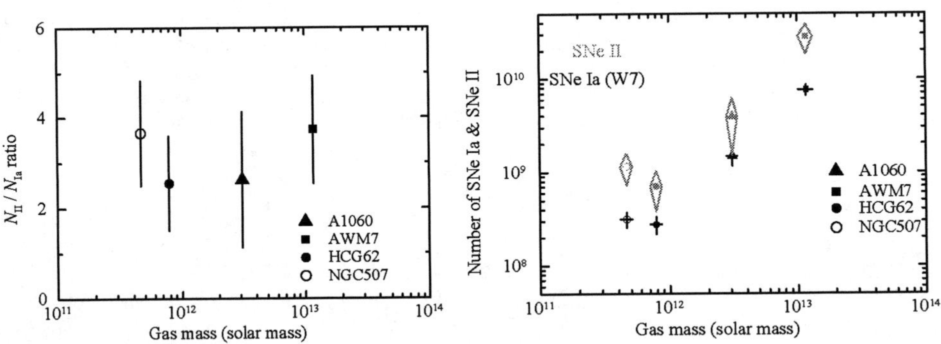

FIGURE 2. *Left*: Integrated numbers of SNe Ia (W7) and SNe II plotted against gas mass within the observed region. *Right*: Number ratio of SNe II to SNe Ia with W7 model.

of the total are not considered.

SNE II PER UNIT GALAXY LUMINOSITY

Based on the luminosity function at high redshift, the star formation history of the universe has been studied. Madau et al. [5] derived a simple stellar evolution model for

TABLE 1. Number of SNe II ($< 0.1\ r_{180}$) per volume ($N_{\rm II}/V$), K-band galaxy luminosity per volume ($L_{\rm K}/V$), and $N_{\rm II}/L_{\rm K}$ in HDF and clusters.

	$N_{\rm II}/V$ (Mpc^{-3})	$L_{\rm K}/V$ ($L_\odot$ Mpc^{-3})	$N_{\rm II}/L_{\rm K}$ ($L_\odot^{-1}$)
HDF	$< 10^5$	4.0×10^8	2.5×10^{-3}
AWM 7*	2.2×10^{11}	5.4×10^{13}	4.1×10^{-3}
A 1060*	8.2×10^{10}	5.1×10^{13}	1.6×10^{-3}

* For AWM 7 and A 1060, the number of SNe II is estimated from the metals of the ICM, not including the member galaxies and stars.

field galaxies in the Hubble Deep Field (HDF). The SFR rises sharply, by about an order of magnitude, from the present to a peak value in the range 0.12–0.17 $M_\odot$ yr^{-1} Mpc^{-3} at $z \sim 1.5$.

If we assume the maximum SFR is 0.1 $M_\odot$ yr^{-1} Mpc^{-1} and integrate it over a Hubble time, the total produced stellar mass is $10^9\ M_\odot$ Mpc^{-1}. Supposing a Salpeter IMF between 10–50 $M_\odot$, estimated number of SNe II per unit volume is $< 10^5$ Mpc^{-3} for field galaxies. When we scale our cluster results only by volume, the normalized numbers of SNe II within $0.1\ r_{180}$ for AWM 7 and A 1060 are 2.2×10^{11} and 8.2×10^{10} Mpc^{-3}, respectively. These factors of $\sim 10^5$ discrepancies are primarily caused by the difference in the number of galaxies between the field and clusters. We therefore scale these results with the K-band luminosity density as summarized in Table 1. Using an average luminosity density $L_{\rm K}/V = (5.74\pm0.86)\times10^8\ h_{100}\ L_\odot$ Mpc^{-3} in Cole et al. [1], the normalized number of SNe II per unit galaxy luminosity is $2.5\times10^{-3}\ L_\odot^{-1}$ for field galaxies. The galaxy luminosity densities of AWM 7 and A 1060 are taken from the Two Micron All Sky Survey.[1] As shown in Table 1, the numbers of SNe II per unit luminosity in the clusters are quite consistent with the above value independently estimated from the star formation history. However, this estimate considers only the ICM metals, and inclusion of stellar metals would roughly double the required SNe II rate. Even though systematic uncertainty in this simple estimation may account for most of the factor of two difference, it is interesting that X-ray metals suggest incompleteness in the optical detection of distant star formation. Also, an environmental difference between fields and clusters may be of relevance.

MASS-TO-LIGHT RATIOS WITH B-BAND LUMINOSITY

We introduce the parameter of cluster metal mass-to-light ratio and estimate its value for both the ICM and the stars. We calculated the iron mass-to-light ratio (IMLR), oxygen mass-to-light ratio (OMLR), and magnesium mass-to-light ratio (MMLR) in unit of $M_\odot/L_\odot$ as shown in Figure 3. We utilized the observed redshift to estimate the

[1] The database address: `http://www.ipac.caltech.edu/2mass/`

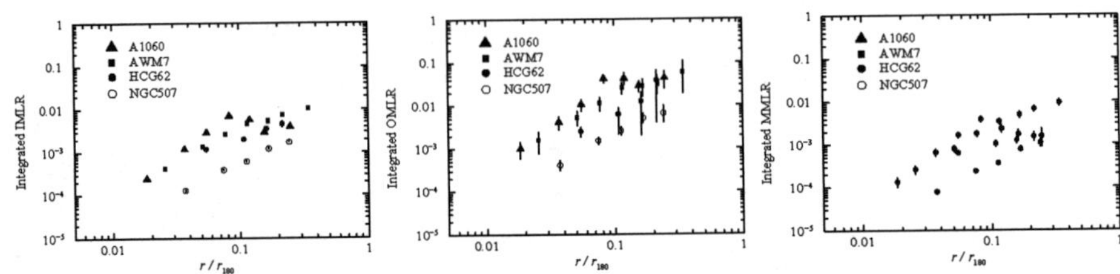

FIGURE 3. These panels show the radial profiles of the integrated IMLR (left), OMLR(middle), and MMLR (right) for AWM 7, A 1060, NGC 507, and HCG 62.

3-dimensional distribution of the galaxy, assuming the spherical symmetry. Although the estimated optical luminosity and gas mass profile are both subjected to large errors, the errors in Figure 3 included only the statistical errors of the metal abundances.

All the integrated IMLR, OMLR, and MMLR show a steep increase up to a radius of $\sim 0.1\ r_{180}$, and seem to reach almost the maximum at 0.1-0.2 r_{180} as shown in Figure 3. The steep increase in the $r < 0.1\ r_{180}$ region suggests that the gas with Fe, O, and Mg ions which were synthesized in the central regions have been distributed to a wide region in the intra-cluster space, and/or that the member galaxies have been more concentrated than the ICM.

REFERENCES

1. Cole, S., et al. 2005, MNRAS, 362, 505
2. de Plaa, J., Werner, N., Bleeker, J. A. M., Vink, J., Kaastra, J. S., & Méndez, M. 2007, A&A, 465, 345
3. Iwamoto, K., Brachwitz, F., Nomoto, K., Kishimoto, N., Umeda, H., Hix, W. R., & Thielemann, F.-K. 1999, ApJS, 125, 439
4. Koyama, K., et al. 2007, PASJ, 59, 23
5. Madau, P., Pozzetti, L., & Dickinson, M. 1998, ApJ, 498, 106
6. Markevitch, M., et al. 1998, ApJ, 503, 77
7. Matsushita, K., et al. 2007a, PASJ, 59, 327
8. Nomoto, K., Tominaga, N., Umeda, H., Kobayashi, C., & Maeda, K. 2006, Nuclear Physics A, 777, 424
9. Sato, K., Tokoi, K., Matsushita, K., Ishisaki, Y., Yamasaki, N. Y., Ishida, M., & Ohashi, T. 2007, ApJL, 667, L41
10. Sato, K., et al. 2007, PASJ, 59, 299
11. Sato, K., Matsushita, K., Ishisaki, Y., Yamasaki, N. Y., Ishida, M., Sasaki, S., & Ohashi, T. 2007, PASJ, 60, S333
12. Tokoi, K., et al. 2007, PASJ, 60, S317
13. Tsujimoto, T., Nomoto, K., Yoshii, Y., Hashimoto, M., Yanagida, S., & Thielemann, F.-K. 1995, MNRAS, 277, 945

15. STELLAR and METEORITIC ABUNDANCES

Presolar Stardust in the Solar System: Implications for Nucleosynthesis and Galactic Chemical Evolution

Larry R. Nittler

Department of Terrestrial Magnetism, Carnegie Institution of Washington, Washington DC, 20015, USA

Abstract.
Primitive meteorites and interplanetary dust particles contain presolar grains: pristine solid samples of stellar material. These grains of stardust are a new source of information on nuclear astrophysics, complementary to traditional astronomical observations. Identified phases include silicates, oxides, SiC, graphite (including sub-grains of metal and carbides), and silicon nitride. Most grains can be attributed to either asymptotic giant branch (AGB) stars or Type II supernovae, based on their isotopic compositions. Detailed high-precision isotopic data for the grains provide important and unique constraints on nuclear processes occurring in these stars.

Keywords: presolar grains, interstellar dust, nucleosynthesis, galactic chemical evolution, isotopes
PACS: 96.30.Za, 96.30.Vb, 97.10.Cv, 98.35.Bd

INTRODUCTION

Presolar stardust grains are tiny mineral grains that condensed in outflows and explosions of previous generations of stars and survived processing in the interstellar medium and early Solar System [1, 2, 3]. They are a trace component (<ppb to a few hundred ppm) of primitive meteorites and interplanetary dust particles (tiny extraterrestrial samples collected in the stratosphere). They were discovered in 1987, following a long search for carriers of isotopically anomalous noble gases in meteorites (see [4] for a historical review). They are identified by their highly unusual isotopic compositions, relative to all other materials in the Solar System. These compositions reflect both Galactic chemical evolution (GCE) and nuclear processing in the parent stars. Because the presolar grains can be studied in great detail in modern microanalytical laboratories, they can provide high-precision constraints on nuclear astrophysics, complementary to traditional astronomical observations.

The discovery and increasingly detailed characterization of presolar grains has been made possible by technological advances in micro- and nano-analytical instrumentation. For example, modern secondary ion mass spectrometers (SIMS) can determine, with high sensitivity, isotopic ratios of many elements in sub-micrometer solid samples, allowing for identification of presolar grains with astrophysically relevant sizes. Fig. 1 shows secondary electron and oxygen isotopic ratio images, produced by a Cameca Instruments NanoSIMS 50L ion microprobe, of a small region of the primitive carbonaceous chondrite Alan Hills 77307. Most materials in the Solar System have highly uniform isotopic composition (<10% variation in O isotopes for instance) due to large-

CP1016, *Origin of Matter and Evolution of Galaxies*,
edited by T. Suda, T. Nozawa, A. Ohnishi, K. Kato, M. Y. Fujimoto, T. Kajino, and S. Kubono

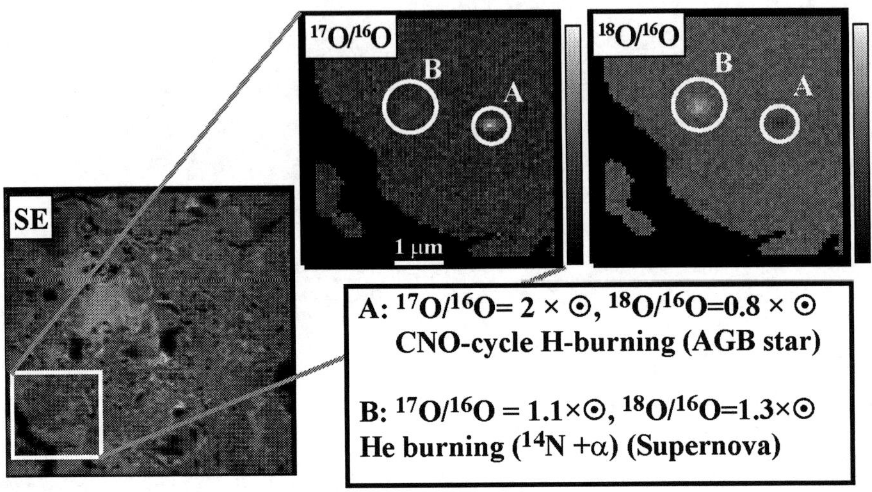

FIGURE 1. Scanning electron (SE) and O-isotopic ratio images of a small region of primitive meteorite ALHA 77307 (Nguyen and Nittler, unpub. data). Two grains (A and B, with indicated highly non-solar isotopic compositions) are identified as presolar silicate grains.

scale homogenization processes in the interstellar medium and early solar system and reflected in the homogeneous gray-scale of most pixels in the isotopic ratio images in Fig. 1. In contrast, two sub-μm grains in this region have relatively huge isotopic variations in both $^{17}O/^{16}O$ and $^{18}O/^{16}O$ ratios. These variations are much larger than can be explained by any known chemical fractionation process and point to nuclear reactions occurring in stars. Indeed, the isotopic patterns exhibited by the grains can be associated with specific nucleosynthetic processes, as indicated on the Figure.

Once identified as stardust, a presolar grain can be further analyzed by a range of techniques. For example, additional isotopic signatures might be determined using SIMS, noble-gas mass spectrometry [5], or resonance ionization mass spectrometry (RIMS, [6, 7]). Detailed chemical and mineralogical investigations can be carried out by scanning and transmission electron microscopy and/or Auger spectroscopy (e.g., [8, 9, 10, 11]). Because each grain is a sample of a *specific* place in a *specific* star at a *specific* time, with very little or no processing since its formation, presolar stardust provides important information about a range of astrophysical processes. In particular, the ability to precisely determine the isotopic composition of multiple elements in a single presolar grain places unprecedented quantitative constraints on stellar evolution and nucleosynthesis models (e.g. [12]). Mineralogical and microstructural studies provide detailed information on grain formation processes in stars (e.g., [8, 13, 10]).

A detailed review of presolar grains and their applications in astrophysics and space science is beyond the scope of this paper. Here I will focus on a few examples that illustrate how meteoritic stardust provides new insights and quantitative constraints on nuclear processes in the Galaxy. Several reviews of the field have been published in recent years [1, 2, 3] and the interested reader is referred to these and the current

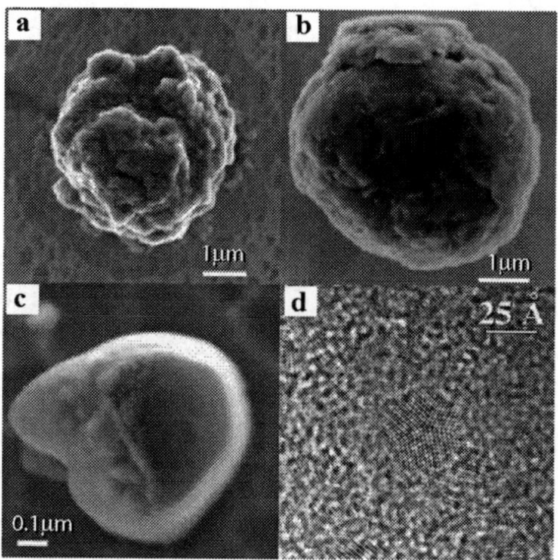

FIGURE 2. Electron micrographs of presolar grains: a) SiC; b) Graphite; c) Al_2O_3; d) nanodiamond. Reprinted from [3].

literature for additional information.

TYPES OF PRESOLAR GRAINS

A large number of presolar phases have now been identified (Table 1); example micrographs of a few are shown in Fig. 2. Most of the carbonaceous grains (and Si_3N_4) have been identified in acid residues of meteorites from which the dominant phases (silicates, metal, sulfides) have been removed. O-rich phases have been identified primarily by automated techniques both *in situ* [14, 15, 16] and in acid residues [17, 18]. Some phases (refractory carbides, metal) have been identified as sub-grains within larger presolar graphite grains [8, 9, 19]. Note that although several hundred presolar silicate grains have now been identified, very few have had detailed mineralogical identifications. Chemical analysis indicates a wide range of silicate compositions. Moreover, more than half that have been analyzed by transmission electron microscopy have proven to be amorphous, non-stoichiometric phases. Note also that the origin for the meteoritic nanodiamonds is unsettled; it is posible that only a tiny fraction are presoalr grains [20].

IMPLICATIONS FOR NUCLEOSYNTHESIS

Because the parent stars of presolar grains ended their lives more than 4.5 Gyr ago, one must use an iterative approach to identify the type of star that produced any given grain. For example, Fig. 3 shows the C and N isotopic ratios of presolar SiC grains. The data fall into distinct groupings. Comparison with spectroscopic observations and theoretical models indicates that these families represent different types of stellar sources, as

TABLE 1. Types of presolar grains in meteorites and interplanetary dust particles (IDPs), after [2]. AGB=Asymptotic Giant Branch stars, SNe=Supernovae, RG=Red Giant stars.

Phase	Abundance (ppm)	Size	Source(s)
Nanodiamond	1400	2 nm	SNe(?)
Silicates (olivine, pyroxene, Ca-, Al-rich, glass ...)	≈500 (IDPs) ≈100 (meteorites)	0.1–1 μm	AGB, SNe
SiC	10	0.1–20 μm	AGB, SNe, J-stars, novae(?)
Graphite	1–2	1–20 μm	AGB, SNe
TiC, ZrC, MoC, RuC, FeC, Fe-Ni	(sub-grains in graphite)	5–220 nm	AGB, SNe
Silicon Nitride (Si_3N_4)	>0.002	∼1 μm	SNe
Oxides (Al_2O_3, $MgAl_2O_4$, $CaAl_{12}O_{19}$, TiO_2, $(Mg,Fe)Cr_2O_4$)	>10	0.1–3 μm	RG, AGB, SNe

indicated on the Figure. For example, the C isotopic distribution of the dominant "Mainstream" population is remarkably similar to that observed in C-rich asymptotic giant branch (AGB) stars [21], pointing to these as sources. Many other isotope signatures in the grains point to an AGB source as well and the high-precision data obtainable on the grains can thus be used to constrain AGB models. Similarly, the "X-grains" have signatures pointing to an origin in Type II supernovae and can provide unique information about such explosions of massive stars. Analogous groupings of O isotopic ratios of presolar oxide and silicate grains have been used to identify red giants, AGB stars and SNe as the sources of these.

Low- and Intermediate-Mass Stars

There is now a large body of evidence that most SiC grains originated in low-mass ($<2M_{\odot}$) AGB stars: the mainstream grains from roughly solar-metallicity stars, the rare Y and Z grains from lower-metallicity stars. One of the strongest pieces of evidence for this comes from isotopic measurements of heavy trace elements in single grains, made possible by resonance ionization mass spectrometry (RIMS). RIMS measurements of single presolar SiC grains reveal almost-pure *s*-process isotopic signatures of many elements, including Mo, Zr, Sr, Ba, and Ru (*e.g.*, [12]). In fact, the relatively high precision of the measurements provides for quantitative constraints on *s*-process models. For example, a free parameter in AGB nucleosynthesis calculations [23, 12] is the amount of ^{13}C present in the region between the AGB He- and H-buring shells, since the primary source of neutrons for the *s*-process in low-mass AGB stars is the $^{13}C(\alpha,n)$ reaction. A recent comparison of models with data acquired for multiple elements in single SiC grains [24] constrains this parameter to a narrow range around that required to explain the solar *s*-process abundance distribution.

Presolar grains also provide important new information about mixing processes in low-mass AGB stars. Spectroscopic observations indicate that an "extra" mixing process, not predicted by standard 1-d models of stellar evolution occurs in low-mass red

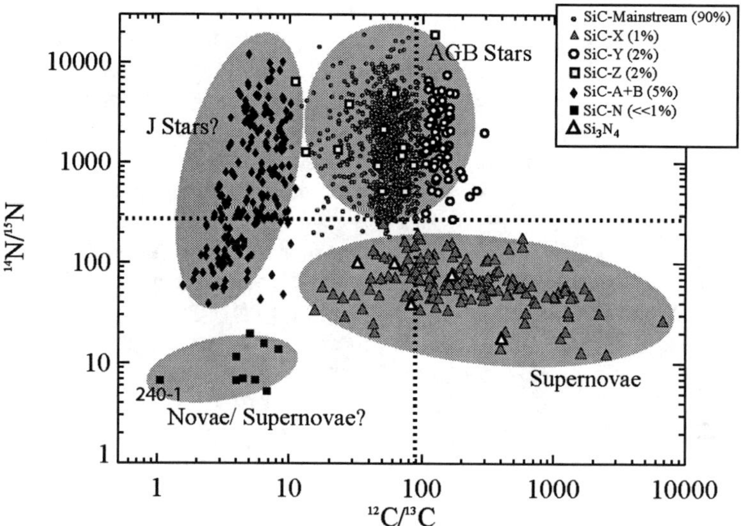

FIGURE 3. C and N isotopic ratios measured in individual presolar SiC grains (see [2] for data sources). Different groupings of isotopic compositions can be attributed to different stellar sources, as indicated. Grain 240-1 [22] is the best candidate yet identified for a nova condensate.

giants [25]. C and N isotopic ratios in presolar SiC grains [26, 27, 28] and O isotopes and inferred $^{26}Al/^{27}Al$ ratios in many presolar oxide grains [17] point to a similar process occurring in AGB stars. The physical mechanism of the mixing (often called "cool bottom processing" or CBP, [29]) is unknown, but it is likely related to rotation. Again, the high-precision grain data can help constrain parameters of mixing models, for example the mixing rate and temperature reached by circulating material [27]. Interstingly, the mixing parameters required to explain the SiC data and the oxide data appear to be quite different, indicating that CBP operates differently in O-rich versus C-rich AGB stars.

There is less compelling evidence for grains from intermediate-mass (IM) stars (2–8 $M_\odot$). Lugaro et al. [30] argued for an IM-AGB origin for an unusual presolar $MgAl_2O_4$ grain, on the basis of O and Mg isotopes. However, this appears to be inconsistent with more recent determinations of key nuclear reaction rates (M. Lugaro, pers. comm.) and the origin of this grain is currently ambiguous.

Supernovae and Novae

A small fraction of presolar SiC ("X grains"), the very rare Si_3N_4 grains and a larger fraction of presolar graphite grains are believed to originate in the cooling ejecta of Type II supernovae (SNe). The most compelling evidence for this origin is the observation of large ^{44}Ca excesses, unaccompanied by anomalies in other stable Ca isotopes, in many grains [31, 32, 33]. This signature points to *in situ* decay of ^{44}Ti, which has a half-life of ≈60 y and is produced solely in SNe. Moreover, extinct ^{49}V observed in some grains [34] also indicates a supernova origin and requires that they formed within one year of the parental explosions. A large number of additional isotope signatures observed in the grains point to a SNe origin as well, and, as for the AGB-derived grains, the data can provide new astrophysical information. A key observation of the supernova grain data is the apparent necessity to heterogeneously mix material from different zones in order to quantitatively reproduce grain isotopic compositions using detailed SN nucleosynthesis calculations [32, 35, 36]. Although extensive macroscopic mixing is both observed in and predicted for SN ejecta, the detailed microscopic mixing required by the grains poses challenges to our understanding of SNe (*e.g.*, [37]).

The SN grains have provided several interesting and important insights into supernova nucleosynthesis. For example, the isotopic patterns observed for heavy elements Mo, Zr, Sr, Ba and Fe in SiC X grains are completely distinct from those observed in the AGB grains, and also differ from expectations for *r*- or *p*-processes associated with supernovae. Rather, these point to a "neutron-burst" nuclear process occurring in a massive star's He shell during the explosion [38]. Moreover, the grains are uniformly more ^{15}N-rich (Fig. 3) than can be explained by the SN mixing models. This result and some astronomical observations [39] indicate that Type II SNe are a major producer of ^{15}N, even though this is not predicted by state-of-the-art calculations (e.g., [40]). As in the case of mixing in low-mass AGB stars (previous section), the solution to this discrepancy between theory and observation may lie in multiple dimensions: models of rotating massive stars appear to produce larger amounts of ^{15}N during core He-burning than 1-d models predict [41], but clearly this problem requires much more attention. Finally, extremely unusual Ca and Ti isotopic compositions were recently reported in presolar SN graphite grains [42]. It remains to be seen whether these can be adequately explained by nucleosynthesis models or if they will point to new insights about massive star evolution.

In addition to SN C- and N-rich phases, a small fraction of presolar oxide and silicate grains also likely formed in SNe. Interestingly, although the most abundant product of Type II SNe is ^{16}O, only one ^{16}O-rich SN grain has been found [43]. The remainder of O-rich grains believed to have a SN origin are ^{18}O-rich [44, 18], reflecting a contribution from the He-burning zone of the massive star (Fig. 1). That a majority of presolar SN O-rich grains are apparently from outer layers of the parent stars might reflect preferential destruction of dust grains in deeper layers by reverse shocks [45].

Amari et al. [46] proposed a nova origin for a handful of isotopically unusual SiC and graphite grains, based mostly on very low $^{12}C/^{13}C$ and $^{14}N/^{15}N$ ratios (Fig. 3), but also on ^{30}Si enrichments. However, a nova origin for these grains is somewhat problematic because huge dilution of the pure nova nucleosynthetic signatures is required to explain the data. Moreover, Nittler and Hoppe [47], reported a SiC grain with very similar C and N isotopes to the putative nova grains, but with Si, Mg, Ca and Ti isotopic ratios strongly indicative of a supernova origin. This result shows that SNe can produce compositions with lower $^{12}C/^{13}C$ and $^{14}N/^{15}N$ ratios than predicted and calls into question whether the other grains formed in novae or SNe. More recently, an Al-rich SiC grain was identified (240-1, Fig. 3, [22]) with $^{12}C/^{13}C=1$, low $^{14}N/^{15}N$ and high inferred $^{26}Al/^{27}Al$. This grain's composition agrees well with nova models [48], without any dilution and is the strongest candidate found to date for a nova condensate in a meteorite.

Galactic Chemical Evolution

The isotopic composition of a given star (and hence grains that condense from it) depends both on the initial composition with which it was born and any nuclear processing that occurs within during its evolution. The initial compositions are set by Galactic chemical evolution (GCE). Stars born at different times and places have different isotopic compositions due to the different nucleosynthetic origins of different isotopes. This is reflected in the isotopic compositions of some elements in presolar grains and thus the grains can also provide constraints on GCE. For example, most of the Si and Ti isotopic ratios of presolar SiC grains appear to be dominated by GCE effects [49, 26]. Although there are many puzzles, the grain data have been used to argue that the Sun might have an atypical isotopic composition for its age [26, 50], that the interstellar medium is well mixed [51], and that the formation of the presolar grain parent stars might have been triggered by a collision between the Milky Way and a dwarf galaxy [52]. Similarly, GCE clearly played a role in setting the Mg and Ca isotopic compositions measured in presolar oxide grains [53, 18].

ACKNOWLEDGMENTS

The author would like to express his deep gratitude to Professor Yoshi Yurimoto and the organizing committee for the invitation to this stimulating meeting and for travel support.

REFERENCES

1. L. R. Nittler, *Earth Planet. Sci. Lett.* **209**, 259–273 (2003).
2. E. Zinner, "Presolar Grains," in *Treatise on Geochemistry vol 1: Meteorites, Comets, and Planets*, edited by A. M. Davis, 2004 (updated 2007), pp. 17–39.
3. D. D. Clayton, and L. R. Nittler, *Ann. Rev. Astron. Astrophys.* **42**, 39–78 (2004).
4. E. Anders, and E. Zinner, *Meteoritics* **28**, 490–514 (1993).
5. P. R. Heck, K. K. Marhas, P. Hoppe, R. Gallino, H. Baur, and R. Wieler, *Astrophys. J.* **656**, 1208–1222 (2007).
6. G. K. Nicolussi, A. M. Davis, M. J. Pellin, R. S. Lewis, R. N. Clayton, and S. Amari, *Science* **277**, 1281–1283 (1997).
7. M. R. Savina, A. M. Davis, C. E. Tripa, M. J. Pellin, R. Gallino, R. S. Lewis, and S. Amari, *Science* **303**, 649–652 (2004).
8. T. J. Bernatowicz, R. Cowsik, P. C. Gibbons, K. Lodders, B. Fegley, S. Amari, and R. S. Lewis, *Astrophys. J.* **472**, 760–782 (1996).
9. T. K. Croat, T. Bernatowicz, S. Amari, S. Messenger, and F. J. Stadermann, *Geochim. Cosmochim. Acta* **67**, 4705–4725 (2003).
10. R. M. Stroud, L. R. Nittler, and C. M. O'D. Alexander, *Science* **305**, 1455–1457 (2004).
11. C. Floss, F. J. Stadermann, J. P. Bradley, Z. R. Dai, S. Bajt, G. Graham, and A. S. Lea, *Geochim. Cosmochim. Acta* **70**, 2371–2399 (2006).
12. M. Lugaro, A. M. Davis, R. Gallino, M. J. Pellin, O. Straniero, and F. Käppeler, *Astrophys. J.* **593**, 486–508 (2003).
13. T. J. Bernatowicz, O. W. Akande, T. K. Croat, and R. Cowsik, *Astrophys. J.* **631**, 988–1000 (2005).
14. S. Messenger, L. P. Keller, F. J. Stadermann, R. M. Walker, and E. Zinner, *Science* **300**, 105–108 (2003).
15. A. N. Nguyen, and E. Zinner, *Science* **303**, 1496–1499 (2004).

16. K. Nagashima, A. Krot, and H. Yurimoto, *Nature* **428**, 921–924 (2004).
17. L. R. Nittler, C. M. O'D. Alexander, X. Gao, R. M. Walker, and E. Zinner, *Astrophys. J.* **483**, 475–495 (1997).
18. L. R. Nittler, C. M. O'D. Alexander, R. Gallino, P. Hoppe, A. Nguyen, F. Stadermann, and E. K. Zinner, *Astrophys. J.* **submitted.**
19. T. K. Croat, F. J. Stadermann, and T. J. Bernatowicz, *Astrophys. J.* **631**, 976–987 (2005).
20. Z. R. Dai, J. P. Bradley, D. J. Joswiak, D. E. Brownlee, H. G. M. Hill, and M. J. Genge, *Nature* **418**, 157–159 (2002).
21. D. L. Lambert, B. Gustafsson, K. Eriksson, and K. H. Hinkle, *Astrophys. J. Supp.* **62**, 373–425 (1986).
22. L. R. Nittler, C. M. O'D. Alexander, and A. N. Nguyen, *Meteoritics and Plan. Sci.* **41 (Supp)**, Abstract #5316 (2006).
23. R. Gallino, C. Arlandini, M. Busso, M. Lugaro, C. Travaglio, O. Straniero, A. Chieffi, and M. Limongi, *Astrophys. J.* **497**, 388–403 (1998).
24. J. G. Barzyk, M. R. Savina, A. M. Davis, R. Gallino, F. Gyngard, S. Amari, E. Zinner, M. J. Pellin, R. S. Lewis, and R. N. Clayton, *Meteoritics and Plan. Sci.* **42**, 1103–1119 (2007).
25. A. I. Boothroyd, and I.-J. Sackmann, *Astrophys. J.* **510**, 232 (1999).
26. C. M. O'D. Alexander, and L. R. Nittler, *Astrophys. J.* **519**, 222–235 (1999).
27. K. M. Nollett, M. Busso, and G. J. Wasserburg, *Astrophys. J.* **582**, 1036–1058 (2003).
28. E. Zinner, L. R. Nittler, R. Gallino, A. I. Karakas, M. Lugaro, O. Straniero, and J. C. Lattanzio, *Astrophys. J.* **650**, 350–373 (2006).
29. G. J. Wasserburg, A. I. Boothroyd, and I.-J. Sackmann, *Astrophys. J.* **447**, L37–L40 (1995).
30. M. Lugaro, A. I. Karakas, L. R. Nittler, C. M. O'D. Alexander, P. Hoppe, C. Iliadis, and J. C. Lattanzio, *Astron. Astrophys.* **461**, 657–664 (2007).
31. L. R. Nittler, S. Amari, E. Zinner, S. E. Woosley, and R. S. Lewis, *Astrophys. J.* **462**, L31–L34 (1996).
32. P. Hoppe, R. Strebel, P. Eberhardt, S. Amari, and R. S. Lewis, *Meteoritics and Plan. Sci.* **35**, 1157–1176 (2000).
33. A. Besmehn, and P. Hoppe, *Geochim. Cosmochim. Acta* **67**, 4693–4703 (2003).
34. P. Hoppe, and A. Besmehn, *Astrophys. J.* **576**, L69–L72 (2002).
35. C. Travaglio, R. Gallino, S. Amari, E. Zinner, S. Woosley, and R. S. Lewis, *Astrophys. J.* **510**, 325–354 (1999).
36. T. Yoshida, *Astrophys. J.* **666**, 1048–1068 (2007).
37. E.-A. Deneault, D. D. Clayton, and A. Heger, *Astrophys. J.* **594**, 312–325 (2003).
38. B. S. Meyer, D. D. Clayton, and L.-S. The, *Astrophys. J.* **540**, L49–L52 (2000).
39. Y. Chin, C. Henkel, N. Langer, and R. Mauersberger, *Astrophys. J.* **512**, L143–L146 (1999).
40. T. Rauscher, A. Heger, R. D. Hoffman, and S. E. Woosley, *Astrophys. J.* **576**, 323–348 (2002).
41. N. Langer, A. Heger, S. E. Woosley, and F. Herwig, in *Nuclei in the Cosmos V*, edited by N. Prantzos, Editions Frontiéres, 1998, pp. 129–135.
42. M. Jadhav, S. Amari, K. K. Marhas, E. Zinner, T. Maruoka, and R. Gallino, in *Lunar Planet. Sci. Conf. XXXVIII*, 2007, Abstract #2256.
43. L. R. Nittler, C. M. O'D. Alexander, J. Wang, and X. Gao, *Nature* **393**, 222 (1998).
44. S. Messenger, L. P. Keller, and D. S. Lauretta, *Science* **309**, 737–741 (2005).
45. T. Nozawa, T. Kozasa, A. Habe, E. Dwek, H. Umeda, N. Tominaga, K. Maeda, and K. Nomoto, *Astrophys. J.* **666**, 955–966 (2007).
46. S. Amari, X. Gao, L. R. Nittler, E. Zinner, J. José, M. Hernanz, and R. S. Lewis, *Astrophys. J.* **551**, 1065–1072 (2001).
47. L. R. Nittler, and P. Hoppe, *Astrophys. J.* **631**, L89–L92 (2005).
48. J. José, M. Hernanz, S. Amari, K. Lodders, and E. Zinner, *Astrophys. J.* **612**, 414–428 (2004).
49. F. X. Timmes, and D. D. Clayton, *Astrophys. J.* **472**, 723–741 (1996).
50. D. D. Clayton, and F. X. Timmes, *Astrophys. J.* **483**, 220–227 (1997).
51. L. R. Nittler, *Astrophys. J.* **618**, 281–296 (2005).
52. D. D. Clayton, *Astrophys. J.* **598**, 313–324 (2003).
53. E. Zinner, L. R. Nittler, P. Hoppe, R. Gallino, O. Straniero, and C. M. O'D. Alexander, *Geochim. Cosmochim. Acta* **69**, 4149–4165 (2004).

AGB stars as an origin of dust and gas in the interstellar medium of galaxies

M. Matsuura*, A.A. Zijlstra†, P.R. Wood**, G.C. Sloan‡, M.A.T. Groenewegen§, E. Lagadec†, J.Th. van Loon¶, P.A. Whitelock||, J. Bernard-Salas‡, J.W. Menzies||, M.-R.L. Cioni††, M.W. Feast‡‡ and G.J. Harris§§

*National Astronomical Observatory of Japan, Osawa 2-21-1, Mitaka, Tokyo 181-8588, Japan
†School of Physics and Astronomy, University of Manchester, Sackville Street, P.O. Box 88, Manchester M60 1QD, United Kingdom
**Research School of Astronomy & Astrophysics, Mount Stromlo Observatory, Australian National University, Cotter Road, Weston ACT 2611, Australia
‡Astronomy Department, Cornell University, 610 Space Sciences Building, Ithaca, NY 14853-6801, USA
§Instituut voor Sterrenkunde, KU Leuven, Celestijnenlaan 200B, 3001 Leuven, Belgium
¶Astrophysics Group, School of Physical and Geographical Sciences, Keele University, Staffordshire ST5 5BG, United Kingdom
||South African Astronomical Observatory, P.O.Box 9, 7935 Observatory, South Africa
††Science & Technology Research Institute University of Hertfordshire College Lane, Hatfield AL10 9AB, United Kingdom
‡‡Astronomy Department, University of Cape Town, 7701 Rondebosch, South Africa
§§Department of Physics and Astronomy, University College London, Gower Street, London WC1E 6BT, United Kingdom

Abstract. We have obtained infrared spectra of carbon stars in four nearby galaxies – the Large and Small Magellanic Clouds, and the Fornax dwarf spheroidal galaxy. Our primary aim is to investigate mass-loss rate and gas compositions of these stars as a function of metallicity, by comparing AGB stars in several galaxies with different metallicities. These stars were observed using the Infrared Spectrometer (IRS) onboard the *Spitzer Space Telescope* which covers 5–35 μm region, and the Infrared Spectrometer And Array Camera (ISAAC) on the Very Large Telescope which covers the 2.9–4.1 μm region. HCN, CH and C_2H_2 molecular bands, as well as SiC and MgS dust features are identified in the spectra. We find no evidence that mass-loss rates depend on metallicity. Carbon stars are strongly affected by carbon production during the AGB phase; primarily mass-loss of carbon-rich stars are driven by amorphous carbon dust grains, that explains the little metallicity dependence of mass-loss rate for carbon-rich stars. We found that C_2H_2 bands are prominent features at 3–15 μm among extragalactic carbon stars, which is not always the case for Galactic carbon stars. We argue that carbon produced in AGB stars dominate the gas and dust chemistry in these stars in low metallicities.

Keywords: stars: AGB and post-AGB – stars: atmospheres – stars:mass-loss stars: carbon –
PACS: 97.10.Me: 97.10.Tk: 97.10.Cv: 97.20.Li: 97.10.Fy: 97.30.Jm: 95.85.Hp: 95.85.Gn

INTRODUCTION

Stars with low and intermediate initial mass (from 1 to 8 $M_\odot$) loose their atmospheres towards the end of their lives (1–10 Gyr after their birth). This mass loss is intense; approximately 50–80 % of the stellar mass is lost during the Asymptotic Giant Branch

CP1016, *Origin of Matter and Evolution of Galaxies*,
edited by T. Suda, T. Nozawa, A. Ohnishi, K. Kato, M. Y. Fujimoto, T. Kajino, and S. Kubono

(AGB) phase. Understanding this process and the composition of the gas lost from AGB stars is important for understanding both stellar and galactic evolution. First, the stellar wind removes material from AGB stars, reducing their mass and influencing their evolution. Secondly, AGB stars are among the primary sources of metal-enriched gas and dust grains in the interstellar medium of galaxies [14]. In particular, carbon-rich dust grains are formed solely in carbon-rich AGB stars [3]. Theoretical work has suggested that AGB mass-loss rates depend on metallicity [2, 28], because the stellar wind is triggered by radiation pressure on dust grains and the dust is made up of astronomical metals, such as oxygen, carbon, silicon, iron and aluminium. Therefore a study of extra-galactic AGB stars is vital, both because it allows us to study mass loss at low metallicity, and because the distances to the parent galaxies are known, so that stellar parameters can be derived with some precision. High sensitivity near- and mid-infrared instrument are required for such a mass-loss study; at present this is provided by the *Very Large Telescope* (VLT) and the *Spitzer Space Telescope* (SST).

Most nearby galaxies have lower metallicities than our Galaxy. Low (initial) metallicity influences mass-loss rates and the chemical composition of the atmospheres of AGB stars. If all elemental abundances of extragalactic AGB stars could be scaled to solar abundances according to the metallicity represented by [Fe/H], there would be fewer molecules (except H_2) in the atmosphere and fewer dust grains in circumstellar envelope at low metallicity. However, there is additional effect to be considered. At lower initial metallicity the elemental abundance is more sensitive to the atoms formed inside AGB stars [e.g. 15]. In addition, the efficiency of the third dredge-up increases at lower metallicity [26]. Thus, abundances of elements in AGB stars are not simply scaled from the iron abundance, which is often used as the measure of metallicity. For example, Lattanzio & Forestini [13], Izzard et al. [9] predicted higher C/O ratios in carbon stars at the end of the AGB phase at lower metallicity for 4 $M_\odot$ initial mass stars. A higher C/O ratio leads to more excess carbon available to form carbon-bearing molecules and carbonaceous dust grains. We investigate this metallicity dependence of molecules and mass-loss rate based on infrared observations.

OBSERVATIONS AND RESULTS

We highlight our series of studies [17, 18, 19, 20, 23, 25, 31, 11, 10] of the 3–35 μm spectra of extragalactic carbon stars based on observations with the Infrared Spectrometer And Array Camera [ISAAC; 22] and the VLT spectrometer and imager for the mid-infrared [VISIR; 12] on the VLT and the Infrared Spectrometer [IRS; 8] on the SST [27]. Our targets are located in three nearby galaxies, namely the Large Magellanic Cloud (LMC), the Small Magellanic Cloud (SMC), and the Fornax dSph galaxy. Properties of the host galaxies and targets of our observations are summarised in Table 1.

Figure 1 shows the spectrum of a star in our Galaxy obtained by the Infrared Space Observatory (ISO) and spectra of LMC and SMC stars observed with ISAAC and IRS. Major molecular bands and dust excess are indicated in the figure. C_2H_2 bands are found at 3.8, 7, and 13 μm. The 3.1 μm absorption is due to HCN and C_2H_2. The 3.8 μm absorption is most likely of photospheric origin, at least among the warmer stars [18, 25], while the 13 μm absorption has a contribution from circumstellar C_2H_2 [19]. There are

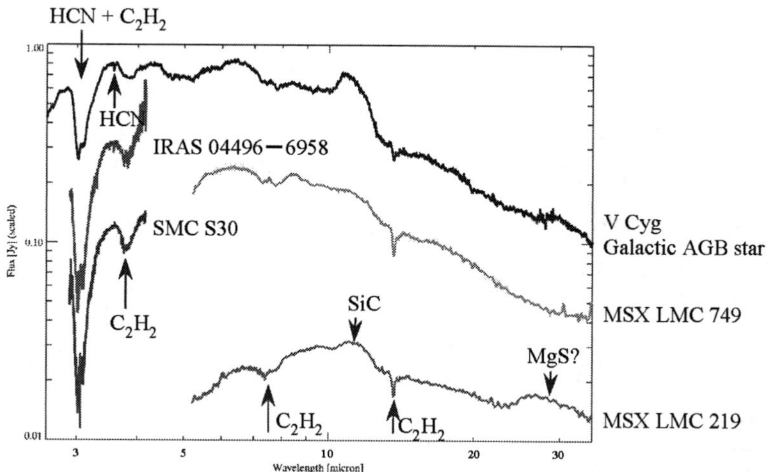

FIGURE 1. Infrared spectra of Galactic, LMC and SMC AGB stars

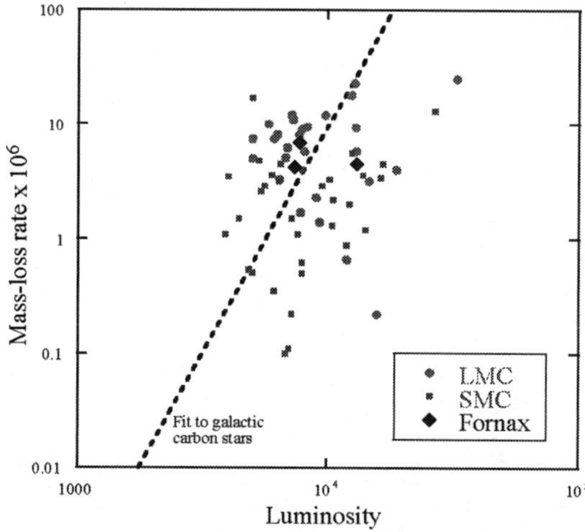

FIGURE 2. Mass-loss rate ($M_\odot$ yr^{-1}) as a function of luminosity ($L_\odot$) for our sample and for the LMC and SMC samples [6]. Fornax stars show high mass-loss rates for their luminosities. The dotted line is the fit to the luminosity vs mass-loss rate relation for Galactic carbon stars [4].

two dust features found, SiC at 11.3 μm and MgS at $\sim$30 μm.

First we measured the equivalent widths of the molecular bands, so as to evaluate the metallicity dependence of these features. The equivalents width of the C_2H_2 molecular bands are measured following the method of Zijlstra et al. [31]. Figs. 3 shows 7.5 μm C_2H_2 equivalent width as a function of infrared colour [6.4]−[9.3]. The [6.4] and [9.3] values are calculated from the Spitzer/IRS spectra, and the definition of these magnitudes

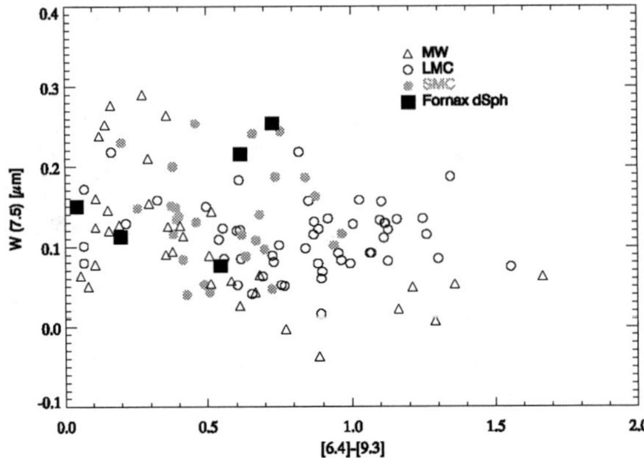

FIGURE 3. The equivalent width of 7.5 μm C_2H_2 as a function of infrared colour [6.4]−[9.3]. Symbols show the host galaxies, MW representing the Milky Way.

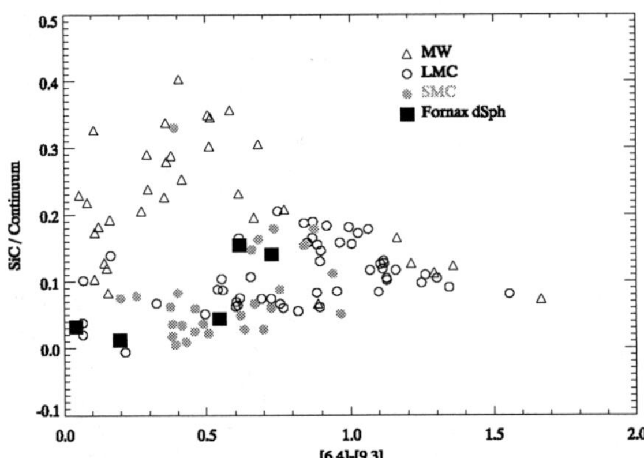

FIGURE 4. The intensity ratio of SiC excess with respect to the pseudo-continuum is plotted against [6.4]−[9.3]; a ratio of 0.0 indicates no SiC excess (continuum only).

is also given by Zijlstra et al. [31]. Within $0.6 < [6.4] - [9.3] < 1.0$ range, a high value of W(7.5), with respect to infrared colour, is found for stars in Fornax and the SMC while a low W(7.5) is found in our Galaxy. This infrared colour measures effective temperature of stars for blue stars, but also circumstellar dust excess for red stars. C_2H_2 equivalent width increases towards lower metallicity,

Secondly, to derive mass-loss rates, the spectral energy distributions (SEDs) are fitted by using a radiative transfer code. Groenewegen et al. [6] and Matsuura et al. [20] give details of the analysis and resultant mass-loss rates. A gas-to-dust ratio of 200 is assumed for both works. The fractions which give the best fit are 5 % SiC and 95 % amorphous

TABLE 1. Properties of host galaxies and the number of observed AGB stars. Number of carbon stars are from Groenewegen [5]. The metallicity of intermediate age stars are discussed in our studies. The number of targets are projects from Zijlstra et al. [31], Sloan et al. [23] and Matsuura et al. [20]. The numbers within the parentheses with † show the number of stars to be analysed in future.

	[Fe/H]	Distance (kpc)	Number of carbon stars			
			Known	Observed with IRS/Spitzer	Observed with ISAAC/VLT	Observed with VISIR/VLT
LMC	−0.3	50	>1000	30	35	
SMC	−0.6	59	∼800	32	2	
Fornax dSph	−1.0	138	104	5 (+6)†	(3)†	2

carbon for Fornax dSph galaxy. Groenewegen et al. [6] gives SiC and amorphous ratio of 2 % in the LMC and the SMC, but these values depend on the optical constants of dust grains. In our galaxy, SiC is typically 10 %. This fraction seems to decrease towards the lower metallicity. This result is consistent with smaller SiC excess, which is plotted in Figure 4.

Figure 2 shows the derived mass-loss rate as a function of luminosity. Little metallicity dependence of mass-loss rates is found for carbon-rich stars. The stars in Fornax dSph galaxy are at the upper end of the SMC mass-loss rates at a given luminosity. LMC stars appear to reach a higher mass-loss rate than the SMC and Fornax stars at a given luminosity.

DISCUSSION

We found little evidence of metallicity dependence of mass-loss rates (Fig. 2). The dust mass is dominated by amorphous carbon dust grains. The quantity of amorphous carbon grains is affected by two factors, i.e. C/O ratio [7] and initial metallicity [32, 24]. The lack of evidence for reduced mass-loss rates at lower metallicity requires a higher C/O ratio towards low metallicity. As new carbon atoms reach the surface of these AGB stars by self production and the third dredge-ups, the C/O ratio increases and the amount of carbon available to form grains also increases. A strong dependence of mass-loss rate on C/O ratio is reported from hydrodynamical models for carbon-rich stars [21]. This is a consequence of the stability of the CO molecule; in carbon stars all available oxygen atoms are locked into carbon monoxide and excess carbon atoms are available to form carbon-bearing molecules and amorphous carbon dust grains.

This result, i.e., the lack of dependence of mass-loss rates on metallicity for carbon-rich stars, is in contrast to that found for oxygen-rich stars, where there is an metallicity dependence [29]. This is because dust driven winds in oxygen-rich stars are dominated by the amount of silicate, and ultimately by the intrinsic silicon elemental abundance [2, 28]. The expansion velocities of oxygen-rich shells are also smaller at low metallicities [29, 30, 16]. A careful interpretation from conversion from the measured OH maser intensity to mass-loss rate is required [24, 16]. In contrast, carbon stars are strongly affected by carbon production during the AGB phase; thus the different dependence of mass-loss rate on metallicity can be understood.

C_2H_2 formation relies on excess carbon atoms after all oxygen atoms are locked into carbon monoxide. Thus the higher abundance of C_2H_2 at lower metallicity is due to the higher C/O ratio [18] caused by carbon synthesised in AGB stars. Figure 13 in [18] shows that the fraction of C_2H_2 in the atmosphere increases drastically up to C/O $\sim$ 1.6. C_2H_2 is thought to be a parent molecule in the formation of PAHs, as indicated by chemical models [1]. It is still unknown whether PAHs are formed during the AGB phase or afterwards. If PAHs are indeed formed during the AGB phase, an over-abundance of C_2H_2 in a low metal environment will affect the growth of these important molecules.

REFERENCES

1. Allamandola L.J., Tielens G.G.M., Barker J.R., 1989, ApJS 71, 733
2. Bowen G.H. & Willson L.A., 1991, ApJ 375, L53
3. Dwek E., 1998, ApJ 501, 643
4. Groenewegen M.A.T., Whitelock P.A., Smith C.H., Kerschbaum F., 1998, MNRAS 293, 18
5. Groenewegen M.A.T., 2006 'Planetary nebulae beyond the Milky Way', Proceedings of the ESO workshop, Edited by Letizia Stanghellini, J. R. Walsh & N.G. Douglas, Berlin, Springer, p.108
6. Groenewegen M.A.T., Wood P.R., Sloan G.C., et al., 2007 MNRAS, 376, 313
7. Habing H.J., Tignon J., Tielens A.G.G.M., 1994, A&A 286, 523
8. Houck J.R., Roellig T.L., van Cleve J. et al., 2004, ApJS 154, 18
9. Izzard R.G., Tout C.A., Karakas A.I., Pols, O.R., 2004, MNRAS 350, 407
10. Lagadec E., Zijlstra A.A., Matsuura M., Menzies J.W., van Loon J.Th., Whitelock P.A., 2008, MNRAS 383, 399
11. Lagadec E., Zijlstra A.A., Sloan G.C., et al. 2007, MNRAS, 376, 1270
12. Lagage, P. O., et al. 2004, The Messenger, 117, 12
13. Lattanzio J. & Forestini M., 1999, in IAU Symp. 191, Asymptotic Giant Branch Stars, ed. T. Le Bertre, A. Lèbre, & C. Waelkens (ASP), p. 31
14. Maeder, A. 1992, A&A, 264, 105
15. Marigo P., 2001, A&A, 370, 194
16. Marshall J.R., van Loon J.Th., Matsuura M., Wood P.R., Zijlstra A.A., Whitelock P.A., 2004, MNRAS 355, 1348
17. Matsuura M., Zijlstra A.A., van Loon J.Th., et al., 2002, ApJ, 580, L133
18. Matsuura M., Zijlstra A.A., van Loon J.Th., et al., 2005, A&A, 434, 691
19. Matsuura M., Wood P.R., Sloan G.C., et al., 2006, MNRAS, 371, 415
20. Matsuura M., Zijlstra A.A., Bernard-Salas J., et al. 2007, MNRAS 382, 1889
21. Mattsson L., Wahlin R., Höfner S., 2007 Proceedings of IAU Symp 241, ed. A. Vazdekis et al. (astro-ph/07052809)
22. Moorwood A., Cuby J.-G., Biereichel P., et al., 1998, Msngr 94, 7
23. Sloan G.C., Kraemer K.E., Matsuura M., et al., 2006, ApJ, 645, 1118
24. van Loon J.Th., 2000, A&A 354, 125
25. van Loon J.Th., Marshall J.R., Cohen M., et al., 2006, A&A, 447, 971
26. Vassiliadis E. & Wood P.R. 1993, ApJ, 413, 641
27. Werner M.W., Roellig T.L., Low F.J. et al., 2004, ApJS154, 1
28. Willson L.A., 2006, Planetary nebulae beyond the Milky Way, Proceedings of the ESO workshop held in Garching, Germany, 19-21 May 2004, Edited by Letizia Stanghellini, J. R. Walsh, and N.G. Douglas, Berlin, Springer, 2006, p.99
29. Wood P.R., Whiteoak J.B., Hughes S.M.G., Bessell M.S., Gardner F.F., Hyland A.R., 1992, ApJ 397, 552
30. Zijlstra A.A., Loup C., Waters L.B.F.M., Whitelock P.A., van Loon J.Th., Guglielmo F., 1996, MNRAS 279, 32
31. Zijlstra A.A., Matsuura M., van Loon J.Th., et al. 2006, MNRAS, 370, 1961
32. Zuckerman B., & Dyck H.M., 1989, A&A 209, 119

Outstanding Problems of Presolar Diamond in Meteorites

Sachiko Amari[a]

[a]*Laboratory for Space Sciences and the Physics Department, Washington University, One Brookings Dr., St. Louis, MO 63130, USA (sa@wuphys.wustl.edu)*

Abstract. Diamond is the first mineral type of presolar grains that were isolated from meteorites, yet it is one of the least understood presolar grain types. An isotopically anomalous component Xe-HL is carried by diamond and is characterized by excesses in both the light, p-process only isotopes (124 and 126) and the heavy, r-process only isotopes (134 and 136). These excesses are always correlated although physical settings of these two processes are quite different and there is little reason to always correlate each other. Furthermore, the r-process Xe and the p-process Xe in diamond inferred from Xe-HL are quite different from what is derived from the solar system abundance as well as stellar models. Further studies are needed to investigate these outstanding problems for over three decades.

Keywords: presolar diamond, isotopic anomalies, meteorites, supernovae, p-process, r-process
PACS: 26.20.Np; 26.30.Hj; 96.30.Za; 97.60.Bw

INTRODUCTION

Meteorites formed 4.6 billion years ago along with other solar system objects. Until the late sixties, it was believed that the solar nebula, from which the planets and meteorites formed, was completely homogenized and that the Solar System was isotopically uniform. This idea began to be questioned when Black and Pepin [1] analyzed Ne in a fragment from the Orgueil meteorite by stepwise heating and found that $^{20}Ne/^{22}Ne$ ratios in high temperature fractions (800 – 1000°C) were much lower (down to 4, air: 9.8) than those commonly observed in meteorites, indicating the presence of a ^{22}Ne-rich component. This component was named Ne-E because alphabets from A to D had already been taken for other Ne components. In subsequent studies, upper limits of $^{20}Ne/^{22}Ne$ of Ne-E became even lower, and this convinced many that some meteorites contain stardust that survived the solar system formation. Other isotopically anomalous noble gas components were also found in meteorites. Xe-HL shows enrichment in both *l*ight (124 and 126) and *h*eavy (134 and 136) isotopes [2]. Kr-S and Xe-S are characterized by enrichment in the s-process only isotopes [3]. A quest for the minerals that contain these noble gas components ultimately led the discovery of presolar grains, opening up a new field of astronomy [e.g., 4].

Presolar grains are defined as stardust that formed in the stellar outflow or ejecta, and was later incorporated in meteorites. This implies that presolar grains are older

CP1016, *Origin of Matter and Evolution of Galaxies*,
edited by T. Suda, T. Nozawa, A. Ohnishi, K. Kato, M. Y. Fujimoto, T. Kajino, and S. Kubono

than the Solar System, a very reason we call *pre*solar grains. Presolar grains identified to date include diamond [5], SiC [6, 7], graphite [8], oxides [9], silicates [10-12], Si_3N_4 [13] and refractory carbides [14]. Their abundances range from a few hundred ppm (diamond, silicates) to a few ppb (Si_3N_4). Of them diamond was the first mineral type that was isolated and identified from meteorites [5], yet it is the least understood carbonaceous type of presolar grains. One of the reasons is that diamonds are very small (average 3nm [15]) and only bulk (≡aggregates of grains) analysis is possible.

In this paper, we will review outstanding problems of presolar diamonds relevant to nucleosynthetic processes in stars.

PRESOLAR DIAMOND

Outstanding Problems of Presolar Diamond

Diamond is the carrier of Xe-HL that is characterized by enrichment in the p-process only isotopes, 124 and 126, and the r-process only isotopes, 134 and 136 (Fig. 1). Excesses in the light isotopes and those in the heavy isotopes are separately called Xe-L and Xe-H, respectively. Interestingly, Xe-L and Xe-H are always positively correlated: the higher the former, the higher the latter. This correlation has been observed in all diamond separates from any kinds of meteorites.

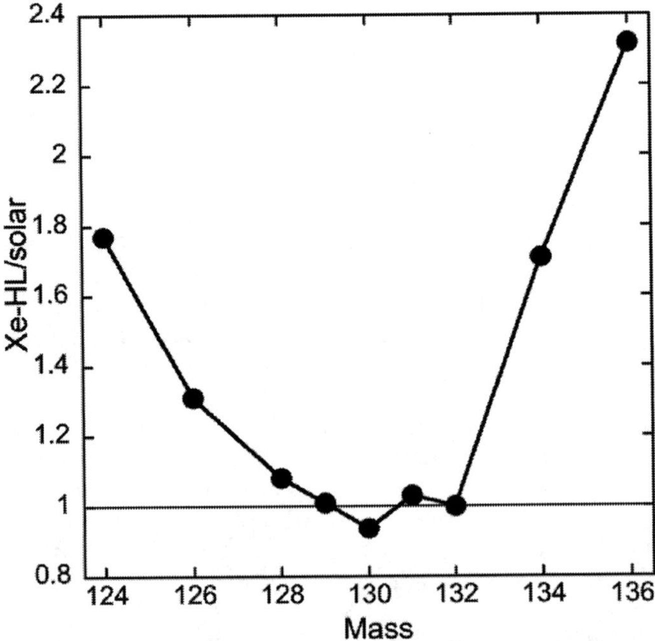

FIGURE 1. The Xe-HL pattern [16]. Xenon isotopes are normalized by ^{132}Xe and the solar ratios. Excesses in the p-process only isotopes and the r-process only isotopes are always observed together.

The close association of the two different nucleosynthetic processes is indeed odd. The p-process is considered to occur in the O/Ne zone in supernovae [17]. It is photo-disintegration processes [(γ,n), (γ,p) and (γ,α)] of pre-existing neutron-rich nuclei under high temperature ($T > \sim 2\times10^9$ K). The r-process is thought to occur in neutrino-driven winds from the neutron star of core-collapse supernovae [18], although the "hot bubble (high entropy)" region which is believed to appear just after the creation of a newly-born proto-neutron star in the core-collapse supernova explosion, neutron star – neutron star binaries, and neutron star – black hole binaries are other candidates. Thus, the isotopes produced by the two completely different processes, which are expected to happen in different physical settings, do not always have to be correlated as what has been observed in diamonds.

Furthermore, the r-process Xe and the p-process Xe inferred from Xe-HL are not what are expected either from the solar system abundance or from stellar models. Assuming that Xe-HL comprises the p-process Xe, the r-process Xe and solar Xe (Xe-HL contains 128 and 130 that are s-process only isotopes. Therefore, Xe-HL must contain a component other than those originated from the p-process and the r-process), Ott [19] inferred the 134/136 ratio of the r-process Xe in diamond to be 0.699 and the 124/126 ratio of the p-process Xe in diamond to be 2.205. These ratios are markedly different from the p-process and the r-process Xe in the Solar System: the r-process 134/136 is derived to be 1.207 after a small correction for the s-process Xe in SiC (although 134 is dominantly produced by the r-process, a small amount of 134 is produced by the s-process) [19]. The p-process Xe in the Solar System is 1.157, taking the ratio of 124 and 126 from the solar abundance. Model calculations of supernovae also predict ratios close to the p-process Xe in the Solar System: Rayet et al. [17] derived 124/126 ratios of stars with mass between 13 and 25 M_{sun} range from 0.952 to 1.188.

Studies on Presolar Diamond

Theoretical Studies

A first attempt to quantitatively explain the Xe-H pattern was made by Howard et al. [20]. They invoked a neutron burst (where the neutron density is lower than that of the r-process), selected one neutron irradiation history, and successfully reproduced Xe-H. Later, Ott [19] proposed the "rapid separation scenario" where unstable precursors with different half-lives are separated from Xe at a proper time to produce the Xe-H pattern: ^{136}Te ($T_{1/2}$ = 17.5 s) and ^{136}I ($T_{1/2}$ = 1.39 m) decay to ^{136}Xe, while ^{134}Te ($T_{1/2}$ = 42 m) and ^{134}I ($T_{1/2}$ = 53 m) decay to ^{134}Xe. The precursors of ^{136}Xe have shorter half-lives than those of ^{134}Xe. If Te and I are separated from Xe before ^{134}Te and ^{134}I completely decay, ^{136}Xe becomes more abundant than ^{134}Xe. The "correct" time for the separation is 8000 seconds after the r-process takes place. Ott [19] applied the same approach to explain Xe-L. This scenario, however, requires very precise timing of the separation. Diamonds in meteorites are undoubtedly formed in more than one supernova. According to this scenario, the separation should have taken place at the "right" time in most supernovae that produced diamonds in meteorites. It is not probable, if not impossible, that most supernovae experienced the same "separation" time.

Another scenario proposed to explain Xe-H is to mix material experienced neutron-burst and solar Xe [21]. However, Xe-L left unexplained in this scenario.

In any case, it has not been satisfactorily explained either why Xe-L and Xe-H are always correlated or why the p- and r-process Xe in diamond are different from those derived from the solar system abundance or stellar models.

Experimental Studies

Compared with huge isotopic anomalies of Xe, those of other heavy elements in diamond are bewilderingly small. Lewis et al. [22] reported Ba and Sr analysis of an Allende diamond separate, using a procedure of cold combustion of diamonds. The highest isotopic anomaly found in Ba is 6.0±1.5 ε in $^{137}Ba/^{138}Ba$ (both isotopes are s- and r-process isotopes){$\varepsilon \equiv [(^{137}Ba/^{138}Ba)_{diamond}/(^{137}Ba/^{138}Ba)_{solar} - 1]\times 10000$} and that in Sr is 120±1 ε in $^{87}Sr/^{88}Sr$ (^{87}Sr: s-process-only isotope, ^{88}Sr: s- and r-processes isotope). Since the Sr and Ba concentrations in the separate was very low, they dismissed the possibility that extremely anomalous Sr and Ba are diluted with a huge amount of normal Sr and Ba because of their low concentrations.

Tellurium isotopic ratios in presolar diamonds were analyzed by Richter et al. [23] and Mass et al. [24]. The r-process only isotopes, ^{128}Te and ^{130}Te show slight excesses [$\delta(^{128}Te/^{124}Te) = 4.0 \pm 1.5$ ‰, $\delta(^{130}Te/^{124}Te) = 9.3 \pm 2.8$ ‰]{$\delta(^{128}Te/^{124}Te)$ ‰ $\equiv [(^{128}Te/^{124}Te)_{diamond}/(^{128}Te/^{124}Te)_{solar} - 1] \times 1000$} [24]. Palladium also shows an elevated $\delta(^{110}Pd/^{104}Pd)$ value of 9.4 ± 5.7 ‰ (^{110}Pd is a r-process only isotope, while ^{104}Pd is an s-process only isotope) [24]. These anomalies are much smaller than those of Xe. Yet they are qualitatively consistent with the rapid separation model as well as the neutron burst model.

CONCLUSIONS

Diamond is the carrier of an isotopically anomalous Xe component, Xe-HL, and the first mineral type of presolar grains that was isolated from meteorites in 1987. Yet it is one of the least understood presolar grain types. A close association of excesses in the p-process only isotopes, 124 and 126, and those in the r-process only isotopes, 134 and 136, is puzzling. Furthermore, the r-process and the p-process in diamond are different from what is derived either from the solar system abundance or from stellar models. Further studies are needed to unlock these outstanding problems.

ACKNOWLEDGMENTS

This work is supported by NASA grant NNX08AG56G.

REFERENCES

1. D. C. Black and R. O. Pepin, *Earth Planet. Sci. Lett.*, **6**, 395-405 (1969).

2. R. S. Lewis, B. Srinivasan and E. Anders, *Science,* **190**, 1251-1262 (1975).
3. B. Srinivasan and E. Anders, *Science,* **201**, 51-56 (1978).
4. K. Lodders and S. Amari, *Chem. Erde,* **65**, 93-166 (2005).
5. R. S. Lewis, M. Tang, J. F. Wacker, E. Anders and E. Steel, *Nature,* **326**, 160-162 (1987).
6. T. Bernatowicz, G. Fraundorf, M. Tang, E. Anders, B. Wopenka, E. Zinner and P. Fraundorf, *Nature,* **330**, 728-730 (1987).
7. M. Tang and E. Anders, *Geochim. Cosmochim. Acta,* **52**, 1235-1244 (1988).
8. S. Amari, E. Anders, A. Virag and E. Zinner, *Nature,* **345**, 238-240 (1990).
9. L. R. Nittler, C. M. O'D. Alexander, X. Gao, R. M. Walker and E. Zinner, *Astrophys. J.,* **483**, 475-495 (1997).
10. S. Messenger, L. P. Keller, F. J. Stadermann, R. M. Walker and E. Zinner, *Science,* **300**, 105-108 (2003).
11. A. N. Nguyen and E. Zinner, *Science,* **303**, 1496-1499 (2004).
12. K. Nagashima, A. N. Krot and H. Yurimoto, *Nature,* **428**, 921-924 (2004).
13. L. R. Nittler et al., *Astrophys. J.,* **453**, L25-L28 (1995).
14. T. J. Bernatowicz, R. Cowsik, P. C. Gibbons, K. Lodders, B. Fegley, Jr., S. Amari and R. S. Lewis, *Astrophys. J.,* **472**, 760-782 (1996).
15. T. L. Daulton, D. D. Eisenhour, T. J. Bernatowicz, R. S. Lewis and P. R. Buseck, *Geochim. Cosmochim. Acta,* **60**, 4853-4872 (1996).
16. G. R. Huss and R. S. Lewis, *Meteoritics,* **29**, 791-810 (1994).
17. M. Rayet, M. Arnould, M. Hashimoto, N. Prantzos and K. Nomoto, *Astron. Astrophys.,* **298**, 517-527 (1995).
18. S. Wanajo, T. Kajino, G. J. Mathews and K. Otsuki, *Astrophys. J.,* **554**, 578-586 (2001).
19. U. Ott, *Astrophys. J.,* **463**, 344-348 (1996).
20. W. M. Howard, B. S. Meyer and D. D. Clayton, *Meteoritics,* **27**, 404-412 (1992).
21. U. Ott, *New Astronomy Reviews,* **46**, 513-518 (2002).
22. R. S. Lewis, G. R. Huss and G. Lugmair, *Lunar Planet. Sci.,* **XXII**, 807-808 (1991).
23. S. Richter, U. Ott and F. Begemann, *Nature,* **391**, 261-263 (1998).
24. R. Maas, R. D. Loss, K. J. R. Rosman, J. R. DeLaeter, R. S. Lewis, G. R. Huss and G. W. Lugmair, *Meteorit. Planet. Sci.,* **36**, 849-858 (2001).

Exchange Frequency of Oxygen Isotope Reservoirs in The Early Solar System

S. Itoh[a]* and H. Yurimoto[a]

[a] *Department of Natural History Sciences, Hokkaido University, Sapporo, Hokkaido 060-0810 Japan*

Abstract. High precision Mg isotope measurements have been performed to determine radiogenic ^{26}Mg of coarse-grained Ca-Al-rich inclusions (CAIs) in two carbonaceous chondrites by secondary ion mass spectrometry using faraday cup multi-collection system. Minerals in each CAI show individual internal isochrones of ^{26}Al-^{26}Mg with respect to each own oxygen isotopic composition of minerals. This indicates that ^{16}O-rich and ^{16}O-poor oxygen isotopic environments existed more than 0.45 Myrs in the solar nebula. The switching time scale between ^{16}O-rich and ^{16}O-poor gaseous environments surrounding CAIs was faster than 10 kyrs.

Keywords: CAI, Early solar system, Radiogenic, Magnesium Isotope, Oxygen Isotope, SIMS
PACS: 91.65.Sn, 91.67.Qr

INTRODUCTION

It is believed that Ca-Al-rich inclusions (CAIs) are the oldest materials in the solar system (4567Ma) [1]. Heterogeneous oxygen isotope distribution is observed in coarse-grained CAIs. Spinel and fassaite are usually rich in ^{16}O, whereas melilite and anorthite are poor in ^{16}O [e.g., 2]. Experimental crystallization sequence of CAI liquid [3] is in the order of spinel, melilite, fassaite and anorthite. To explain the heterogeneous oxygen isotopic distribution among the minerals, a simple crystallization of ^{16}O-rich CAI liquid in ^{16}O-poor gas is not applicable because melilite is ^{16}O-rich and fassaite is ^{16}O-poor.

Recently, CAIs bearing ^{16}O-rich melilite have been reported [4, 5]. For example, a type A CAI 7R19-1(a) from Allende CV3 chondrite experienced two heating events: Initially, this CAI crystallized from ^{16}O-rich CAI liquid. ^{16}O-rich spinel, melilite and fassaite were crystallized from the liquid. Secondly, this CAI experienced partial melting in ^{16}O-poor gas. Oxygen isotopic exchange occurred between ^{16}O-rich liquid and ^{16}O-poor gas, and ^{16}O-poor melilite and fassaite crystallized from the ^{16}O-poor liquid. However, time-scales between heating events and time-scales for oxygen isotope switching between ^{16}O-rich and ^{16}O-poor environment of CAI forming region have not been established.

The ^{26}Al-^{26}Mg dating using a short-lived radionuclide of ^{26}Al (a half-life of 0.73Ma) is useful to determine the time scales because CAIs contained live ^{26}Al. Recently, multi-collector secondary ion mass spectrometry (SIMS) has made possible high precision magnesium isotope analyses that provide chronologically useful data on low-Al/Mg phases in Ca-Al-rich inclusions (CAIs) in primitive chondrites [e.g., 6]. If

CP1016, *Origin of Matter and Evolution of Galaxies*,
edited by T. Suda, T. Nozawa, A. Ohnishi, K. Kato, M. Y. Fujimoto, T. Kajino, and S. Kubono

the radiogenic ^{26}Mg for individual minerals in CAIs could be determined with the precision of less than 0.2‰, time resolution of early solar system chronology would be achieved to 100 kyr. Because each CAI experienced multiple melting events, each internal ^{26}Al-^{26}Mg isochron in a CAI corresponds to individual melting and provide a time scale between heating events. This approach is independent of spatial heterogeneities of $^{26}Al/^{27}Al$ in the early solar system.

In this study, we report a high precision Al-Mg isotopic study of coarse-grained CAIs from Y81020 CO chondrite and Allende CV chondrites in order to determine time scale of ^{16}O-rich and -poor isotope exchange events.

EXPERIMENT METHODS

The sample used in this study is a type A 7R-19-1(a) CAI [4, 5], a type B HN3-1 CAI [e.g., 7] from Allende CV chondrite and a type A Y20a CAI from Y81020 CO chondrite. A carbon evaporation film of about 30 nm was coated on the thin section in order to reduce electrostatic charging during high-energy electron and ion bombardments for chemical and isotopic analyses. Oxygen isotopic studies of 7R-19-1 (a) and HN3-1 have been reported elsewhere [4, 5, 7]. Oxygen isotope measurements for minerals of Y20a CAI have been performed by the Hokudai isotope microscope system (Cameca ims-1270 SIMS system and SCAPS ion imager at Hokkaido University). The analytical conditions described elsewhere [8, 9].

For Al-Mg isotope measurements, a Cameca ims-1270 of Hokkaido University has been used in this study. A 13 keV $^{16}O^-$ primary ion beam focused to 10-20 μm diameter was used. Measurements were done at the high mass resolving power of ~2000, sufficient to resolve all molecular ion interferences (e.g., ^{24}MgH, $^{48}Ca^{2+}$). The effect of all molecular ion interferences to Mg isotopes was smaller than 0.1‰ levels. The primary ion currents were adjusted to 5-38 nA in order to set that the secondary ^{24}Mg ion counts for each standard mineral were obtained as $\sim 10^8$ cps.

Secondary ions were collected by four faraday cups of the multi-collection system for ^{24}Mg for L'2, ^{25}Mg for C, ^{26}Mg for H1 and ^{27}Al for H'2. Total of 60 cycle measurements for one measurement spot were done in 10 minutes. The last 20 cycles were used to calculate the Mg isotope ratios because the secondary ion intensity ratios between isotopes drifted systematically in the first tens cycles, which would be due to changes of electrostatic charging conditions on sample surface by starting of sputtering. After each sample measurement, we measured background noises of faraday cup detectors to compensate the background drifts to reduce the systematic error.

Russian spinel, Takashima augite, synthetic melilite-glass ($åk_{\sim 40}$) and synthetic fassaite-glass standards were used to correct instrumental mass fractionation for Mg isotopes. The excess ^{26}Mg calculated using the exponential law with a slope of 0.5140 derived from evaporated CAI-like materials [6] after correcting instrumental mass fractionation by matrix effect for CAI minerals [10]. The internal and external error of excess ^{26}Mg is less than 0.1‰ for the standard minerals. Representative error values (‰) of $\delta^{26}Mg_{excess}$ could be calculated as 0.13‰ (2σ). The detail experimental procedures are described elsewhere [10].

RESULTS AND DISCUSSION

Constituent minerals in the 7R-19-1(a) CAI are classified into ^{16}O-rich spinel, ^{16}O-rich melilite, ^{16}O-rich fassaite, ^{16}O-poor melilite and ^{16}O-poor fassaite. The ^{16}O-rich minerals plotted along a straight line in the Al-Mg isotope diagram. The ^{16}O-poor minerals also seem to be plotted to the other straight line. We calculate an initial $^{26}Al/^{27}Al$ ratio of each isochron (Fig. 1): ^{16}O-rich spinel, fassaite and melilite, $(5.83\pm0.27) \times 10^{-5}$ (2σ); ^{16}O-poor melilite and fassaite, $(4.71\pm0.43) \times 10^{-5}$ (2σ) (forced through the origin). Al-Mg isotopic distribution in each phase of 7R-19-1 CAI indicates that the chronological sequence of crystallization is consistent with that of oxygen isotopic petrography [4, 5]. The timing of heating events for ^{16}O-rich and ^{16}O-poor phases are clearly distinguishable and the time interval is about 0.23±0.11Myr.

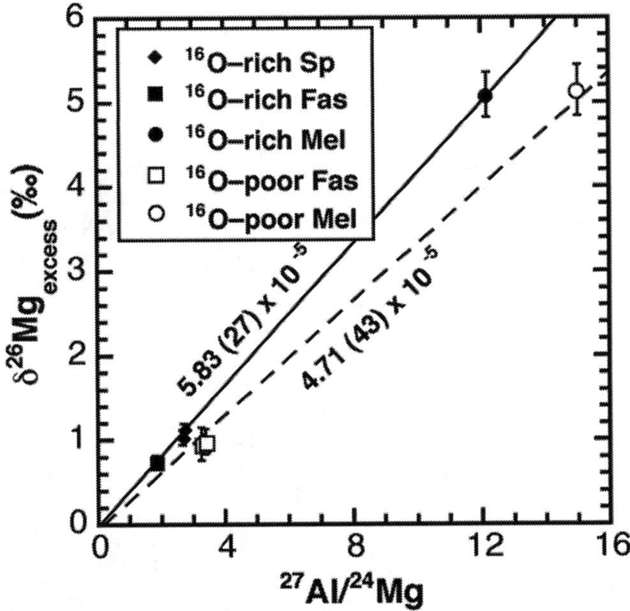

FIGURE 1. Al-Mg evolution diagram of 7R-19-1 (a) CAI. Solid symbols and a line denote ^{16}O-rich phases. Open symbols and a dashed line denote ^{16}O-poor phases. Each line and the number correspond to the internal isochron. Sp, spinel; Mel, melilite; Fas, fassaite.

The Y20a CAI consists of melilite, fassaite, and Wark-Lovering (WL) rim. The melilite contains small fassaite and spinel grains. Fassaite shows fringe texture resulting from the partial melting. Oxygen isotopes are homogeneously distributed in fassaite and WL rim. The oxygen isotopic compositions are enriched in ^{16}O ($\delta^{17,18}O$=~-40‰). Oxygen isotopes are heterogeneously distributed in melilite. The distribution is bimodal, i.e., ^{16}O-rich ($\delta^{17,18}O$=~-40‰) and ^{16}O-poor ($\delta^{17,18}O$=~0‰). From the oxygen isotopic compositions and petrographic texture, the constituent minerals of this CAI are classified into ^{16}O-rich fassaite, ^{16}O-rich interior melilite, ^{16}O-rich exterior melilite, ^{16}O-poor interior melilite and ^{16}O-rich WL rim. Minerals of each classification consist of individual isochrones (Fig. 2). The ^{16}O-rich fassaite shows the highest δ^{26}Mg excess ((6.06±0.01) x 10^{-5} (2σ)), and followed as, ^{16}O-rich interior melilite ((5.24±0.22) x 10^{-5} (2σ)), ^{16}O-rich exterior melilite ((4.60±0.02) x 10^{-5} (2σ)) and ^{16}O-poor interior melilite ((4.59±0.01) x 10^{-5} (2σ)), and WL-rim spinel ((3.94±0.30) x 10^{-5} (2σ)) (forced through the origin). These isochrones indicate that heating events occurred at least five times during this CAI formation of ~0.45Myr. The oxygen isotopic environment surrounding the CAI switched from ^{16}O-rich to ^{16}O-poor, and then to ^{16}O-rich. The relative age of ^{16}O-poor interior melilite and ^{16}O-rich exterior melilite in Y20a shows that the oxygen isotopic exchange events occurred within analytical uncertainty (less than 10kyrs).

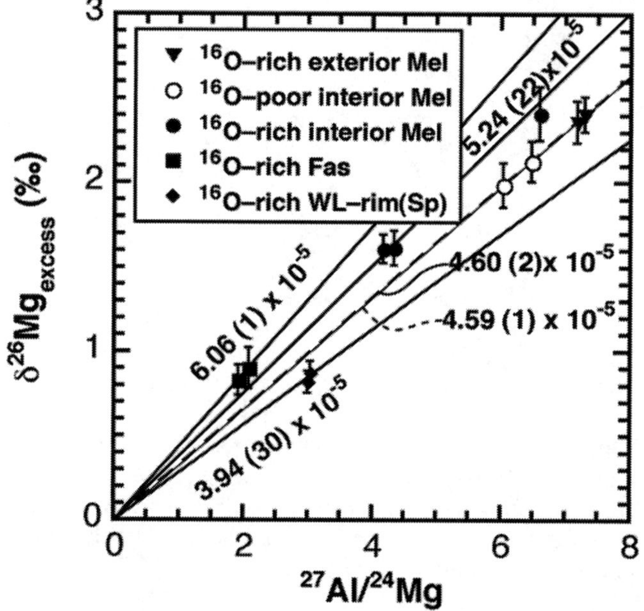

FIGURE 2. Al-Mg evolution diagram of Y20a CAI. Solid symbols and lines denote ^{16}O-rich phases. Open symbols and a dashed line denote ^{16}O-poor phases. Each line and the number correspond to the internal isochron. Mel, melilite; Fas, fassaite; WL-rim (Sp), Wark-Lovering rim (spinel).

Constituent minerals in the HN3-1 CAI are classified into ^{16}O-rich spinel, ^{16}O-poor melilite and ^{16}O-rich fassaite. We calculate an initial ^{26}Al/^{27}Al ratio of each phase with different oxygen isotopic compositions (Fig. 3), following as, ^{16}O-rich spinel and fassaite; (5.63±0.64) x 10^{-5} (2σ) and that of ^{16}O-poor melilite and fassaite; (5.50±0.04) x 10^{-5} (2σ) (forced through the origin). Although the timing of heating events for ^{16}O-rich and ^{16}O-poor phases may not be clearly distinguishable, switching between ^{16}O-rich and ^{16}O-poor environments was less than 0.03±0.12 Ma.

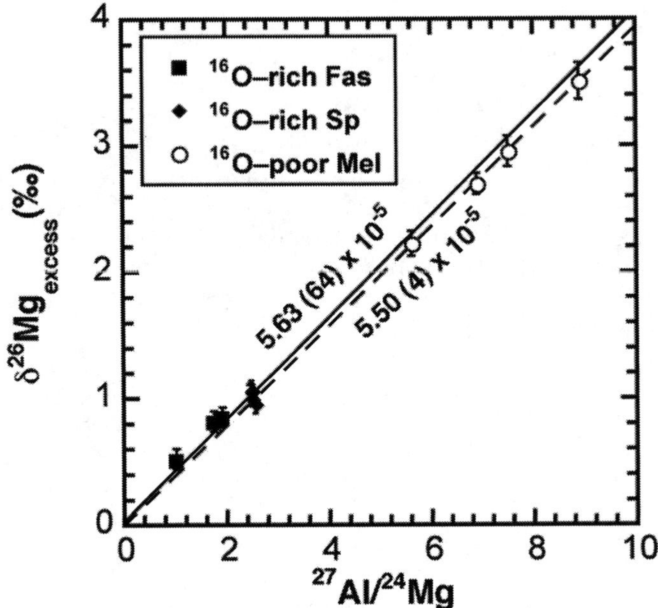

FIGURE 3. Al-Mg evolution diagram of HN3 CAI. Solid symbols and a line denote ^{16}O-rich phases and open symbols and a dashed line denote ^{16}O-poor phases. Each line and numbers corresponds to the internal isochron. Mel, melilite; Fas, fassaite; Sp, spinel.

The Al-Mg model age in this study indicates that these CAIs have been experienced multiple heating events accompanied by oxygen isotopic changing of the solar nebula. The oxygen isotopic compositions surrounding these CAIs changed from ^{16}O-rich to ^{16}O-poor or vice versa. The oxygen isotope switching abruptly occurred as fast as less than 10 kyrs in some case. These two oxygen isotopic environments existed at least 0.45 Myrs in the solar nebula, although it is unclear whether the two oxygen isotopic environments coexisted continuously or appeared frequently in the early solar system.

CONCLUSIONS

High precision Al-Mg dating is applied to individual minerals in three coarse-grained CAIs from Y81020 CO chondrite and Allende CV chondrites having O isotopic heterogeneity. Al-Mg isotopic systematics of internal isochron in these CAIs determined the time scale of multiple heating events and changing of oxygen isotope environments in the solar nebula. Al-Mg model age from each CAI shows that time interval of O isotopic exchange between ^{16}O-rich and ^{16}O-poor environments are less than 10 kyrs. The time duration of each CAI formation is variable, and in the formation of Y20a CAI lasted 0.45Myrs.

ACKNOWLEDGMENTS

This work was partly supported by the grant of Monka-sho (S.I. and H.Y.).

REFERENCES

1. Y. Amelin et al., *Science* 297, 1678 (2002)
2. R. N. Clayton, *Ann. Rev. Earth Planet. Sci.* 21, 115 (1993).
3. E. Stolper, *GCA* 46, 2159 (1982).
4. H. Yurimoto et al., *Science* 182, 1874 (1998).
5. M. Ito et al., *GCA* 68, 2905 (2004).
6. A. Davis et al., *LPSC XXXVI* , No. 2334 (2005).
7. M. Ito. et al., *LPSC XXIX* No. 1556 (1998).
8. S. Itoh and H. Yurimoto, *Nature* 423, 728 (2003).
9. H. Yurimoto et al., *Appl. Surf. Sci.,* 203-204, 793 (2003).
10. S. Itoh et al., *Appl. Surf. Sci.,* (2008) in press.

POSTER SESSION

Big Bang Nucleosynthesis: Impact of Nuclear Physics Uncertainties on Baryonic Matter Density Constraints

Michael S. Smith*, Blake D. Bruner†, Raymond L. Kozub†, Luke F. Roberts*, David Tytler**, George M. Fuller**, Eric Lingerfelt*,‡, W. Raphael Hix* and Caroline D. Nesaraja*,‡

Physics Division, Oak Ridge National Laboratory, Oak Ridge, Tennessee, 37831-6354, USA
†Physics Dept., Tennessee Technological Univ., Cookeville, Tennessee, 38505, USA
**Dept. of Physics, Univ. California San Diego, La Jolla, CA, 92093, USA
‡Dept. of Physics & Astronomy, Univ. of Tennessee, Knoxville, Tennessee, 37996-1200, USA

Abstract.
We performed new Big Bang Nucleosynthesis simulations with the **bigbangonline.org** suite of codes to determine, from the nuclear physics perspective, the highest achievable precision of the constraint on the baryon-to-photo ratio η given current observational uncertainties. We also performed sensitivity studies to determine the impact that particular nuclear physics measurements would have on the uncertainties of predicted abundances and on the η constraint.

Keywords: cosmology, big bang nucleosynthesis, thermonuclear reactions, reaction rates
PACS: 98.80.-k, 98.80.Ft, 98.80.Bp, 98.80.Cq, 98.80.Es, 26.35.+c, 27.10.+h, 27.20.+n, 29.87.+g

INTRODUCTION

In the standard model of Big Bang Nucleosynthesis (BBN), the creation of the light elements ^{2}H, ^{4}He, and ^{7}Li in the early universe depends on the rates of thermonuclear reactions such as p(n,γ)d as well as on the baryon-to-photon ratio η (or equivalently the ratio Ω_b of the density of baryonic matter to the closure density). The value of η can be constrained [1] by comparing each of the primordial abundances of ^{2}H, ^{4}He, and ^{7}Li (inferred from present-day observations) to those predicted by BBN theory. Disagreements between these constraints (e.g., using ^{7}Li) and the η constraint derived independently from Cosmic Microwave Background Radiation (CMBR) observations [2] can indicate, and constrain, potential new physics beyond the standard model [3, 4].

The precision of BBN η constraints depends both on the uncertainties in primordial abundances and on the uncertainties in BBN theory (abundance predictions). The latter are determined by Monte Carlo BBN calculations in which *input* thermonuclear reaction rate uncertainties are translated into *output* abundance prediction uncertainties [5, 6].

PRECISION LIMIT OF BBN CALCULATIONS

We performed Monte Carlo BBN simulations with the following estimates of the uncertainties of the input thermonuclear reaction rates: neutron lifetime [$\pm$0.8 seconds],

CP1016, *Origin of Matter and Evolution of Galaxies*,
edited by T. Suda, T. Nozawa, A. Ohnishi, K. Kato, M. Y. Fujimoto, T. Kajino, and S. Kubono

$p(n,\gamma)d$ [$\pm$7%], $d(p,\gamma)^3He$ [$\pm$6%], $d(d,n)^3He$ [$\pm$3%], $d(d,p)^3H$ [$\pm$3%], $^3He(n,p)^3H$ [$\pm$4%], $^3H(d,n)^4He$ [$\pm$4%], $^3He(d,p)^4He$ [$\pm$4%], $^3H(\alpha,\gamma)^7Li$ [$\pm$4%], $^3He(\alpha,\gamma)^7Be$ [$\pm$6%], $^7Li(p,\alpha)^4He$ [$\pm$5%], and $^7Be(n,p)^7Li$ [$\pm$5%]. These uncertainties are larger than those found in a recent BBN study [7] which relied substantially on renormalizing cross section values from different measurements. The primordial abundance determinations we chose were: D/H = $2.78 \pm 0.40 \times 10^{-5}$; Y_p [^{4}He] = 0.242 ± 0.002; and ^{7}Li/H = 1.23 [+0.68/-0.02]$\times 10^{-10}$. Combining the predicted abundances as a function of η from the Monte Carlo BBN simulation with the observations gives the following "canonical" constraints on (allowable ranges for) $\eta_{10} = \eta \times 10^{10}$: 5.26 - 6.56 from D/H [a full-width precision of 22%]; 3.45 - 5.33 from ^{4}He [42%], and 1.54 - 4.38 from ^{7}Li/H [96%]. All the calculations in this work were made with the new online suite of codes freely available at **bigbangonline.org**.

Next, we performed Monte Carlo BBN simulations wherein all reaction rate uncertainties were reduced to their smallest *reasonable* limit, in order to determine the smallest reasonable uncertainties in BBN theory predictions. We chose input uncertainties of $\pm$ 3% for the strong interactions based on our experience with high precision nuclear reaction cross section measurements. We chose $\pm$ 0.5 seconds for the smallest reasonable neutron lifetime uncertainty using current neutron bottle technologies [8] – although magnetic bottles [9] have the potential to go as low as 0.1 sec in the next decade [10]. The Monte Carlo BBN calculations were run with these small, but realistically achievable, nuclear reaction rate uncertainties as input. A comparison of the output with the observations leads to the following allowable ranges for η_{10} : 5.28 - 6.52 from D/H [full width precision of 21%]; 3.48 - 5.28 from ^{4}He [41%], and (Fig. 1) 1.56 - 2.38 and 2.86 - 4.22 from ^{7}Li/H [42% and 38%, respectively]. The range from ^{7}Li is bifurcated because the predicted abundances now drop below the lower observational limit of ^{7}Li /H = 1.21×10^{-10}.

A comparison of these results to the CMBR η constraint [2] (η_{10} = 5.87 - 6.3, a full-width precision of 7.2%) makes it clear that: (a) the BBN η_{10} ranges from ^{4}He and ^{7}Li/H (Fig. 1) disagree with the η_{10} range from CMBR studies, and the disagreement is slightly worsened with more precise rates; (b) the BBN η_{10} range using D/H agrees with that from the CMBR; and (c) *with current uncertainties on the primordial abundances*, reasonable improvements in nuclear physics are not able to give constraints on η with the same precision as that obtainable by CMBR studies. To achieve a comparable precision, uncertainties in primordial abundance observations must be significantly reduced – and the extent of this reduction is currently under study. The primary motivations for improved nuclear reaction rates are, at this time, better defining the central values of the disagreement between the CMBR η constraint and that from BBN using ^{4}He and ^{7}Li – and investigating the bifurcation of the constraint using ^{7}Li (Fig. 1).

SENSITIVITY STUDIES

To determine which reaction rates have the largest impact on reducing the predicted abundance uncertainties and the uncertainty of the BBN η constraint, we performed a series of Monte Carlo calculations with a systematic reduction in the uncertainty of each individual reaction rate (from the current value down to zero). The goal of these

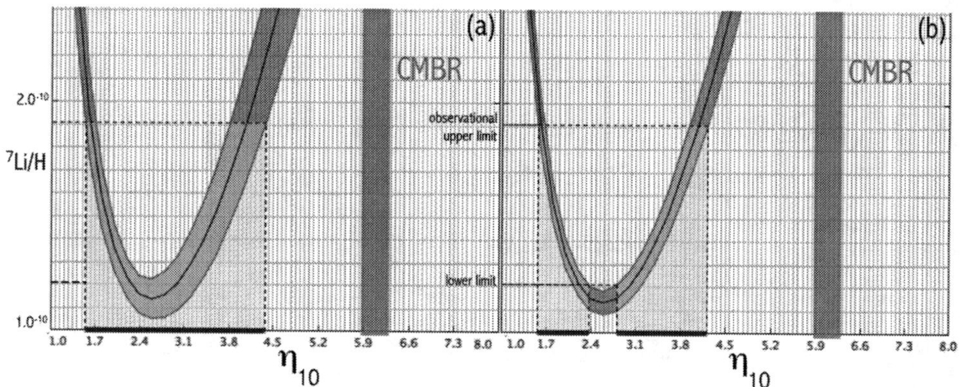

FIGURE 1. BBN predictions vs. baryon-to-photon ratio for ^{7}Li abundance with nuclear reaction rate uncertainties at their (a) current and (b) smallest reasonable values. Light shaded region indicates allowable η values from BBN; allowable range from CMBR is also shown.

calculations was to single out a subset of BBN thermonuclear reactions for further experimental and theoretical attention. We found that reducing the uncertainty of some reactions had essentially no (less than 5%) impact on the uncertainty of predicted abundances or η constraints. However, the D/H predictions were significantly sharpened by improvements in the d(p,γ)^{3}He, d(d,n)^{3}He, d(d,p)^{3}H, and p(n,γ)d reactions (in order of importance). For ^{4}He, the neutron lifetime and the p(n,γ)d reaction gave significant improvements, while the p(n,γ)d, ^{3}He(α,γ)^{7}Be, and d(p,γ)^{3}He reactions were important for improving the precision of the ^{7}Li predictions. Combining these results, we recommend improvements in the neutron lifetime and the p(n,γ)d, d(p,γ)^{3}He, d(d,n)^{3}He, d(d,p)t, and ^{3}He(α,γ)^{7}Be reaction rates to sharpen the cosmological predictions of BBN theory.

ACKNOWLEDGMENTS

ORNL is managed by UT-Battelle, LLC for the U.S. Department of Energy under contract DE-AC05-00OR22725.

REFERENCES

1. J. Yang, D. N. Schramm, G. Steigman, R.T. Rood, *Astrophys. J.* **227** (1979) 697.
2. D.N. Spergel et al., *Astrophys. J. Suppl.* **170** (2007) 377.
3. K. Jedamzik, *Phys. Rev.* D 70 (2004) 063524.
4. M. Kusakabe, T. Kajino and G. J. Mathews, *Phys. Rev.* D 74 (2006) 023526.
5. L.M. Krauss, P. Romanelli, *Astrophys. J.* **358** (1990) 47.
6. M.S. Smith, L.H. Kawano, R.A. Malaney, *Astrophys. J. Suppl.* **85** (1993) 219.
7. P. Descouvemont et al., *Atomic Data Nuclear Data Tables* **88** (2004) 203.
8. A. Serebov et al., *Phys. Lett.* **B605** (2005) 72.
9. P.R. Huffman et al., *Nature* **403** (2000) 62.
10. G. Greene, private communication (2008).

Periodic light variations from the triple-disk system around supermassive binary black holes

K.Hayasaki and S. Mineshige

Yukawa Institute for Theoretical Physics, Kyoto University Kitashirakawa, Oiwake-cho, Sakyo-ku, Kyoto 606-8502, Japan

Abstract.
We investigate accretion flows around supermassive binary black holes (BBHs) with the orbital eccentricity $e = 0.5$, the semi-major axis $a = 0.01\,\mathrm{pc}$, and the low mass ratio of the secondary black hole (BH) to the primary BH $q = 0.1$. In the simulations we consider a triple-disk system composing of two accretion disks around BHs and one circumbinary disk surrounding the two. The circumbinary disk works as a mass reservoir. We confirm that a non-axisymmetric accretion disk is formed around each BH. The X-ray luminosity of the secondary BH exhibits the double peaks every binary orbit, whereas that of the primary BH shows a single peak. Such properties can not be seen in the case of equal mass BBHs and therefore provide a potentially important observational signature of supermassive BBHs with low mass ratios.

Keywords: black hole physics – accretion, accretion disks – binaries:general – galaxies:nuclei

INTRODUCTION

In the standard paradigm of structural formation of the Universe, the galaxies experience multiple hierarchical mergers during their lifetimes. There are growing observational evidences that the supermassive black holes (BHs) have co-evolved with their host galaxies [1, 2]. These strongly suggest that a BH growth is mainly caused by a BH merger and a subsequent accretion of gas as a result of a galaxy merger. Then, supermassive binary black holes (BBHs) with a sub-parsec scale separation are inevitably formed before the BHs merge by emitting gravitational radiation[3]. It is likely that a rotating gas disk (i.e., circumbinary disk) forms around the supermassive BBHs in the course of the gas rich galaxy merger. Then, what will happen as a result of the interaction between the supermassive BBHs and the circumbinary disk?

Hayasaki et al. [4, 5], using a smoothed particle hydrodynamics (SPH) code[6], studied accretion flows from the circumbinary disk (hereafter, the CBD) to the supermassive BBHs with the equal BH masses. In principle, all the processes, including the mass transfer from the circumbinary disk to the individual BHs, and the accretion flow onto each BH, should be taken account of to understand the dynamics of a triple disk system, however, such a simulation would require an enormous computational time. Therefore, they divided all the processes into two steps: In the first-stage simulation, they focus on the mass transfer process from the CBD to the individual BHs. They found that the mass transfer from the CBD to each BH modulates with orbital phase when the supermassive BBHs are on an eccentric orbit. In the second-stage simulation, they studied the accretion flow onto each BH by assuming the mass transfer rate from the CBD. It was, then, confirmed that two accretion disks are formed around the BHs in the case of equal mass

CP1016, *Origin of Matter and Evolution of Galaxies*,
edited by T. Suda, T. Nozawa, A. Ohnishi, K. Kato, M. Y. Fujimoto, T. Kajino, and S. Kubono

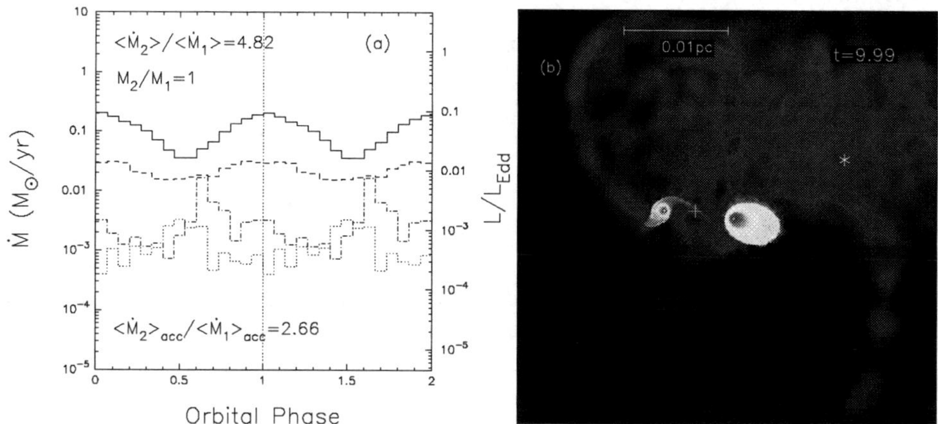

FIGURE 1. *(a)* Orbital phase dependence of the mass transfer rate from the CBD to the effective common gravitational radius of each BH, and of the mass accretion rate at the inner simulation boundary $r_{\text{in}} = 5.0 \times 10^{-3}a$, of the BHs. The solid-line and dashed line show the mass transfer rate to the secondary BH and that to the primary BH, respectively. The dotted line and the dash-dotted line show the mass accretion rate to the secondary BH and that to the primary BH, respectively. The right axis shows the bolometric luminosity L_{bol} corresponding to the mass transfer rate and the mass accretion rate with the energy conversion $\eta = 0.1$, normalized by the Eddington luminosity L_{Edd} for a total BH mass $M_{\text{bh}} = 1.0 \times 10^8\ M_{\odot}$, where η is defined by $L_{\text{bol}} = \eta \dot{M}_{\text{BH}} c^2$. In the panel, $\langle \dot{M}_2 \rangle / \langle \dot{M}_1 \rangle$, M_2/M_1, and $\langle \dot{M}_2 \rangle_{\text{acc}} / \langle \dot{M}_1 \rangle_{\text{acc}}$ show the ratio of the mass transfer rates, the mass ratio, and the ratio of the mass accretion rates, respectively. *(b)* Density maps of the two accretion disks around the supermassive BBH rotating with $P_{\text{orb}} \sim 9.4\,\text{yr}$, $e = 0.5$, and $q = 0.1$ at the periastron. Panel shows on a logarithmic scale the surface density contours over a range of 5 orders of magnitude. The white cross and the white asterisk indicate the positions of the mass input by the boundary condition. The supermassive BBHs are rotating in a counterclockwise direction. The time and a length scale are shown in the top-right and the top-left of the panel, respectively. The total number of SPH particles is 112,297 in the end of the simulation.

BHs. If the BH masses are much different, however, it is little known how the accretion disks are formed around BHs and what signatures they will exhibit. We have, therefore, performed the two-stage simulation of accretion flows as mentioned above but for the case of the supermassive BBHs with $a = 0.01\,\text{pc}$, eccentricity $e = 0.5$, and the low mass ratio $q = 0.1$.

DISTINCT SIGNATURES FROM SUPERMASSIVE BBHS

In the first-stage simulation, we calculate the mass transfer rate from the CBD to the effective gravitational radius of each BH as the averaged values over $40 \leq t \leq 100$, where the unit of time is $P_{\text{orb}} \simeq 9.4\,\text{yr}$. The orbital phase dependence of the mass transfer rate of the primary BH (hereafter, the primary transfer rate), and of the mass transfer rate of the secondary BH (the secondary transfer rate) are shown by the dashed line and the solid line in Figure 1*a*, respectively.

Next, we have performed the second-stage simulation based on the first-stage simulation. Here, each BH has an accretion radius $5 \times 10^{-3}a$. The mass accretion rates at this radius are also shown in Figure 1*a*, where the dotted line shows the mass accretion rate to the primary BH (hereafter, the primary accretion rate), and the dash-dotted line shows the same but to the secondary BH (the secondary accretion rate). Figure 1*b* shows the snapshot of the accretion flow around the supermassive BBHs at the time of periastron passage of the 10th binary orbit. The white cross and the white asterisk in the density map indicate the mass supply points from the CBD to the primary BH and to the secondary BH, respectively.

Our main conclusions are summarized as follows:

- We confirm the formation of the two accretion disks around BHs. These disks are significantly non-axisymmetric.
- The accretion disk around the primary BH is much larger than the accretion disk around the secondary BH.
- While the averaged secondary transfer rate during one orbital period is about 4.8 times higher than that of the primary transfer rate, the averaged secondary accretion rate are about 2.7 times higher than that of the primary accretion rate. These indicate that the system does not reach the quasi-steady state, yet.
- Unexpectedly the secondary accretion rate shows double peaks during one orbital period, whereas the primary accretion rate has a single peak. These features are unique to supermassive BBHs with low mass ratios, thus offering a basic signature of the presence of supermassive BBHs.

ACKNOWLEDGMENTS

The simulations reported here were performed using the facility at the Centre for Astrophysics & Supercomputing at Swinburne University of Technology, Australia and at YITP in Kyoto University. This work has been supported in part by the Grants-in-Aid of the Ministry of Education, Science, Culture, and Sport and Technology (MEXT; 30374218 K.H., 14079205 K.H. & S.M.), and by the Grant-in-Aid for the 21st Century COE Scientific Research Programs on "Topological Science and Technology" and "Center for Diversity and Universality in Physics" from MEXT.

REFERENCES

1. Ferrarese, L. & Merritt, D., *ApJL*, 539, L9-12, (2000).
2. Gebhardt, K., et al., *ApJL*, 539, L13-16, (2000).
3. Begelman, M. C., Blandford, R. D. & Rees, M. J., *Nature*, 287, 307-309, (1980).
4. Hayasaki, K., Mineshige, S. & Sudou, H., *PASJ*, 59, 427, (2007).
5. Hayasaki, K., Mineshige, S. & Ho, C.L., *ApJL*, submitted
6. Bate, M. R., Bonnel, I. A. & Price, N. M., *MNRAS*, 277, 362-376, (1995).

Oxygen Isotopic Evolution in the Early Solar Nebula: Validation of the H_2O Transport Model

Takashi Fukui and Kiyoshi Kuramoto

Department of Cosmosciences, Hokkaido University, Sapporo 060-0810, Japan

Abstract. We perform a numerical simulation on the global transport process of ^{16}O-poor H_2O and silicate in a protoplanetary disk to examine whether the process is responsible for the oxygen isotopic systematics observed in chondritic components without conflicting with observation of protoplanetary disks and theory of planet formation.

Keywords: Chondrites, Chondrules, CAIs, Oxygen isotope, Protoplanetary disks
PACS: 96.30.Za, 97.82.Jw

INTRODUCTION

Oxygen isotopic systematics observed in chondritic components provides one of the best clues to elucidate chemical and physical processes that occurred during the formation of the solar system [1]. Chondrules show the oxygen isotopic composition similar to those of terrestrial materials with little variation. In contrast, CAIs, the oldest minerals in the solar system and formed $\sim 10^6$ yr prior to chondrules [2], have anomalously ^{16}O-rich composition.

It is recently suggested that the systematics would reflect the oxygen isotopic evolution of the inner solar nebula gas associated with enhancement of ^{16}O-poor H_2O produced in the parent molecular cloud [3]. The enhanced H_2O is supplied by evaporation of icy content in dust aggregates drifted from the cold outer nebula, losing their orbital angular momentum due to the nebula gas drag. Therefore, the mean oxygen isotopic composition of the inner solar nebula is expected to evolve from ^{16}O-rich (CAIs like) to relatively ^{16}O-poor (chondrules and terrestrial like) one, which is, at least qualitatively, consistent with the systematics and chronology.

However, the degree and duration of H_2O enhancement, which depend on the size of dust aggregates and the properties of the nebula (such as mass, radius, the intensity of turbulence, etc.), have been poorly studied. In addition, although the model relies on substantial infall of dust aggregates, a sufficient amount of solid components to form planets must be retained in the nebula at the same time.

In this study, we perform a set of numerical simulation on the global transport process of the ^{16}O-poor H_2O and silicate under the typical properties of observed protoplanetary disks to examine whether the process is capable of reproducing the oxygen isotopic systematics and chronology of chondritic components without any conflicts with theory of planet formation.

CP1016, *Origin of Matter and Evolution of Galaxies*,
edited by T. Suda, T. Nozawa, A. Ohnishi, K. Kato, M. Y. Fujimoto, T. Kajino, and S. Kubono

MODEL

We numerically solve one-dimensional transport equations for the nebula gas (containing CO), H_2O and silicate in solid and gas phases. The equations for H_2O and silicate include the exchange term between the both phases associated with evaporation and condensation at each evaporation front. Transport in gas phase is driven by the disk gas turbulence, whereas transport in solid phase is also contributed by inward drift due to the nebula gas drag. The difference in the transport rate between gas and solid phases is responsible for the enhancement of H_2O inside the evaporation front of ice (hereafter, we call it "the ice line").

To solve the equations, we need to determine the typical size of dust aggregate, which affects the drift velocity, at each location in the disk. Mechanisms which control the typical size of dust aggregates are currently unsettled. Here we assume that collisional agglomeration is suspended at the aggregate size much smaller than m-size because they hardly stick each other above a threshold collision velocity. Our description is supported by collision experiments of dust aggregates [4]. Theoretical analysis of collisional mechanics suggests that the threshold velocity for silicate is an order of magnitude lower than that of H_2O ice [5]. This implies that silicate aggregates inside the ice line are smaller and therefore drift slower than icy aggregates.

The temperature of the disk midplane, which affects the intensity of turbulence and the location of the ice line, is calculated with a simple radiative transfer model including irradiation from the central star and viscous dissipation as heat source.

We consider a disk with the initial mass and dust-gas ratio of $0.1M_\odot$ and 0.01 which surrounds the central star with mass, radius and surface temperature of 1 $M_\odot$, $2R_\odot$ and 4000 K, respectively. The porosity of dust aggregate is 0.9. The initial disk radius $r_{\rm init}$, the dimensionless parameter α which is related to the intensity of turbulence, and the threshold collision velocity for H_2O ice $V_{\rm crit,ice}$ are parameters ranging 100–1000 AU, 10^{-2}–10^{-3} and 0.3–3 m s^{-1}, respectively. The relative abundance and the initial fractionation of H_2O, silicate and CO [3,6] are summarized in Table 1.

RESULT AND DISCUSSION

Figure 1 shows the result for $r_{\rm init} = 300$ AU, $\alpha = 3 \times 10^{-3}$ and $V_{\rm crit,ice} = 1$ m s^{-1}. In this case, the ice line locates around 3 AU. After the onset of disk accretion, solid components (H_2O and silicate) begin to drift inward from the outer region of the disk. This leads to enhancement of water vapor inside the ice line because inward transport of H_2O is decelerated after sublimation. Silicate is also decelerated and concentrated in the inner disk, because the adhesivity and thus the size of dust aggregates decrease after

TABLE 1. Properties of oxygen carriers

	H_2O	Silicate	CO
relative O abundance	4	3	7
$\delta^{17,18}O_{\rm solar}$ (‰)	+105	0	−60

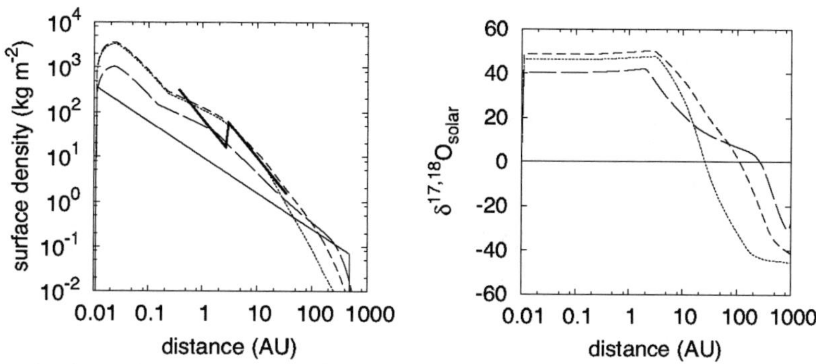

FIGURE 1. Radial profiles of the surface density of solid components (left) and the local mean oxygen isotopic composition (right). The thin solid, long-dashed, short-dashed and dotted curves represents $t = 0$, 3×10^5, 10^6 and 2×10^6 yr, respectively. The thick solid curve in the left panel corresponds to the minimum mass solar nebula model [7].

removal of the icy mantle. The individual silicate aggregate inside the ice line is $\sim$ mm or less in size, which is comparable with the typical chondrule size.

As a result of the H_2O enhancement, the mean oxygen isotopic composition of the inner region of the disk evolves toward ^{16}O-poor. In this case, $\delta^{17,18}O_{solar}$ increases by 40 ‰ within $\sim 3 \times 10^5$ yr. Thus, the difference in oxygen isotopic composition between CAIs and chondrules is reproduced in this model. Until $t \sim 2 \times 10^6$ yr, the oxygen isotopic composition is temporally and spatially almost constant. This may explain the little variation in the oxygen isotopic composition among chondrules. During this period, a sufficient mass of solid components is available for planet formation. The duration of the H_2O enhancement inside the ice line depends on r_{init}, α and $V_{crit,ice}$. The dependence on the three parameters is given by;

$$\tau \sim 2 \times 10^6 \left(\frac{r_{init}}{300\ \mathrm{AU}}\right)^{1/2} \left(\frac{\alpha}{3 \times 10^{-3}}\right) \left(\frac{V_{crit,ice}}{1\ \mathrm{m\ s^{-1}}}\right)^{-2} \mathrm{yr}.$$

We conclude that the H_2O transport model is capable of reproducing the oxygen isotopic evolution of the inner solar nebula gas consistently with the oxygen isotopic systematics and chronology of chondritic components, observation of protoplanetary disks and theory of planet formation.

REFERENCES

1. R. N. Clayton, *Ann. Rev. Earth Planet. Sci.* **21**, 115–149 (1993).
2. Y. Amelin et al., *Science* **297**, 1678–1683 (2002).
3. H. Yurimoto and K. Kuramoto, *Science* **305**, 1763–1766 (2004).
4. J. Blum and G. Wurm, *Icarus* **143**, 138–146 (2000).
5. C. Dominik and A. G. G. M. Tielens, *Astrophys. J.* **480**, 647–673 (1997).
6. K. Lodders, *Astrophys. J.* **591**, 1220–1247 (2003).
7. C. Hayashi, *Prog. Theor. Phys. Suppl.* **70**, 35–53 (1981).

Identification of Silicate and Carbonaceous Presolar Grains in the type 3 Enstatite Chondrites

Shingo Ebata and Hisayoshi Yurimoto

Department of Natural History Sciences, Hokkaido University, Sapporo 060-0810, Japan

Abstract. We surveyed presolar grains in primitive enstatite chondrites by isotopography using the HokuDai isotope microscope system. The mineral identification has been conducted by X-ray analysis with scanning electron microscopy. The chemical compositions are determined for eight silicate and ten carbonaceous presolar grains. Presolar grains of pyroxene compositions are dominant in the enstatite chondrites. This suggests that presolar silicates of enstatite composition were selectively survived in the enstatite chondrite parent body or the enstatite chondrite formation area in the solar nebula.

Keywords: presolar grain, silicate, carbon, meteorites, SIMS, isotope imaging
PACS: 96.30.Za, 32.10.Bi, 68.49.Sf, 82.80.Ms, 91.65.Dt, 97.10.-q, 81.05.Uw

INTRODUCTION

Primitive meteorites, interplanetary dust particles (IDPs) and Antarctic micro meteorites (AMMs) contain presolar silicate grains that predate the formation of our solar system [1-7]. The silicate grains have been found by O-isotope ion imaging with a NanoSIMS [e.g. 1-3,7] and with an isotope microscope (Cameca ims-1270 + SCAPS [8]) [e.g. 4-6]. Although about 200 presolar silicates were discovered, the chemical compositions of only 27 presolar silicate grains have been studied in carbonaceous chondrites (CCs) [e.g. 2-4], IDPs [e.g. 1] and AMMs [e.g. 7]. The spatial resolution and the quantitative analysis capability of the conventional ion imaging technique have been difficult to determine the mineral species of sub-micron presolar silicate grains because of the small size and abundant silicate grains formed in the solar system. Among the 27 presolar silicate grains identified so far, eight have compositions similar to olivine, eleven are pyroxene-like, and seven are GEMS (glass with embedded metal and sulfides)-like compositions. The sizes of presolar silicates seem to be between 100 and 1000 nm with most grains less than 300 nm.

Ebata et al. [9] have discovered silicate and carbonaceous presolar grains from three type 3 enstatite chondrites (ECs) by in-situ measurements. However, grain characterizations were limited to about 17 % of all presolar grains found and the characterized mineral species were only pyroxene compositions. It seems that presolar silicate species of ECs may be different from those of CCs, IDPs and AMMs. Here we report further in situ studies of silicate and carbonaceous

CP1016, *Origin of Matter and Evolution of Galaxies*,
edited by T. Suda, T. Nozawa, A. Ohnishi, K. Kato, M. Y. Fujimoto, T. Kajino, and S. Kubono

presolar grains in three primitive EH3 chondrites, Yamato-691, ALHA81189 and Sahara 97072.

EXPERIMENTAL

We surveyed presolar grains by isotopography using the HokuDai isotope microscope system (Cameca ims-1270 + SCAPS [8]). We applied two homogeneous irradiation conditions of primary beam for about 70 μm across field; the rastered ion beam of small size and the static broad ion beam. Total integration time for each field was ~1 hour for rastered beam condition and ~1.5 hours for static beam condition. We used a 50 μm contrast aperture (CA) except for C isotopes. A 150 μm CA was used for C isotopes. The primary beam intensity was adjusted to ~0.5 nA for both irradiation conditions. The sputtering depth was less than 100 nm for the sequence. The digital image processing methods, the selection criterion for distinguishing presolar grains and estimation of the analytical errors are the same as [4].

For mineral identification of presolar grains, mineralogical and petrographical characterization of matrix areas containing isotopic anomalous grains has been conducted using a field emission type scanning electron microscope (JEOL JSM-7000F) equipped with energy dispersive X-ray spectrometer (Oxford INCA).

RESULTS AND DISCUSSION

Presolar silicates were identified by oxygen isotopographies: 3 grains from areas of about 61,000 μm^2 for Y-691 (the volume abundance: 4 ppm); 19 from about 99,000 μm^2 for ALHA81189 (17 ppm), whereas no presolar silicate grains were identified grains from areas of about 30,000 μm^2 for SAH 97072 (<3 ppm). Presolar carbonaceous grains were also identified by carbon isotopographies: 14 grains from areas of about 63,000 μm^2 for Y-691 (20 ppm); 13 from about 96,600 μm^2 for ALHA81189 (12 ppm); and 3 from about 32,000 μm^2 for SAH 97072 (8 ppm). The abundance of presolar silicates is much smaller in these EH3 chondrites than in primitive CCs and IDPs [e.g. 1, 5].

The oxygen isotopic compositions of presolar silicates in the EH3 chondrites are shown in Figure 1. The most grains (~ 90 %) have excesses in ^{17}O with nearly normal $^{18}O/^{16}O$ ratios. These grains are categorized into the so-called group 1, which likely formed around O-rich red giant or asymptotic giant branch (AGB) stars [10]. The rest belongs to the group 4, which has nearly normal $^{17}O/^{16}O$ ratios and excesses in ^{18}O. The origin of group 4 grains is considered as AGB stars or super novae.

The sizes of presolar silicates are 0.2-1.1 μm (average: 0.49 μm) and carbonaceous grains are 0.1-1.2 μm (average: 0.40 μm). The average size of presolar silicate grains of the ECs is larger than those of CCs and IDPs [e.g. 1, 2, 3]. In the case of presolar carbonaceous grains, the average size is smaller than those from CCs [11].

The chemical compositions are determined for eight presolar silicate grains (e.g. Figure 2 (a, c)); Enstatite: 3, Fe-rich hypersthene pyroxene (En_{50}): 1, Fe-rich olivine (Fo_{30}): 1, SiO_2: 1, aggregates of pyroxene-like compositions: 2. The SiO_2 grain and aggregates may be amorphous. In the case of carbonaceous grains, the chemical compositions are determined for ten grains (e.g. Figure 2 (b,d)); graphite: 4, SiC: 6. Presolar grains of pyroxene compositions are more dominant than other presolar silicate species in the ECs, whereas olivine, pyroxene and GEMS are equally distributed CCs [e.g. 2-4], IDPs [e.g. 1] and AMMs [7]. This suggests that presolar silicates of enstatite composition were selectively survived in the ECs parent body or the enstatite chondrite formation area in the solar nebula.

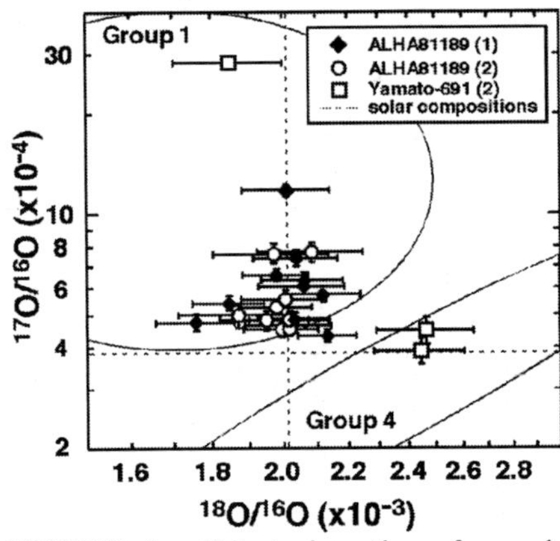

FIGURE 1. O-isotopic ratios of presolar silicate grains from Y-691 and ALHA 81189. Error bars are 2σ. Ten grains belong to Group 1 and two grains belong to Group 4 defined by [11]. (1) and (2) in the legend correspond to the primary beam conditions of static and rastered beam, respectively.

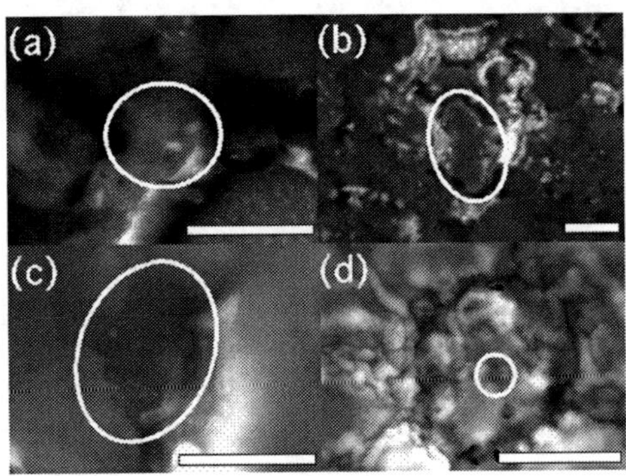

FIGURE 2. Examples of identified presolar grains in this study (indicated by circles). (a) enstatite grain from ALHA81189. (b) SiC grain from ALHA81189. This grain is surrounded by FeS. (c) enstatite grain from Y-691. (d) Graphite grain form Y-691. This grain is surrounded by Fe metal. Scale bars are 1 μm.

ACKNOWLEDGMENTS

We thank T. McCoy and L. Welzenbach for loaning us the thin section of ALHA81189. We also thank S. Kobayashi, N. Sakamoto, T. J. Fagan and S. Itoh for the many useful discussions and their help with solving numerous technical problems.

REFERENCES

1. Messenger et al. (2003) *Science* **300**, 105-108.
2. Nguyen and Zinner (2004) *Science* **303**, 1496-1499.
3. Mostefaoui and Hoppe (2004) *Astrophysical Jarnal* **613**, L149-L152.
4. Nagashima et al. (2004) *Nature* **428**, 921-924.
5. Kobayashi et al. (2005) *Antarct. Meteor.* **XXIX**, 30-31.
6. Tonotani et al. (2006) *Lunar and Planetary Sci.* **XXXVII**, 1539.
7. Yada et al. (2006) *Lunar and Planetary Sci.* **XXXVII**, 1470.
8. Yurimoto et al. (2003) *Appl. Surf. Sci.* **203-204**, 793.
9. Ebata et al. (2006) *Lunar and Planetary Sci.* **XXXVII**, 1619.
10. Nittler et al. (1997) *Astrophysical Jarnal* **483**, 475-495.
11. Amari (1994) *Geochim. Cosmochim. Acta* **58**. 459-470.

Waiting Points in Nova and X-ray burst Nucleosynthesis

Tomomi Sunayama*,†, Michael S. Smith**, Eric Lingerfelt**,‡, Kim Buckner**,‡, W. Raphael Hix** and Caroline D. Nesaraja**,‡

*Knox College, 2 East South Street, Galesburg, Illinois, 61401-4999, USA
†Oak Ridge Institute for Science Education, Oak Ridge, Tennessee 37831-0117
**Physics Division, Oak Ridge National Laboratory, Oak Ridge, Tennessee, 37831-6354, USA
‡Dept. of Physics & Astronomy, Univ. of Tennessee, Knoxville, Tennessee, 37996-1200, USA

Abstract. In nova and X-ray burst nucleosynthesis, waiting points are nuclei in the reaction path which delay the nuclear flow towards heavier nuclei, typically because of a weak proton capture reaction and a long β^+ lifetime. Waiting points can influence the energy generation and final abundances synthesized in these explosions. We have constructed a systematic, quantitative set of criteria to identify rp-process waiting points, and use them to search for waiting points in post-processing simulations of novae and X-ray bursts. These criteria have been incorporated into the **Computational Infrastructure for Nuclear Astrophysics**, online at **nucastrodata.org**, to enable anyone to run customized searches for waiting points.

Keywords: nova, X-ray burst, nucleosynthesis, waiting point, reaction network
PACS: 26.30.-k, 26.30.Ca, 26.50.+x, 97.10.Cv, 97.30.Qt, 97.80.Gm, 97.80.Jp

INTRODUCTION

Novae and X-ray bursts are stellar explosions driven by sequences of thermonuclear reactions on proton-rich unstable nuclei - including the Hot CNO cycle, the rp-process, and the αp-process [1]. These reaction sequences, often visualized as a "flow" up the nuclide chart from lighter to heavier nuclei [Fig. 1], directly determine the observable luminosity as well as the abundances synthesized in these explosions.

To more thoroughly understand the nuclear reaction sequences in these explosions, it is important to identify their waiting points for further experimental and theoretical investigations. Waiting points are nuclei in the reaction path which delay the reaction flow towards heavier nuclei. They are so named because they typically have limited reaction flux through the proton capture reaction and a long β^+ lifetime, and the reaction flow must "wait" for the long decay before heavier nuclei are synthesized.

There is, unfortunately, no systematic, quantitative definition of rp-process waiting points. A variety of criteria have been sporadically applied in the literature, including (p,γ) reaction Q-value, β^+ lifetime τ_β, change in abundance, and others. Many of these, unfortunately, have serious limitations. For example: Q-values do not always correlate with reaction rates; some waiting points have lower abundances than nuclei with similar Z and N; β lifetime does not take into account the total destruction lifetime both by decay and by strong interactions; and stable nuclei with no β decay can be waiting points. The lack of systematic criteria together with the large number of nuclei often forces searches to rely heavily on preconceptions (e.g., even-even $N = Z$ nuclei) which

CP1016, *Origin of Matter and Evolution of Galaxies*,
edited by T. Suda, T. Nozawa, A. Ohnishi, K. Kato, M. Y. Fujimoto, T. Kajino, and S. Kubono

are in some cases misleading and in most cases incomplete.

WAITING POINT CRITERIA

Our approach is to first specify a time window within a simulation, then to apply a set of tests to each nuclear species – comparing its properties [abundance, destruction lifetime, flux through creation and destruction reactions, and others] during that time window to those of "neighboring" nuclei in the $N-Z$ plane. Some tests are utilized to reject a nucleus as a waiting point; if it passes, more stringent tests are used to decide whether or not it is a waiting point during the specified time window.

Our waiting point rejection tests are: total destruction half life should be larger than a minimum threshold [with a default of 0.2 seconds]; the maximum creation flux should be larger than the maximum destruction flux; the abundance should be larger than a minimum value [1.0×10^{-6}]; and the total integration of flux over the time window is positive. The more stringent tests ("acceptance tests") are: the average over the time window of the total destruction lifetime (β decay plus strong interactions) is larger than τ_β; total destruction half life is larger than a chosen value [1.0 second]; average total flux is larger than half the maximum total flux; abundance is larger than a chosen value [1.0×10^{-4}]; and the total integrated flux is greater than a chosen value [1.0×10^{-4}].

Our current decision logic is that a nucleus must pass all rejection tests and at least one of the acceptance tests for it to be considered a waiting point. The numerical parameters chosen in these criteria must be adjusted for different simulations; the default values are show in brackets above. These criteria have been incorporated into the **Computational Infrastructure for Nuclear Astrophysics**, a suite of nuclear astrophysics codes available online at `nucastrodata.org`. With this new software tool, users can now specify a simulation, a time window, and values of parameters that define waiting points, and the system will systematically test all nuclei and determine which ones satisfy these criteria. These waiting points can then be visualized on a nuclide chart along with the reaction flux and/or the abundance of the nuclei in the simulation.

FIRST RESULTS

We used our criteria to search for waiting points in post-processing simulations of novae and X-ray bursts, and compared our results to those found in the literature. We searched for waiting points in 7 different time windows in a single-zone X-ray burst simulation similar to that in [2], and found ^{14}O and ^{15}O are waiting points at very early times [(-33,-28) and (-22,-18), respectively, where (t_1, t_2) are the start and stop times in seconds and 0 seconds is the peak temperature of the burst]. However, we find additional waiting points during these windows not reported in [2]: ^{15}O, ^{19}Ne, ^{22}Mg, ^{23}Mg, and ^{32}S in (-33,-28); ^{14}O, ^{26}Si, ^{30}S, and ^{32}S in (-22,-18). In later time windows, we confirmed some waiting points, found some new ones, rejected others, and found that some (e.g., ^{56}Ni) were waiting points in different windows (e.g., (2,5) instead of (-3,0) in [2]). We confirmed all of our waiting points by detailed examination of the abundance vs. time and flux vs. time over the time windows of interest. At later times in an X-ray burst simulation, it

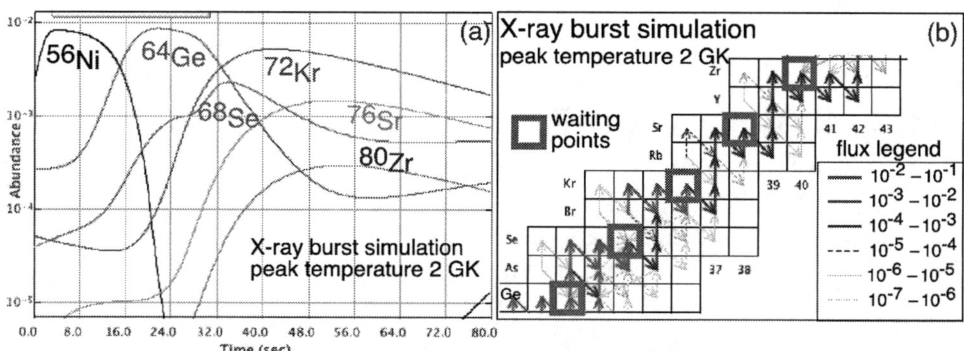

FIGURE 1. (a) Abundances vs. time for even-even, $Z = N$ nuclei in the late times of an X-ray burst; (b) corresponding nuclear reaction flux diagram with waiting points indicated.

is thought that ^{56}Ni and the even-even, $N = Z$ nuclei above ^{64}Ge are waiting points [3]. We confirm this in windows such as (2,5), where we find ^{56}Ni, ^{64}Ge, ^{68}Se, and ^{76}Sr are waiting points. In later time windows, higher masses are synthesized, then decay away and the flow moves to even heavier nuclei [Fig. 1]. We confirmed this, finding that ^{72}Kr, ^{74}Sr, ^{76}Sr, ^{80}Zr, ^{84}Mo, ^{88}Ru, ^{96}Cd are waiting points in (27,32). In this window, we also find three additional waiting points (^{75}Sr, ^{79}Zr, and ^{81}Zr) all *adjacent* to even-even, $N = Z$ nuclei that would likely be missed by searches based on past literature results.

SUMMARY AND FUTURE PLANS

We have developed a quantitative set of criteria for rp-process waiting points that influence nucleosynthesis in novae and X-ray bursts. These criteria have been incorporated into the **Computational Infrastructure for Nuclear Astrophysics** at `nucastrodata.org` to enable rapid searches and visualization of waiting points. Future plans include guiding the choice of time windows, enabling customized search logics, ranking each waiting point impact, and running the system on a variety of element synthesis simulations to search for previously neglected waiting points.

ACKNOWLEDGMENTS

ORNL is managed by UT-Battelle, LLC for the U.S. Department of Energy under contract DE-AC05-00OR22725. TS was partially supported by the ORISE HERE program.

REFERENCES

1. M.S. Smith, K.E. Rehm, *Ann. Rev. Nucl. Part. Sci.* **51** (2001) 91.
2. M. Wiescher *et al.*, *J. Phys. G* **25** (1999) R133-R161.
3. H. Schatz *et al.*, *Phys. Rep.* **294** (1998) 167.

Hydrodynamical simulations of detonations in superbursts.

Claire Noël*, Yves Busegnies*, Miltiadis V. Papalexandris† and Stephane Goriely*

*Institut d'Astronomie et d'Astrophysique, Université Libre de Bruxelles, Campus plaine CP 226, Boulevard du Triomphe, 1050 Bruxelles, Belgium
†Département de Mécanique, Université catholique de Louvain, 1348 Louvain-la-Neuve, Belgium

Abstract. A new hydrodynamical algorithm to study astrophysical detonations is presented. A prime motivation of this development is the description of a carbon detonation in conditions relevant to superbursts, which are thought to result from the propagation of a detonation front around the surface of a neutron star in the carbon layer underlying the atmosphere. The one-dimensional calculations we have performed demonstrate that the carbon detonation at the surface of a neutron star is a multiscale phenomenon. We show that a multi-resolution approach can be used to solve all the reaction lengths. For mixed H/He accreting systems, we have introduced a new reduced network to study the impact of the photodesintegration of the heavy elements on the detonation.

Keywords: Hydrodynamics; Nuclear reactions, nucleosynthesis, abundances; Shock waves; Stars: neutron; X-rays: bursts
PACS: 26.30.Ca, 95.30.Lz, 97.10.Cv, 97.60.Jd, 97.80.Jp

INTRODUCTION

The current view is that superbursts are due to the thermally unstable ignition of ^{12}C at densities of about 10^8 - 10^9 g cm^{-3} [1, 2]. Strohmayer & Brown [2] have shown that for the pure He accreting system 4U 1820-30 the superburst is thermonuclear in origin, fueled by the burning of carbon.For mixed H/He accretors a very small amount of ^{12}C remains after the combustion of H and He via the rp-process [3]. However, Cumming & Bildsten [1] have shown that a small residual amount of ^{12}C in a heavy element bath might be enough to ignite a superburst. Moreover, Schatz et al. [4] have shown that, at the high temperatures reached in superbursts, the photodisintegration of the heavy rp-process ashes influences the energetics of the combustion. In this case, for the calculation of the energy production during the superbursts, the simplified nuclear reaction network usually adopted in astrophysical hydrodynamics simulations [5] has to be extended in order to include the contribution of the rp-process ashes.

NUMERICAL METHOD

We use the hydrodynamical algorithm described in [5]. It solves the adiabatic Euler equations for compressible, non viscous gas dynamics with source terms. It is a finite-volume method of MUSCL type [6].The algorithm avoids dimensional splitting and is parallel. The equation of state accounts for partially degenerate and partially relativistic

CP1016, *Origin of Matter and Evolution of Galaxies,*
edited by T. Suda, T. Nozawa, A. Ohnishi, K. Kato, M. Y. Fujimoto, T. Kajino, and S. Kubono

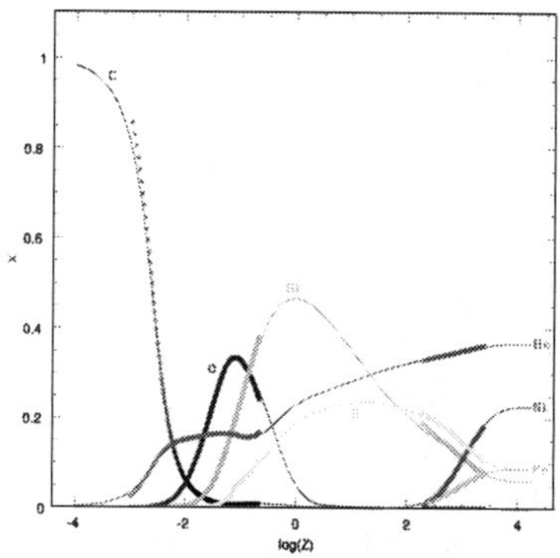

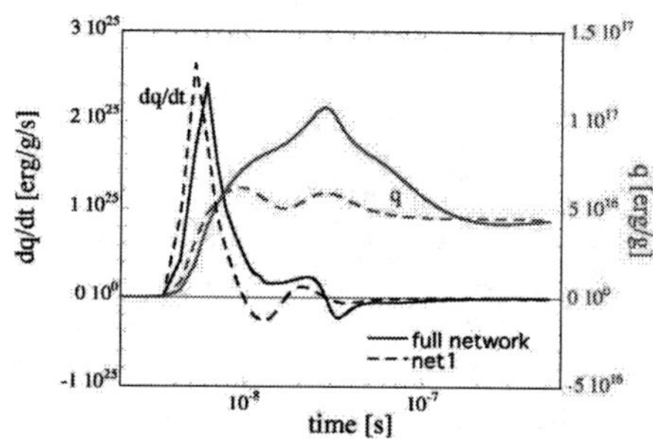

FIGURE 1. Left: Nuclear mass fraction profiles ^{4}He, ^{12}C, ^{16}O, ^{28}Si, ^{32}S, ^{52}Fe and ^{56}Ni for a detonation front in pure ^{12}C at $T = 10^8$ K and $\rho = 10^8\,\mathrm{g\,cm^{-3}}$. Z is the distance to the shock in cm. The thin solid lines give the steady state solution of the ZND model. The heavy dots (the density of the dots is so high that they form thick curves) are obtained with the time-dependent calculations with two different resolutions. Right: Network calculation of the energy production rate (erg g^{-1} s^{-1}) and total energy (erg g^{-1}) for a detonation in a mixture $X_{^{12}\mathrm{C}} = 0.1$, $X_{^{96}\mathrm{Ru}} = 0.9$. for two nuclear reaction networks. Solid lines: full network including 14758 reactions on 1381 nuclides; dashed lines: new reduced network. The time in s is measured since the passage of the shock, assuming a constant velocity $D = 1.09\,10^9$ cm s^{-1}.

electrons and positrons. The ions are treated as a Maxwell-Boltzmann gas, and the radiation follows the Planck law. The selected nuclear reaction network involves 13 nuclides from ^{4}He to ^{56}Ni linked by 27 reactions [5].

PURE HE ACCRETORS

Since superbursts in pure He accreting systems may be due to pure C detonation at high density and temperature we first calculate the ZND profiles [7] for a detonation in pure ^{12}C at a temperature $T = 10^8$ K and a density $\rho = 10^8\,\mathrm{g\,cm^{-3}}$. This provides the main defining features of the detonation, like the characteristic time and length-scales and the reaction-zone structure. The nuclear mass fraction profiles of some of the most abundant species are presented in Fig. 1 (thin solid lines). As can be seen, a large variety of length scales are at work. This suggests that the full resolution of the detonation requires a time-dependent simulation on a domain as large as 10^4 cm with a resolution of 10^{-4} cm. For computational time reason we are unable to reach this resolution. However, following Gamezo et al. [8], we have been able to perform two sets of calculations, one with a resolution of 10 cm on a domain of 10^4 cm, and one with a resolution of 10^{-4} cm on a domain of 1 cm. We have superimposed the profiles of the nuclear mass fractions of some of the most abundant species as a function of the distance to the shock obtained with the ZND algorithm and with the hydrodynamical algorithm. As can be seen in Fig. 1, the time-dependent profiles are very close to the steady state ones. So, as in Gamezo et al. [8], the partial resolution approach can be applied for the simulations of detonations in this system.

MIXED H/HE ACCRETORS

In a mixed H/He accreting system the detonation leading to the superburst is expected to propagate in the ashes of the rp-process which occured in the upper atmosphere. For simplicity, as in Cumming & Bildsten [1], we take ^{96}Ru as a fiducial heavy nucleus to represent these ashes. We simulate the propagation of a detonation in a mixture of ^{12}C and ^{96}Ru with mass fractions $X_{^{96}\mathrm{Ru}} = 0.9$ and $X_{^{12}\mathrm{C}} = 0.1$. Since Schatz et al. [4] show that the heavy rp-process ashes may photodisintegrate during the superbursts, our reduced nuclear reaction network has to be extended in order to include these heavy ashes. In a first step, we complement the reduced network of thirteen species from ^{12}C to ^{56}Ni with an α-chain between ^{64}Ni and ^{96}Ru. We perform the hydrodynamical simulation of the detonation and, using the obtained thermodynamical profiles, a calculation is performed with a full network. The energy production rates deduced from the full and reduced network are compared. The full network predicts an endothermic episode which is not reproduced by the reduced one.A new reduced network is constructed in order to match more closely the full network energy production rate. We adopt an iterative procedure of hydrodynamical simulations, a new reduced network being adjusted at each iteration on the basis of the hydrodynamical conditions provided at the previous step. This is repeated until the energy production by the reduced and full networks converge to a good level of accuracy. After convergence is reached [9], the detonations conditions calculated with the reduced network result from an energy production in very good agreement with the one calculated with the full network, as seen in Fig. 1.

CONCLUSIONS

Using our new hydrodynamical algorithm we have studied the thermodynamical profiles of detonations in conditions relevant to superbursts in pure He accreting systems and in mixed H/He accreting systems. In the first case, we have underlined the large difference between the total reaction length and the length on which some species burn. We have constructed a new reduced network which reproduces quite well the energy generation rate of a reference network including all reactions on all nuclides probably involved in the specific case of the propagation of a detonation in a mixture $X_{^{12}\mathrm{C}} = 0.1$, $X_{^{96}\mathrm{Ru}} = 0.9$. The presence of ^{96}Ru decreases the total energy of the detonation and its velocity, due to an endothermic phase at early stages associated with photodisintegrations.

REFERENCES

1. Cumming, A. & Bildsten, L. 2001, ApJ, 559, L127
2. Strohmayer, T. E. & Brown, E. F. 2002, ApJ, 566, 1045
3. Schatz, H., Aprahamian, A., Barnard, V. & al. 2001, Phys. Rev. Lett., 86, 3471
4. Schatz, H., Bildsten, L. & Cumming, A. 2003, ApJ, 583, L87
5. Noël, C., Busegnies, Y., Papalexandris, M. V. & al. 2007, A&A, 470, 653
6. van Leer, B. 1979, J. Comp. Phys., 21, 101
7. Fickett, W. & Davis, W. C. 1979, Detonation, (Univ. California Press, Berkeley)
8. Gamezo, V.N., Wheeler, J.C., Khokhlov, A.M. & Oran, E.S. 1999, ApJ, 512, 827
9. Noël, C., Goriely, S., Busegnies, Y. & Papalexandris, M. V. 2008, A&A, in press

Interpretation of Extremely Metal-Poor Stars as Candidates of First Generation Stars

Takanori Nishimura*, Masayuki Aikawa†, Nobuyuki Iwamoto**, Takuma Suda*, Masayuki Y. Fujimoto* and Icko Iben Jr.‡

*Department of Physics, Faculty of Science, Hokkaido University, Sapporo 060-0810, Japan
†Hokkaido University OpenCourseWare Hokkaido University, Sapporo 060-0813, Japan
**Japan Atomic Energy Agency, Tokai, Ibaraki 319-1195, Japan
‡Departments of Astronomy and of Physics, University of Illinois at Urbana-Champaign, Urbana, Illinois 61801 USA

Abstract. The evolution of extremely metal-poor (EMP) stars of low-/intermediate-masses is distinct from those of metal-rich stars in that the convection driven by the helium shell flash can extend outward into the hydrogen-rich layer during TP-AGB phase. In the circumstance of $[\mathrm{Fe/H}] < -2.5$, protons are mixed and converted into neutrons in the convective zone to promote nucleosynthesis through neutron and α-captures. We study the nucleosynthesis in the helium-flash convective zone, induced by this hydrogen mixing. In the dearth of the pristine metals, the neutron-recycling reactions, $^{12}\mathrm{C}(n,\gamma)^{13}\mathrm{C}(\alpha,n)^{16}\mathrm{O}$, and in some cases, the subsequent $^{16}\mathrm{O}(n,\gamma)^{17}\mathrm{O}(\alpha,n)^{20}\mathrm{Ne}$, play an important role and catalyze the syntheses of O through Mg and still heavier elements. In particular, it is demonstrated that such peculiar abundance patterns of light elements from C through Al and heavy elements of Sr as observed from the two most iron-deficient stars, HE0107-5240 and HE1327-2326, can well be reproduced in terms of the nucleosynthesis in the metal-free and EMP AGB stars. In addition, the lack of Na and Al enhancement for a carbon-rich giant HE0557-4840 can be interpreted as the absence of neutron-capture reactions because of the negligible amount of proton ingestion. Based on these results, we assign their origin to the Pop III stars, born out of the primordial gas. We also discuss about the surface pollution both via the mass transfer in the binary systems and via the accretion of interstellar gas.

Keywords: stellar evolution, nucleosynthesis, population III stars
PACS: 97.10.Cv, 97.20.Wt

INTRODUCTION

Recently, the studies of EMP stars in the Galactic halo with high resolution spectroscopy reveal that their abundance patterns are different from those known from younger Population I and II stars. Striking examples are the two stars of giant HE0107-5240 [3] and subgiant HE1327-2326 [5, 1], which have the smallest iron abundances among known EMP stars, i.e., of $[\mathrm{Fe/H}] = -5.3$ and -5.46, respectively. The both stars share the peculiar abundance patterns among the light elements; in addition to the huge enrichment of carbon ($[\mathrm{C/Fe}] \simeq 4.0$), the light elements such as N, O, Na, Mg and Al are reported to be largely enhanced [3, 4, 2, 5, 6, 1]. There also exist differences in abundances between these two stars; HE1327-2326 has larger enhancements for N (by 2.2 dex), O (by 0.5 dex), and Na through Al (by $1.2-1.5$ dex) than HE0107-5240. Moreover, the former displays the enrichment of a s-process element, Sr ($[\mathrm{Sr/Fe}] = 1.1-1.3$), but HE0107-5240 does not so ($[\mathrm{Sr/Fe}] < -0.5$). Very recently, the discovery of another gi-

CP1016, *Origin of Matter and Evolution of Galaxies*,
edited by T. Suda, T. Nozawa, A. Ohnishi, K. Kato, M. Y. Fujimoto, T. Kajino, and S. Kubono

ant, HE0557-4840, is reported at the iron abundance of $[Fe/H] = -4.75$, which is also the minority among EMP stars compared with those of $[Fe/H] > -4$ [10].

It is known that the stars with mass of $M < 3.0 M_{\odot}$ and the metallicity of $[Fe/H] < -2.5$ experience the engulfment of protons in the hydrogen burning shell by the extension of helium flash convection during the core and shell helium flash phase [7, 8]. The mixed protons are carried down and captured by ^{12}C to produce ^{13}C via β-decay, which produces neutrons via $^{13}C(\alpha,n)^{16}O$ and can be the source of light elements like Na, Mg, and Al. If these stars belong to the binary system with low-mass secondaies of $M \sim 0.8 M_{\odot}$, they can show the nucleosynthetic signature of the EMP stars or possibly Population III stars. This paper focuses on the origins of the most iron-deficient stars in the standpoint of nucleosynthesis in intermediate mass EMP stars. We elaborate the nucleosynthesis in the helium flash convection with use of the nuclear network code with various range of parameters related to the stellar evolution. Based on the binary scenario, we persue the possibility that the iron-poorest stars observed today can be the survivors of Population III stars.

METHODS AND ASSUMPTIONS

We use the two nuclear reaction networks in combination with the program of M. A. and N. I. The network program by M. A. covers light elements of $Z < 16$ and compute the time evolution of the helium flash by the one zone approximation developed by [12]. The other by N. I. covers $16 < Z < 83$ of heavy elements including up to the edge of s-process elements. Since the hydrogen mixing mechanism has not yet been established in detailed perception, we treat the duration of mixing event and the amount of hydrogen mixing as parameters in free but reasonable range based on some sets of initial stellar models [9]. For simplicity, the amount of matter mixed into the helium flash convection is measured by that of ^{13}C relative to the abundance of ^{12}C instead of that of hydrogen.

RESULTS AND DISCUSSION

The key reactions controlling the final abundance distributions under consideration. The nucleosynthesis in helium flash convection proceeds in step by step as described below. At first stage, the neutron ejected via $^{13}C(\alpha,n)^{16}O$ reaction is consumed by overwhelming ^{12}C to make ^{13}C until the ^{16}O increases to achieve comparable neutron absorption with ^{12}C. The second stage has branchings at ^{17}O and ^{22}Ne attributed by α- and neutron-capture reactions; (1) for ^{17}O, the branching is noncritical to produce heavier elements because the divided flows join together at ^{22}Ne. (2) however, the branching for ^{22}Ne is important to make ^{23}Na with neutron-capture and to make ^{25}Mg and ^{26}Mg with α-capture additionally. At ^{26}Mg the α-capture reactions are no longer effective because of large Coulomb barrier for the temperature under consideration ($\log T < 8.5$). Accordingly the last stage is dominated by neutron-capture reactions for heavier isotopes. In particular, the neutron cross sections of ^{26}Mg and ^{27}Al are critical for a starting point of producing heavier isotopes including s-process elements.

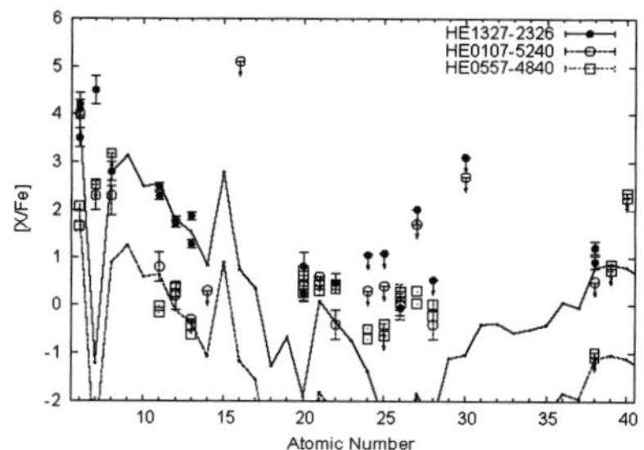

FIGURE 1. The comparisons with observed abundance patterns of three most iron-deficient stars : Note that the calculated abundance distributions are normalized by carbon abundance.

The interpretation of the most iron-deficient stars as Pop.III survivors. We confront our results with abundance patterns of currently known iron-poorest stars on the basis of binary scenario [11] as shown in Fig. 1, where we normalized our results by C abundance to compare with observed data. For HE1327-2326, our result well reproduces the abundances of C, O, Na, Mg, Al, and Sr, while N is supplied by the dredge-up event, He-FDDM [11] with CN-cycle. Since the abundances of iron group elements are not peculiar in HE1327-2326, it is probable that the star is a Pop.III star polluted by interstellar gas. The consideration of the effect of dilution by the third dredge-up, as shown by dashed line in Fig. 1, is consistent with abundances of Na, Mg and Al for HE0107-5240, although those of C, N and O are not satisfactory. This can be also reconciled with the consideration of He-FDDM for N and the third dredge-up for C and O, respectively. Finally, HE0557-4840 denotes only C enhancement relative to Fe and have different ratios for Na, Mg and Al from other two stars. These features are derived from effective third dredge-up and interstellar gas accretion.

To discriminate whether they are Pop.III or the second generation stars born out of gas polluted by the Pop.III supernovae, we need to sophisticate our models of nucleosynthesis as well as those of stellar evolution and binary evolution. It is to be fully investigated how the occurrence of dredge-up mechanisms in model stars depends on its initial mass.

REFERENCES

1. W. Aoki et al. 2006, ApJ, 897
2. M. S. Bessell, N. Christlieb, & B. Gustafsson. 2004, ApJ, 61
3. N. Christlieb et al. 2002, Nature, 904
4. ——. 2004, ApJ, 708
5. A. Frebel et al. 2005, Nature, 871
6. ——. 2006, ApJ, L17
7. M. Y. Fujimoto, I. Iben Jr., & D. Hollowell. 1990, ApJ, 580
8. M. Y. Fujimoto, Y. Ikeda, & I. Iben Jr. 2000, ApJ, L25
9. T. Nishimura, et al. 2008, in preparation
10. J. Norris et al. 2007, ApJ, 670, 774
11. T. Suda et al. 2004, ApJ, 476
12. D. Sugimoto, & M. Fujimoto. 1978, PASJ, 30, 467

Production of light elements on a low-mass secondary in a soft X-ray transient

Shin-ichiro Fujimoto*, Ryuichi Matsuba† and Kenzo Arai†

*Kumamoto National College of Technology, Kumamoto 861-1102, Japan
† Kumamoto University, Kumamoto 860-8555, Japan

Abstract. We examine production of light elements (Li, Be, & B) on the surface of a low-mass secondary in a black hole soft X-ray transient (BHSXT) through the spallation of CNO nuclei by neutrons which are ejected from a hot (> 10 MeV) advection-dominated accretion flow (ADAF) around the black hole. Using updated binary parameters, cross sections of neutron-induced spallation reactions, and mass accretion rates in ADAFs derived from the spectrum fitting of multi-wavelength observations of quiescent BHSXTs, we obtain the equilibrium abundances of Li by equating the production rate of Li and the mass transfer rate through accretion to the black hole. We find that the resulting abundances are found to be in good agreement with the observed values in seven BHSXTs. Moreover, the isotopic ratio ^{6}Li/^{7}Li and the abundance ratios Be/Li and B/Li are calculated to be about 0.7–0.8, 1.0–1.3, and 1.8–2.1, respectively, on the secondaries.

Keywords: Accretion, accretion disks — black hole physics — nuclear reactions, nucleosynthesis, abundances — stars: abundances
PACS: 97.80.Jp

INTRODUCTION

High abundances of Li have been detected in late-type secondaries of black hole soft X-ray transients (BHSXTs) and a neutron star soft X-ray transient (NSSXT) in quiescence [1, 2], though Li would be destructed in a deep convective envelope of a late-type star. The Li enrichment has not, however, been observed on a late-type secondary in a compact binary with a white dwarf [3]. These facts strongly suggest that a production mechanism of Li operates in compact binaries [4, 5] and that the nature of the primaries is crucial for the mechanism.

Multi-wavelength spectra of BHSXTs in quiescence are successfully fitted to the radiation from an advection-dominated accretion flow (ADAF) around the black hole [6, 7]. Density is so low in ADAFs that ions interact inefficiently with electrons. Consequently ions have high temperatures due to viscous heating up to about 30 MeV near the inner edge of ADAFs, $r_{\rm in}$. At such high temperatures, α-α reaction proceeds to synthesize Li inside ADAF [1, 5]. It is necessary that a fraction $10^{-3} - 10^{-4}$ of the accreting gas is transported to the secondary to explain the high abundances of Li observed in BHSXTs. However, such a high fraction is uncertain to be realized due to strong gravity of the black hole and the Coulomb interactions with nuclei inside ADAFs [4]. Helium breaks via spallation with protons to produce neutrons at the inner region of ADAFs. A large fraction of neutrons can be ejected from the ADAF, because they do not interact with nuclei through the Coulomb interactions. Neutrons intercepted by the secondary interact with CNO nuclei through spallation to produce Li on the surface [4]. This scenario is of

CP1016, *Origin of Matter and Evolution of Galaxies*,
edited by T. Suda, T. Nozawa, A. Ohnishi, K. Kato, M. Y. Fujimoto, T. Kajino, and S. Kubono

particular interest, because the Li enrichment is anticipated in secondaries of BHSXTs and NSSXTs. It does not work, however, in a binary with white dwarf primaries due to low disk temperatures for helium breakup into nucleons.

LI PRODUCTION ON THE SECONDARY

We firstly examine abundance distributions in ADAFs. The temperatures and the densities of the ADAF at a given radius r are expressed as a self-similar solution [e.g. 8]. with the viscous parameter, $\alpha_{\rm vis}$, the black hole mass $m = M/M_{\odot}$, and the mass accretion rate $\dot{m}$ in units of the Eddington accretion rate $\dot{M}_{\rm Edd} = 1.4 \times 10^{17} m\,{\rm g\,s^{-1}}$. Once the temperatures, densities and drift timescales are specified, we can follow the abundance evolution in ADAF from the outer boundary $r_{\rm out}$ to $r_{\rm in}$, using a nuclear reaction network [9]. We set $r_{\rm out}$ to be $100 r_g$, where r_g is the Schwarzschild radius of the black hole and the abundances is set to be the solar composition. We find that neutrons are produced significantly via the breakup of ^{4}He at $r < 20 r_g$. Hereafter we fix $\alpha_{\rm vis} = 0.3$ in the present paper [6]. The neutrons produced in ADAF have positive Bernoulli numbers [8], so that a fraction of the neutrons thermally overcomes the deep gravitational well of the black hole before inelastic scattering with protons. We evaluate the ejection fraction of neutrons $f_{\rm ej} \simeq 0.12$, which depends weakly on r [9], and the energy of the ejected neutrons averaged over the radius of the ADAF, as $\langle E_{\rm ej} \rangle \simeq 78\,{\rm MeV}$. We then calculate the abundance of Li on secondaries as Eq. (8) in FMA08, assuming the isotopic ejection of neutrons and equating the production rate of Li through the spallation of CNO nuclei by neutrons from ADAFs and the deposition rates of Li via mass accretion, as in FMA08. It should be emphasized that $Y_{\rm Li,eq}$ depends on $\alpha_{\rm vis}$ and $\dot{m}$ as the combination of $\dot{m}/\alpha_{\rm vis}^2$ through $Y_{\rm n,in}$. We note that for the solar abundance and $E_{\rm n} = \langle E_{\rm ej} \rangle = 78\,{\rm MeV}$, the cross sections for the spallation of CNO nuclei by neutrons are calculated from the Talys nuclear reaction code [10], as $\sigma_{\rm sp6} = 10.5\,{\rm mb}$ and $\sigma_{\rm sp7} = 15.3\,{\rm mb}$ for the production of ^{6}Li and ^{7}Li, respectively, yielding the sum $\sigma_{\rm sp} = \sigma_{\rm sp6} + \sigma_{\rm sp7} = 25.8\,{\rm mb}$. If we adopt the abundance of CNO-processed material, in which all the original ^{12}C are converted to ^{14}N through CNO cycle in the interior of the secondary, the total cross section decreases to be $\sigma_{\rm sp} = 18.0\,{\rm mb}$.

We compare the evaluated and observed abundances of Li in seven BHSXTs, using updated parameters of the binaries, shown in Table 1 of FMA08. The binary separations are calculated from $R_*/a = 0.46(1 + M/M_*)^{-1/3}$ [11]. The observed abundances of Li are adopted from Casares et al. [12]. The mass accretion rates in ADAF are found from the spectrum fitting of the multi-wavelength observations of quiescent BHSXTs to be $\dot{m}_{\rm spe} = 4.3 \times 10^{-3}$ and 2.0×10^{-2} for A0620-003 and V404 Cyg, respectively [6, 7]. For the other objects, where the spectrum fitting has not yet been performed in quiescence, we simply specify the accretion rates from the minimum X-ray luminosities in these systems [13]. These values of $\dot{m}_{\rm spe}$ are given in Table 2 of FMA08. Using these values of parameters all together, we calculate the equilibrium abundances of Li on the secondaries in seven BHSXTs. We show the resulting abundances $A({\rm Li}) = \log(Y_{\rm Li,eq}/Y_{\rm p}) + 12$ against the orbital periods $P_{\rm orb}$ of the binaries by the open circles in Figure 1. It is found that our results are in good agreement with the observed abundances.

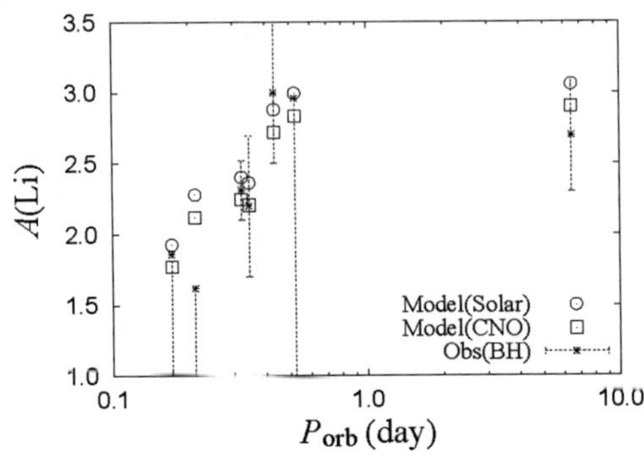

FIGURE 1. Evaluated and observed abundances of Li on the secondaries in seven BHSXTs.

Difference of the Li abundances on secondaries are possibly due to their different mass accretion rates, rather than difference of the operation of Li depletion via the hot-bottom burning [12]. Finally, we evaluate the isotopic ratio ^{6}Li/^{7}Li on the secondaries. It is easily calculated from the cross sections of ^{6}Li and ^{7}Li for the spallation reactions, or $\sigma_{sp6}/\sigma_{sp7}$, to be 0.69 – 0.81, depending on the CNO abundances of the secondaries for $E_n = 78\,\mathrm{MeV}$. We also calculate abundance ratios of Be and B to Li as 1.0–1.3 and 1.8–2.2, respectively. We note that the ^{6}Li/^{7}Li ratio is much larger than 0.12 for a NSSXT Cen X-4 [12] and 0.081 for meteorites. Detection of such a high ^{6}Li/^{7}Li ratio will be an evidence for the production of Li on the secondaries in BHSXTs and for the existence of an ADAF, rather than an accretion disk corona system [e.g. 14].

This work was supported in part by Grant-in-Aid for the Scientific Research from the Ministry of Education, Science and Culture of Japan (No.17540267)

REFERENCES

1. Martin, E. L., Rebolo, R., Casares, J., & Charles, P. A. 1994, ApJ, 435, 791
2. Martin, E. L., Casares, J., Molaro, P., Rebolo, R., & Charles, P. 1996, New Astron. 1, 197
3. Martin, E. L., Casares, J., Charles, P. A., & Rebolo, R. 1995, A&A, 303, 785
4. Guessoum, N., & Kazanas, D. 1999, ApJ, 512, 332
5. Yi, I., & Narayan, R. 1997, ApJ, 486, 363
6. Narayan, R., Barret, D., & McClintock, J. E. 1997, ApJ, 482, 448
7. Quataert, E., & Narayan, R. 1999, ApJ, 520, 298
8. Narayan, R., & Yi, I. 1994, ApJL, 428, L13
9. Fujimoto, S., Matsuba, R., & Arai K. 2008, ApJL, 673, L51 (FMA08)
10. Koning, A. J., Hilaire, S., & Duijvestijn, M. C. 2005, Proc. Int. Conf. on Nuclear Data for Science and Technology, ed. C. Haight et al., AIP Conf., 769, 1145
11. Paczyński, B. 1971, ARA&A, 9, 183
12. Casares, J., Bonifacio, P., González Hernández, J. I., Molaro, P., & Zoccali, M. 2007, A&A, 470, 1033
13. McClintock, J. E., Narayan, R., Garcia, M. R., Orosz, J. A., Remillard, R. A., & Murray, S. S. 2003, ApJ, 593, 435
14. Malzac, J. 2007, Memorie della Societa Astronomica Italiana, 78, 382

The Origin and Evolution of the Extremely Metal-Poor Halo Planetary Nebulae

M. Otsuka*, H. Izumiura*, A. Tajitsu† and S. Hyung**

*Okayama Astrophysical Observatory (NAOJ), Kamogata, Okayama 719-0232, Japan
†Subaru Telescope (NAOJ), 650 North A'ohoku Place, Hilo, Hawaii 96720, U.S.A.
**School of Science Education (Astronomy), Chungbuk National University, 12 Gaeshin-dong Heungduk-gu, CheongJu, Chungbuk 361-763, Korea

Abstract. We present recent results of our research on the extremely metal-poor ([Fe/H] ~ -2) halo planetary nebulae (PNe), K 648, BoBn 1, and H 4–1. Chemical abundance analysis using Subaru/HDS spectra and archive data shows that these halo PNe might have evolved from carbon-enhanced metal-poor (CEMP) stars. In BoBn 1, we have detected two fluorine lines and found enhanced fluorine abundance, [F/H] = +1.1. The [C,N,F/Fe] abundances of BoBn 1 are comparable to those of the CEMP-*s* star, HE1305+0132 ([Fe/H] = –2.5). BoBn 1 might have evolved from a CEMP-*s* star such as HE1305+0132 in the origin and evolution.

Keywords: ISM: planetary nebulae, ISM: abundances, stars: Population II
PACS: 43.35.Ei, 78.60.Mq

INTRODUCTION

Over 1000 planetary nebulae (PNe) have been recognized in the Galaxy thus far. Most of them are members of the galactic disk population, while only 14 PNe have been identified as halo members. The fact that all the halo PNe are metal-poor suggests that the progenitors of the halo PNe are Population II stars. It is widely believed that halo PNe not only provide us with an opportunity to study the evolution of individual metal-poor stars but also make it possible to probe the chemical evolution of the Galaxy at an early period. However, carbon- and nitrogen-rich (C-, N-rich) halo PNe, K 648 (in the globular cluster M 15, [Fe/H] = –2.1), H 4–1 ([Fe/H] = –2.1), and BoBn 1 ([Fe/H] = –2.0) have some unresolved issues on their origin.

For example, the halo PNe can become N-rich by the first dredge-up but not C-rich, if the initial mass of the progenitor is $\sim$0.8 $M_{\odot}$, which is the turn-off star mass of M 15, according to the theoretical models of single metal-poor stars (e.g., Fujimoto *et al.* 2000, Suda *et al.* 2004). To become C-rich PNe, the third dredge-up must take place in late-AGB phase. But, the third dredge-up is more efficient in stars with initial masses $\gtrsim$ 1.2–1.5 $M_{\odot}$. Also, current stellar evolution models employing empirical mass-loss laws predict a too slow post-AGB evolution toward PN for a star of $\sim$0.8 $M_{\odot}$, which does not explain the presence of halo PNe. Note that we have a little knowledge about the duration of thermal pulses and mass-loss laws, which are crucial parameters for the post-AGB evolution. Depending on them, such halo stars may evolve into PNe. The issues on chemical abundances and evolutionary time scale of C- and N-rich halo PNe must be resolved before one can proceed to discuss the evolution of metal-poor stars and the early chemical evolution of the Galaxy through the observation of halo PNe.

CP1016, *Origin of Matter and Evolution of Galaxies*,
edited by T. Suda, T. Nozawa, A. Ohnishi, K. Kato, M. Y. Fujimoto, T. Kajino, and S. Kubono

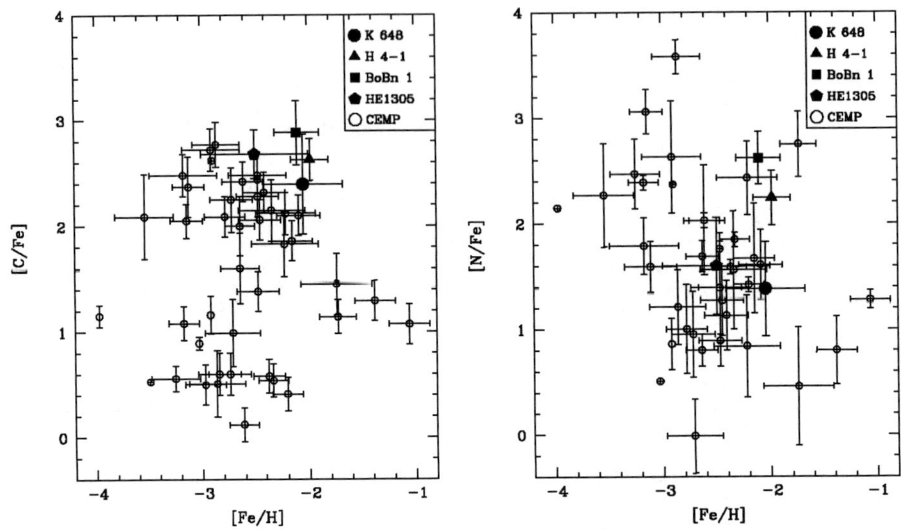

FIGURE 1. The diagrams of [Fe/H] vs. [C/Fe] (left panel) and [Fe/H] vs. [N/Fe] (right panel). For halo PNe, we adopt the averaged C and N derived from forbidden and recombination lines and use Ar instead of Fe. HE1305 stands for HE1305+0132, which is a fluorine enhanced CEMP star (see text).

Chemical abundance analysis of PNe is a key tool to investigate these issues. Comparing the observed abundances with those from theoretical evolution and nucleosynthesis models for metal-poor stars will shed light on the origin and evolution of C- and N-rich halo PNe. In order to solve these issues, we have carried out high-sensitivity, high-resolution, and wide-waveband (3600 Å–7500 Å) spectroscopy using Subaru/HDS and gathered archive data and performed chemical abundance analysis.

RESULTS & DISCUSSION

Our analysis reveals that [C/Fe] and [N/Fe] abundances of K 648, H 4–1, and BoBn 1 are compatible with those of carbon-enhanced metal-poor (CEMP) stars as shown in the [Fe/H]-[C/Fe] and [Fe/H]-[N/Fe] diagrams (Figure 1). In view of their location on the diagrams, these halo PNe might have close connection with CEMP stars in the origin and evolution.

As currently defined, CEMP stars fall into two categories, those that exhibit large enhancements of *s*-process elements (CEMP-*s*) and those that do not have such enhancements (CEMP-*no*). Most of the CEMP-*s* stars show large enhancements of C and N. Low-metallicity stellar evolution models have demonstrated that the carbon and nitrogen overabundances in CEMP-*s* stars can be reproduced by binary model (e.g., Lau *et al.* 2007) and Pre-SNe model (e.g., Hirschi 2007). Binary scenario seems to be plausible explanations for the problems of the evolutionary timescale and chemical abundances for C and N-rich halo PNe, because the binary evolution (mass transfer, coalescence,

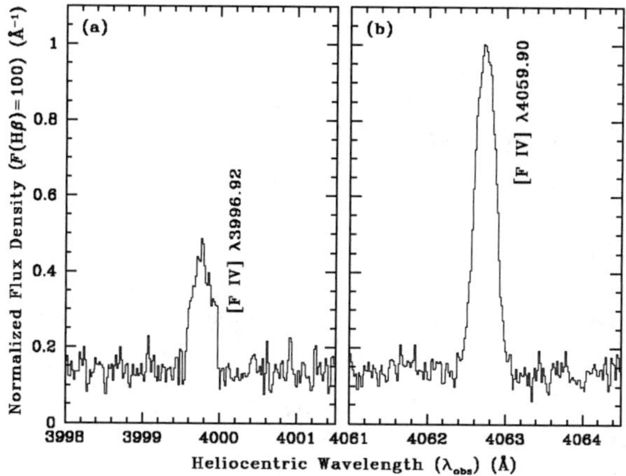

FIGURE 2. The line-profiles of [F IV] λ3996.92 (left panel) and [F IV] λ4059.90 (right panel). The intensity is normalized to the Hβ flux, F(Hβ).

etc.) would shorten the time taken for AGB stars to evolve into PNe. In fact, about 70 % of the CEMP stars are identified as long-period (0.3–100 yr) binaries (Beers & Christlieb 2005). C- and N-rich halo PNe might have evolved from CEMP-*s* stars.

Interestingly, in BoBn 1, we have detected two fluorine lines (Figure 2). From these line, we have estimated [F/H] = +1.1±0.1 and [F/O] = +2.0±0.1. To our knowledge, BoBn 1 is the most metal-poor, fluorine-rich PN among fluorine-detected, H-rich 14 PNe. The [F/H] and [F/O] abundances of BoBn 1 are compatible with the AGB models for 1.5–2.5 $M_\odot$ star, with $Z = 10^{-4}\,Z_\odot$ ([Fe/H] $\simeq -2.3$) by Lugaro *et al.* (2004) and Renda *et al.* (2004). The single metal-poor star models of Suda *et al.* (2004) and Fujimoto *et al.* (2000) have demonstrated that the stars with [Fe/H] $\lesssim -2.5$ and initial masses of 1.2–3.5 $M_\odot$ can become C, N, and *s*-process elements rich by the mixing of the ^{12}C-rich intershell and hydrogen-rich shell and a dredge-up. Such a massive star, however, would not have survived in the Galactic halo up to now.

BoBn 1 might have evolved from a CEMP-*s* star such as HE1305+0132. The [C,N,F/Fe] abundances of BoBn 1 are comparable to those of HE1305+0132 ([Fe/H] = −2.5, [C,N/Fe] > +1.5, [F/Fe] = +3.0; Schuler *et al.* 2007). Schuler *et al.* (2007) predict that *s*-process elements should be enhanced in this CEMP star.

REFERENCES

1. Beers, T. C. & Christlieb, N. 2005, ARAA, 43, 531
2. Fujimoto, M. Y., Ikeda, Y., & Iben, I.Jr. 2000, ApJL, 529, 25
3. Hirschi, R. 2007, A&A, 461, 571
4. Lau, H. H. B., Stancliffe, R. J., & Tout, C. A. 2007, MNRAS, 378, 563
5. Lugaro *et al.* 2004, ApJ, 615, 934
6. Renda, A. *et al.* 2004, MNRAS, 354, 575
7. Schuler, S. C. *et al.* 2007, ApJL, 667, L81
8. Suda, T. *et al.* 2004, ApJ, 611, 476

Light Element Production in Type Ib/c Supernovae

Ko Nakamura

Research Center for the Early Universe, Graduate School of Science, University of Tokyo, Bunkyo-ku, Tokyo 113-0033, Japan

Abstract. Recent observations of halo stars indicate non-primordial ^{6}Li and Be, which are produced through cosmic-ray interactions. Some energetic supernovae can accelerate particles to energies enough to interact with ambient material and produce light elements including ^{6}Li and ^{9}Be. We investigate interactions between Type Ib/c supernova ejecta and the circumstellar matter (CSM) as production sites for the light elements, particularly ^{6}Li and ^{9}Be in the early stages of the Galactic chemical evolution. Considered are energetic Type Ib/c supernova explosions of Wolf-Rayet type stars at low metallicities, embedded in the dense CSM. The energetic portion of the supernova ejecta can interact directly with the circumstellar material and induce light-element production via spallation and α-α fusion reactions. We find that accelerated particles lose their energy through Coulomb collisions with free electrons near the stellar surface where very dense CSM resides, so that most of light elements are produced at the near-surface region. This fact suggests that the amounts of produced light elements and their ratios strongly depend on the chemical compositions of the region. The resulting ^{6}Li and ^{9}Be abundances in the ejecta and the cicumstellar material out of which very metal-poor stars form may show scattering, though it has not yet been confirmed.

Keywords: nuclear reactions — supernovae: general — stars: abundances
PACS: 26.40.+r, 97.10.Cv, 97.10.Fy, 97.10.Me, 97.10.Qh, 97.10.Tk, 97.20.Tr, 97.60.Bw, 98.35.Bd

INTRODUCTION

Recently ^{6}Li has been detected in some very metal-poor stars [1]. The observed ^{6}Li abundance is several orders of magnitude larger than that predicted from the standard big bang nucleosynthesis theory. It supports the idea that the cosmic-ray interactions have played some important roles in the light element production at low metallicities. Previous theoretical studies, however, produce less ^{6}Li than observations.

Nakamura et al. [5] investigated the LiBeB production by the interactions between SN Ic ejecta and the CSM, assuming the thick, dense CSM. Since SNe Ic undergo intense mass loss via stellar wind, some of such SNe should be embedded in the dense matter constructed by the stellar wind material blown in the lifetime of their progenitors. Thus the ejecta reasonably interact with the ambient medium. Based on the evolutionary simulations of massive, rapidly-rotating, very metal-poor stars [e.x., 3], they constructed the progenitor models by replacing the compositions of outer layers of SN 1998bw model with He and N. Accelerated He nuclei on the surface are expected to collide with He in the CSM and produce plenty of Li isotopes. At the same time, much ^{9}Be should be produced via the spallation reactions of nitrogen. In fact, the most metal-poor star among stars from which ^{6}Li is detected, LP815-43, exhibits the enhanced amount of Be that deviates from the Be-Fe relation in stars. Moreover, the N abundance of LP815-43 is also on the enhanced level than that expected from previous

CP1016, *Origin of Matter and Evolution of Galaxies*,
edited by T. Suda, T. Nozawa, A. Ohnishi, K. Kato, M. Y. Fujimoto, T. Kajino, and S. Kubono

measurements of stars with similar metallicities. These observations may suggest that nitrogen nuclei play an important role in Be production sites. Although Nakamura et al. [5] succeeded in reproducing the ^{6}Li and ^{9}Be abundances of LP 815-43, they used metal-rich stellar models to account for the observations of very metal-poor stars. In this paper, I use a stellar model derived from the simulations of the evolution of extremely metal-poor stars. Below the adopted stellar model is explained. Outer layers of the star are accelerated by supernova explosion and produce LiBeB via interactions with the CSM. Results are compared with observations of the very metal-poor star LP 815-43.

CALCULATIONS

Stellar models. Type Ic supernovae, which are considered to be explosions of WR type stars, are likely to be a strong candidate of CNO cosmic-ray accelerators. Hirschi [2] calculated the evolution of a series of fast rotating stars at very low metallicity and his 85 $M_{\odot}$ model undergoes considerable mass loss in the red supergiant stage. It loses about three quarters of its initial mass at the end of the Si burning and becomes a WO type WR star, the core of which is considered to collapse in a short time and induce SN Ic explosion. We adopt his 85 $M_{\odot}$ model ($\sim 20\,M_{\odot}$ at explosion, hereafter HM20 model) as a progenitor. I have simulated the hydrodynamic interactions between the stellar winds and the surrounding matter, using the data of mass-loss history corresponding to the HM20 model. The resultant density distribution of the CSM is proportional to r^{-2} near the stellar surface, therefore the progenitor is surrounded by a dense CSM, through which supernova ejecta pass and produce light elements.

Particle acceleration. We use a hydrodynamic code to calculate supernova explosions. It has been constructed for special relativistic hydrodynamic simulations with Lagrangian coordinates. The explosion energy $E_{\rm ex}$, given in the range between typical 10^{51} ergs and highly energetic 10^{52} ergs corresponding to SN 1998bw, is released at the center of the stellar model, traced the evolution of physical quantities. Details of the code and procedures are described in Nakamura & Shigeyama [4].

Energy loss of ejecta and LiBeB production. To investigate the role of supernova ejecta in light element nucleosynthesis, the modification of the energy distribution of ejecta when they transfer in the CSM need to be considered because the cross sections of spallation and α-α fusion reactions are sensitive to energies of particles. Energetic ejecta accelerated by a supernova explosion lose energy when they collide with and ionize neutral atoms in the ambient medium, or, through the Coulomb scattering. We assume that the CSM is completely ionized, so that ejecta accelerated by SN explosion lose their energy mainly through Coulomb collisions with free electrons in the CSM. The energy loss of ejecta is estimated by Monte-Carlo simulation. The LiBeB yields are calculated simultaneously, using the empirical cross sections given by, for example, Read & Viola [6]. Typically they suddenly become negligible below some thresholds in energy around $\sim 5-30\,\mathrm{MeV\,A^{-1}}$, and have peaks adjacent to the bluff. Energetic explosions can accelerate particles beyond these thresholds.

RESULTS AND CONCLUSIONS

We can estimate the amounts of light elements produced via each reaction as functions of the explosion energy. We find that a significant amount of ^{6}Li (e.g., $\sim 10^{-5}\ M_\odot$ for $E_{\rm ex} = 3 \times 10^{52}$ ergs) is produced through the α-α fusion reaction. We also find that the most effective reaction for the ^{9}Be production is the spallation of C in the CSM with accelerated He. This is caused by the facts that the progenitor has a significant amount of C in the CSM and the cross section of the C + He reaction has a broad peak. On the other hand, hydrogen and nitrogen nuclei in the CSM do not play any important roles as a target of ejecta. This means that most of ejecta lose their energy within the dense near-surface region of the CSM and that the reactions producing LiBeB also terminate in this region. As a result of the localized production of the light elements, their abundance ratios with respect to heavy elements averaged over the CSM and SN ejecta are likely to be inherited by stars of the next generation in the Galactic halo, as pointed out by Shigeyama & Tsujimoto [7]. Thus, it is reasonable to compare the resulting abundance ratios with observations of Pop II stars.

I focus on the abundances of the very metal-poor star LP 815-43 from which ^{6}Li, Be, and N have been detected, all at apparently enhanced levels. As summarized in Nakamura et al. [5], $X_{^6{\rm Li}}/X_{\rm O} \sim 6.88^{+3.08}_{-3.22} \times 10^{-7}$ and $X_{^9{\rm Be}}/X_{\rm O} \sim 1.32^{+0.77}_{-0.49} \times 10^{-8}$ for this star. Here X_i/X_j denotes the mass ratio of element i to j. With the oxygen mass of 12.3 $M_\odot$ in the HM20 model, we can reproduce these ratios when the explosion energy is about 10^{52} ergs. This is not impractical energies for SNe Ic, some of which are associating with gamma-ray bursts. The observed ratio of ^{6}Li to ^{7}Li is about a factor of 10 less than our prediction, ~ 0.4. This is reasonable because ^{6}Li produced in the interacting region should be diluted with the primordial ^{7}Li in the interstellar matter.

In the on-site production of the light elements by supernova ejecta interacting with the CSM, it is revealed here that the yield of LiBeB strongly depends on the chemical compositions of the stellar outermost layer and of the near-surface CSM. This result implies that LiBeB abundances of very metal-poor stars should show scattering depending on the properties of supernovae which induce formation of the stars. More observational data are necessary to elucidate what fraction of the LiBeB abundances seen in halo stars of different metallicities can be explained by the mechanism proposed here.

Acknowledgments. The author is grateful to Raphael Hirschi for the offering of his model and Toshikazu Shigeyama for valuable discussions. This work has been partially supported by the Grant-in-Aid for JSPS Fellows (18·10689).

REFERENCES

1. Asplund, M., Lambert, D. L., Nissen, P. E., Primas, F., & Smith, V. V. 2006, ApJ, 644, 229
2. Hirschi, R. 2007, A&A, 461, 571
3. Meynet, G., & Maeder, A. 2002, A&A, 390, 561
4. Nakamura, K., & Shigeyama, T. 2004, ApJ, 610, 888
5. Nakamura, K., Inoue, S., Wanajo, S., & Shigeyama, T. 2006, ApJL, 643, L115
6. Read, S. M., & Viola, V. E. 1984, Atomic Data and Nuclear Data Tables, 31, 359
7. Shigeyama, T., & Tsujimoto, T. 1998, ApJL, 507, L135

Study of ^{17}O(p,α)^{14}N reaction via the Trojan Horse Method for application to ^{17}O nucleosynthesis

M.L. Sergi[*,†], C. Spitaleri[*,†], A. Coc[**], R.G. Pizzone[*,†], M. Gulino[*,†], V. Burjan[‡], S. Cherubini[*,†], V. Crucillà[*,†], F. Hammache[§], Z. Hons[‡], G. Kiss[¶], V. Kroha[‡], M. La Cognata[*,†], L. Lamia[*,†], S.M.R. Puglia[*,†], G.G. Rapisarda[*,†], S. Romano[*,†], N. de Séréville[§], E. Somorjai[¶], S. Tudisco[*,†] and A. Tumino[*,†]

[*] *Laboratori Nazionali del Sud, Catania, Italy*
[†] *Dipartimento di Metodologie Fisiche e Chimiche per l'Ingegneria, Università di Catania, Catania, Italy*
[**] *CSNSM, CNRS/IN2P3/UPS, Bâtiment 104, 91405 Orsay Campus, France*
[‡] *Nuclear Physics Institute of ASCR, Rez near Prague, Czech Republic*
[§] *IPN, IN2P3-CNRS et Université de Paris-Sud, 91406 Orsay Cedex, France*
[¶] *Institute of Nuclear Research of Hungarian Academy of Sciences (ATOMKI), Debrecen*

Abstract. Because of the still present uncertainties on its rate, the ^{17}O(p,α)^{14}N is one of the most important reaction to be studied in order to get more information about the fate of ^{17}O in different astrophysical scenarios. The preliminary study of the three-body reaction ^{2}H(^{17}O,α^{14}N)n is presented here as a first stage of the indirect study of this important ^{17}O(p,α)^{14}N reaction through the Trojan Horse Method (THM).

Keywords: Nuclear Astrophycsics; Indirect Methods
PACS: 24.10.-i, 24.50.+g

INTRODUCTION

The role of oxygen, in particular of the isotope ^{17}O, is related to various open questions in astrophysics. First of all, it is one of the very few isotopes whose nucleosynthetic origin can be attributed to novae. In novae, ^{17}O is produced in one of the two paths leading to ^{18}F production which is of special interest for gamma ray astronomy [1, 2]. Secondly, the relative abundances of the oxygen isotopes have been observed at the surface of some Red Giant stars [3]. Thirdly, hundreds of presolar grains found in meteorites are composed of oxides with laboratory measured isotopic compositions [4]. Hence more precise evaluations of the nuclear path involving ^{17}O are needed. While the nuclear reaction rate ^{16}O(p,γ)^{17}F(β^+)^{17}O governing the production of ^{17}O is rather well known (NACRE), the experimental status for its destruction by proton capture is much less satisfactory. The ^{17}O(p,γ)^{18}F reaction is important for ^{18}F production in novae while the competitive ^{17}O(p,α)^{14}N reaction represents the dominant channel for the destruction [5]. At nova temperatures ($\sim$10^8 K), several measurements [2, 6] of the E_R^{cm} = 183 keV resonance (E_X=5.790 MeV) both in the (p,γ) and (p,α) channels have drastically reduced the uncertainties on these rates in the context of explosive

CP1016, *Origin of Matter and Evolution of Galaxies*,
edited by T. Suda, T. Nozawa, A. Ohnishi, K. Kato, M. Y. Fujimoto, T. Kajino, and S. Kubono

H-burning. At lower temperatures relevant for hydrostatic H-burning in AGB and RG stars (2-8$\times 10^7$ K), the only direct measurement [7] concerns the E_R^{cm} = 65 keV (p,α) resonance (E_X=5.672 MeV) but additional uncertainties are still present on the available direct data mainly due to the presence of various resonances. In addition some sub-threshold levels could contribute to the total reaction rate and then a further study of this reaction in the energy region relevant for astrophysics is necessary.
However, these energies (few keVs) are much lower than the Coulomb barrier of a few MeV, thus implying the reactions take place via tunnel effect with an exponential decrease of the cross section. Assuming this exponential suppression, the behavior of the direct cross sections at astrophysical energies are usually extrapolated from that at higher energies by using the definition of the astrophysical factor [8]

$$S(E) = E\sigma(E)\exp(2\pi\eta) \tag{1}$$

which, for non resonant reactions, varies smoothly with energies. In order to avoid the uncertainties due to the extrapolation (e.g. presence of unexpected subthreshold resonances or electron-screening effects) in the last twenty years many indirect methods have been developed in order to extract the S(E)-factor. In particular the THM [9, 10] is a powerful tool which selects, under appropriate kinematical conditions, the quasi-free (QF) contribution of a suitable three-body reaction performed at energies well above the Coulomb barrier to extract a charged particle two-body cross section at astrophysical energies, free of Coulomb suppression and electron screening effects.
The present study of the ^{17}O+p$\rightarrow^{14}$N+α reaction in the energy window relevant for astrophysics was performed by selecting the QF-contribution of the ^{2}H(^{17}O,α^{14}N)n three body reaction. The deuteron was used as the "trojan horse nucleus" because of its p$\oplus$n cluster structure: the proton is brought in the nuclear field of ^{17}O while the neutron acts like a spectator to the reaction [11].

DATA ANALYSIS: FROM THE THREE-BODY TO THE TWO-BODY CROSS SECTION

The experiment was performed at the Laboratori Nazionali del Sud in Catania. The SMP Tandem Van de Graaf accelerator provided a 41 MeV ^{17}O beam with a spot size on target of about 1.5 mm and intensities up to 2-3 nA. Deuterated polyethylene targets (CD_2) of about 150 μg/cm^2 were placed at 90° with respect to the beam axis. The detection setup consisted of six Position Sensitive Detectors (PSD) and of two ionization chambers filled with 60 mbar of isobuthane gas as ΔE detector. The angular ranges were chosen in order to cover momentum values p_n of the undetected neutron ranging from -100 MeV/c to 100 MeV/c. This allows then to select the kinematical region where a strong contribution of the Quasi Free (QF) mechanism is expected.
The first step in a THM analysis is to select the correct channel for the three-body reaction. After the energy and position calibration of the two PSD's in telescopes, ^{14}N and α particles were selected with the standard ΔE-E technique. In order to investigate the presence of the QF-mechanism on the three-body coincidence yield, a study of the momentum distribution $|\Phi(\vec{P}_n)|^2$ of the neutron inside the deuteron was made and the

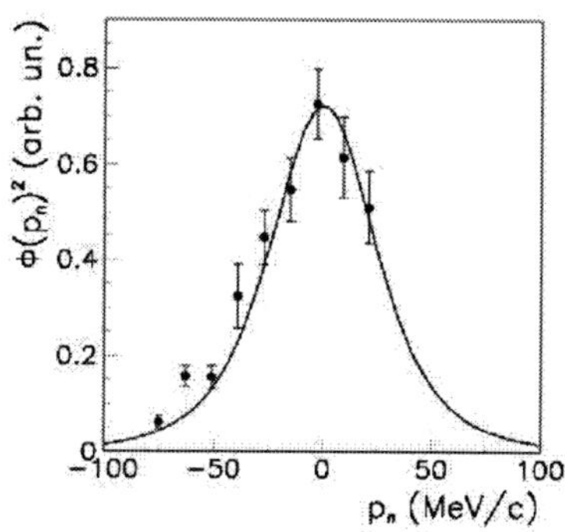

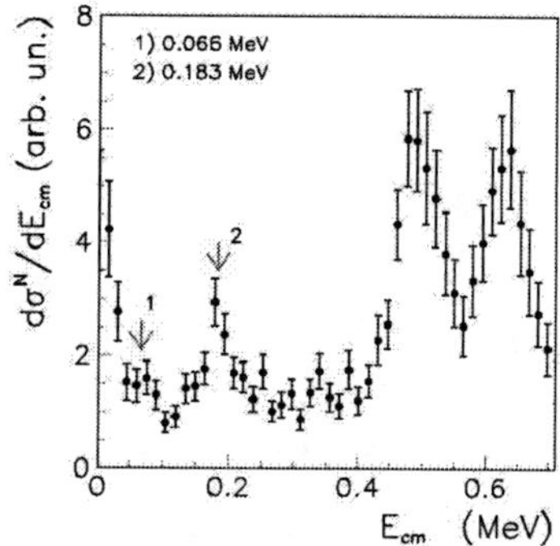

FIGURE 1. Left panel: experimental momentum distribution (points) compared with the theoretical one (full line). Right panel: ^{17}O(p,α)^{14}N nuclear cross section.

results is shown in Fig.1 (left panel). The good agreement between the experimental data and the theoretical Hulthen wave function makes us confident that in the chosen kinematical region it is possible to select the QF-contribution for the three-body reaction ^{2}H(^{17}O,^{14}Nα)n and that no other mechanism are important.

In the framework of the PWIA (Plane Wave Impulse Approximation) [12], the three-body cross section can be factorized into different terms:

$$\frac{d^3\sigma}{dE_{cm}d\Omega_{^{14}N}d\Omega_\alpha} \propto KF \cdot |\Phi(\vec{P}_n)|^2 \cdot \frac{d\sigma^N}{d\Omega} \tag{2}$$

By selecting the condition $|\vec{P}_n|<30$ MeV/c, the behavior of the nuclear cross section for the reaction studied here is shown in right panel of Fig.1. From this preliminary analysis, hints on the excitation function have been obtained. Further progress on data analysis is still required.

REFERENCES

1. A. Coc et al., Astronomy & Astrophysics **357** (2000) 561.
2. A. Chafa et al., PRL **95** (2005) 031101.
3. D.S.P. Dearborn, Physics Report **210** (1992) 367.
4. L. R. Nittler, in Workshop on Oxygen in the Earliest Solar System, LPI Editorial Board, Contribution N.1278, p.33.
5. C. Iliadis, *"Nuclear physics of Stars"*, WILEY-VCH Verlag GmbH & Co. KGaA, 2007.
6. C. Fox et al., PRL **93** (2004) 081102.
7. J.C. Blackmon et al., PRL **74** (1995) 2642.
8. C. Rolfs & W. Rodney, *"Cauldrons in the Cosmos"*, The University of Chicago press, Chicago (1988).
9. C. Spitaleri et al., Phys. Rev. **C60**, 055802 (1999).
10. C. Spitaleri et al., PRC **69** (2004) 55806.
11. M. Zadro et al., Phys. Rev. C **C40**, 181, 1989.
12. G.F.Chew & G.C. Wick, Phys. Rev., **85**, 636, 1952.

The r-Process in Supersonic Neutrino-Driven Winds: The Roll of Wind Termination Shock

Takami Kuroda*, Shinya Wanajo* and Ken'ichi Nomoto*,†

*Department of Astronomy, School of Science, The University of Tokyo
†Institute for the Physics and Mathematics of the Universe, The University of Tokyo

Abstract. Recent hydrodynamic studies of core-collapse supernovae imply that the neutrino-heated ejecta from a nascent neutron star develops to supersonic outflows and forms the wind termination shock. We investigate the effects of the termination shock in neutrino-driven winds and its role on the r-process. We find a couple of effects that can be relevant for the r-process. First is the sudden slowdown of the temperature decrease by the wind termination plays a decisive role to determine the r-process abundance curves. This is due to the strong dependences of the nucleosynthetic path on the temperature during the r-process freezeout phase. Second is the entropy jump by termination-shock heating, up to several $100 N_{\rm A} k$. We, however, find negligible roles of the entropy jump on the r-process. This is a consequence that the sizable entropy increase takes place only at a large shock radius ($\gtrsim$ 10,000 km) where the r-process has already ceased.

Keywords: nuclear reactions, nucleosynthesis, abundances — stars: abundances — stars: neutron — supernovae: general
PACS: 97.60.Bw

TERMINATION SHOCK IN NEUTRINO DRIVEN WINDS

Whether the rapid-neutron capture (r-process) can take place under the circumstance of core-collapse supernovae has been an unresolved problem for a long time (see, [3, 7, 2] for a recent review). The neutrino-heated supernova ejecta (neutrino-driven winds, NDW) from a proto-neutron star has been expected as a suitable astrophysical site, since, it meets the conditions such as high entropy, high temperature and high neutron-to-seed ratio.

Previous studies, however, confirmed that the spherically expanding winds from a typical proto-neutron star (e.g., $1.4 M_\odot$ with a 10 km radius) cannot attain the requisite physical conditions for the r-process. Recently, Arcones et al. [1] have suggested that the reverse shock propagating from the preceding supernova ejecta might have important effects on the r-process nucleosynthesis. However, systematic nucleosynthesis calculations with the termination-shocked, supersonic NDW have been absent.

We, thus, investigate the effects of the wind termination shock on the r-process nucleosynthesis in some detail. For this purpose, the thermodynamic histories of supersonic outflows are calculated with a semi-analytic, spherically symmetric, general relativistic NDW model [5, 6]. The termination-shocked outflows are then obtained by applying the non-relativistic Rankine-Hugoniot relations at arbitrary chosen shock radii. By applying the obtained wind trajectories, the r-process calculations are performed with the

CP1016, *Origin of Matter and Evolution of Galaxies*,
edited by T. Suda, T. Nozawa, A. Ohnishi, K. Kato, M. Y. Fujimoto, T. Kajino, and S. Kubono

extensive nuclear reaction network. For actual methods, see Kuroda, Wanajo & Nomoto [4].

RESULTS

1: The Effects of Termination Shock Heating on the r-Process

In Fig:1, the wind solutions are displayed. Neutrino luminosities are set as $L_{\nu,51} \equiv (L_\nu/10^{51}\text{ergs s}^{-1})$=10 (*dash dotted* lines) and 1 (*dotted* lines). Matters are blown off from the surface of nascent neutronstar, $r = 10$km, with interacting neutrinos at first and then the adiabatic expansion phase starts beyond $r \gtrsim 30 \sim 40$km. Beyond the shock

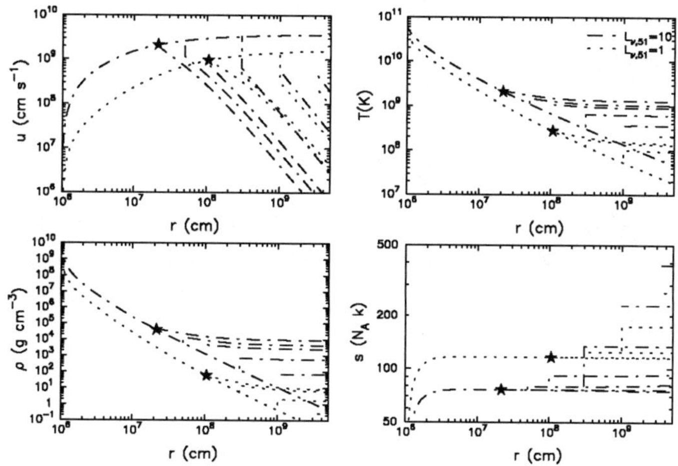

FIGURE 1. The wind solutions (radial velocity u, temperature T, density ρ and entropy s) and the termination-shocked outflows. Mass of the central neutronstar and the neutrino luminosity are set as $M_{\rm NS} = 1.4 M_\odot$ and $L_{\nu,51} = 10(\textit{dash dotted}), 1(\textit{dotted})$. $\star$ marks indicate sonic point.

planes, the radial velocities suddenly decrease while the density and the temperature stay almost constant. The remarkable point is the entropy jump and they increase up to $s(\mathrm{N_A k}) \gtrsim 200$ in some models. We can see that the further the shock plane goes from the sonic point, the larger the entropy jump is.

In Fig:2, the relations between the entropies S_s and the temperature $T_{9,s}$ ($T_9 \equiv T/10^9$K) just above the shock planes are shown. We can see that the significant entropy jump ($\gtrsim 200$) occurs only when the temperature decreases below $T_9 \sim 2.5$ (at which the r-process already begins). We, thus, claim that the entropy jump does not help to enhance the neutron-to-seed ratio prior to the r-process. The high entropy should be attained much earlier than the current situation to influence the neutron-to-seed ratio.

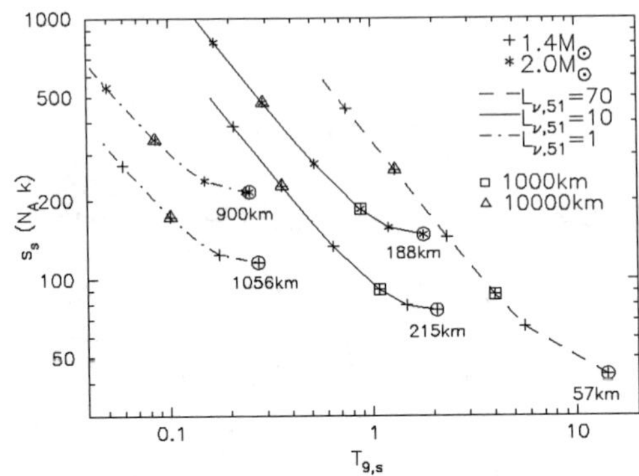

FIGURE 2. The temperature $T_{9,s}$ and the entropy S_s just above the shock plane of various models. Numbers beside circles, triangles and squares represent the shock radii.

CONCLUSIONS

The entropy jumps by the termination-shock heating have negligible effects on the final *r*-process neucleosynthesis. This is a consequence that a sizable entropy jump occurs only when the *r*-process has already ceased or, even if it happens before freezeout, the entropy jump is too small.

REFERENCES

1. Arcones, A., Janka, H. -Th., & Scheck, L. 2006, A&A, 467, 1227
2. Arnould, M., Goriely, S., & Takahashi, K. 2007, Phys. Rep., 450, 97
3. Cowan, J. J.; Thielemann, F.-K. 2004, *Phys. Today*, 57, 47
4. Kuroda,T., Wanajo,S. and Nomoto,K., 2008, ApJ, 672, 1068
5. Wanajo, S., Kajino, T., Mathews, G. J., & Otsuki, K. 2001, ApJ, 554, 578
6. Wanajo, S., Itoh, N., Ishimaru, Y., Nozawa, S., & Beers, T. C. 2002, ApJ, 577, 853
7. Wanajo, S. & Ishimaru, I. 2006, Nucl. Phys. A, 777, 676

Experimental Astrophysical Reaction Rates of Threshold (p,n)-Reactions on Pd Isotopes

Ye. Skakun[a] and T. Rauscher[b]

[a]*Kharkiv Institute of Physics and Technology, Academicheskaja str., 1, 61108 Kharkiv, Ukraine*
[b]*Departement für Physik, Universität Basel, Klingelbergstr, 82, CH-4056 Basel, Switzerland*

Abstract. Astrophysical *S*-factors for (p,n) reactions on stable Pd isotopes derived from experimental cross sections were compared to statistical model predictions with the NON-SMOKER code. Satisfactory agreement was found. Reaction rates were derived by combining experiment and theory.

Keywords: (p,n) reactions, statistical model, reaction rates, p-process, nucleosynthesis.
PACS: 25.40.Kv, 24.60.Dr, 26.30.Ef.

Nucleosynthesis models studying the so-called *p*-isotopes in the *p*- or γ-process require the knowledge of neutral- and charged-particle induced reactions and their inverses on stable as well as unstable isotopes [1, 2]. Therefore, predictions based on Hauser-Feshbach statistical theory [3] have to be combined with cross section measurements involving stable targets to improve reaction models and the understanding of branchings. Measurements of proton captures in the A=70-120 mass region (see, e.g., [4-8]) have been performed for that purpose. The typical γ-process temperatures T=(1.8-3.3) GK correspond to a proton energy region of several MeV, often reaching the threshold of (p,n)-reactions. The steep increase of the (p,n) excitation functions leads to competition with the (p,γ)-channel. Thus, (p,n)-reactions may also affect abundances in the γ-process. Moreover, current databases for astrophysics [3, 9] cover the temperature range up to T=10 GK, required for full reaction networks for stellar nucleosynthesis. This translates into a proton energy region up to ~10 MeV for nuclei in the A=100-120 mass region.

Earlier [10], we measured the cross sections of ^{104}Pd(p,n)^{104m,g}Ag, ^{105}Pd(p,n)^{105m,g}Ag, ^{106}Pd(p,n)^{106m,g}Ag, ^{108}Pd(p,n)^{108g}Ag, and ^{110}Pd(p,n)^{110m,g}Ag in the energy range 4-9 MeV. The reactions were selected because the residual nuclei have half-lives suitable for off-line γ-detection in a low background area. The total cross sections ($\sigma=\sigma_m+\sigma_g$) were determined for all reactions except for ^{108}Pd(p,n) because of the long half-life of the ^{108m}Ag isomer. Here, we compare experimental astrophysical *S*-factors and laboratory reaction rates with statistical model predictions [3] by the NON-SMOKER code [11] (Fig. 1).

The energy dependence of experimental and theoretical *S*-factors for the reaction ^{104}Pd(p,n)$^{104m+g}$Ag (panel *a*) is the same but the absolute values are underestimated by theory by ~15%. Because of Q=-5.06 MeV the data is covered by the Gamow window only at the temperature T≥3.8 GK. The theoretical temperature dependence of the rate is given by the solid curve on panel *b* while the points correspond to a combination of

CP1016, *Origin of Matter and Evolution of Galaxies*,
edited by T. Suda, T. Nozawa, A. Ohnishi, K. Kato, M. Y. Fujimoto, T. Kajino, and S. Kubono

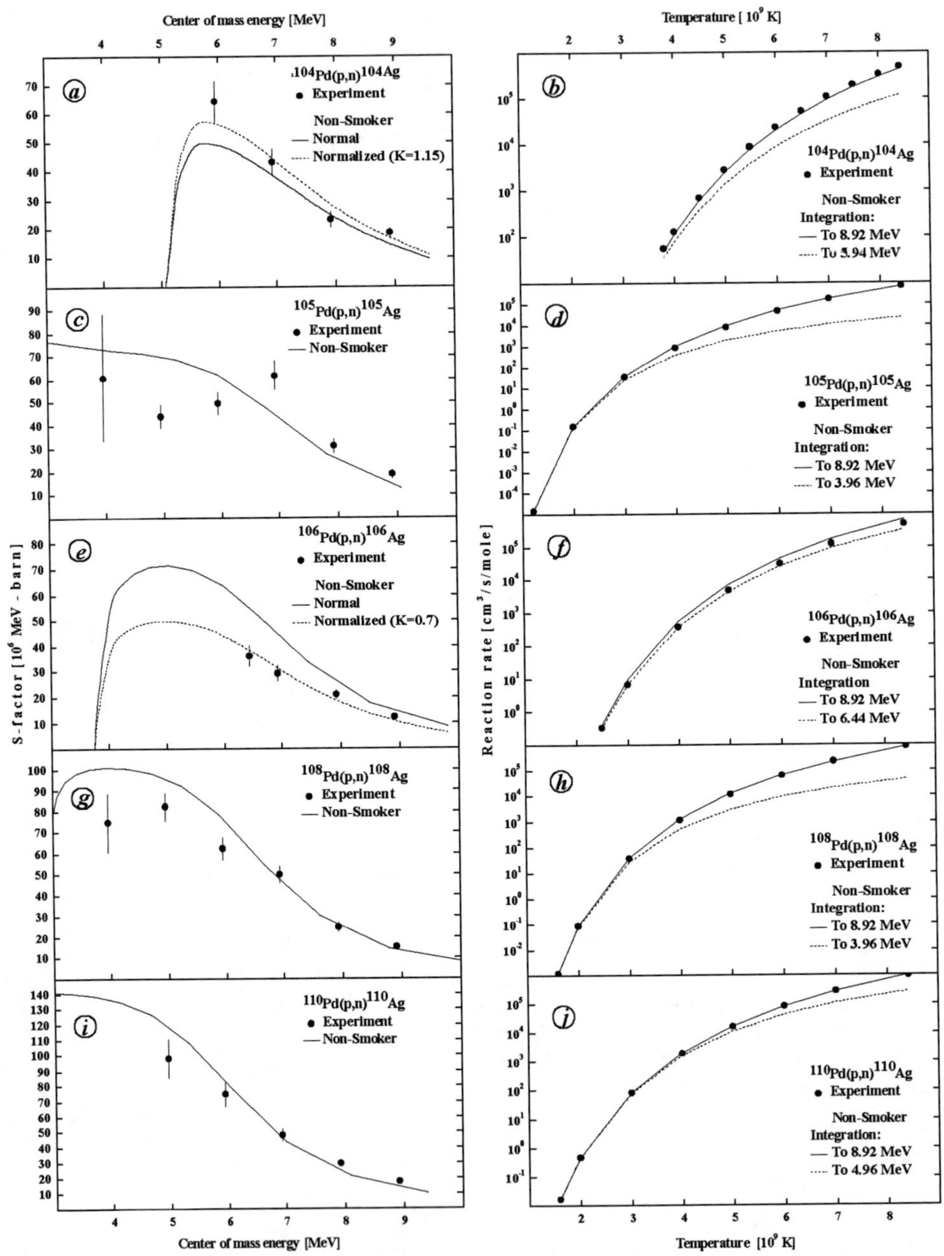

FIGURE 1. Astrophysical *S*-factors (left column) and reaction rates (right column): Points are experimental values, curves are theory predictions. Laboratory rates were derived from combinations of experimental data and theory (see text).

experiment and theory, utilizing renormalized theoretical S-factors. The upper temperature limit (T=8.4 GK) for the rate is determined by the available data. The contribution of the unmeasured energy interval to the rate (the dashed curve on panel b) is about 65% and 24% at the temperatures T=3.8 GK and 8.4 GK, respectively. The original theoretical rates underestimate the combined values by 16÷17 %.

Although scattering of the experimental points is observed for the reaction ^{105}Pd(p,n)$^{105m+g}$Ag (panel c), the agreement with theory is satisfactory on the whole. The rate (panel d) can be determined for $T \geq 1.2$ GK because Q=-2.13 MeV. The largest disagreement (22%) between the original theoretical and the combined rates is found in the temperature range T=(4-5) GK. The contribution of the unmeasured interval decreases from almost 100% at the lowest temperature to ~4% at T=8.4 GK.

The population of the low-spin ground states dominates the S-factors of ^{106}Pd(p,n)$^{106m+g}$Ag (Q=-3.75 MeV) and ^{108}Pd(p,n)$^{108m+g}$Ag (Q=-2.70 MeV) at low projectile energies. The decays of the residuals ^{106g}Ag and ^{108g}Ag lead to the formation of the p-process nuclei ^{106}Cd (branching ratio <1%) and ^{108}Cd (branching ratio 97.15%) [12], respectively. The experimental S-factors of ^{106}Pd(p,n) (panel e) are in good agreement with normalized (K=0.7) NON-SMOKER calculations, the ones of ^{108}Pd(p,n)^{108g}Ag (panel g) with the original predictions for total cross sections. Since the experimental data for ^{106}Pd(p,n) only cover a small interval, the contribution of normalized NON-SMOKER values to the rate is already large (~67%) at T=8.4 GK and further increasing with decreasing T. For ^{108}Pd(p,n) the corresponding theoretical contribution amounts to ~5% at T=8.4 GK. The original theoretical rate for the reaction ^{106}Pd(p,n)$^{106m+g}$Ag (panel f) exceeds the combined values by not more than 47% and for ^{108}Pd(p,n)^{108g}Ag (panel h) by not more than 9%.

The reaction ^{110}Pd(p,n)^{110}Ag has a low threshold (Q=-1.67 MeV) and therefore commences at T=0.8 GK. The experimental and theoretical S-factors and rates (panels i and j) are in good agreement. Thus, the difference between predicted and combined rates is ≤2% in the whole temperature range although the contribution of the theoretical low energy interval amounts to ~28% for T=8.4 GK. The astrophysical relevance of this reaction lies in the decay of its product nucleus ^{110}Ag (branching ratio 99.7% [12]) to the stable ^{110}Cd which is an *s-only* isotope.

REFERENCES

1. M. Arnould and S. Goriely, *Phys. Rep.* **384**, 1 (2003).
2 T. Rauscher, *Phys. Rev. C* **73**, 015804 (2006).
3. T. Rauscher, F.-K. Thielemann, *At. Data Nucl. Data Tables.* **75**, 1 (2000); **79**, 1 (2001).
4. J. Bork, H. Schatz, and F. Käppeler, *Phys. Rev.* **C58**, 524 (1998).
5. F.R. Chloupek et al. *Nucl. Phys.* **A652**, 391 (1999).
6. N. Özkan et al. *Nucl. Phys.* **A710**, 469 (2002).
7. S. Harissopulos et al, *Nucl. Phys.* **A719**, 115c (2003).
8. G. G. Kiss et al. *Phys. Rev. C* **76**, 055807 (2007).
9. http://nuclear-astrophysics.fzk.de/kadonis/.
10. V.G. Batij, E.A. Skakun, Yu.N. Rakivnenko, O.A. Rastrepin, *Yad. Fiz.* **47**, 609 (1988); EXFOR Records A0304004, A0304007, A0304011, A0304013, A0304016.
11 T. Rauscher and F.-K. Thielemann, in *Stellar Evolution, Stellar Explosions, and Galactic Chemical Evolution*, ed. A. Mezzacappa (IOP, Bristol 1998), p. 519.
12. S.Y.F. Chu, L.P. Ekström and R.B. Firestone, 1999. The Lund/LBNL Nuclear Data Search, http://nucleardata.nuclear.lu.se/nucleardata/toi/.

Cryogenic gas target system for intense RI beam productions in nuclear astrophysics

Y. Wakabayashi*, H. Yamaguchi*, S. Hayakawa*, Y. Kurihara*, G. Amadio*, H. Fujikawa*, D.N. Binh*,†, J.J. He**, A. Kim‡ and S. Kubono*

*Center for Nuclear Study, University of Tokyo, Japan
†Institute of Physics and Electronics, Vietnam Academy of Science and Technology, Vietnam
**School of Physics, The University of Edinburgh, U.K.
‡Department of Physics, Ewha Womans University, Korea

Abstract.
A cryogenic gas target system was newly developed to produce intense RI beams at the low-energy in-flight radio-isotope beam separator (CRIB) of the University of Tokyo. The main features of the cryogenic gas target system are the direct cooling of the target cell by a liquid N_2 finger and the circulation of the target gas that goes through the liquid N_2 tank. Hydrogen gas was cooled down to 85-90 K by liquid nitrogen and used as a secondary beam production target which has a thickness of 2.3 mg/cm^2 at the gas pressure of 760 Torr. Intense RI beams, such as a ^{7}Be beam of 2×10^8 particles per second, were successfully produced using the target.

Keywords: RI beam production, gas target
PACS: 29.25.-t, 29.27.-a

INTRODUCTION

Intense radio-isotope (RI) beams are the key to investigate directly astrophysical nuclear reactions under explosive conditions. To produce RI beams, light gases, such as H_2, D_2, and ^{3}He, have often been used as the RI beam production target at CRIB[1, 2].

A thick and thermostable gas target is necessary to produce intense RI beams for measurements of low cross-section reactions. However, the gas cell for CRIB cannot be elongated much more than 80 mm, because of a significantly worse collection efficiency and larger beam images at the downstream focal planes. We have also experienced that the foils of the water-cooled gas target were broken by high-current primary beams.

To solve these problems at the same time, we have constructed a cryogenic gas target system. Such a use of the cryogenic gas target for the RI beam production has already been made at other facilities[3]. The design and basic test results of the new cryogenic gas target system are described below.

DESIGN

Fig. 1 shows the whole structure of the target system. The target gas cell is shown at the bottom of the figure. On the beam direction sides of the cell, flanges with Havar foils of 2.5 μm were attached to seal the cell. Above the target cell, there is a tank filled

CP1016, *Origin of Matter and Evolution of Galaxies*,
edited by T. Suda, T. Nozawa, A. Ohnishi, K. Kato, M. Y. Fujimoto, T. Kajino, and S. Kubono

with liquid nitrogen. The liquid nitrogen is filled also in the vertical duct in between the tank and the gas cell to cool the cell directly. There is a gas circulation line through the target cell and the tank. A diaphragm pump was installed in the line for making a gas circulation. The main features of the cryogenic gas target system are as follows:

1. Direct cooling by liquid nitrogen
   The target cell is cooled by liquid nitrogen. If the target is ideally cooled to the liquid nitrogen temperature (77 K), it becomes about 4 times thicker compared to the room temperature target with the same pressure. Low temperature is also good for having a stability against high-current beams.
2. Forced gas circulation
   A forced flow of the target gas can be made, by circulating the gas using a diaphragm pump and valves in an enclosed system including the target cell.
3. Oxygen concentration monitoring
   An oxygen analyzer is installed for monitoring the concentration to avoid the explosion of the hydrogen gas during the circulation.

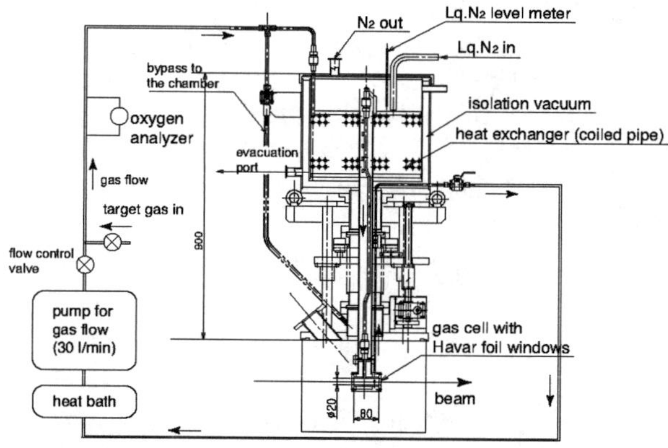

FIGURE 1. Design of the cryogenic gas target system.

TEST FOR THE BASIC FEATURES OF THE TARGET SYSTEM

We have performed some basic tests of the cryogenic target system producing an RI beam at CRIB. The primary beam of $^{7}Li^{2+}$ at 5.6 MeV/u with current up to 2.8 eμA bombarded the cryogenic gas target of 1-atm hydrogen. The secondary beam of ^{7}Be was produced via the ^{7}Li(p,n) reaction for various beam intensities and gas circulation settings. We had almost 100% pure ^{7}Be beam at the final focal plane.

Fig. 2 shows the production rates of the ^{7}Be beam, measured for various primary-beam currents and circulation rates of the target gas. The production rates were normalized for the momentum acceptance of $\pm 3\%$, and we confirmed that we can obtain an intense ^{7}Be beam of 2×10^{8} per second in the best cases.

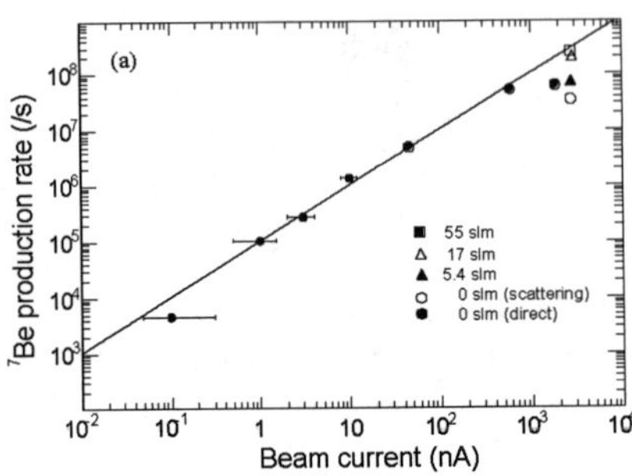

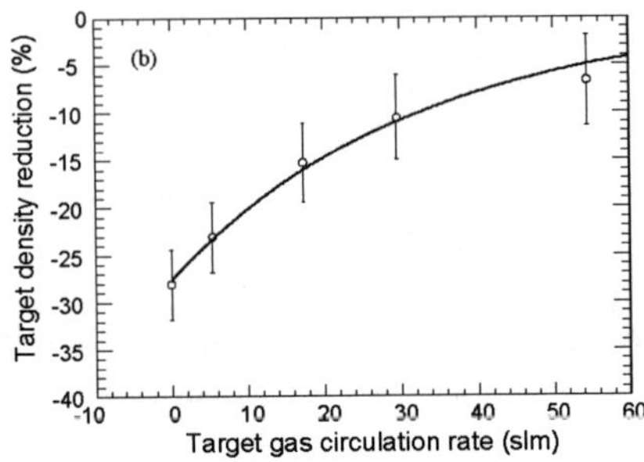

FIGURE 2. (a)Production rate of the ^{7}Be beam and (b) Density reduction of the target gas

The result in Fig. 2 (a) shows a good linearity between the production rate and the primary beam current less than 1 eμA. However, with higher intensities, the production rate deviated from the proportional line, and the beam purity also started to decrease. This is regarded as the density-reduction effect caused by the heat.

Target density reduction was measured by the $^{7}Li^{2+}$ beam of 2.8 eμA. The reduction is denoted as the deviation from the standard thickness measured with a low-current beam. The result in Fig. 2 (b) illustrates that the density reduction effect can be minimized by circulating the target gas. We succeeded in reducing this reduction effect to around –5% by making the gas circulation of 55 slm (standard litter per minute). The approximated curve is given by

$$-0.28 \times \exp(-0.031 \times \sigma) \tag{1}$$

where σ is the circulation rate in slm.

CONCLUSION

A cryogenic gas target system was developed for the RI beam production at CRIB. The target system was used for the production of ^{7}Be and we obtained the beam of 2×10^{8} particles per second, with the target of 2.3 mg/cm^{2}-thick hydrogen gas at 85–90 K. The target gas circulation with the rate of 55 slm can almost eliminate the target density reduction effect.

ACKNOWLEDGMENTS

We want to thank all staffs at RIKEN and CNS accelerator for their help.

REFERENCES

1. S. Kubono *et al.*, Eur. Phys. J. A **13** (2002) 217.
2. Y. Yanagisawa *et al.*, Nucl. Instrum. Methods Phys. Res., Sect. A **539** (2005) 74.
3. A. Azhari *et al.*, Phys. Rev. C **63** (2001) 055803.

Description of three-body scattering for astrophysics

Y. Kikuchi*, T. Myo†, M. Takashina**, K. Katō* and K. Ikeda‡

*Division of Physics, Graduate School of Science, Hokkaido University, Sapporo 060-0810, Japan
†Research Center for Nuclear Physics (RCNP), Osaka University, Ibaraki 567-0047, Japan
**Institute for Theoretical Physics (YITP), Kyoto 606-8502, Japan
‡The Institute of Physical and Chemical Research (RIKEN), Wako 351-0198, Japan

Abstract. A new and simple discription of the three-body scattering states, which is based on the Lippmann-Schwinger equation and the complex scaling method, is developed. As applications of this method, the energy distribution of the $E1$ transition strength for ^{6}He is calculated. It is also shown that the correlations of subsystems in ^{6}He are well reproduced.

Keywords: Complex scaling method, Three-body scattering
PACS: 25.60.Gc

INTRODUCTION

The three-body nuclear reactions are important in nucleosyntheses, as in the triple-α reaction in ^{12}C production. [1] The description of scattering states is essential to estimate the relevant cross sections and reaction rates. However, in many cases, it is complicated to describe the scattering states of three-body systems. In this work, we develop a new and simple method to describe the three-body scattering states with the help of the Lippmann-Schwinger equation and the complex scaling method (CSM).

COMPLEX SCALED LIPPMANN-SCHWINGER EQUATION

To describe the three-body scattering states, we use the formal solution of the complex scaled Lippmann-Schwinger (CSLS) equation. This solution is given as follows:

$$|\Psi^{(+)}(\mathbf{k},\mathbf{K})\rangle = |\mathbf{k},\mathbf{K}\rangle + \sum_i U^{-1}(\theta)|\Phi_i^\theta\rangle \frac{1}{E-E_i^\theta}\langle\tilde{\Phi}_i^\theta|U(\theta)\hat{V}|\mathbf{k},\mathbf{K}\rangle, \qquad (1)$$

where $|\mathbf{k},\mathbf{K}\rangle$ is a asymptotic solutions for three-body systems, and $\mathbf{k}$ and $\mathbf{K}$ indicate the momenta of subsystems. The operator $U(\theta)$ is a complex scaling operator. The eigenstates and eigenvalues of the complex scaled total Hamiltonian $\hat{H}^\theta$ are represented as Φ^θ and E^θ, respectively. In the present work, we use the numerical solutions within the variational method so-called the hybrid-TV model [2] for Φ^θ and E^θ to extract Green's function.

In Eq. (1), the boundary conditions are included by eigenvalues of $\hat{H}^\theta$. The energy eigenvalues in CSM are given as complex numbers, and their imaginary parts impose

CP1016, *Origin of Matter and Evolution of Galaxies*,
edited by T. Suda, T. Nozawa, A. Ohnishi, K. Kato, M. Y. Fujimoto, T. Kajino, and S. Kubono

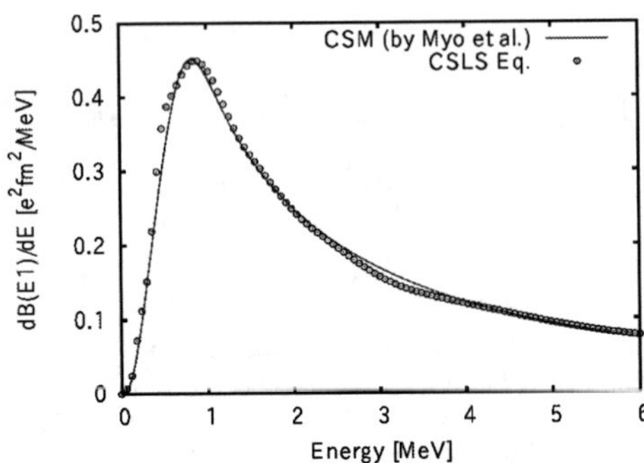

FIGURE 1. Energy distributions of the $E1$ transition strength of ^{6}He. The points and line show the results of the CSLS Eq. and CSM + responce function method [6], respectively.

the boundary condition for each states. [3] These eigenvalues in CSM make it possible to take the contributions from each decay mode into Green's function.

In addition, the CSLS Eq. has advantage that the physical quantities such as an S-matrix are treated separately, as devided into the contibutions from the initial states and the final state interaction (FSI). The matrix element of an arbitrary operator $\hat{O}$ can be expressed as

$$\langle \Phi_{\text{in.}} | \hat{O} | \Psi^{(+)}(\mathbf{k},\mathbf{K}) \rangle = \langle \Phi_{\text{in.}} | \hat{O} | \mathbf{k},\mathbf{K} \rangle + \sum_i \langle \Phi_{\text{in.}} | \hat{O} U^{-1}(\theta) | \Phi_i^\theta \rangle \frac{1}{E - E_i^\theta} \langle \Phi_i^\theta | U(\theta) \hat{V} | \mathbf{k},\mathbf{K} \rangle \tag{2}$$

, where $\Phi_{\text{in.}}$ is an initial state wave function. The first term in Eq. (2) presents a contributions of direct breakup components from the initial state, and the second term indicates the effect of FSI.

APPLICATION TO COULOMB BREAKUP REACTION OF ^{6}HE

To test the validity of CSLS Eq., we first calculate the energy distributions of the dipole transition of ^{6}He [4, 5] and show the result in FIG. 1. The points and line show the present result with the CSLS Eq., compared tp the one obtained by the preocedure given in Ref. [6], respectively. Two result shows a reasonable agreement implying the reliability of the present method.

We also calculate the two-demensional energy distributions of the $E1$ strength as shown in FIG. 2. In the left panel, which shows the strength from all the contributions, the large enhancement of the transition strength is seen around $E(\alpha\text{-}n) \sim 0.6$ MeV. This peak can be understood as the effect of the ^{5}He$(3/2^-)$ resonance. On the other hand, the distribution in the right panel, which shows the strength from the direct breakup only, has much smaller contibution than the total distribution. These results indicate that the

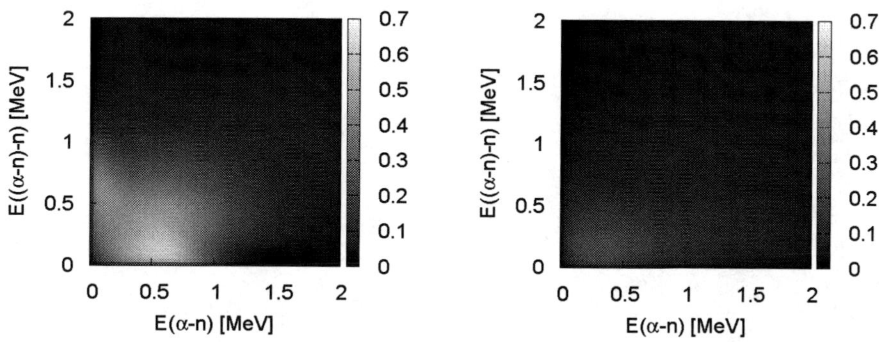

FIGURE 2. Two-demensional energy distributions of the $E1$ transition strength. The left and right panel show the total distributions and the contribution of the direct breakup from the initial states, respectively.

constributions of direct breakup components from the initial state is much weak, and masked by the FSI of the ^{5}He resonance.

These results show that the CSLS Eq. well describes the correlations of subsystems as well as the total system. The use of the CSLS Eq. allows us to understand the reaction mechanisms in cluster nucleosyntheses. The other application to the neutron capture reaction of ^{16}O may be shown by Yamamoto *et al.* [7]

SUMMARY

In this work, we develop a simple discription of the three-body scattering states. We calculate the energy distributions of the $E1$ transition strength by using the CSLS Eq., and find that the CSLS Eq. can describe the correlations of both total system and subsystems. We also find that the contributions of the direct breakup from the initial states is masked in ^{6}He by the strong FSI coming from the ^{5}He resonance.

REFERENCES

1. F. Hoyle, Astrophys. J., Suppl. **1**, 121 (1954).
2. Y. Tosaka, Y. Suzuki and K. Ikeda, Prog. Theor. Phys. **83**, 1140 (1990).
3. J. Aguilar,and J.M. Combes, Commun. Math. Phys. **22**, 269 (1971).
   E. Balslev and J.M. Combes, Commun. Math. Phys. **22**, 280 (1971).
4. T. Aumann *et al.*, Phys. Rev. C **59**, 1252 (1999).
5. T. Nakamura *et al.*, Phys. Rev. Lett. **96**, 252502 (2006).
6. T. Myo, K. Katō, S. Aoyama and K. Ikeda, Phys. Rev C **63**, 054313 (2001).
7. K. Yamamoto, H. Masui, T. Wada, K. Katō and M. Ohta, OMEG'07 oral presentation.

Neutron capture and inelastic scattering cross sections for ^{186}Os, ^{187}Os, and ^{189}Os and the Re-Os chronology

M. Segawa[a], Y. Nagai[b], T. Masaki[b], Y. Temma[b], T. Shima[b], K. Mishima[b], M. Igashira[c], S. Goriely[d], A. Koning[e] and S. Hilaire[f]

[a]*Quantum Beam Science Directorate, Japan Atomic Energy Agency, Tokai, Ibaraki 319-1195, Japan*
[b]*Research Center for Nuclear Physics, Osaka University, Osaka 567-0047, Japan*
[c]*Research Laboratory for Nuclear Reactors, Tokyo Institute of Technology, Tokyo 152-8550, Japan*
[d]*Institut d'Astronomie et d'Astrophysique, Université Libre de Bruxelles, Campus de la Plaine, CP226, 1050 Brussels, Belgium*
[e]*Nuclear Research and Consultancy Group, P.O. Box 25, NL-1755 ZG Petten, The Netherlands*
[f]*CEA/DAM Ile-de-France, DPTA/Service de Physique Nucléaire, BP 12, 91680 Bruyères-le-Châtel, France*

Abstract. We measured the neutron capture cross sections of ^{186,187,189}Os taking for the first time their pulse height spectra for neutrons between 5 and 90 keV by means of an anti-Compton NaI(Tl) spectrometer. The neutron inelastic scattering cross section for ^{187}Os as well as the neutron elastic scattering cross sections for ^{186,187}Os were also observed with use of ^{6}Li-glass scintillation detectors with a small systematic uncertainty.

Keywords: neutron capture reaction, cosmo-chronology
PACS: 25.40.Lw, 26.20.+f, 27.70.+q, 29.30.Kv

1. INTRODUCTIION

It has been considered that Re-Os pair can be one of the good cosmochronometers, since it has unique features as discussed below [1]. First, ^{187}Re is produced by only *r*-process and the half life is quite long 42.3± 1.3Gyr [2]. Second, ^{186}Os is the *s*-only isotope. ^{187}Os is known to be produced not only by the decay of ^{187}Re but also by slow neutron capture process for ^{186}Os, and it is depleted by neutron capture reactions of the ground and the first excited states of ^{187}Os at a stellar temperature. In order to extract the abundance of ^{187}Os, which can be attributable to the ^{187}Re decay, therefore, one must know both the production and depletion rates of ^{187}Os. The production rate of ^{187}Os and the depletion rate of ^{187}Os via its ground state could be obtained by measuring the neutron capture cross section of ^{186}Os, $\sigma_\gamma(^{186}\mathrm{Os})$, and ^{187}Os, respectively. On the other hand, the depletion rate via the excited state should be calculated theoretically, since it is not possible to measure the neutron capture cross section for the 9.75 keV first excited state of ^{187}Os using a current technique. In order to construct reliable theoretical models to calculate the excited state neutron capture cross section, the measurements of the inelastic scattering cross section off the ground state (J^π= 1/2$^-$

CP1016, *Origin of Matter and Evolution of Galaxies,*
edited by T. Suda, T. Nozawa, A. Ohnishi, K. Kato, M. Y. Fujimoto, T. Kajino, and S. Kubono

) of ^{187}Os to its excited 9.75 keV state (J^{π}= 3/2^{-}), $\sigma_{n,n'}$(^{187}Os), were suggested. From mentioned points of view, σ_{γ}(^{A}Os) for ^{186,187,189}Os, and $\sigma_{n,n'}$(^{187}Os) were extensively measured by several groups [3-8]. However, previous data differ between different data sets and/or have a large uncertainty, and therefore they are hardly used to discriminate existing theoretical models.

Hence, in the present study, we have measured σ_{γ}(^{A}Os) for ^{186,187,189}Os accurately and $\sigma_{n,n'}$(^{187}Os) in the neutron energy range from 10 to 70 keV.

2. EXPERIMENTS

We used neutrons, which were produced by the ^{7}Li(p,n) reaction using a proton beam provided from the pelletron accelerator of the Research Laboratory for Nuclear Reactors at Tokyo Institute of Technology.

2.1 Cross section measurements of (n,γ) reactions on Os isotopes

The measurements of the neutron capture reactions have been carried out by detecting a prompt γ-ray by means of an anti-Compton NaI(Tl) spectrometer [9]. The σ_{γ}(^{A}Os) data for ^{186,187,189}Os were obtained by comparing the γ-ray yield of the ^{186,187,189}Os(n,γ) reactions to that of the ^{197}Au(n,γ) reaction, whose cross section is well known within an error of 3 %. The obtained σ_{γ}(^{A}Os) of ^{186,187,189}Os are shown in Fig.1 together with previous experimental data and a theoretical one calculated using the TALYS code as described in refs.[10,11]. The obtained cross sections decrease quite smoothly with increasing the neutron energy, and therefore the present keV neutron capture reaction is considered to dominantly proceed via an s-wave neutron capture process. The present result for ^{187}Os is ~20 % larger than previous data, while the results for ^{186}Os and ^{189}Os are in good agreement with the previous result taken by Browne et al. Details are described in ref. 9.

2.2 Cross section measurement of (n,n') reactions on Os isotopes

The measurements of the neutron elastic scattering cross sections of ^{186}Os and ^{187}Os, and of $\sigma_{n,n'}$(^{187}Os) were performed by detecting neutrons scattered by the samples with four ^{6}Li-glass detectors with a TOF method [12]. In order to subtract the elastically scattered neutrons from total neutrons scattered by ^{187}Os to obtain the yield due to the inelastic scattering by ^{187}Os, we measured the neutron yield Y_n(^{186}Os) due to the elastic scattering by ^{186}Os. Using the obtained neutron yield due to the inelastic scattering by ^{187}Os, we could determine $\sigma_{n,n'}$(^{187}Os) by referring to the neutron elastic scattering cross section for ^{12}C, which is known accurately within an uncertainty of 4 %. The detailed analysis for the elastic as well as inelastic scattering cross sections for Os isotopes mentioned is in progress.

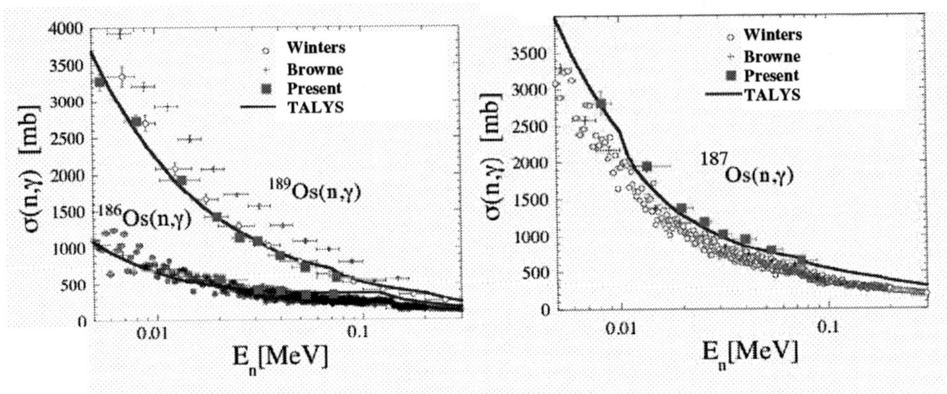

Fig.1 The obtained neutron capture cross sections for ^{186,187,189}Os together with previous data (circle and cross) and theoretical calculation (solid line) [3-9].

3. CONCLUSION

The first successful measurement of continuum as well as discrete γ-ray spectra from the neutron capture by ^{186,187,189}Os at $5 < E_n < 90$ keV enabled us to accurately determine the cross sections with a small systematic uncertainty. We also succeeded to measure the inelastically scattered neutrons spectrum for ^{187}Os accurately. Using the present results together with the TALYS code [10], $\sigma_\gamma(^{187}$Os) by the first excited state of ^{187}Os is calculated.

ACKNOWLEDGMENTS

We would like to thank K. Tosaka for his reliable operation of the Pelletron accelerator. This work was supported by a Grant-in-Aid for Scientific Research of the Japan Ministry of Education, Culture, Sports, Science, and Technology.

REFERENCES

1. D. D. Clayton, *Astrophys. J.* **139**, 637 (1964).
2. M. Linder et al, *Geochem. Consochem. Acta.* **53**, 1593 (1989).
3. R. R. Winters and R. L. Macklin, *Phys. Rev. C* **25**, 208 (1982).
4. R. R. Winters and R. L. Macklin, *Astron. Astrophys.* **171**, 9 (1987)
5. J. C. Browne and B. L. Berman, *Phys. Rev. C* **23**, 1434 (1981)
6. R. L. Macklin, R. R. Winters et al, *Astrophs. J.***274**, 408 (1983)
7. R. L. Hershberger, R. L. Macklin et al, *Phys. Rev. C* **28,** 2249 (1983)
8. M. T. McEllistrem, R. R. Winters et al, *Phys. Rev. C* **40**, 591 (1989)
9. M. Segawa, T. Masaki, Y. Nagai et al, *Phys. Rev. C* **76**, 022802(R) (2007).
10. A. J. Koning, S. Hilaire, and M. C. Duijvestijn, NRG report 21297/04, 62741/P (2004).
11. S. Hilaire and S. Goriely, *Nucl. Phys. A* **779**, 63 (2006).
12. M. Segawa, Y. Temma, Y. Nagai et al, *Nucl. Instr. and Meth. A* **564**, 370 (2006).

$E0$ transition strength and cluster structure in ^{13}C

T. Yoshida*, N. Itagaki† and T. Otsuka†

*Meme Media Laboratory, Hokkaido University 060-8628 Sapporo, Japan
†Department of physics, University of Tokyo, Hongo, 113-0033 Tokyo, Japan

Abstract. We study the structure of low-lying states of ^{13}C with a microscopic cluster model. The second 0^+ state of ^{12}C, which is known as the Hoyle state and is important for astrophysical reactions, has been clarified to have dilute α-cluster structure. On the basis of the α-cluster model, we discuss the effect of one valence neutron in ^{13}C on the α-cluster configuration of the Hoyle state. For these purpose, we investigate the iso-scalar $E0$ transition probability from the ground $1/2^-$ state to this excited states.

Keywords: nuclear structure, α-cluster, $E0$ transition strength
PACS: 43.35.Ei, 78.60.Mq

INTRODUCTION

Various microscopic cluster models have been successfully applied to light nuclei [1]. Recently, systematic analyses have been extensively performed in neutron-rich nuclei such as α-α cluster structure in the Be isotopes. As the next step of such studies, we focus on the 3α-cluster structure of C isotopes. The second 0^+ state of ^{12}C has been well known to have a developed 3α-cluster structure, which is recently reinterpreted as the α-condensed state [2]. Therefore, it is intriguing to see how the structure changes when valence neutrons are added. However, the clusters may change their geometric shape in neutron rich C isotopes. For instance, an equilateral-triangular shape of 3α surrounded by excess neutrons is suggested in the second 3^- state of ^{14}C based on the molecular-orbit model [3]. Also, the presence of the cluster states in ^{13}C has been experimentally suggested by measuring the iso-scalar $E0$ transitions from the ground $1/2^-$ state induced by the ^{13}C(α,α')^{13}C reaction[4]. The obtained B(E0) values of 55±6 fm^4($1/2_2^-$), 35±4 fm^4 (third $1/2_3^-$)[4]) are much larger than those by the shell-model calculation. However, they are significantly smaller than the value for the $0_1^+ \rightarrow 0_2^+$ transitions in ^{12}C. In order to investigate the property of these states, we calculate the eigen states and the $E0$ transition strength from the ground $1/2^-$ state to the excited $1/2^-$ states in ^{13}C.

RESULTS AND DISCUSSION

The total wave function is fully antisymmetrized and is given by a superposition of the Slater determinants. Projection onto a good angular momentum is performed by the projection operator, and the coefficients of this wave function are determined by diagonalizing the Hamiltonian matrix after this projection. Each Slater determinant (Ψ_k)

CP1016, *Origin of Matter and Evolution of Galaxies*,
edited by T. Suda, T. Nozawa, A. Ohnishi, K. Kato, M. Y. Fujimoto, T. Kajino, and S. Kubono

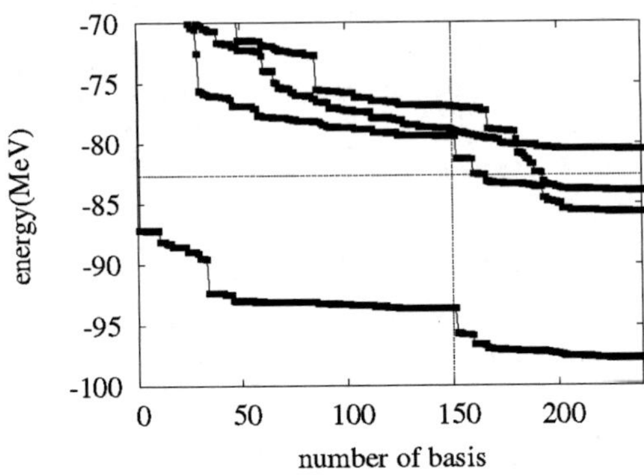

FIGURE 1. Convergence of energy for the $1/2^-$ states. The horizontal axis show the number of trial Slater determinants. The Φ_1 and Φ_2 configurations (noted in the text) correspond to base number 1-150 and 151-240 respectively. The dotted line shows the α-α-α-n threshold energy.

consists of A nucleons $\Psi_k = \mathscr{A}[(\psi_1\chi_1)(\psi_2\chi_2)\cdots]$, and each nucleon ($\psi_i\chi_i \;\; i = 1 \sim A$) has a Gaussian form the same as many conventional cluster models, and the oscillator parameter ($b = 1/\sqrt{2\nu}$ = 1.46 fm) is common for all nucleons to exactly remove the center-of-mass kinetic energy. As for the Hamiltonian, we use the Volkov No.2 (V2) effective $N-N$ potential[5] for the central part and G3RS potential[6] for the spin-orbit part. To reproduce the scattering phase shifts of the $\alpha+n$ and $\alpha+\alpha$ systems, M (Majorana parameter of V2) = 0.6 and V_0 (strength of G3RS) = 2000 MeV should be adopted. However, we use V_0=1500 MeV for ^{13}C[7]. Also, B(Bartlet) $=$ H(Heisenberg) $= 0.125$ is introduced for V2 to reproduce the binding energy of deuteron and to remove the bound state of neutron-neutron system. In order to achieve sufficient description for ^{12}C including gas-like nature of the second 0^+ state, we include deviation from the equilateral triangular shape of the 3α core configurations [8] (noted as Φ_1). In addition, the breaking effect of one of the α-cluster is taken into account by using simplified method to include the spin-orbit contribution (SMSO) [9] (noted as Φ_2). Here, the imaginary part of each gaussian center for an α-cluster is introduced by parameter Λ, the strength of α-cluster breaking effect. As for the valence neutron, we distribute its Gaussian center parameter randomly around the core. For the selection of the basis states, we use the idea of the stochastic variational method.

The energy convergence of the $1/2^-$ states is shown in Fig. 1, where the horizontal axis shows the number of trial Slater determinants introduced. The base number 1 to 150 (151 to 240) corresponds to the Φ_1 (Φ_2) configuration. As we can see, the convergence is reasonable. We also obtained large E0 transition strength by using these eigen states. The calculated $B(E0, 1/2^-_1 \rightarrow 1/2^-_{2,3})$ values are ~ 0, 150 fm^4, respectively. On the other hands, the result for gas-like state of ^{12}C become $\sim$200 fm^4 with the same method.

Here, we discuss the mechanism of this large $E0$ value by using simpler model space. Three α configuration of equilateral triangular shape with the relative $\alpha-\alpha$ distance of $D = 4$fm (noted as $\Phi_{D=4}$), and ^{8}Be+α like configuration (noted as Φ_Λ) where one α-cluster breaking effect is taken into account through Λ are considered here. After

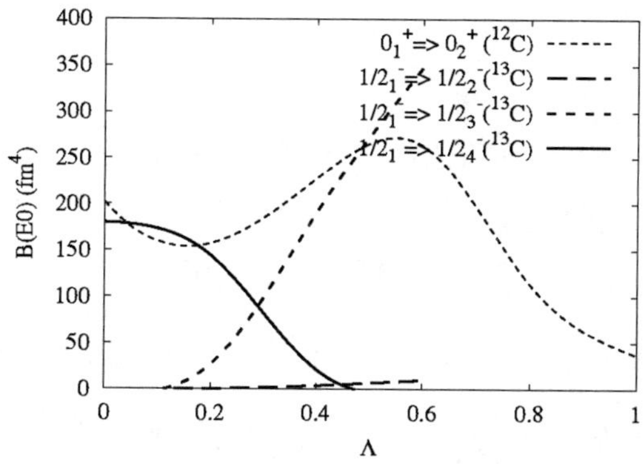

FIGURE 2. $E0$ transition strength for ^{12}C ($0_1^+ \to 0_2^+$) where we use two simplified models; $\Phi_{D=4}$ and Φ_Λ configuration. The thick dashed lines show the $E0$ transition strength for ^{13}C ($1/2_1^- \to 1/2_{2,3,4}^-$). The horizontal axis shows the Λ parameter of SMSO.

solving the wave function of one valence neutron, we calculate the eigen states in the same method as that mentioned previously. The results of $E0$ transition have the same patterns as we can see in Fig. 2. Although the order of the level may not correspond to the experiments exactly due to the smallness of the model space, the large $E0$ value is obtained for $B(E0, 1/2_1^- \to 1/2_4^-)$. Some part of the large $E0$ value can be understood by this cluster configuration ($\Phi_{D=4}$). The value is weakened when α-cluster breaking effect due to the Λ parameter.

In our calculation, the large $E0$ value has been reproduced, which is somehow small compared to that for the $B(E0, 0_1^+ \to 0_2^+)$ value of ^{12}C, which suggests the shrinkage of 3α in the excited states of ^{13}C.

REFERENCES

[1] Y. Fujiwara, H. Horiuchi, K. Ikeda, M. Kamimura, K. Katō, Y. Suzuki, and E. Uegaki, Prog. Theor. Phys. Suppl. **68**, 60 (1980).

[2] A. Tohsaki, H. Horiuchi, P. Schuck, and G. Röpke, Phys. Rev. Lett. **87**, 192501 (2001)

[3] N. Itagaki, T. Otsuka, K. Ikeda and S. Okabe, Phys. Rev. Lett. **92** 142501 (2004).

[4] Y. Sasamoto et al., Mod. Phys. Lett. A**21**, 2393 (2006).

[5] A.B. Volkov, Nucl. Phys. **74**, 33 (1965).

[6] R. Tamagaki, Prog. Theor. Phys. **39**, 91 (1968).

[7] T. Otsuka, T. Matsuo, D. Abe, Phys. Rev. Lett. **97** 162501 (2006).

[8] E. Uegaki, S. Okabe, Y. Abe and H. Tanaka Prog. Theor. Phys. **57**, 1262 (1977).

[9] N. Itagaki, H. Masui, M. Ito, and S. Aoyama, Phys. Rev. C **71**, 064307 (2005).

Neutrino-Mass Effects on the CMB Anisotropies in a Cosmological Model with a Primordial Magnetic Field

K. Kojima*, D. G. Yamazaki† and T. Kajino†,*

*Department of Astronomy, The Graduate School of Science, University of Tokyo, 7-3-1 Hongo, Bunkyo-ku, Tokyo 113-0033, Japan

†Division of Theoretical Astrophysics, National Astronomical Observatory, 2-21-1 Osawa, Mitaka, Tokyo 181-8588, Japan

Abstract. In recent years neutrino oscillation experiments have showed that neutrinos have finite mass. One can determine the neutrino mass from neutrinoless double-beta decay experiments, but the exact mass of neutrinos is still unknown. However, we can also constrain the neutrino mass from a precise analysis of the CMB data, which constrains the mass to be smaller than 2 eV (2σ) in a standard ΛCDM model [1]. In recent studies it has beddn shown that a primordial magnetic field (PMF) exists, and it makes important effects on the CMB anisotropies. In this work, we constrain the total neutrino mass from the CMB anisotropies with the PMF effect. We thus a obtained a smaller upper limit on the neutrino mass, i.e. $< 1.3\text{eV}(2\sigma)$. We will also discuss the dependence on the power law spectral index of the PMF and on the vector CMB anisotropies.

Keywords: Neutrino, CMB, Primordial Magnetic Field
PACS: 98.70.Vc, 98.62.En, 14.60.Pq

NEUTRINO MASS AND STRUCTURE FORMATION

Neutrinos are one of the dark matter candidates, and massive finite mass neutrinos have a large effect on the formation of large scale structure (LSS) as a hot dark matter. The change of CMB anisotropies by a massive neutrino makes it possible to constrain the neutrino mass from cosmology. Fukugita et al. [1] showed that $\sum m_\nu < 2\text{eV}(2\sigma)$ within a standard ΛCDM model. Another constraint is taken from the cosmological study of [2] which used the WMAP-3yr data and the Lyman-α (Ly-α) forest power spectrum from SDSS. They obtained a very severe constrait, $\sum m_\nu < 0.17\text{eV}(2\sigma)$. However, the analysis of the Ly-α data makes the effective number of neutrinos $N_\nu > 3$ and the best fit value is around $N_\nu = 5.3$, which does not agree with the result from the CMB power spectrum. In this work, we try to constrain the total neutrino mass using only the WMAP-3yr data by assuming three neutrino species.

The effect of the neutrino was studied in [3], and it is known to change the CMB power spectrum for $\ell > 300$. Neutrinos cause two different effects: The first is the free-streaming effect. The typical scale affected by this is smaller than the conformal time when massive neutrinos become non-relativistic. The multipoles affected by free-streaming are about $\ell > 300$ when $\sum m_\nu \sim 2\text{eV}$. This effect decays the gravitational potential and the curvature rapidly and enforces the acoustic oscillation. This process upshifts the zero-point of the oscillation. The second effect is a shift to the horizontal

CP1016, *Origin of Matter and Evolution of Galaxies*,
edited by T. Suda, T. Nozawa, A. Ohnishi, K. Kato, M. Y. Fujimoto, T. Kajino, and S. Kubono

peak in the CMB power spectrum. It is caused by an increase of the sound horizon [4]. These neutrino effects on the power spectrum make it possible to constrain the neutrino mass from the analysis of observed anisotropies of the CMB. This method is expected to constrain $\sum m_\nu < 2$eV under standard ΛCDM model [1] with WMAP-3yr data [5].

PRIMORDIAL MAGNETIC FIELD

Magnetic fields in clusters of galaxies have been observed with a strength of 0.1–1.0 μG at $\sim$ 1Mpc. The existense of a primordial magnetic field (PMF) of order 1 nG, whose field lines collapses as structure forms, is one possible explanation for such magnetic fields in galactic clusters. Also the PMF could influence a variety of phenomena in the early universe such as the CMB and the matter density field. Therefore, in order to constrain $\sum m_\nu$ more precisely, we should take into consideration the PMF effect.

The treatment of the primordial magnetic field was studied in [6, 7]. We parametarize the magnetic field B_λ, which is the smoothed magnetic field strength over a scale $\lambda = 1$Mpc. We introduce several assumptions which are usually taken when calculating the PMF. First, electrons and protons distribute uniformly in the early universe. It means that the conductance of the universe is infinite, and the electric field is approximately 0. This is the so called MHD approximation. Second, the magnetic field has power law spectrum with a spectral index n_B. In refs. [8, 7] the spectral index is severely constrained from the gravitational waves produced by the magnetic field, namely the constraint varied depending on the epoch when the PMF is generated. In this work, we set $n_B = -2.9$, which is valid no matter when the PMF was generated.

The effect of the PMF can be separated into three modes. The scalar mode [9] increases anisotropies dramatically for $\ell < 10$. This is because of the Integrated Sachs-Wolfe effect. The vector mode enforces the change in the power spectrum for $\ell > 1000$, which can be neglected in the WMAP region, $\ell < 800$. The tensor mode is much smaller than the other two modes, so we also neglect it here.

CALCULATED RESULT

We constrain $\sum m_\nu$ from CMB anisotropies by taking account of only the scalar PMF effect. We used a Markov chain Mote Carlo method (COSMOMC [10]) in the statistical analysis and calculated the likelihood with 600,000 samples with the WMAP-3yr data. Free parameters are the density parameters for cold dark matter, baryonic matter, and massive neutrinos, the hubble parameter, the optical depth, the scalar spectral index of primordial density perturbations, their amplitude, and the magnetic field strength B_λ. The fixed parameter is the spectral index of the PMF, $n_B = -2.9$. The constraint obtained in this calculation is $\sum m_\nu < 1.35$eV, which is stronger than that of the previous work. We know that $\sum m_\nu$ and B_λ have almost no correlation from the likelihood contour on the $B_\lambda^2 - \sum m_\nu$ plane displayed at the left hand side of Fig. 1. This is because neutrinos and the scalar PMF have effects at separate regions of mutipolarity ℓ. The neutrino effect appears for $\ell > 300$, while the effect of a scalar PMF appears for $\ell < 10$. This applies only to the multipole region $\ell < 800$ for which the WMAP-3yr data are available. The

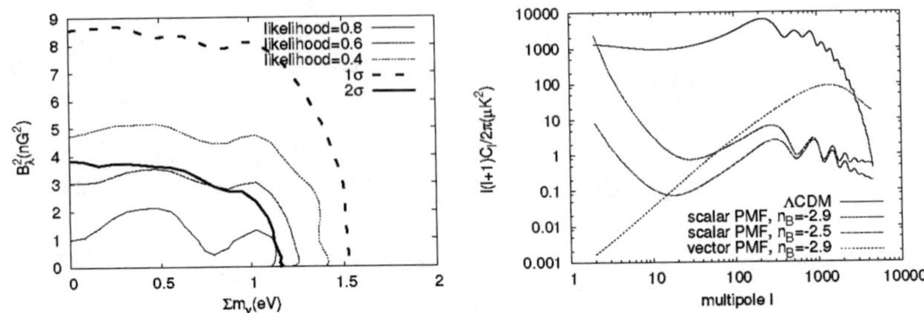

FIGURE 1. Left figure: The likelihood contour in the $B_\lambda^2 - \sum m_\nu$ plane. The narrow solid, long-dashed and short-dashed curves represent likelihood contours for 0.8, 0.6 and 0.4, respectively. The thick solid and dotted curves are 1σ C.L. and 2σ C.L., respectively. Right figure: The solid line is the primary CMB power spectrum. The dashed line and dash-dotted lines are anisotropies produced a by scalar PMF where $n_B = -2.9$ and $n_B = -2.5$ are used, respectively. The thin dashed line is produced by the vector mode of the PMF where $n_B = -2.9$.

more interesting effect is found for $\ell > 1000$, as we will discuss below.

In this work although we set $n_B = -2.9$, we need to vary n_B to get a more general result. We here discuss the dependence of the neutrino mass constraint on n_B at least qualitatively. If n_B is larger than -2.9, the effect of the primordial magnetic field in lower ℓ becomes smaller (r.h.s. of Fig. 1). It means that the spectrum looks similar to the primary spectrum. As a result the constraint increases to 2eV.

In the near future, the PLANCK mission will reveal the CMB anisotropies very pricisely up to $\ell \sim 2500$, which should make it possible to better constrain neutrino masses. When we calculate the neutrino mass from such higher multipoles, the vector PMF becomes very important as indicated on the r.h.s. of Fig. 1. Neutrinos and the PMF may have a correlation at higher multipoles because both massive neutrinos and the vector PMF have a large effect on the CMB anisotropies there. At present, only the CBI and ACBAR data are available for high multipoles. Our goal before the PLANCK mission is to calculate the neutrino mass with the scalar and vector PMF using other observational data, such as CBI and ACBAR, and by varing all parameters including n_B. This work is now under way.

REFERENCES

1. M. Fukugita, K. Ichikawa, M. Kawasaki, and O. Lahav, *Phys. Rev. D* **74**, 027302 (2006).
2. U. Seljak, A. Slosar, and P. McDonald, *JCAP* **10**, 14 (2006).
3. K. Ichikawa, M. Fukugita, and M. Kawasaki, *Phys. Rev. D* **71**, 043001 (2005).
4. S. Dodelson, E. Gates, and A. Stebbins, *ApJ* **467**, 10 (1996).
5. D. N. Spergel, R. Bean, O. Doré, M. R. Nolta, C. L. Bennett, J. Dunkley, G. Hinshaw, N. Jarosik, E. Komatsu, L. Page, H. V. Peiris, L. Verde, M. Halpern, R. S. Hill, A. Kogut, M. Limon, S. S. Meyer, N. Odegard, G. S. Tucker, J. L. Weiland, E. Wollack, and E. L. Wright, *ApJS* **170**, 377–408 (2007).
6. A. Mack, T. Kahniashvili, and A. Kosowsky, *Phys. Rev. D* **65**, 123004 (2002).
7. D. G. Yamazaki, K. Ichiki, T. Kajino, and G. J. Mathews, *ApJ* **646**, 719–729 (2006).
8. C. Caprini, and R. Durrer, *Phys. Rev. D* **65**, 023517 (2002).
9. M. Giovannini, *Classical and Quantum Gravity* **23**, 4991–5025 (2006).
10. A. Lewis, and S. Bridle, *Phys. Rev. D* **66**, 103511 (2002).

The Magnetorotational Explosion of Core-Collapse Supernovae with Initially Weak Magnetic Field

Takami Kuroda and Hideyuki Umeda

Department of Astronomy, School of Science, The University of Tokyo

Abstract. Core-collapse supernovae (CCSNe) are the final fate of the massive stars, but their explosion mechanisms are still uncertain. One of the clues to the solution of the explosion mechanism is to examine the asymmetric effects. This is because most of observed CCSNe are asymmetric explosions. One of the factors to the asymmetric explosions are the magnetorotational effects. The magnetic fields are amplified intensively along the rotational axsis during the collapse, and it leads to the bipolar outflows which may eject outer mantle. To understand the role of magnetorotational effects during CCSNe, we have developed a new multidimensional magnetohydrodynamic(MHD) code and calculate collapse of a $25M_{\odot}$ star with various magnetic field and rotational velocity.

Keywords: magnetorotational explosion, — supernovae: general — supernovae: MHD
PACS: 97.60.Bw

INTRODUCTION AND OUR MHD CODE

Recent MHD simulations of CC-SNe (e.g. [2, 4, 6]) report that the strongly amplified magnetic field along the rotational axis via the rotational wrapping launch bipolar outflows which may trigger the explosions. This scenario is called "Magnetorotational Explosion". However, in [2, 4, 6], successful explosion models (i.e. explosion energy, $E_{\rm exp}$, achieves $\sim 10^{51}$ergs) employ strong magnetic field as the initial values ($B_Z \sim 10^{11-13}$) which are considered to be unrealistically strong at the pre-collapse phase.

On the other hand, no clear explosions are reported with initially weak magnetic field, $B_Z \sim 10^9$ (e.g. [4, 5]). Their explosion energies are far less than the typical explosion energies (for instance, $E_{\rm exp} \sim 10^{48}$ergs in Takiwaki et al. [5]). They suggested the failed explosions may be due to the fact that the computational resolutions are not fine enough and the magnetic field is not amplified sufficiently to affect the explosion dynamics. To overcome this problem we have developed a new MHD code. The features of our code are 1) MHD Roe's approximate Riemann solver is employed; 2) Adaptive Mesh Refinement; 3) The general equation of states can be treated and 4) Self gravitation is included.

With our new MHD code, we calculated collapse of $25M_{\odot}$ star with various initial rotational velocity and magnetic field strength. Initial rotational velocity and magnetic field are set as the following forms.

$$\Omega_{\rm ini}(R,Z) = \Omega_0 \frac{R_0^2}{R^2+R_0^2} \tag{1}$$

CP1016, *Origin of Matter and Evolution of Galaxies*,
edited by T. Suda, T. Nozawa, A. Ohnishi, K. Kato, M. Y. Fujimoto, T. Kajino, and S. Kubono

$$\mathbf{B}(R,Z) = (B_r, B_\phi, B_z) = (0,0,B_Z) \tag{2}$$

Where Ω_0 and B_Z are free parameters and R_0 is set as 10^8cm. We employed the same equation of state as in Ardeljan et al. [1].

RESULTS

We calculated various models with changing Ω_0 and B_Z. In Hirschi et al. [3] it is reported that $25M_\odot$ star with solar metallicity($Z = 0.02$) has $\Omega_0 \sim 1.0$ rad s^{-1} at the end of silicon burning stage. Meanwhile, in Yoon & Langer [7], they calculated stellar evolution with including the effects of magnetic torques and they suggested local specific angular momentum is lowered by approximately one order of magnitude compared to nonmagnetic field case. However these results are still not conclusive ones and thus we choose various Ω_0 from $0.0 \sim 6.0$ rad s^{-1}. For initial magnetic field B_Z, we used $B_Z = 10^{6,9,12}$G and confirmed similar results to previous works [2, 4–6] in models with $B_Z = 10^{12}$G.

We find all models are divided into two types, "*fast*" and "*slow*" rotational models and only fast rotational models ($\Omega_0 \geq 1.9$rad s^{-1}) exceed $\sim 10^{51}$ergs in our study. On the other hand, initially slow rotational models ($\Omega_0 \leq 1.5$rad s^{-1}) do not exceed 10^{51}ergs at the end of the calculations. These models accrete too much matter onto the central object, due to stronger effective gravitational force, and it makes harder to explode. Hereafter we only focus on the successful explosion models which have fast rotations.

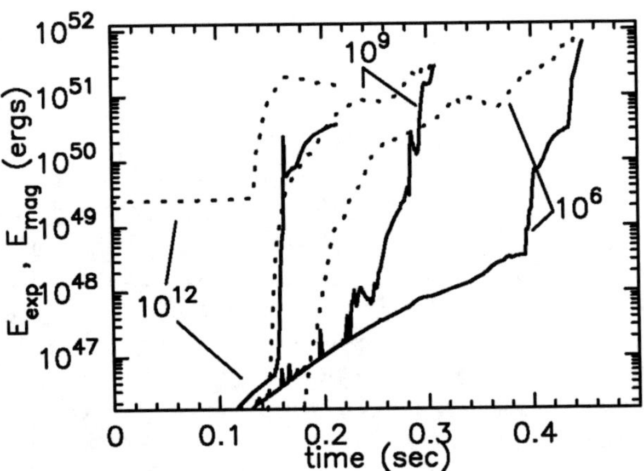

FIGURE 1. Time evolutions of explosion energies $E_{\rm exp}$ (*solid* line) and magnetic energies $E_{\rm mag}$ (*dotted* line) in the computational domain of different models. All models are with $\Omega_0 = 4.0$rad s^{-1} and the numbers which are pointing lines correspond to B_Z. All models enter the core bounce time at $t \sim 130$ms.

In Fig:1, it is shown time evolutions of explosion energies $E_{\rm exp}$ and magnetic energies in the computational domain of different models. As seen, $E_{\rm exp}$ with $B_Z = 10^{12}$G rapidly increases after the core bounce, $t \sim 130$ms, since the magnetic field has been already amplified sufficiently strong at that time ($E_{\rm mag} \sim 10^{51}$ergs). This is a similar result of previous works (e.g. [2, 4–6]).

Meanwhile models with $B_Z = 10^{6,9}$G have much weaker magnetic field immediately after the core bounce, though, they are amplified subsequently and reach $E_{\rm mag} \sim 10^{51}$ergs, finally. We confirmed turbulent flows in the central region ($r \lesssim 100$km) and, thus, we consider it is the dynamo mechanism that amplifies the magnetic field. The dynamo mechanism works until the exhaustion of rotational free energy (i.e. difference between the rotational energy with shearing motion and with rigid rotation with same total angular momentum) which limits the saturated magnetic energy. In these models the explosions occur when the magnetic fields are amplified sufficiently ($E_{\rm mag} \sim 10^{51}$ergs), too. It requires longer time to amplify the seed magnetic field and produces time lag, $\sim 0.1-0.3$s, till explosion as seen in Fig:1. Due to the time lag, the explosion energies become larger and larger as the explosions occur later and later. This is because, as time passes, there are less matters along the rotational axis due to mass falling and centrifugal force.

CONCLUSIONS

I) We have developed new MHD-code which implement Poisson's solver for self-gravity, Adaptive Mesh Refinement scheme and general EOS.

II) We calculated core-collapse of a $25M_\odot$ star with various Ω_0 and B_Z and found,

1. Faster rotational models $\Omega_0 \geq 1.9$rad s^{-1} can produce typical size explosion energy of core-collapse SNe ($\sim 10^{51}$ergs) independently on the initial magnetic field, meanwhile slower rotational models $\Omega_0 \leq 1.5$rad s^{-1} do not in our calculations.
2. Magnetic field is amplified sufficiently strong in both cases, $B_Z = 10^6, 10^9$G, via the dynamo mechanism and affects the explosion dynamics, finally. This is mainly due to our fine mesh size thanks to AMR (~ 1.2km at "requisite" location). Previous works, e.g. [4, 5], do not report such amplification and they thus did not report any explosions.
3. Initially weaker magnetic field requires longer time to be amplified and produces time delay until the explosion. Delayed explosion causes less matter along the rotational axis and the explosion energy becomes larger.

REFERENCES

1. Ardeljan, N.V., Bisnovatyi-Kogan, G.S. and Moiseenko, S.G., 2003, MNRAS, 359, 333
2. Burrows, A., Dessart, L., Livne, E., Ott, C.D. and Murphy, J., 2007, ApJ, 664, 416
3. Hirschi, R., Meyner, G. and Maeder, A., 2004, A&A, 425, 649
4. Sawai, H., Kotake, K., and Yamada, S, 2005, ApJ, 631, 446
5. Takiwaki, T., Kotake, K., Nagataki, S. and Sato, K., 2004, ApJ, 616, 1086
6. Yamada, S. and Sawai, H., 2004, ApJ, 608, 907
7. Yoon, S.-C. and Langer, N., 2005, A&A, 443, 643

Neutron Stars as a Source of the Short-Lived Nuclides in Ap-star Atmospheres

Vera F. Gopka[a], Oleg M. Ulyanov[b], Sergey M. Andrievsky[a]

[a] Department of Astronomy, Odessa National University, T.G. Shevchenko Park, Odessa 65014, Ukraine (e-mail: gopka.vera@mail.ru; scan@deneb1.odessa.ua)

[b] Institute of Radio Astronomy of National Academy of Sciences of Ukraine, 4 Chervonopraporna str., Kharkov 61002, Ukraine (e-mail: oulyanov@rian.kharkov.ua)

Abstract. We propose a new explanation of some magnetic chemically peculiar (MCP) star anomalies, which is based on an assumption that such stars be the close binary systems with a secondary component being a neutron star. Within this hypothesis one can naturally explain the main anomalous features of MCP stars: first of all, an existence of the short-lived radioactive isotopes detected in some stars (like Przybylski's star (PS) and HR465), and some others peculiarities. Also we can assume the presence of the electron-positron annihilation emission lines (0.511 MeV) in the gamma spectrum of some MCP stars.

Key words: MCP star, binary system, electron-positron plasma, neutron star, pulsar, annihilation.
PACS: 95.30.Cq

INTRODUCTION

From the past century various aspects of the MCP stars have been studied: variability of the absorption line intensities, magnetism, brightness, radial velocities, an emission in the radio, X, IR ranges. However, the nature of these stars is still enigmatic. The energy source needed to continuously support the abundance of short-lived isotopes in MCP star atmospheres is not understood even now. We propose a new hypothesis that MCP stars can be the close binary systems with a secondary component being a neutron star (NS) interacting with a primary component. It allows one to identify the origin of the peculiarity of the chemical abundances. The MCP star phenomenon could be explained by an interaction of the ultra-relativistic electron-positron plasma emitted by NS with gamma factor $\gamma>20$ with the gas of the upper atmosphere of a companion star. In particular, such an interaction produces the bremsstrahlung gamma quanta within the energy range of the photonuclear resonance (2-30 MeV).

THE START-POINT OF HYPOTHESIS IS PRZYBYLSKY'S STAR

The roAp star HD101065 is in fact an unique astrophysical laboratory for understanding and exploring the extreme phenomena of the stellar evolution. The existence in the PS atmosphere of such short-lived isotopes as ^{61}Pm [1, 2, 3], or trans-bismuth elements [4, 5], neutron rich isotopes of Ca and Li abundance [6, 7] testifies the peculiar physical processes that lead to these anomalies. Now it is generally accepted that detected overabundances of some elements do not reflect the chemical composition of the entire star, but only of its photosphere. In some cases overabundance of some elements can reach the values up to 6 dex. Nevertheless, the mechanism that is responsible for such uncommon processes was not identified at that time. Recently, Goriely [8] showed that nucleosynthesis of the heavy elements can

CP1016, *Origin of Matter and Evolution of Galaxies*,
edited by T. Suda, T. Nozawa, A. Ohnishi, K. Kato, M. Y. Fujimoto, T. Kajino, and S. Kubono

occur in the star atmosphere due to the high-energy particles entering the atmosphere of the star. At the same time, the origin of such particles was not clearly specified. The r-process has been frequently mentioned in relation to the abundance anomalies observed in Ap stars. The work [9] indicates the fact that an explosion remnant of a more massive star can be a possible explanation for the origin of the peculiarity. But there is a problem: a single SN explosion that took place in a binary system long time ago, and the following pollution of the nearby star atmosphere with synthesized isotopes cannot explain an existence of the short-lived elements in the atmosphere of this star observed at present time (the most stable isotope ^{145}Pm has half-life time 17.7 years, the isotope ^{252}Es - only 471.7 days). In order to explain several anomalous features of some MCP stars and especially their extremely peculiar chemical composition we propose the following hypothesis. Let us assume that PS is a close binary system with a non-visible companion being the NS. For this system an orbital plane is near perpendicular to the line of sight. The rapid wind, generated by NS, which consists of the electron-positron plasma, is accelerated to the speed of light and hit the PS atmosphere. The electron-positron plasma falling on the PS must be ultra-relativistic, so that the kinetic energy of each particle $E \gg m_e c^2$. The gamma-factor of the "fast" electrons/positrons, which are necessary for starting the photonuclear reactions, must exceed 20. This estimation is completely realistic, and this is supported by an example of pulsar PSR B0531+21. The high-energy electrons can also generate free neutrons via direct interaction with the hydrogen nuclei in the PS atmosphere ($p + e^- \rightarrow n + \nu$). Such free neutrons are necessary for the r-process to occur in the considered medium. Thus, the nuclei of heavy elements in the PS atmosphere can be synthesized as a result of two processes: the photonuclear reaction and the neutron capture by the seed nuclei of lighter element. In both cases the source of the energetic particles that trigger the nuclear reactions in the PS atmosphere is associated with the NS. The binary system PS-NS must be a close one. At present there is a large uncertainty in the mass estimations of the PS. If we take two extreme estimates we obtain that the mass of this star falls in the range $M_{PS} \approx (1.5\text{–}8)M_{\odot}$, and the mass of NS is $M_{NS} \approx (0.7\text{–}1.4)M_{\odot}$. It is possible to estimate the parabolic velocity for this system. Taking into account that the range of characteristic tangential velocities for the radio pulsars is $100 < V_{PSRs} < 500$ km/s, we obtain the range of the most probable parameters. It can be seen that the distances, estimated in that way, ranges from 0.035 AU up to 1.0 AU. Knowing the parallax of PS $\pi \approx 7.95 \pm 1.07$ mas, one can estimate its distance (D=125.8±15.7 pc), angular size (~0.17mas), and orbital period (1–270 days).

In the frame of our hypothesis one can expect the presence of the strong emission line in the gamma range (0.511 MeV) resulting from the electron-positron annihilations in two photons $e^+ + e^- \rightarrow 2\gamma$. The main reason of this line existence is complete loses of kinetic energy of electron-positron plasma as a result of the bremsstrahlung in the MCP star atmosphere. A presence of the annihilation line will be the strongest test of our hypothesis. Actually if this line will be found we can estimate proper radial velocities, local temperature and magnetic field strength in the MCP atmospheres.

ABOUT OBSERVATIONS THAT SUPPORT PRESENCE OF NS AS A COMPANION OF MCP STAR IN THE BINARY SYSTEM

The majority of mentioned above features can be explained in the framework of our hypothesis when active area of NS is touching the surface of a companion star. Such interaction can explain the following main characteristics of MCP stars:

1. Periodic variations of the light, spectrum, magnetic field, the spots formation on the stellar surface; the profile variability of PrIII, NdIII, NdII lines, the rapid non-radial oscillations of roAp stars [10, 11]. The variations of some MCP stars of the SrCrEu subgroup in the infrared with the same period as the visible light, spectrum and magnetic field variations [12].
2. The correlation between the effective temperature, magnetic field strength and radio luminosities of MCP stars; radio emission detected in about 25% MCP stars; moderate circular polarization of MCP stars [13], and 100% polarization of the radio emission for CU Vir at 1.4 GHz [14].
3. An anisotropic stellar wind of MCP stars.
4. The X-ray emission detected in some MCP stars (ROSAT All-Sky Survey).
5. The anomalous chemical composition of some MCP stars.

CONCLUSIONS

The proposed hypothesis based on the assumption that some MCP stars can be the binary stellar systems containing as a secondary component NS enables one to explain in a natural way some peculiarities associated with these stars. Among them: anomalous chemical composition, an existence of the short-lived radioactive isotopes, short-time and long-period variations of the light and magnetic field, the X-ray and radio-emission detected in some MCP stars. The most important criterion supporting our hypothesis should be the detection of the electron-positron annihilation line.

ACKNOWLEDGMENTS

O.M. Ulyanov thanks V.V. Ilyushin, V.V. Zakharenko and Grant J. Mathews for helpful discussions; V.F. Gopka was supported by the research fund of Ghonbuk National University, Korea, by Dr. C. Kim.

REFERENCES

1. G. Wegner, A.D. Petford, *MNRAS* **168**, 557-575 (1974).
2. C. Cowley, W.P. Bidelman et al., *Astron. & Astrophys.* **419**, 1087-1093 (2004).
3. V. Fivet et al., *MNRAS* **380**, 771-780 (2007).
4. V. Gopka, A. Yushchenko et al., AIP Conference Proceedings **843**, Tokyo, Japan, 2005, pp. 389-391.
5. W.P. Bidelman, ASP Conference Series **336**, 309 (2005).
6. C. Cowley, S. Hubrig, *Astron. & Astrophys.* **196**, 21-24 (2005).
7. A.V. Shavrina, N.S. Polosukhina et al., *Astron. & Astrophys.* **409**, 707-713 (2003).
8. S. Goriely, *Astron. & Astrophys.* **466**, 619-627 (2007).
9. G.R. Burbidge, Proc. of the IUA, Symp. No 22, 418 -419 (1965).
10. V. Gopka, O. Ulyanov, S. Andrievsky, http://adsabs.harvard.edu/abs/2007arXiv0712.2409G
11. O. Kochukhov, T. Ryabchykova et al., *MNRAS* **376**, 651-672 (2007).
12. F.A. Catalano, F. Leon, R. Kroll, *Astron. & Astrophys. Suppl. Ser.* **129**, 463-477 (1998).
13. S.A. Drake, J.L. Lynsky et al., *Ap. J.* **420**, 387-391 (1994).
14. C. Trigilio et al., *Astron. & Astrophys.* **362**, 281-288 (2000).

Measurement of the half-life of ^{60}Fe for a Nearby Supernova Source

K. Takahisa[a], T. Shima[a], Y. Nagai[a] and N. Takahasi[b]

[a]*Research Center for Nuclear Physics, Osaka University*
[b]*Department of Chemistry, Faculty of Science, Osaka University*

Abstract. A nearby supernova (SN) explosion in the past can be confirmed by the detection of radioisotopes on Earth. The largest error of this estimate is attributed to the half-life of ^{60}Fe (18%). We thus propose a precise measurement of the half-life of ^{60}Fe.

Keywords: Supernova, half-life, ^{60}Fe.
PACS: 20

INTRODUCTION

A nearby supernova (SN) explosion in the past can be confirmed by the detection of radioisotopes on Earth that were produced and ejected by the SN. Recently, a well resolved time profile of ^{60}Fe concentration in a deep-sea ferromanganese crust was measured, and its highly significant increase was found for the period of about 2.8 Myr ago[1].

The well defined time of the SN explosion makes it possible to search plausible correlations with other events in Earth's history. The cosmic rays (CR) flux enhancement due to an expanding supernova remnants (SNR) is estimated to be around 15% for a few 100kyr (for an interstellar medium (ISM) density of 0.5 atoms cm^{-3}. The ^{60}Fe fluence can be calculated as $(2.9 \pm 1.0) \times 10^6$ atoms cm^{-2}. The error comes mainly from the stastical one in AMS measurement, i.e., uncertainly in the half-life of ^{60}Fe (18%), that in the ^{10}Be dating (assumed to be 10%), and the 5% error for the crust's density and its iron content, respectively.

In this case, the largest error of this estimate is attributed to the half-life of ^{60}Fe (18%). We thus propose a measurement of the half-life of ^{60}Fe with higher accuracy (less than 10% error).

The half-life of ^{60}Fe was measured only by two groups. Roy and Kohman reported $T_{1/2} = 3 \times 10^5$year, with uncertainly by factor of 3 in 1957. Kutschera et al. reported $T_{1/2} = (1.49 \pm 0.27) \times 10^6$ year in 1984[2]. The specific activity of ^{60}Fe in ^{nat}Fe was measured through the grow-in of the 1.332 MeV gamma-ray line of the ^{60}Co daughter activity.

CP1016, *Origin of Matter and Evolution of Galaxies*,
edited by T. Suda, T. Nozawa, A. Ohnishi, K. Kato, M. Y. Fujimoto, T. Kajino, and S. Kubono

EXPERIMENT

We plan to measure the half-life of ^{60}Fe. So far, a new irradiation equipment was installed at the WN course at the Research Center for Nuclear Physics (RCNP), Osaka University in 2006. A copper disk of 4.5 cm diameter and 0.55 cm thickness was irradiated with protons of 200 MeV incident energy at the viewer point inside the WN course in May 2006.

The beam intensity was monitored and integrated by the SEC (Secondary electron monitor). The efficiency of the SEC was measured by a Faraday cup. The efficiency of the SEC was about 5%. The beam intensity was about 1 micro **A** and one day beam time.

After bombarding, the target was melted by concentrated nitric in July 2007. The concentrated nitric was evaporated. After residual thing was adjusted by Hydrochloric acid (6mol/l), the ion exchange was started. The iron ions were separated from the other materials by using the ion exchange method.

RESULS AND OUTLOOK

We have measured the grow-in of the 1.332 MeV gamma-ray line of the ^{60}Co daughter activity in this sample by using a Germanium detector enclosed by lead block in the low background laboratory at under floor of RCNP. The efficiency of this Germanium detector was about 0.1%. The room temperature of this laboratory is homoeothermic by air conditioning system.

Figure 1 shows the activity of the ^{60}Co nuclei in this sample during 150 day after separation from the other materials.

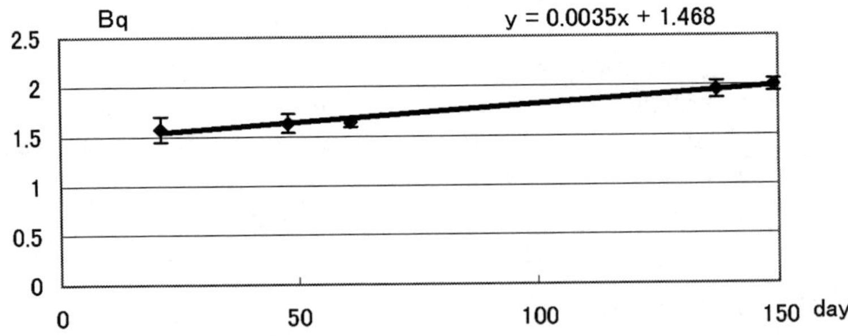

FIGURE 1. The activity of the ^{60}Co nuclei during 150 day after separation from the other materials. Vertical axis (y) means Bq and Horizontal axis (x) means day. 1.468Bq means initial activity of ^{60}Co nuclei and 0.0035 means grow-in of ^{60}Co nuclei depend on the ^{60}Fe nuclei.

In this result, the grow-in of the 1.332 MeV gamma-ray line of the ^{60}Co daughter activity shows that 4×10^{14} ^{60}Fe nuclei were produced when the half life of the ^{60}Fe is $T_{1/2} = 1.5\times10^{6}$ year.

This sample of iron will be measured to extract the ratio of ^{59}Fe to ^{60}Fe by using the Inductively Coupled plasma Mass Spectroscopy (ICP-MS).

We will measure the 59 keV gamma-ray from the first excitation level of the ^{60}Co by using of the Germanium detector.

We are able to determine the half-life of ^{60}Fe by using the combination of the grow-in of ^{60}Co daughter activity and information of the mass separation after a few years.

ACKNOWLEDGMENTS

This experiment was performed under the program E281 at RCNP. We thank RCNP cyclotron crew for their supports before and during and after the experiment.

REFERENCES

1. K. Knie et al., *Phys.Rev.Lett.* **93** (2004) 4171103.
2. W. Kutschera et al., *Nucl.Ins.Meth. Phys.Res./Sect.* **5** (1984) 430.

Thermonuclear Reaction Rate Libraries and Software Tools for Nuclear Astrophysics Research

Michael S. Smith*, Richard Cyburt†, Hendrik Schatz†, Michael Wiescher**, Karl Smith†, Scott Warren†, Ryan Ferguson†, Eric Lingerfelt*,‡, Kim Buckner*,‡ and Caroline D. Nesaraja*,‡

Physics Division, Oak Ridge National Laboratory, Oak Ridge, Tennessee, 37831-6354, USA
†Dept. of Physics & Astronomy, & National Superconducting Cyclotron Lab., Michigan State Univ., East Lansing, Michigan, 48824-1321, USA
**Dept. of Physics, Univ. of Notre Dame, Notre Dame, Indiana, 46556-5770, USA
‡Dept. of Physics & Astronomy, Univ. of Tennessee, Knoxville, Tennessee, 37996-1200, USA

Abstract. Thermonuclear reaction rates are a crucial input for simulating a wide variety of astrophysical environments. A new collaboration has been formed to ensure that astrophysical modelers have access to reaction rates based on the most recent experimental and theoretical nuclear physics information. To reach this goal, a new version of the REACLIB library has been created by the Joint Institute for Nuclear Astrophysics (JINA), now available online at **http://www.nscl.msu.edu/~nero/db**. A complementary effort is the development of software tools in the **Computational Infrastructure for Nuclear Astrophysics**, online at **nucastrodata.org**, to streamline, manage, and access the workflow of the reaction evaluations from their initiation to peer review to incorporation into the library. Details of these new projects will be described.

Keywords: reaction rate, nucleosynthesis, stellar modeling, software, nuclear data, evaluations
PACS: 24.10.-i, 21.10.-k, 26.20.-f, 26.30.-k, 26.30.Ca, 26.50.+x, 97.10.Cv, 97.30.Qt, 97.80.Gm, 97.80.Jp

INTRODUCTION

Simulations of the element synthesis and nuclear energy generation in astrophysical environments require libraries containing hundreds to thousands of thermonuclear reaction rates. These rates are based on laboratory measurements and theoretical calculations. It is challenging, but essential, to keep libraries updated with rates based on the latest nuclear physics information. One of the most popular libraries, REACLIB [1], has grown to a collection of over 62,000 rates and is used by researchers around the world. However, the lack of resources and mechanisms to regularly update and easily disseminate this library have prevented it from reaching its true potential as a community resource. There has not been a public release of REACLIB with updated experimental and theoretical rates since 1995, although numerous proprietary, partially updated versions of the library exist. Furthermore, there is no unified effort to evaluate reactions for input into REACLIB, nor efforts to streamline the process of getting the latest results into the library. A new collaboration [2, 3] has been formed to address these and other related issues. Two of the first activities of this effort are (1) the creation of a new, publicly-available version of REACLIB, and (2) software tools to help ensure that the library

CP1016, *Origin of Matter and Evolution of Galaxies*,
edited by T. Suda, T. Nozawa, A. Ohnishi, K. Kato, M. Y. Fujimoto, T. Kajino, and S. Kubono

stays up to date with the most recent nuclear physics information.

NEW VERSION OF RATE LIBRARY REACLIB

A new, public, web-based version of REACLIB has been created by the Joint Institute for Nuclear Astrophysics (JINA) and is available online at `http://www.nscl.msu.edu/~nero/db`. Data, stored in the standard REACLIB format, are continually updated as new rate assessments become available. A versioning system has been adopted to keep track of new rates, and rates are independently verified for quality control. Recommended rate libraries representing "snap shots" of the live database are stored for researchers who need a fixed/unchanging set of rates. The categorization of reactions is slightly changed from the original REACLIB formulation to accommodate reverse rate calculations, exporting to different file formats, and future developments. A nuclide-based search is available that enables searches of just the nuclides that you are interested in. Plotting and commenting on rates is also available. The database contains multiple versions of many rates, with one recommended rate that is continuously updated and documented on a status/discussion page. The rates are stored in a MySQL database with a PHP-driven web interface. The library is complete for explosive H and He burning, and rates for r-, s-, and p-processes are being added. The ground state weak rates are taken from the Nuclear Wallet Cards [4] when possible, and from experiment or theory for other weak rates.

NEW SOFTWARE TOOLS FOR EVALUATION WORKFLOW

It is essential that the new REACLIB has rates based on the most recent experimental information. This will require significant effort by many researchers to combine the latest measurements, previous results, and theoretical calculations to determine the "best" possible rate of any particular reaction. Because there are so many rates to evaluate and such limited manpower, it is imperative to design software tools to speed up, streamline, standardize, and automate many reaction evaluation tasks. Furthermore, to ensure quality control, reaction evaluations should be peer reviewed before inserting the corresponding rates into the library. Finally, it is advantageous to make the workflow of the evaluation and reviewing processes as transparent and accessible to the research community as possible.

The **Computational Infrastructure for Nuclear Astrophysics**, a suite of nuclear astrophysics codes available online at `nucastrodata.org`, already has a number of tools that aid evaluations. These include: uploading cross sections; rescaling and gain matching them; extending experimental cross sections with theoretical calculations; performing numerical integrations to determine reaction rates; and parameterizing rates in the standard REACLIB format. The suite also has easy-to-use tools to visualize, search, comment on, and compare reactions in rate libraries [REACLIB and others]. To streamline the *new* evaluation activities in support of the new REACLIB library, we are adding a set of *workflow management tools* to our online suite. We identified four different categories of Users – Scientific Contributors, Evaluators, Referees, and Editors

– each of which can perform a variety of specific tasks. These include, for example: uploading measurements or calculations (by a Scientific Contributor); assigning a rate to an Evaluator (Editor); initiating and completing an evaluation (Evaluator); initiating and completing a review of an evalution (Referee); and accepting and distributing a reviewed evaluation for inclusion in REACLIB (Editor). Anyone can view graphically the status of the process from the initiation of an evaluation to the final decision of the Editor, making the workflow as transparent as possible. Additional software tools to further aid the calculations and workflow of evaluations are also in development.

SUMMARY

Two new projects are designed to significantly enhance the ability of astrophysical modelers to get easy access to reaction rates based on the most up-to-date experimental and theoretical nuclear physics information. The first is the release of a new, public, web-based REACLIB with updated reaction rates and a set of recommended rates. The second is the development of new software tools at **nucastrodata.org** to streamline the process of getting the latest and best information into this new library. With this work, we hope to ensure the availability of the highest quality thermonuclear reaction rates that are essential for simulations of a wide variety of astrophysics phenomena.

ACKNOWLEDGMENTS

ORNL is managed by UT-Battelle, LLC for the U.S. Department of Energy under contract DE-AC05-00OR22725. MSU NSCL and JINA are funded by the National Science Foundation.

REFERENCES

1. F.-K.Thielemann, M. Arnould, J.W. Truran, in *Advances in Nuclear Astrophysics*, ed. E. Vangioni-Flam et al. (Editions Frontiere, Gif sur Yvette, 1987), p.525.
2. http://www.physik.unibas.ch/~nic9sat/
3. http://www.jinaweb.org/events/ect07/
4. http://www.nndc.bnl.gov/wallet/

Fission modes of neutron-rich nuclei in the r-process nucleosynthesis

S. Tatsuda, K. Yamamoto, T. Asano, M. Ohta, T. Wada[1], S. Chiba[2], H. Koura[2], T. Maruyama[2], T. Tachibana[3], T. Kajino[4], K. Sumiyoshi[5], K. Otsuki[6]

Department of Physics, Konan University, 8-9-1 Okamoto, Kobe658-8501, Japan
[1]Kansai Univ., [2] JAEA, [3] Waseda Univ., [4] NAO,
[5] Numazu College of Technology, [6] Univ. of Chicago

Abstract. The fission fragments mass distribution (FFMD) which is the important nuclear information in the study of the r-process nucleosynthesis is estimated for the neutron-rich nuclei ($Z>85$) according to the theoretical investigation of the potential energy surface. The details for determining FFMD are discussed. In this paper, the network calculation on the r-process nucleosynthesis is also performed including the data of FFMD. The comparison of the results of the network calculation, with and without the fission processes, is shown.

Keywords: r-process nucleosynthesis, mass distribution, fission
PACS: 25.85.-w, 26.30.+k

INTRODUCTION

The fission fragments mass distribution (FFMD) is one of the most important nuclear information in the r-process nucleosynthesis, together with nuclear mass, (n, γ) and (γ, n) reaction rates, β-decay rates and various fission rates. For the reasons, the fission processes before and after the freeze-out of neutron density may affect strongly the element abundance patterns. For example, fission may affect the termination of the r-process, and also before the neutron freeze-out, the fission products from transuranium region may rejoin the r-process nucleosynthesis (so-called fission cycling). In the r-process nucleosynthesis, the following fission processes may play a significant role to the element abundance patterns; β-delayed fission process, neutron induced fission process, fission after β-delayed neutron emission process and neutrino induced fission process. The experimental information of FFMD for the neutron-rich nuclei hardly exists; then, there are only a few groups which perform the network calculation taking into accounts of FFMD[1][2]. Furthermore, the theoretical estimation of FFMD has never been established. In the present paper, the attention is focused on FFMD and the role of fission in the r-process nucleosynthesis.

ANALYSIS OF FFMD

FFMD is estimated by searching for the local minimum of PES near the scission point, and examining the valley depth and its location in the deformation parameter

CP1016, *Origin of Matter and Evolution of Galaxies*,
edited by T. Suda, T. Nozawa, A. Ohnishi, K. Kato, M. Y. Fujimoto, T. Kajino, and S. Kubono

space. PES is calculated by using the liquid drop model with the microscopic correction of two-center shell model[3][4][5] in the three-dimensional parameter space which is composed of the distance between the centers of two-harmonic oscillators z, the deformation of the fragments δ and the mass asymmetric parameter α.[6] The value of the potential depth is defined relatively to the one for the spherical mononucleus.

FFMD corresponding to the single fission mode is assumed to be given by Gaussian distribution whose dispersion is $\sigma^2=49$ (taken from Langevin calculation by Asano et al.[7]) and mean values are A_H and A_L, where $A_H=A(1+\alpha)/2$ and $A_L=A(1-\alpha)/2$. If the symmetric fission ($\alpha=0$) and the asymmetric fission ($\alpha\neq 0$) are mixing, FFMD is determined by superposing the two components with the factor w_{sym} and w_{asym}, respectively. The weight factors are defined as $w_{sym}=-0.16(\Delta V-3.0)$ for $\Delta V\leq 3.0$ [MeV], $w_{sym}=0$ for $\Delta V>3.0$[MeV] and $w_{asym}=1-w_{sym}$, where $\Delta V=V_{sym}-V_{asym}$, V_{sym} and V_{asym} are the potential depth located at $\alpha=0$ and $\alpha\neq 0$ respectively.

From this analysis, it turned out that all neutron-rich nuclei, except the nuclei near Z~100 and N~164 (producing the double magic fragments), decayed into the asymmetric fragments via fission. Figure 1 shows samples of FFMD.

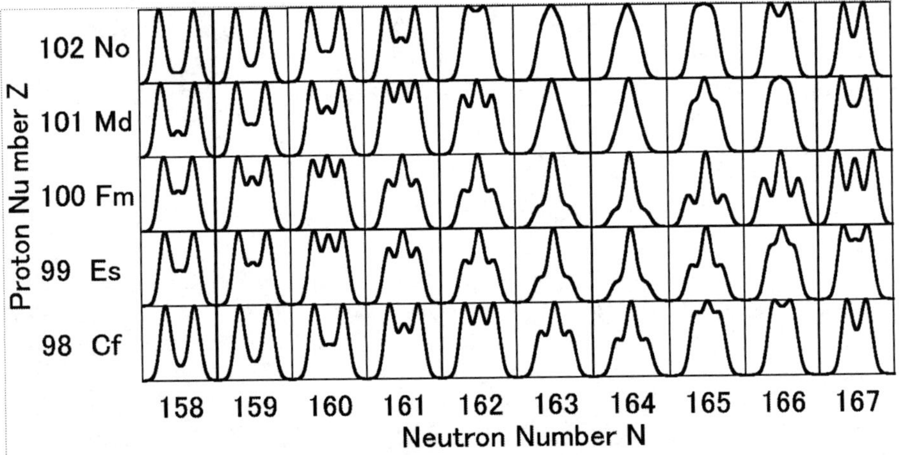

Figure 1. This figure shows FFMD of the nuclei in the region $158 \leq N \leq 167$ and $98 \leq Z \leq 102$.

NETWORK CALCULATION

The network calculation in the r-process nucleosynthesis is carried out in the three cases; (i) without the fission effects, (ii) with the fission effects taking account of the realistic FFMD and (iii) with the fission effects accompanying the symmetric fission. In this paper, only the β-delayed fission processes are taken into account. The β-delayed fission rates are defined as the half value of the β-decay rates taken from Klapdor, Metzinger and Oda (1984). For the astrophysical condition, we use the exponential model[8] which leads to the trajectory of Terasawa[9]. The modified version[9] of the nuclear mass table from Hilf, von Groote and Takahashi[10] is used for the nuclear masses. Figure 2 displays the results of the network calculation.

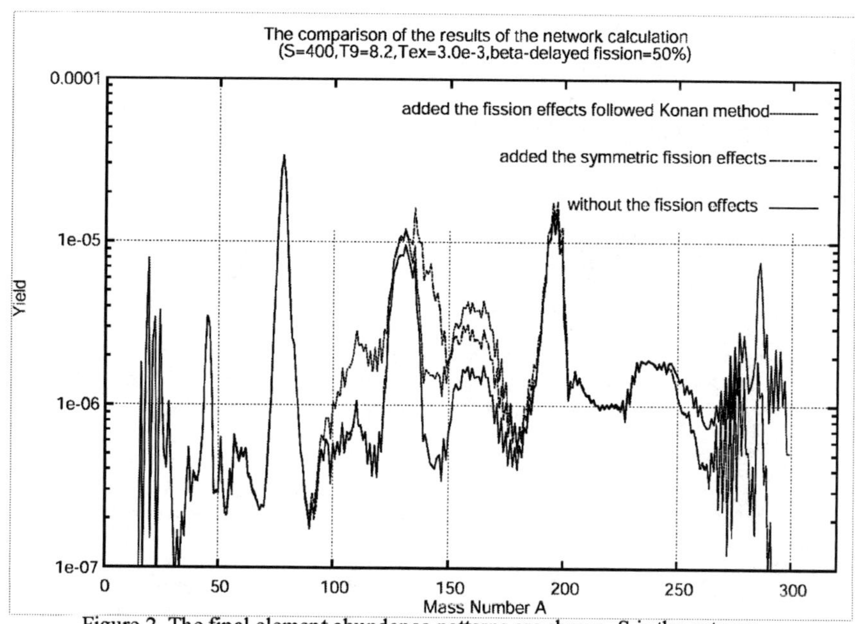

Figure 2. The final element abundance patterns are shown. S is the entropy and T_9 is temperature [10^9K].

SUMMARY

From the study of FFMD, it turns out as follows. Except for the nuclei in the neighborhood of ^{264}Fm (Z=100 and N=164), the asymmetric valleys ($\alpha \neq 0$) are usually deeper than the symmetric valley ($\alpha=0$). In the special case that the nucleus may decay into identical double magic nuclei, the symmetric valley becomes the deepest by the shell structure, so the symmetric fission becomes dominant. Therefore, all neutron-rich nuclei, except the nuclei near Z~100 and N~164 (twice of double magic), decay into asymmetric fragments through fission.

In the present results of the network calculation, the final element abundance patterns are much affected by FFMD. Consequently, it is important to analyze FFMD in detail by taking account of the dynamics from saddle to scission.

REFERENCES

1. I. V. Panov et al. / Nucl. Phys. A 747 (2005) 633-654
2. J. Benlliure et al. / Nucl. Phys. A 628 (1998) 458-478
3. J. Maruhn and W. Greiner / Z. Phys.251, 431(1972)
4. S. Suekane, A. Iwamoto, S. Yamaji and K. Harada / JAERI-memo 5918(1974)
5. A. Iwamoto, S. Yamaji, S. Suekane and K. Harada / Prog. Theor. Phys. 55, 115(1976)
6. S. Tatsuda, M. Ohta and et al. / *Proceedings of Tours Symposium on Nuclear Physics VI (Tours 2006)*, France (2006), p. 423
7. T. Asano, M. Ohta and et al. / Journal of Nucl. and Radiochemical Sciences, 5, 1 (2004) 1-5
8. K. Otsuki et al. / New Astro. 8 (2003) 767
9. M. Terasawa et al. / APJ 562 (2001) 470-479
10. H. Klapdor, J. Metzinger and T. Oda / At. Data Nucl. Data Tables 31 (1984) 81

Nucleosynthesis in Surface Detonation Type Ia Supernovae

Ivo R. Seitenzahl[*,†], Casey A. Meakin[**], Dean M. Townsley[‡,*], James W. Truran[‡,§] and Don Q. Lamb[‡,**]

[*]*Joint Institute for Nuclear Astrophysics*
[†]*Department of Physics, Enrico Fermi Institute, University of Chicago, Chicago, IL 60637*
[**]*Center for Astrophysical Thermonuclear Flashes, University of Chicago, Chicago, IL 60637*
[‡]*Department of Astronomy and Astrophysics, University of Chicago, Chicago, IL 60637*
[§]*Physics Division, Argonne National Laboratory, Argonne, L 60439*

Abstract. A common method for calculating nucleosynthetic yields for Type Ia supernovae (SNe Ia) is the post-processing of a large number of passively advected tracer particles with a nuclear reaction network. In this paper we present a novel and more computationally efficient method for calculating nucleosynthetic yields for material burned to nuclear statistical equilibrium (NSE). The method exploits a tight correlation that exists between the final entropy and neutron excess for material processed by a detonation. We generate a table of freeze out abundances by integrating a 443 nuclide nuclear reaction network along synthetic thermodynamic trajectories at a fixed entropy and neutron excess. Final yields are calculated by first interpolating in the table and then decaying short lived radionuclides back to stability.

We apply this method to obtain Fe-peak yields for a suite of 2D axisymmetric SNe Ia simulations in the framework of the gravitationally confined detonation (GCD) explosion mechanism developed at the Flash Center. Our models produce ^{56}Ni masses that are comparable to observed high luminosity SNe Ia and have a reasonable isotopic composition of other Fe-peak elements.

Keywords: supernovae:general — nuclear reactions:nucleosynthesis, abundances — white dwarfs
PACS: 26.30.-k,26.50.+x,97.60.Bw

INTRODUCTION

Type Ia supernovae (SNe Ia) are believed to be thermonuclear explosions of accreting C/O white dwarfs near the Chandrasekhar limit, powered by the energy liberated by the fusion of the initial composition to nuclear statistical equilibrium (NSE). They rank among the most important contributing sites to the production of Fe-peak elements in the Universe and can be used distance indicators to probe the cosmology of the universe. This importance of SNe Ia to both cosmology and galactic chemical evolution (in addition to being already an inherently interesting problem) has long attracted the attention of theorists and lead in time to more and more sophisticated explosion models and computer simulations. To obtain a light curve and spectrum from a model simulation and to quantify the contribution to galactic nucleosynthesis, it is necessary to calculate the detailed make-up of the nuclear composition of the ejected material. This is commonly done via the tracer particle method [1, 2], in which a large number (typically $\sim 10^6$) of massless tracer particles is placed proportional to the mass density in the initial stellar model. The tracer particles are then passively advected by the hydrodynamics and keep a record of the time history of thermodynamic state variables along their path. Each tracer

CP1016, *Origin of Matter and Evolution of Galaxies*,
edited by T. Suda, T. Nozawa, A. Ohnishi, K. Kato, M. Y. Fujimoto, T. Kajino, and S. Kubono

particle is identified with a Lagrangian mass, the composition of which is calculated by integrating a nuclear reaction network along the trajectory after the simulation. Regardless of being a trivially parallelizable problem, the post-processing a large number of tracers with a reaction network is computationally expensive.

Additionally, in spite of being standardizable candles, SNe Ia exhibit large variations of their peak luminosity and spectral properties. And even though the explosion mechanism(s) is (are) still unknown and hotly debated, it is foreseeable that a large number of simulations will be required to explain the observed diversity. The expected large number of simulations further makes an efficient method for calculating nucleosynthetic yields very desirable. In the next section, we proceed to describe a novel method based on a look-up scheme of tabulated yields for calculating Fe-peak yields for surface detonation SNe Ia simulations that circumvents costly post-processing of each individual trajectory. The last section summarizes results of a first application of this method to SNe Ia simulations that explode within the framework of the gravitationally confined detonation (GCD) scenario [3, 4].

METHODOLOGY

We have developed a computationally very efficient method to calculate the Fe-peak yields of SNe Ia that explode via detonations ignited near the surface. At the heart of the method lies a tight correlation between the final entropy S_f and the final electron fraction $Y_{e,f}$ of material processed in the detonation. This correlation arises because the detonation Mach number $M \sim M(\rho)$ is mainly a function of the density ρ. The final entropy of a zone, consisting of the sum of the entropy generated in the shock, the initial entropy, and the entropy generated in the burn of the initially cold fuel is therefore also mainly a function of ρ. Since the time rate of change of the electron fraction $\dot{Y}_e \sim \dot{Y}_e(\rho)$ is also mainly a function of density, it follows that $S_f \sim S_f(Y_{e,f})$.

We generate a one parameter (logarithmically spaced in Y_e) lookup table of the freeze out abundances by integrating a nuclear reaction network over a suite of analytic trajectories for a fixed entropy and Y_e as determined by a fit to their correlation. These trajectories are constructed by imposing an analytic temperature profile $T_9(t) = 5.5\exp(-t/0.4s)$, and solving for the density at every time step using an equation of state and adiabaticity. We initialize the network with NSE mass fractions corresponding to the chosen values for $S_f, Y_{e,f}$ and $T_9(0) = 5.5$ (the temperature was chosen to be just high enough to guarantee that the reactions are in equilibrium). The results are not very sensitive to the exact value of the decay constant and a representative value was chosen. We would like to emphasize that this method is by construction only applicable for material that reached temperatures high enough during the burn that the freeze out proceeds from the NSE state. Final yields are obtained by first interpolating in the table and then decaying the short-lived radionuclides back to stability. The yields obtained by integrating along the analytic trajectories compare very well yields obtained from post-processing Lagrangian tracer particles, whereas NSE based freeze out estimates give a poor estimates of many Fe-peak isotopes.

APPLICATION TO SIMULATIONS

We use the adaptive mesh reactive-hydrodynamics code FLASH [5] to evolve massive WD models from ignition to homologous expansion in 2D axisymmetric geometry. Nuclear flames in these simulations are propagated using an advection-diffusion-reaction flame model [4] and the nuclear energetics of [6, 7]. We calculate a suite of 2D axisymmetric single off center ignition point supernova models for a range of offsets [8], a large fraction of which undergoes surface detonations. The Fe-peak yields are then calculated with the method outlined above by using a reaction progress variable [4] as a marker for complete burning to NSE. The models have explosion energies around 1.5×10^{51} ergs, produce ^{56}Ni masses comparable to observed high luminosity SNe Ia [9] and have a reasonable isotopic composition of other Fe-peak elements. The total amount of ^{56}Ni produced is interestingly almost constant ($\sim 1.08\ M_{\odot}$) for all of the models that detonate. For example, models with larger offset lead to less pre-expansion and more NSE material synthesized, but the detonation occurs at higher densities near the center of the star and neutronization is more vigorous leading to a smaller fraction of the material ending up as ^{56}Ni.

ACKNOWLEDGMENTS

This work is supported by the U.S. Department of Energy under contract B523820 to the ASC Alliances Center for Astrophysical Flashes and by the National Science Foundation under grant PHY 02-16783 for the Frontier Center "Joint Institute for Nuclear Astrophysics" (JINA). JWT acknowledges support from Argonne National Laboratory, operated under contract No. W-31-109-ENG-38 with the DOE. IRS acknowledges support from the AAS international travel grant.

REFERENCES

1. C. Travaglio, W. Hillebrandt, M. Reinecke, and F.-K. Thielemann, *A&A* **425**, 1029–1040 (2004), arXiv:astro-ph/0406281.
2. E. F. Brown, A. C. Calder, T. Plewa, P. M. Ricker, K. Robinson, and J. B. Gallagher, *Nuclear Physics A* **758**, 451–454 (2005), arXiv:astro-ph/0505417.
3. T. Plewa, A. C. Calder, and D. Q. Lamb, *Astrophys. J. Lett.* **612**, L37–L40 (2004), astro-ph/0405163.
4. D. M. Townsley, A. C. Calder, S. M. Asida, I. R. Seitenzahl, F. Peng, N. Vladimirova, D. Q. Lamb, and J. W. Truran, *ApJ* **668**, 1118–1131 (2007), arXiv:0706.1094.
5. B. Fryxell, K. Olson, P. Ricker, F. X. Timmes, M. Zingale, D. Q. Lamb, P. MacNeice, R. Rosner, J. W. Truran, and H. Tufo, *Astrophs. J. Supp.* **131**, 273–334 (2000).
6. A. C. Calder, D. M. Townsley, I. R. Seitenzahl, F. Peng, O. E. B. Messer, N. Vladimirova, E. F. Brown, J. W. Truran, and D. Q. Lamb, *ApJ* **656**, 313–332 (2007), arXiv:astro-ph/0611009.
7. I. R. Seitenzahl, D. M. Townsley, F. Peng, and J. T. Truran, *Atomic Data and Nuclear Data Tables* **accepted for publication** (2008).
8. C. A. Meakin, I. R. Seitenzahl, D. M. Townsley, J. W. Truran, and D. Q. Lamb, *ApJ* **submitted** (2008).
9. M. Stritzinger, P. A. Mazzali, J. Sollerman, and S. Benetti, *A&A* **460**, 793–798 (2006), arXiv:astro-ph/0609232.

OMEG07: Program

The 10th Int. Symp. on Origin of Matter and Evolution of Galaxies
- From the Dawn of Universe to the Formation of Solar System -

December 4-7, Hokkaido University, Sapporo, Japan
http://nucl.sci.hokudai.ac.jp/~omeg07/

Symposium Subjects

1. Big Bang Cosmology and Particle Astrophysics
2. Cosmic and Galactic Chemical Evolution and Structure Formation
3. Stellar and Meteoritic Abundances
4. Nucleosynthesis in Stars, Novae, and Supernovae
5. Nuclear Structure and Reactions for Astrophysics
6. Weak Interaction and Neutrino Physics
7. Explosion Mechanism of Supernovae
8. Neutron Star and High Density Matter
9. X-Ray, Gamma Ray, Cosmic Ray, and Meteorites
10. Nuclear Data for Astrophysics

Symposium Schedule

12/03 (Mon.)	17:30-	Registration and Welcome Party at Enrei-sou
12/04 (Tue.)	AM	Opening Session Big Bang Cosmology and Particle Astrophysics Special Session: First Stars (1)
	PM	Special Session: First Stars (2) Nucleosynthesis (1), (2)
	Evening	Short Poster Presentation
12/05 (Wed.)	AM	Nuclear Data Nuclear Physics (1), (2) (in parallel)
	PM	Explosion Mechanism of Supernovae Neutron Star and High Density Matter Poster Session
	Evening	Banquet
12/06 (Thu.)	AM	Special Session: RI Beams Weak Interaction and Neutrino Physics
	PM	Nuclear Abundances in Supernovae Stellar and Meteoritic Abundances Closing
	Evening	Excursion
12/07 (Fri.)	-15:30	Excursion

Plenary and Parallel Sessions

Invited speakers are marked with *.
Allocated time includes the time for discussions (30=25+5, 20=16+4, 15=12+3).

12/03 (Mon.)

- **Registration at Enrei-sou** (17:30-18:30)
- **Welcome Party at Enrei-sou** (18:00-20:00)

12/04 (Tue.)

- **Opening**
  (9:00-9:15, *chaired by K. Kato*)

Masayuki Y. Fujimoto (Hokkaido U.) (5m)
Opening Address
Keizou Yamaguchi (Dean, Graduate School of Science, Hokkaido U.) (5m)
Welcome address

- **Big Bang Cosmology and Particle Astrophysics**
  (9:15-10:15, *chaired by T. Kajino*)

- **Special Session: First Stars (1) Nuclear Abundances in First Stars**
  (10:30-12:00, *chaired by M. Limongi*)

12/05 (Wed.)

12/06 (Thus.)

Kenzo Ishikawa (Hokkaido U.) (30m*)
Present Status of Neutrino Mixing and Possible Effects of Wave Packet Width . . . . (6-2-2)

Yoshitaka Fujita (Osaka U.) (15m)
Gamow-Teller Transitions in Proton Rich pf-shell Nuclei - Relevance to Supernova Explosion - . . . . (6-2-3)

Yong-Yeon Keum (National Taiwan U.) (15m)
Neutrino Mass Bounds from the Nuetrinoless Double Beta Decays and Large Scale Structures . . . . (6-2-4)

■ **Nuclear Abundances in Supernovae**
(13:30-14:45, *chaired by T. Kozasa*)

Lawrence Rudnick (U. Minnesota) (30m*)
Gas and Dust Layers from Cas A's Explosive Nucleosynthesis . . . . (6-3-1)

Katsuji Koyama (Kyoto U.) (30m*)
Suzaku Results of SN 1006: Chemical Abundances of the youngest Galactic Type Ia Supernova Remnant . . . . (6-3-2)

Kosuke Sato (Tokyo U. of Sci.) (15m)
Type Ia and II Supernovae Contributions to the Metal Enrichment in Intra-Cluster Medium Observed with Suzaku . . . . (6-3-3)

■ **Stellar and Meteoritic Abundances**
(15:00-16:30, *chaired by H. Yurimoto*)

Larry R Nittler (CIW) (30m*)
Presolar Stardust in the Solar System: Implications for Nucleosynthesis and Galactic Chemical Evolution . . . . (6-4-1)

Mikako Matsuura (NAOJ) (15m)
Asymptotic Giant Branch Stars as an Origin of Material in the ISM of Galaxies . . . . (6-4-2)

Sachiko Amari (Washington U.) (15m)
Presolar Diamond in Meteorites . . . . (6-4-3)

Shoichi Itoh (Hokkaido U.) (15m)
Fluctuation Frequency of Oxygen Isotope Reservoirs in the Early Solar System . . . . (6-4-4)

Naoya Sakamoto (Hokkaido U.) (15m)
Origin of ^{17,18}*O-rich Materials from Acfer 094* . . . . (6-4-5)

■ **Closing** (16:45-17:00, *chaired by T. C. Beers*)

Shigeru Kubono (CNS, U. Tokyo) (10m)
Concluding Remarks

■ **Excursion: Jozankei Spa** (18:00)

12/07 (Fri.)

■ **Excursion: Otaru City** (9:00-15:30)

Poster Session

(P-01) **Michael Scott Smith** (ORNL)
Big Bang Nucleosynthesis: Impact of Nuclear Physics Uncertainties on Baryonic Matter Density Constraints (subj. 1)
(P-02) **Lawrence Rudnick** (U. Minnesota)
Did a Giant Void Cause the WMAP Cold Spot? (subj. 1)
(P-03) **Kimitake Hayasaki** (YITP, Kyoto U.)
Periodic Light Variations from the Triple-Disk System around a Supermassive Binary Black Hole (subj. 2)
(P-04) **Takashi Fukui** (Hokkaido U.)
Oxygen Isotopic Evolution of the Early Solar Nebula: H_2O Transport Theory and its consistency with observation of Protoplanetary Disks (subj. 3)
(P-05) **Sachio Kobayashi** (Hokkaido U.)
Searching for Presolar Silicate Grains from Petrologic type 3.0-3.2 Carbonaceous Chondrites by in-situ Isotope Imaging (subj. 3)
(P-06) **Shingo Ebata** (Hokkaido U.)
Identification of Silicate and Carbonaceous Presolar Grains in the type 3 Enstatite Chondrites (subj. 3)
(P-07) **Tomomi Sunayama** (Knox College)
Waiting Points in Explosiove rp-Process Nucleosynthesis (subj. 4)
(P-08) **Claire Noel** (ULB)
Hydrodynamical Simulations of Detonations in Superbursts. (subj. 4)
(P-09) **Takanori Nishimura** (Hokkaido U.)
Interpretation of Extremely Metal-Poor Stars as Candidates of First Generation Stars (subj. 4)
(P-10) **Shin-ichiro Fujimoto** (Kumamoto CT)
Production of Light Elements on a Low-Mass Secondary in a Soft X-ray Transient (subj. 4)
(P-11) **Masaaki Otsuka** (Okayama, NAOJ)
The Origin and Evolution of the Extreme Metal-Poor Halo Planetary Nebulae (subj. 4)
(P-12) **Ko Nakamura** (U. Tokyo)
Light Element Production in Type Ib/c Supernovae (subj. 4)
(P-13) **Maria Letizia Sergi** (U. Catania)
The Role of Oxygen in Astrophysical Domains (subj. 4)
(P-14) **Takami Kuroda** (U. Tokyo)
The r-Process in Supersonic Neutrino-Driven Winds: The Role of Wind Termination Shock (subj. 4)
(P-15) **Ivo Rolf Seitenzahl** (Chicago)
Nucleosynthesis in Surface Detonation Type Ia Supernovae (subj. 4)
(P-16) **Yevgen Skakun** (Nat. Sci. Center Kharkiv Inst.)
Experimental and Theoretical Astrophysical S-factors and Reaction Rates of Threshold (p,n)-Reactions on Pd Isotopes (subj. 5)
(P-17) **Yasuo Wakabayashi** (CNS, U. Tokyo)
Cryogenic Gas Target System for Intense RI Beam Productions in Nuclear Astrophysics (subj. 5)
(P-18) **Yuma Kikuchi** (Hokkaido U.)
Description of Three-Body Scattering for Astrophysics (subj. 5)

(P-19) **Mariko Segawa** (JAEA)

Neutron Capture and Inelastic Scattering Cross Sections for ^{186}Os, ^{187}Os, and ^{189}Os and the Re-Os Chronology (subj. 5)

(P-20) **Tooru Yoshida** (Hokkaido U.)

E0 Transition Strength and Cluster Structure in ^{13}C (subj. 5)

(P-21) **Kazuhiko Kojima** (U. Tokyo/NAOJ)

Neutrino-Mass Effect on the CMB Anisotropies in a Cosmological Model with Primordial Magnetic Field (subj. 6)

(P-22) **Takami Kuroda** (U. Tokyo)

The Magnetorotational Explosion of Core-Collapse Supernovae with Initially Weak Magnetic Field (subj. 7)

(P-23) **Oleg M. Ulyanov** (NASU)

Neutron Stars as Source of the Short-Lived Nuclides in Ap-star Atmospheres (subj. 8)

(P-24) **Keiji Takahisa** (RCNP, Osaka U.)

Measurement of the Half-Life of ^{60}Fe for a Nearby Supernova Source (subj. 9)

(P-25) **Michael Scott Smith** (ORNL)

Thermonuclear Reaction Rate Libraries and Software Tools for Nuclear Astrophysics Research (subj. 10)

(P-26) **Sayuki Tatsuda** (Konan U.) *Fission Modes of Neutron-Rich Nuclei in the r-Process Nucleosynthesis* (subj.10)

List of Participants

Amari, Sachiko
Physics Department
Washington University
One Brookings Dr.
St. Louis, MO 63130
USA
sa@wuphys.wustl.edu

Ando, Ryosuke
Division of Physics
School of Science
Hokkaido University
Kita 10 Nishi 8, Kita-ku
Sapporo 060-0810
Japan
ando@nucl.sci.hokudai.ac.jp

Aoki, Wako
National Astronomical Observatory of Japan
2-21-1 Osawa, Mitaka
Tokyo 181-8588
Japan
aoki.wako@nao.ac.jp

Aumann, Thomas
GSI
Planckstr.1
D-64291 Darmstadt
Germany
t.aumann@gsi.de

Beers, Timothy C.
Department of Physics and Astronomy
Joint Institute for Nuclear Astrophysics
Michigan State University
E. Lansing, MI 48824
USA
beers@pa.msu.edu

Blinnikov, Sergei I.
Institute for Theoretical and Experimental Physics (ITEP), and University of Tokyo
Moscow 117218
Russia
sergei.blinnikov@itep.ru

Carollo, Daniela
Research School of Astronomy and Astrophysics, The Australian National University, Mount Stromlo Observatory, and INAF
Cotter Road, Weston, ACT 2611
Australia
carollo@mso.anu.edu.au

Cheoun, Myung-Ki
Dept. of Physics
Soongsil University
511 Sangdo-dong, Dongjak-Gu
Seoul 156-743
Korea
cheoun@ssu.ac.kr

Cherubini, Silvio
Dipartimento di Metodologie Fisiche e Chimiche per l'Ingegneria
Universita' di Catania and INFN-LNS
c/o INFN - LNS, Via S. Sofia 62
Catania 95125
Italy
cherubini@lns.infn.it

Chiba, Satoshi
Advanced Science Research Center
Japan Atomic Energy Agency
2-4 Shirakata-Shirane, Tokai, Naka
Ibaraki 319-1195
Japan
chiba.satoshi@jaea.go.jp

Coc, Alain
CSNSM, CNRS/IN2P3
University Paris Sud
UMR 8609, Batiment 104
F-91405 Orsay Campus
France
Alain.Coc@csnsm.in2p3.fr

Dillmann, Iris
Institut fuer Kernphysik
Forschungszentrum Karlsruhe
Postfach 3640
D-76021 Karlsruhe
Germany
iris.dillmann@ik.fzk.de

Ebata, Shingo
Department of Natural History Sciences
Graduate School of Science
Hokkaido University
Kita 10 Nishi 8, Kita-ku
Sapporo 060-0810
Japan
ebashin@ep.sci.hokudai.ac.jp

Ethugal Pedige, Berni Ann Thushari
Department of Physics
Kyushu University
4-2-1 Ropponmatsu, Chuo-ku
Fukuoka 810-8560
Japan
berni@gemini.rc.kyushu-u.ac.jp

Fischer, Tobias
Department Physics
Basel University
Klingelbergstrasse 82
CH-4056 Basel
Switzerland
tobias.fischer@unibas.ch

Fujimoto, Masayuki
Department of Cosmosciences
Graduate School of Science
Hokkaido University
Kita 10 Nishi 8, Kita-ku
Sapporo 060-0810
Japan
fujimoto@astro1.sci.hokudai.ac.jp

Fujimoto, Shin-ichiro
Department of Electronic Control
Kumamoto National College of Technology
2659-2, Suya, Koshi
Kumamoto 861-1102
Japan
fujimoto@ec.knct.ac.jp

Fujita, Yoshitaka
Department of Physics
Osaka University
Machikaneyama 1-1, Toyonaka
Osaka 560-0043
Japan
fujita@rcnp.osaka-u.ac.jp

Fukui, Takashi
Department of Cosmosciences
Graduate School of Science
Hokkaido University
Kita 10 Nishi 8, Kita-ku,
Sapporo 060-0810
Japan
ftakashi@ep.sci.hokudai.ac.jp

Gulino, Marisa
Dipartimento di Metodologie Fisiche e Chimiche per l'Ingegneria
Universita' di Catania and INFN-LNS
c/o INFN - LNS, Via S. Sofia 62
Catania 95125
Italy
gulino@lns.infn.it

Habe, Asao
Department of Cosmosciences
Graduate School of Science
Hokkaido University
Kita 10 Nishi 8, Kita-ku
Sapporo 060-0810
Japan
habe@astro1.sci.hokudai.ac.jp

Hashimoto, Takashi
Advanced Science Research Center
Japan Atomic Energy Agency
2-4 Shirakata-Shirane, Tokai, Naka
Ibaraki 319-1195
Japan
hashimoto.takashi62@jaea.go.jp

Hayakawa, Takehito
Kansai Photon Science Institute
Japan Atomic Energy Agency
8-1 Kunimidai, Kizu
Kyoto 619-0215
Japan
hayakawa.takehito@jaea.go.jp

Hayasaki, Kimitake
Yukawa Institute for Theoretical Physics
Kyoto University
Kitashirakawa, Oiwake-cho
Sakyo-ku, Kyoto 606-8502
Japan
kimitake@yukawa.kyoto-u.ac.jp

Hirabayashi, Yoshiharu
Information Initiative Center
Hokkaido University
Kita 11 Nishi 5, Kita-ku
Sapporo 060-0811
Japan
hirabay@iic.hokudai.ac.jp

Isaka, Masahiro
Division of Physics
School of Science
Hokkaido University
Kita 10 Nishi 8, Kita-ku
Sapporo 060-0810
Japan
isaka@nucl.sci.hokudai.ac.jp

Ishikawa, Kenzo
Department of Cosmosciences
Graduate School of Science
Hokkaido University
Kita 10 Nishi 8, Kita-ku
Sapporo 060-0810
Japan
ishikawa@particle.sci.hokudai.ac.jp

Ishimaru, Yuhri
Academic Support Center
Kogakuin University
2665-1, Nakanomachi, Hachioji
Tokyo 192-0015
Japan
kt13121@ns.kogakuin.ac.jp

Ishizuka, Chikako
Department of Cosmosciences
Graduate School of Science
Hokkaido University
Kita 10 Nishi 8, Kita-ku
Sapporo 060-0810
Japan
chikako@nucl.sci.hokudai.ac.jp

Ito, Makoto
RIKEN Nishina Center
2-1 Hirosawa, Wako
Saitama 351-0198
Japan
itomk@riken.jp

Ito, Takashi
Division of Physics
School of Science
Hokkaido University
Kita 10 Nishi 8, Kita-ku
Sapporo 060-0810
Japan
ito@astro1.sci.hokudai.ac.jp

Itoh, Shoichi
Department of Natural History Sciences
Graduate School of Science
Hokkaido University
Kita 10, Nishi 8, Kita-ku
Sapporo 060-0810
Japan
sitoh@ep.sci.hokudai.ac.jp

Iwamoto, Nobuyuki
Nuclear Data Center
Japan Atomic Energy Agency
2-4 Shirakata-Shirane, Tokai, Naka
Ibaraki 319-1195
Japan
iwamoto.nobuyuki@jaea.go.jp

Kajino, Toshitaka
Division of Theoretical Astrophysics
National Astronomical Observatory
of Japan, and University of Tokyo
2-21-1 Osawa, Mitaka
Tokyo 181-8588
Japan
kajino@nao.ac.jp

Kanzawa, Hiroaki
Department of Pure and Applied
Physics, Science and Engineering
Waseda University
3-4-1 Okubo, Shinjuku-ku
Tokyo 169-8555
Japan
kann-sein@toki.waseda.jp

Kasagi, Jirohta
Laboratory of Nuclear Science
Tohoku University
Mikamine 1-2-1, Tihaku-ku
Sendai 982-0826
Japan
kasagi@lns.tohoku.ac.jp

Kato, Kiyoshi
Department of Cosmosciences
Graduate School of Science
Hokkaido University
Kita 10 Nishi 8, Kita-ku
Sapporo 060-0810
Japan
kato@nucl.sci.hokudai.ac.jp

Kawanomoto, Satoshi
Optical and Infrared Astronomy
Division, National Astronomical
Observatory of Japan
2-21-1 Osawa, Mitaka
Tokyo 181-8588
Japan
kawanomoto.satoshi@nao.ac.jp

Keum, Yong-Yeon
Department of Physics
National Taiwan University
Da-An Gu, Roosevelt Road
Taipei 10672
Taiwan (R.O.C.)
yykeum@phys.ntu.edu.tw

Kikuchi, Yuma
Division of Cosmosciences
Graduate School of Science
Hokkaido University
Kita 10 Nishi 8, Kita-ku
Sapporo 060-0810
Japan
yuma@nucl.sci.hokudai.ac.jp

Kiss, Gabor G.
Institute of Nuclear Research
(ATOMKI)
Bem ter 18/c
Debrecen H-4026
Hungary
ggkiss@atomki.hu

Kobayashi, Sachio
Department of Natural History Sciences
Graduate School of Science
Hokkaido University
Kita 10 Nishi 8, Kita-ku
Sapporo 060-0810
Japan
sachio@ep.sci.hokudai.ac.jp

Kojima, Kazuhiko
National Astronomical Observatory
of Japan, and University of Tokyo
2-21-1 Osawa, Mitaka
Tokyo 181-8588
Japan
kojima@th.nao.ac.jp

Komiya, Yutaka
Department of Astronomy
Graduate School of Science
Tohoku University
Aza-Aoba 6-3, Aramaki
Aoba-ku, Sendai 980-8578
Japan
komiya@astr.tohoku.ac.jp

Koyama, Katsuji
Department of Physics
Kyoto University
Oiwake-cho, Kita-Shirakawa
Sakyo-ku, Kyoto 606-8502
Japan
koyama@cr.scphys.kyoto-u.ac.jp

Kozasa, Takashi
Department of Cosmosciences
Graduate School of Science
Hokkaido University
Kita 10 Nishi 8, Kita-ku
Sapporo 060-0810
Japan
kozasa@mail.sci.hokudai.ac.jp

Kubono, Shigeru
Center for Nuclear Study
University of Tokyo
2-1 Hirosawa, Wako
Saitama 351-0198
Japan
kubono@cns.s.u-tokyo.ac.jp

Kuramoto, Kiyoshi
Department of Cosmosciences
Graduate School of Science
Hokkaido University
Kita 10 Nishi 8, Kita-ku
Sapporo 060-0810
Japan
keikei@ep.sci.hokudai.ac.jp

Kuroda, Takami
Department of Astronomy
School of Science
The University of Tokyo
6-18-5 Kitayamada, Tsuzuki-ku
Yokohama 224-0021
Japan
kuroda@astron.s.u-tokyo.ac.jp

Kusakabe, Motohiko
Division of Theoretical Astronomy
National Astronomical Observatory
of Japan, and University of Tokyo
2-21-1 Osawa, Mitaka
Tokyo 181-8588
Japan
kusakabe@th.nao.ac.jp

Limongi, Marco
INAF - Osservatorio Astronomico di
Roma
Via Frascati 33, I-00040
Monteporzio Catone (Roma)
Italy
marco@oa-roma.inaf.it

Makii, Hiroyuki
Japan Atomic Energy Agency
2-4 Shirakata Shirane
Tokai-mura, Naka-gun
Ibaraki 319-1195
Japan
makii.hiroyuki@jaea.go.jp

Makinaga, Ayano
Department of Physics
Konan University
8-9-1 Okamoto, Higashinada
Kobe 658-8501
Japan
dn521001@center.konan-u.ac.jp

Mathews, Grant J.
Center for Astrophysics
Department of Physics
University of Notre Dame
Notre Dame, IN 46556
USA
gmathews@nd.edu

Matsui, Kana
Department of Cosmosciences
Graduate School of Science
Hokkaido University
Kita 10 Nishi 8, Kita-ku
Sapporo 060-0810
Japan
kmatsui@astro1.sci.hokudai.ac.jp

Matsumiya, Hiroshi
Department of Cosmosciences
Graduate School of Science
Hokkaido University
Kita 10 Nishi 8, Kita-ku
Sapporo 060-0810
Japan
matsumiya@nucl.sci.hokudai.ac.jp

Matsushita, Masafumi
Department of Physics
Rikkyo University
3-34-1 Nishi Ikebukuro, Toshima-ku
Tokyo 171-8501
Japan
matsushi@ribf.riken.jp

Matsuura, Mikako
National Astronomical Observatory
of Japan
2-21-1 Osawa, Mitaka
Tokyo 181-8588
Japan
mikako@optik.mtk.nao.ac.jp

Minamidani, Tetsuhiro
Department of Cosmosciences
Graduate School of Science
Hokkaido University
Kita 10 Nishi 8, Kita-ku
Sapporo 060-0810
Japan
minamidani@astro1.sci.hokudai.ac.jp

Minato, Futoshi
Department of Physics
Nuclear Theory Group
Tohoku University
6-3 Aoba, Aramaki, Aoba-ku
Sendai 980-8578
Japan
minato@nucl.phys.tohoku.ac.jp

Miwa, Shinichi
Department of Cosmosciences
Graduate School of Science
Hokkaido University
Kita 10 Nishi 8, Kita-ku
Sapporo 060-0810
Japan
miwa@astro1.sci.hokudai.ac.jp

Miyatake, Hiroaki
IPNS, KEK
2-4 Shirakata Shirane, Tokai, Naka
Ibaraki 319-1195
Japan
hiroaki.miyatake@kek.jp

Mizukawa, Rei
Department of Cosmosciences
Graduate School of Science
Hokkaido University
Kita 10 Nishi 8, Kita-ku
Sapporo 060-0810
Japan
rei@nucl.sci.hokudai.ac.jp

Moller, Peter
Theoretical Division
Los Alamos National Laboratory
P. O. Box 1663 MS B243
Los Alamos, NM 87545
USA
moller@lanl.gov

Motobayashi, Tohru
RIKEN Nishina Center
2-1 Hirosawa, Wako
Saitama 351-0198
Japan
motobaya@riken.jp

Murakami, Takaomi
Department of Cosmosciences
Graduate School of Science
Hokkaido University
Kita 10 Nishi 8, Kita-ku
Sapporo 060-0810
Japan
takaomi@nucl.sci.hokudai.ac.jp

Nagai, Yasuki
Japan Atomic Energy Agency
2-4 Shirakata Shirane
Tokai-mura, Naka-gun
Ibaraki 319-1195
Japan
nagai@rcnp.osaka-u.ac.jp

Nakamura, Ko
Research Center for the Early Universe
Graduate School of Science
University of Tokyo
7-3-1 Hongo, Bunkyo-ku
Tokyo 113-0033
Japan
nakamura@resceu.s.u-tokyo.ac.jp

Nakazato, Ken'ichiro
Department of Physics
Waseda University
3-4-1 Okubo, Sinjuku-ku
Tokyo 169-8555
Japan
nakazato@heap.phys.waseda.ac.jp

Namekata, Daisuke
Department of Cosmosciences
Graduate School of Science
Hokkaido University
Kita 10 Nishi 8, Kita-ku
Sapporo 060-0810
Japan
name@astro1.sci.hokudai.ac.jp

Nishimura, Nobuya
Department of Physics
School of Science
Kyusyu University
Jonan-ku Tashima 1-2-1
Fukuoka 814-0113
Japan
nobuya@gemini.rc.kyushu-u.ac.jp

Nishimura, Takanori
Department of Cosmosciences
Graduate School of Science
Hokkaido University
Kita 10 Nishi 8, Kita-ku
Sapporo 060-0810
Japan
nishimura@astro1.sci.hokudai.ac.jp

Nittler, Larry R
Department of Terrestrial Magnetism
Carnegie Institution of Washington
5241 Broad Branch Rd NW
Washington DC 20015
USA
lnittler@ciw.edu

Noel, Claire
Institut d'Astronomie et
d'Astrophysique
University Libre de Bruxelles, IAA
Campus plaine, CP 226
Boulevard du Triomphe, 1050 Bruxelles
Belgium
cnoel@ulb.ac.be

Nozawa, Takaya
Department of Cosmosciences
Graduate School of Science
Hokkaido University
Kita 10 Nishi 8, Kita-ku
Sapporo 060-0810
Japan
tnozawa@mail.sci.hokudai.ac.jp

Ohnishi, Akira
Department of Cosmosciences
Graduate School of Science
Hokkaido University
Kita 10 Nishi 8, Kita-ku
Sapporo 060-0810
Japan
ohnishi@nucl.sci.hokudai.ac.jp

Ohta, Masahisa
Department of Physics
Konan University
8-9-1 Okamoto, Higashinada
Kobe 658-8501
Japan
masaota@konan-u.ac.jp

Oogi, Taira
Department of Cosmosciences
Graduate School of Science
Hokkaido University
Kita 10 Nishi 8, Kita-ku
Sapporo 060-0810
Japan
oogi@astro1.sci.hokudai.ac.jp

Otsuka, Masaaki
Okayama Astrophysical Observatory
National Astronomical Observatory
of Japan
Honjo 3037-5, Kamogata-cho
Asakuchi-shi, Okayama 719-0232
Japan
masaaki@oao.nao.ac.jp

Otsuka, Takaharu
Department of Physics
University of Tokyo, and CNS
7-3-1 Hongo, Bunkyo-ku
Tokyo 113-0033
Japan
otsuka@phys.s.u-tokyo.ac.jp

Otuka, Naohiko
Nuclear Data Center
Japan Atomic Energy Agency
2-4 Shirakata Shirane, Tokai, Naka
Ibaraki 319-1195
Japan
ohtsuka@nucl.sci.hokudai.ac.jp

Pizzone, Rosario Gianluca
Dipartmento di Metodologie Fisiche e
Chimiche per l'Ingegneria
Universita' di Catania, and INFN-LNS
via santa sofia 62
Catania 95100
Italy
rgpizzone@lns.infn.it

Rehm, Ernst K.
Physics Division
Argonne National Laboratory
9700 South Cass Av.
Argonne, IL 60439
USA
rehm@anl.gov

Rudnick, Lawrence
Department of Astronomy
University of Minnesota
Minneapolis, MN 55455
USA
larry@astro.umn.edu

Sakamoto, Naoya
Department of National History Science
Graduate School of Science
Hokkaido University
Kita 10 Nishi 8, Kita-ku
Sapporo 060-0810
Japan
naoya@ep.sci.hokudai.ac.jp

Sato, Kosuke
Department of Physics
Tokyo University of Science
1-3 Kagurazaka, Shinjuku-ku
Tokyo 162-8601
Japan
ksato@rs.kagu.tus.ac.jp

Segawa, Mariko
Quantum Beam Science Directorate
Japan Atomic Energy Agency
2-4 Shiraka Shirane
Naka-gun, Tokai-mura
Ibaraki 319-1195
Japan
segawa.mariko@jaea.go.jp

Seitenzahl, Ivo R.
Department of Physics
Enrico Fermi Institute
University of Chicago
Chicago, IL 60637
USA
irs@uchicago.edu

Sergi, Maria L.
Dipartmento di Metodologie Fisiche e
Chimiche per l'Ingegneria
Universita' di Catania, and INFN-LNS
Via santa sofia 62
Catania 95100
Italy
sergi@lns.infn.it

Shibuya, Takatoshi
Division of Physics
School of Science
Hokkaido University
Kita 10 Nishi 8, Kita-ku
Sapporo 060-0810
Japan
shibuya@astro1.sci.hokudai.ac.jp

Shima, Tatsushi
Research Center for Nuclear Physics
Osaka University
10-1 Mihogaoka, Ibaraki
Osaka 567-0047
Japan
shima@rcnp.osaka-u.ac.jp

Shimizu, Hirohito
Division of Physics
School of Science
Hokkaido University
Kita 10 Nishi 8, Kita-ku
Sapporo 060-0810
Japan
shimizu@astro1.sci.hokudai.ac.jp

Skakun, Yevgen
National Science Center
Kharkiv Institute of Physics and
Technology
Academicheskaja str.
1, 61108 Kharkiv
Ukraine
skakun@kipt.kharkov.ua

Smith, Michael S.
Physics Division
Oak Ridge National Laboratory
MS-6354, Bldg. 6025, PO Box 2008
Oak Ridge, TN 37831-6354
USA
smithms@ornl.gov

Sorai, Kazuo
Department of Cosmosciences
Graduate School of Science
Hokkaido University
Kita 10 Nishi 8, Kita-ku
Sapporo 060-0810
Japan
sorai@astro1.sci.hokudai.ac.jp

Spitaleri, Claudio
Dipartmento di Metodologie Fisiche e
Chimiche per l'Ingegneria
Catania University
Via S. Sofia 62
Catania 95100
Italy
spitaleri@lns.infn.it

Suda, Takuma
Department of Cosmosciences
Graduate School of Science
Hokkaido University
Kita 10 Nishi 8, Kita-ku
Sapporo 060-0810
Japan
suda@astro1.sci.hokudai.ac.jp

Sumiyoshi, Kohsuke
Numazu College of Technology
Ooka 3600, Numazu
Shizuoka 410-8501
Japan
sumi@numazu-ct.ac.jp

Sumayama, Tomomi
Oak Ridge Institute for Science
Education, K-1473 Knox College
2 East South Street
Galesburg, IL 61401-4999
USA
tsunayam@knox.edu

Suzuki, Toshio
Department of Physics
College of Humanities and Sciences
Nihon University
Sakurajosui 3-25-40, Setagaya-ku
Tokyo 156-8550
Japan
suzuki@phys.chs.nihon-u.ac.jp

Tachibana, Takahiro
Waseda University Senior High School
1-31-3 Kamishakujii, Nerima-ku
Tokyo 167-0044
Japan
ttachi@waseda.jp

Takahisa, Keiji
Research Center for Nuclear Physics
Osaka University
10-1 Mihogaoka, Ibaraki
Osaka 567-0047
Japan
takahisa@rcnp.osaka-u.ac.jp

Takano, Masatoshi
Research Institute for Science and
Engineering
Waseda University
3-4-1 Okubo, Shinjuku-ku
Tokyo 169-8555
Japan
takanom@waseda.jp

Takekoshi, Tatsuya
Department of Physics
School of Science
Hokkaido University
Kita 10 Nishi 8, Kita-ku
Sapporo 060-0810
Japan
takekoshi@astro1.sci.hokudai.ac.jp

Tanaka, Kazunori
Research Institute for Science and
Engineering
Waseda University
3-4-1 Okubo, Shinjuku-ku
Tokyo 169-8555
Japan
tanaka-kznr@aoni.waseda.jp

Tanaka, Masaomi
Department of Astronomy
Graduate School of Science
University of Tokyo
7-3-1 Hongo, Bunkyo-ku
Tokyo 113-0033
Japan
mtanaka@astron.s.u-tokyo.ac.jp

Tatsuda, Sayuki
Department of Physics
Konan University
8-9-1 Okamoto, Higashinada
Kobe 658-8501
Japan
mn621010@center.konan-u.ac.jp

Terada, Kentaro
Department of Earth and Planetary
Systems Science, Graduate School of
Science, Hiroshima University
1-3-1 Kagamiyama, Higashi-Hiroshima
Hirosima 739-8526
Japan
terada@sci.hiroshima-u.ac.jp

Teranishi, Takashi
Kyushu University
6-10-1 Hakozaki, Higashi-ku
Fukuoka 812-8581
Japan
teranishi@nucl.phys.kyushu-u.ac.jp

Togano, Yasuhiro
Department of Physics
Rikkyo University
3-34-1 Nishi-Ikebukuro, Toshima-ku
Tokyo 171-8501
Japan
toga@ne.rikkyo.ac.jp

Tominaga, Nozomu
Department of Astronomy
Graduate School of Science
University of Tokyo
7-3-1 Hongo, Bunkyo-ku
Tokyo 113-0033
Japan
tominaga@astron.s.u-tokyo.ac.jp

Trache, Livius M.
Cyclotron Institute
Texas A&M University
MS 3366, College Station
TX 77843-3366
USA
livius_trache@tamu.edu

Tsubakihara, Kohsuke
Department of Cosmosciences
Graduate School of Science
Hokkaido University
Kita 10 Nishi 8, Kita-ku
Sapporo 060-0810
Japan
tsubaki@nucl.sci.hokudai.ac.jp

Ulyanov, Oleg M.
Institute of Radio Astronomy of
National Academy of Sciences of
Ukraine
4 Cervonopraporna str.
Kharkov 61002
Ukraine
oulyanov@rian.kharkov.ua

Utsunomiya, Hiroaki
Department of Physics
Konan University
8-9-1 Okamoto, Higashinada
Kobe 685-8501
Japan
hiro@center.konan-u.ac.jp

Wada, Takahiro
Department of Pure and Applied
Physics, Kansai University
3-3-35 Yamate-cho
Suita 564-8680
Japan
wadataka@ipcku.kansai-u.ac.jp

Wakabayashi, Yasuo
Center for Nuclear Study
University of Tokyo
2-1 Hirosawa, Wako
Saitama 351-0198
Japan
bayashi@cns.s.u-tokyo.ac.jp

Wanajo, Shinya
Department of Astronomy
School of Science
University of Tokyo
7-3-1 Hongo, Bunkyo-ku
Tokyo 113-8654
Japan
wanajo@astron.s.u-tokyo.ac.jp

Wano, Atsushi
Division of Physics
School of Science
Hokkaido University
Kita 10 Nishi 8, Kita-ku
Sapporo 060-0810
Japan
wano@nucl.sci.hokudai.ac.jp

Watanabe, Shigeto
Department of Cosmosciences
Graduate School of Science
Hokkaido University
Kita 10 Nishi 8, Kita-ku
Sapporo 060-0810
Japan
shw@ep.sci.hokudai.ac.jp

Watanabe, Yoshimasa
Department of Cosmosciences
Graduate School of Science
Hokkaido University
Kita 10 Nishi 8, Kita-ku
Sapporo 060-0810
Japan
nabe@astro1.sci.hokudai.ac.jp

Yamaguchi, Hidetoshi
Center for Nuclear Study
University of Tokyo
2-1 Hirosawa, Wako
Saitama 351-0198
Japan
yamag@cns.s.u-tokyo.ac.jp

Yamamoto, Kazuyuki
Department of Physics
Konan University
8-9-1 Okamoto, Higashinada
Kobe 658-8501
Japan
dn621003@center.konan-u.ac.jp

Yamamoto, Yu
Department of Physics
Waseda University
Aobaku, Matsukazedai 41-13
Yokohama 227-0067
Japan
yamamoto@heap.phys.waseda.ac.jp

Yoneda, Ken-ichiro
RIKEN Nishina Center
2-1 Hirosawa, Wako
Saitama 351-0198
Japan
kyoneda@riken.jp

Yoshida, Naoki
Department of Physics
Nagoya University
Furocho, Chikusa
Nagoya 464-8602
Japan
nyoshida@a.phys.nagoya-u.ac.jp

Yoshida, Tooru
Meme Media Laboratory
Hokkaido University
Kkita-ku, Sapporo 060-8628
Japan
tooru@nucl.sci.hokudai.ac.jp

Yurimoto, Hisayoshi
Department of Natural History Sciences
Graduate School of Science
Hokkaido University
Kita 10 Nishi 8, Kita-ku
Sapporo 060-0810
Japan
yuri@ep.sci.hokudai.ac.jp

Author Index

H

I

J

K

L

M

N

O

P

R

S

T

U

V

W

Y

Z